Maximum-Entropy and Bayesian Methods in Science and Engineering

Volume 2: Applications

Fundamental Theories of Physics

An International Book Series on The Fundamental Theories of Physics: Their Clarification, Development and Application

Maximum-Entropy and Bayesian Methods in Science and Engineering

Volume 2: Applications

edited by

Gary J. Erickson

Department of Electrical Engineering,
Seattle University, Seattle, Washington, U.S.A.

and

C. Ray Smith

Advanced Sensors Directorate
Research, Development and Engineering Center,
US Army Missile Command, Redstone Arsenal,
Alabama, U.S.A.

KLUWER ACADEMIC PUBLISHERS

DORDRECHT / BOSTON / LONDON

Library of Congress Cataloging in Publication Data

ISBN-13: 978-94-010-9056-8 e-ISBN-13: 978-94-010-9054-4
DOI: 10.1007/978-94-010-9054-4

Published by Kluwer Academic Publishers,
P.O. Box 17, 3300 AA Dordrecht, The Netherlands.

Kluwer Academic Publishers incorporates
the publishing programmes of
D. Reidel, Martinus Nijhoff, Dr W. Junk and MTP Press.

Sold and distributed in the U.S.A. and Canada
by Kluwer Academic Publishers,
101 Philip Drive, Norwell, MA 02061, U.S.A.

In all other countries, sold and distributed
by Kluwer Academic Publishers Group,
P.O. Box 322, 3300 AH Dordrecht, The Netherlands.

In honour of E. T. Jaynes

CONTENTS

PREFACE ix

SETI, RADON TRANSFORMS, AND OPTIMIZATION
Stanley R. Deans 1

INDUCTIVE INFERENCE BY NEUTRAL NETWORKS
David Hestenes 19

ON THE EFFICIENCY OF A CLASS OF MAXIMUM ENTROPY ESTIMATORS
C.C. Rodriguez 33

QUANTUM STATISTICAL MECHANICS IN PHASE SPACE AND THE CLASSICAL LIMIT
Y. Tikochinsky and D. Shalitin 51

SUPERPOSITION EFFECTS IN DIFFERENTIAL ENTROPY AND KULLBACK-LEIBLER INFORMATION
A.K. Rajagopal, P.J. Lin-Chung and S. Teitler 83

SUPER VARIATIONAL PRINCIPLES
L.H. Schick 89

EINSTEIN'S REVERSAL OF THE BOLTZMANN PRINCIPLE AND PARTICLE STATISTICS
A.K. Rajagopal and S. Teitler 105

CLASSICAL ENTROPY OF A COHERENT SPIN STATE: A LOCAL MINIMUM
C.T. Lee 111

LEAST MAXIMUM ENTROPY AND MINIMUM UNCERTAINTY COHERENT STATES
A.K. Rajagopal and S. Teitler 121

MAXIMUM ENTROPY SPECTROSCOPY - DIMES AND MESA
John Skilling 127

INFORMATION AND ENTROPY OF PATTERNS IN GENETIC SWITCHS
Thomas Dana Schneider 147

MAXIMUM ENTROPY AND THE PHASE PROBLEM IN PROTEIN CRYSTALLOGRAPHY
R.K. Bryan 155

CONTRAST TRANSFER FUNCTION CORRECTION IN ELECTRON MICROSCOPY
R.K. Bryan 171

CLIMATICALLY INDUCED CYCLIC VARIATIONS IN UNITED STATES CROP PRODUCTION: IMPLICATIONS IN ECONOMIC AND SOCIAL SCIENCE
Robert Guinn Currie 181

A MAXIMUM ENTROPY METHOD FOR EXPERT SYSTEM CONSTRUCTION
Alan Lippman 243

STOCHASTIC RELAXATION METHODS FOR IMAGE RESTORATION AND EXPERT SYSTEMS
Stuart Geman 265

TOWARDS A METHOD OF CORRECTING BIAS INTRODUCED BY THE USE OF THE ERROR FITTING METHOD IN CONJUNCTION WITH A MAXIMUM ENTROPY PROCESSING ALGORITHM
N.A. Farrow and F.P. Ottensmeyer 313

MAKING MAXIMUM ENTROPY COMPUTATIONS EASIER BY ADDING EXTRA CONSTRAINTS (EXTENDED ABSTRACT)
Sally A. Goldman and Ronald L. Rivest 323

IMAGE RESTORATION AND RECONSTRUCTION USING ENTROPY AS A REGULARIZATION FUNCTIONAL
Ali Mohammad-Djafari and Guy Demoment 341

APPLICATION OF LIKELIHOOD AND ENTROPY FOR TOEPLITZ CONSTRAINED COVARIANCE ESTIMATION
Michael I. Miller 357

THE CONCEPT OF EPOCH ENTROPY IN COMPLEX SYSTEMS
K.L. Ngai, A.K. Rajagopal and S. Teitler 363

A GENERAL THEORY OF INHOMOGENEOUS SYSTEMS
S.A. Trugman 371

MAXIMUM ENTROPY AND CRACK GEOMETRY IN GRANITIC ROCKS
P.M. Doyen 381

MAXIMUM ENTROPY ANALYSIS OF LIQUID DIFFRACTION DATA
John H. Root, P.A. Egelstaff and B.G. Nickel 395

RANDOM ARRAY BEAMFORMING
Keith H. Norsworthy and Paul N. Michels 409

DECISION MAKING WITH BARELY ANY INFORMATION: THE ROLE OF MIXED STRATEGIES
P.B. Kantor and M.J. Kantor 421

COMPARISON OF BAYESIAN AND DEMPSTER'S RULES IN EVIDENCE COMBINATION
Yizong Cheng and R.L. Kashyap 427

Subject Index 435

PREFACE

This volume has its origin in the Fifth, Sixth and Seventh Workshops on "Maximum-Entropy and Bayesian Methods in Applied Statistics", held at the University of Wyoming, August 5-8, 1985, and at Seattle University, August 5-8, 1986, and August 4-7, 1987. It was anticipated that the proceedings of these workshops would be combined, so most of the papers were not collected until after the seventh workshop. Because most of the papers in this volume are in the nature of advancing theory or solving specific problems, as opposed to status reports, it is believed that the contents of this volume will be of lasting interest to the Bayesian community.

The workshop was organized to bring together researchers from different fields to critically examine maximum-entropy and Bayesian methods in science and engineering as well as other disciplines. Some of the papers were chosen specifically to kindle interest in new areas that may offer new tools or insight to the reader or to stimulate work on pressing problems that appear to be ideally suited to the maximum-entropy or Bayesian method.

These workshops and their proceedings could not have been brought to their final form without the support or help of a number of people. Professor Alwyn van der Merwe, the Editor of Fundamental Theories of Physics, and Dr. D. J. Larner of Kluwer, provided encouragement and friendship at critical times. Others who have made our work easier or more rewarding include Professor Paul D. Neudorfer of Seattle University, Mr. Robert M. Braukus, P.E., Director of Telecommunications of Puget Sound Power and Light Co., Dr. J. M. Loomis of the Radar Technology Branch of MICOM's Research, Development, and Engineering Center, and Dr. Rabinder Madan of the Office of Naval Research.

Partial support of the fifth and seventh workshops was provided by the Office of Naval Research under Grants No. N00014-G-0219 and N0001487-G-0231.

DEDICATION

In commemoration of the thirtieth anniversary of his first papers (published in the Physical Review) on maximum-entropy, the 1987 workshop and these proceedings are proudly dedicated to Edwin T. Jaynes. May his contributions continue for at least another thirty years.

SETI, Radon Transforms, and Optimization*

Stanley R. Deans†
NASA–Ames Research Center 229-8
Moffett Field, CA 94035

1. Introduction

My purpose today is to explain the connections indicated in the title, tell you about a recent successful observation by the NASA–SETI team using prototype SETI hardware at Goldstone, outline a data analysis problem facing the Microwave Observing Project, and give some preliminary results that may prove useful in solution of the data analysis problem. You will not hear anything about how Bayesian probability theory and maximum entropy are being used to solve signal processing problems in SETI. One of my purposes in attending this conference is to learn how SETI might profit by using these methods for processing a prodigious amount of data in real time. Your suggestions will be greatly appreciated.

2. NASA–SETI Program

There are several excellent sources that provide history and background for the NASA–SETI Program and for SETI in general [1]. A recent report on "Signal Processing in SETI" by Cullers, Linscott and Oliver [2] is especially relevant to the discussion here. Hardware and software developments are described along with some important results of a field test trial of a functional prototype of the automated SETI electronic system as the design has been developed thus far by the NASA–SETI R&D program.

Another important report describes a series of SETI Science Workshops [3], conducted as part of a two-year feasibility study, chaired by Philip Morrison of the Massachusetts Institute of Technology and supported by the NASA Office of Space Science. Four major conclusions of the Workshops were:

---

*SETI is an acronym for Search for Extraterrestrial Intelligence.

†National Research Council Associate. Permanent university address: Department of Physics, University of South Florida, Tampa, FL 33620.

G. J. Erickson and C. R. Smith (eds.),
Maximum-Entropy and Bayesian Methods in Science and Engineering (Vol. 2), 1–17.

1. It is both timely and feasible to begin a serious search for extraterrestrial intelligence.
2. A significant SETI program with substantial potential secondary benefits can be undertaken with only modest resources.
3. Large systems of great capability can be built if needed.
4. SETI is intrinsically an international endeavor in which the United States can take a lead.

Conclusion 2 has had a profound influence on the direction of the SETI effort. It is now clear that a significant SETI program can be carried out without having to build new special purpose radio telescopes. Powerful new instrumentation, utilizing the most recent VLSI technology, placed at existing radio telescope sites will provide the opportunity, at modest cost, to conduct significant new searches with high sensitivity and broad sky and frequency coverage. There are many aspects of the proposed observing program that represent enormous improvements over all previous searches. New technological developments make it possible to process tens of millions of frequency channels simultaneously during each second of observation time. Hence it will be possible for the first time to explore, systematically, a significant portion of the microwave spectrum. Frequencies from 1 to 10 GHz will be analyzed, in some cases with a resolution as narrow as 1 Hz. Increased sensitivity will be achieved with low noise, tunable, wide band, cryogenic receivers. SETI signal detection processors using sophisticated algorithms will search for a variety of artificial signals. These algorithms are being designed to do the signal processing in real time, due to the enormous amount of data that will be coming in each second. Finally, the overall signal identification process will be controlled by expert systems.

3. Pioneer 10 received on the MCSA

A 74,000-channel prototype of the multi-channel spectrum analyzer (MCSA), to be used in the planned ten year microwave observing program, is now being field-tested at Goldstone, California. Pioneer 10, now outside the solar system over 3.3 billion miles away, is still radiating its 1 watt carrier from an antenna with 32.6 dB gain. The signal, received on the 26 meter DSS 13 antenna at Goldstone, is too weak for the station to achieve lock. However, using a frequency synthesizer as the local oscillator, the signal is clearly visible on the display terminal of the prototype SETI MCSA as a bright straight line trace, illustrated in Figure 1. The same clearly visible trace would be produced using Arecibo as the receiver and a 64 meter transmitting antenna radiating 600 kilowatts at a distance of 100 light years [2].

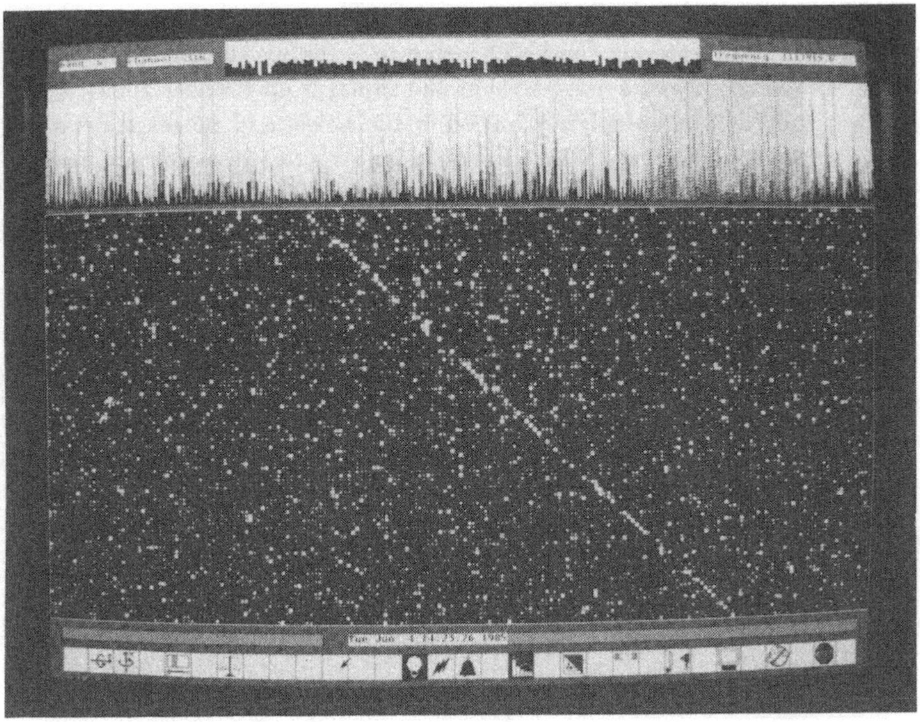

Figure 1. Prototype SETI hardware detects Pioneer 10. The top window displays the noise power in 144 bands each 576 Hz wide. The center window shows the noise power in 1024 bins of one of these bands; each bin is 0.5 Hz wide. The large lower window displays as brightness (or pixel area) the power in 206 adjacent bins. Each of the 100 raster lines is a spectral sample, the most recent being at the bottom. Every two seconds a new spectrum is added and the display scrolls up. Thus time increases downward. The frequency scale is reversed, the higher frequency bins are at the left. The slanting trace shows the received carrier frequency to be dropping about 0.2 Hz s^{-1}. The drift is caused by the Earth's rotation. The figures at the bottom are a menu of icons selected by cursor and used to modify the display. Photograph courtesy of: NASA–Ames Research Center, Moffett Field, California.

4. The Radon Transform

If any signals are detected in the SETI search, it is highly unlikely that they will be as far above the average noise level as the signal from Pioneer 10. It is more likely that they will be completely buried in the noise, and pulses are certainly as likely as continuous-wave (CW) carrier signals since less average power is required. Detection of both CW and drifting pulsed signals is discussed in [2]. Here the discussion is confined to power spectra for CW signals that may or may not be drifting in frequency. Simulation of such a signal is illustrated in Figure 2. In this case the signal power, in a bin where the signal is centered, is about twice the average noise power. If you look closely, you can find the drifting signal.

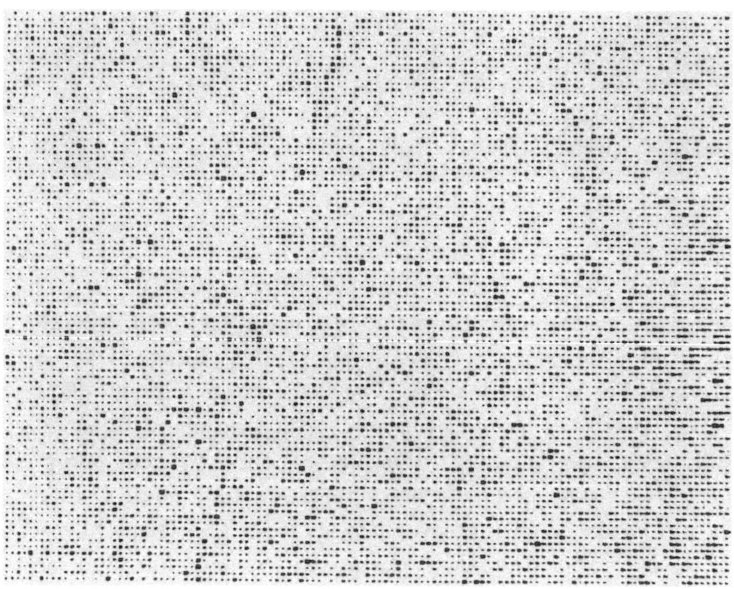

Figure 2. Simulated signal in noise field. In this simulated power spectrum the signal power is 2.0 times the average noise power; the drift rate is 0.27 Hz s^{-1}. As in Figure 1, frequency is horizontal and time is vertical.

In Figure 3 the noise field is the same, but the signal power is only half the average noise power. Not many people can find this signal by visual means. However, by integrating along the actual signal path the signal-to-noise ratio (SNR) can be increased enough to make it possible to identify the location and drift rate of the signal. The development of algorithms for the detection of linearly drifting signals leads naturally to a consideration of the Radon transform [4,5]. (Note: If the sending antenna and receiving antenna are accelerating relative to each other at a constant rate, then the received frequency will drift at a constant rate and thus show up as a line in a frequency-time plot, such as Figures 1–3. Moreover, astrophysical considerations lead to the conclusion that the most likely drift rates would lie between ± 1 Hz s^{-1}.)

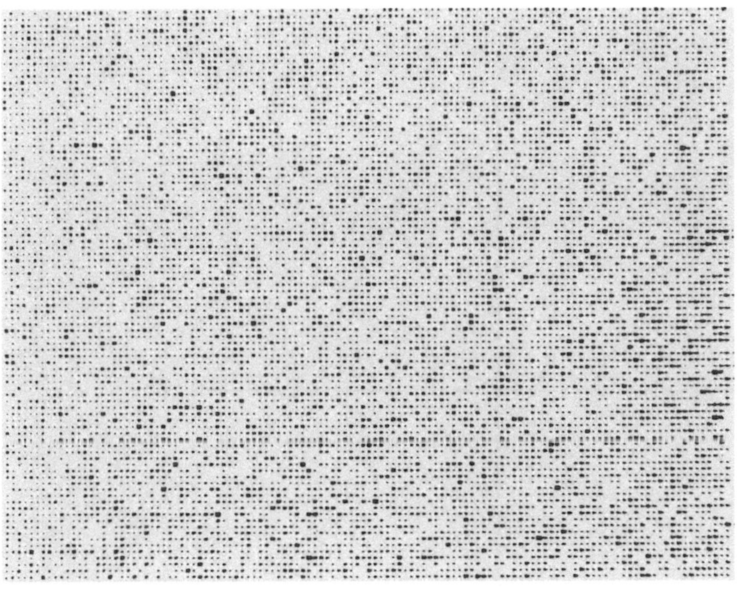

Figure 3. Simulated signal buried in noise field. This power spectrum is the same as Figure 2, except that the signal power is only half the average noise power.

The Radon transform of a function of two independent variables, x and y in Figure 4, can be written as

$$\check{f}(p,\phi) = \mathcal{R}\, f(x,y) = \int_{-\infty}^{\infty}\int_{-\infty}^{\infty} f(x,y)\,\delta(p - x\cos\phi - y\sin\phi)\,dx\,dy\,.$$

The Dirac delta function is one-dimensional. Thus the integral over the xy plane reduces to a line integral along the line defined by

$$p = x\cos\phi + y\sin\phi\,.$$

Clearly, one way to locate a linearly drifting signal in a noisy ensemble is to examine $\check{f}$ as a function of p and ϕ. This has been done and the results are just fine for simulations, but the complexity of the resulting algorithm is such that it is not a very good candidate for a detection algorithm that operates in real time with as many as 10^7 channels per second.

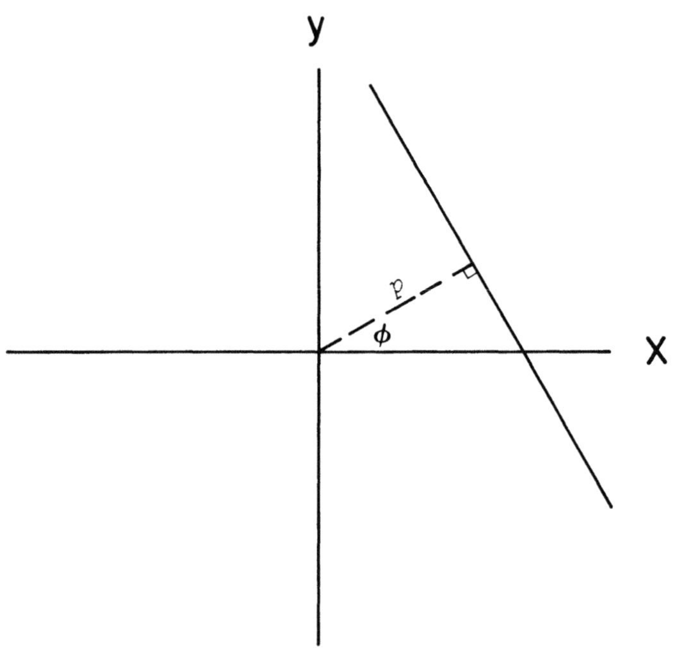

Figure 4. Coordinates for the Radon transform.

One way to speed up the calculation without sacrificing very much sensitivity is simply to add the contributions made by entire pixels along an assumed line rather than actually integrating along the assumed line. However, as we shall see, this is still not satisfactory, due to memory constraints and the necessity of doing the analysis in real time.

5. Optimization

For purposes of reference we say ***the optimum detection strategy*** is to test all drift rates and not discard any data. For example, a 1000 second observation with one spectrum per second would require approximately 2000 drift rates per frequency bin for a careful scan of all drift rates between -1.0 and $+1.0$ Hz s^{-1}. The general goal is to reduce the memory requirements as much as possible and at the same time keep the loss of detection sensitivity as low as possible. One very attractive way to accomplish this is by dividing the observation period into various stages with decisions made at the end of each stage. With these restrictions (minimum memory, maximum sensitivity, multiple stages) the detection problem can be formulated as a constrained optimization problem. Before illustrating a possible approach using two or three stages it will prove useful to quantify the concept of *sensitivity* and introduce what we will call (for lack of a better name) a ***memory*** *cost factor.*

Sensitivity

A measure of sensitivity for the 1000 second observation mentioned above can be computed by finding the signal power to average noise power ratio r one could expect to detect, given the overall probability of false alarm P_{fa} and overall detection probability P_d. (These probabilities are defined in *Appendix A*.) For example, $r = 0.15796$ corresponds to $P_d = 0.5$ and $P_{fa} = 10^{-6}$. To accomodate those who think in terms of decibels, this signal-to-noise ratio (SNR) is given by $10 \log_{10} r = -8.0145\,\text{dB}$. For the given values of P_{fa} and P_d this represents the best sensitivity one could expect. This optimum occurs for a ***one-stage procedure.*** It will be compared to the best one can hope to achieve by using a two-stage procedure and a three-stage procedure.

Memory cost factor

The memory cost factor per frequency bin is computed by insisting that the spacing between tested and stored drift rates is proportional to the reciprocal of the number of spectra in the stage under consideration. If there are N_1 spectra in the first stage the spacing is ϵ/N_1, where ϵ is the proportionality factor. Reasonable values for ϵ are in the interval $1 \leq \epsilon \leq 2$. Throughout this discussion the number of stored drift rates per frequency bin is assumed to be

$$\frac{2N_1}{\epsilon} + 1 .$$

A simple example may be useful. Suppose there are two stages with

$$\epsilon = 2 , \qquad N_1 = 10 , \qquad N_2 = 100 .$$

During the first stage drift rates

$$-1.0, \ -0.8, \ -0.6, \ \cdots , \ 0.8, \ 1.0$$

are tested, a total of 11. This factor 11 is the memory cost factor for the first stage. Suppose there is a "hit" at 0.4; during the second stage, drift rates

$$0.30, \ 0.32, \ 0.34, \ \cdots , \ 0.48, \ 0.50$$

are tested. Once again, this is a total of 11. In this example the memory cost factor is the same for each stage.

6. Two–stage procedure

Suppose the observations are made in two stages: N_1 seconds in the first stage with false alarm probability P_1, and N_2 seconds in the final stage with false alarm probability P_2. The overall false alarm probability $P_{fa} = P_1 P_2$, the total number of spectra $N = N_1 + N_2$, and the detection probability P_d are held constant as in the one-stage procedure,

$$N = 1000 , \qquad P_{fa} = 10^{-6} , \qquad P_d = 0.5 \tag{1}$$

The memory cost factor per frequency bin in the first stage is given by

$$C = \frac{2N_1}{\epsilon} + 1$$

and the average number of hits coming in from the first stage that must be investigated in the second stage is CP_1. If the spacing in the second stage is proportional to the reciprocal of N_2 with proportionality factor ϵ, then an interval of length ϵ/N_1 where the hit occurred must be subdivided into intervals of length ϵ/N_2. This means it is necessary to examine

$$\frac{\dfrac{\epsilon}{N_1}}{\dfrac{\epsilon}{N_2}} + 1 = \frac{N_2}{N_1} + 1$$

drift rates per hit. It is important to observe that this result is not dependent on ϵ. (Clearly, there are reasons why the signal might be missed if ϵ is too large, but they do not change the conclusions about the optimization discussed here.)

The memory requirement in stage two is the number of hits coming from stage one times the number of drift rates per hit,

$$CP_1 \left(\frac{N_2}{N_1} + 1 \right) .$$

To keep the memory requirement in the second stage from being greater than the requirement in the first stage, it is essential that

$$CP_1 \left(\frac{N_2}{N_1} + 1 \right) \leq C .$$

The C cancels and it is convenient to define

$$k = P_1 \left(\frac{N_2}{N_1} + 1 \right) .$$

The essential requirement is

$$k \leq 1 . \tag{2}$$

The constrained optimization problem is to find the best values of N_1, N_2, and P_1 such that r is as low as possible and equations (1) and (2) hold. More details are given in *Appendix A*.

The results of the two-stage optimization show that the best values occur then the equality in (2) holds. In this case the same amount of memory is used in the second stage as in the first stage. A careful examination of the region with $k = 1$ reveals that the optimum parameters are:

$$r = 0.16816 , \qquad P_1 = 0.166 , \qquad N_1 = 166 , \qquad N_2 = 834 .$$

This best sensitivity corresponds to a SNR of $10 \log_{10} r = -7.7428\,\text{dB}$. When compared to the one-stage procedure it follows that the loss in sensitivity is about 0.272 dB. This loss in sensitivity yields a saving in memory requirements. If $\epsilon = 1$ the one-stage requirement is $2N + 1 = 2001$, and the two-stage requirement is $2N_1 + 1 = 333$. The two-stage requirememt is about 16% of the one-stage requirement. Another way to think about this is that an 84% saving in memory costs about 0.272 dB loss in sensitivity.

7. Three–stage procedure

Suppose the observations are made in three stages: N_1 seconds in the first stage, N_2 seconds in the second stage, and N_3 seconds in the third stage,

$$N_1 + N_2 + N_3 = N\,. \tag{3}$$

Let the false alarm probabilities in the first and second stages be denoted by P_1 and P_2, respectively. Then the false alarm probability in the third stage is determined,

$$P_3 = \frac{P_{fa}}{P_1 P_2}\,. \tag{4}$$

For convenience, define

$$P_2 = xP_1\,. \tag{5}$$

The memory requirements provide further conditions on these variables for the three-stage procedure.

Memory requirememts

In the first stage the memory cost factor per frequency bin is

$$C = \frac{2N_1}{\epsilon} + 1\,.$$

The memory requirement in stage two is

$$CP_1\left(\frac{N_1}{N_2} + 1\right)\,.$$

It is possible to select N_2 such that the memory requirement in stage two is the same as in stage one,

$$CP_1\left(\frac{N_2}{N_1} + 1\right) = C\,. \tag{6}$$

The constant C cancels and

$$N_2 = \frac{N_1(1 - P_1)}{P_1}\,. \tag{7}$$

Each hit from stage two, CP_1P_2, must be investigated in stage three and increments related to $1/N_3$. It may be too restrictive to require an equality on the memory needs in the last stage; rather than an equation similar to (6) the requirement is

$$CP_1P_2\left(\frac{N_3}{N_2} + 1\right) \leq C\,.$$

Once again, the constant cancels. By using (5) the inequality is

$$xP_1^2\left(\frac{N_3}{N_2}+1\right)\leq 1\,. \tag{8}$$

Three–stage optimization

Selection of the best values for x and P_1 is not a trivial matter. Given (x, P_1) it is necessary to find the value of N_1 that yields the lowest possible SNR consistent with the constraints, equations (1), (3), (4), (5), (7), and (8). For more details, see *Appendix A* and *Appendix B*. For convenience, define

$$K = xP_1^2\left(\frac{N_3}{N_2}+1\right)\,, \tag{9}$$

so the constraint equation is simply $K \leq 1$. It is possible to compute an optimum set of parameters for points in the (x, P_1) plane, and verify that the equality $K = 1$ yields the best constrained optimum. Moreover, the parameter r varies slightly along the $K = 1$ curve in the (x, P_1) plane. The optimum parameters along this curve in the neighborhood of minimum r are given in Table 1.

Table 1. Parameters along the $K = 1$ curve.

r	x	P_1	N_1	N_2	N_3
0.1755	0.47	0.50	105	105	790
0.1751	0.50	0.49	103	107	790
0.1744	0.60	0.46	98	114	788
0.1741	0.72	0.43	91	120	789
0.1742	0.81	0.41	86	124	790
0.1745	0.92	0.39	82	128	790
0.1749	1.00	0.38	81	132	787

From this table the best parameters for the three-stage procedure are along the line where $r = 0.1741$, set apart with the double horizonatal lines. This optimum is illustrated in more detail in Figure 5, where the detection probability is shown as a function of N_1. It is important to note the broad maximum in this figure. This is characteristic of the region where the optimum occurs. To show the effect more dramatically, suppose the region with $(x, P_1) = (1.0\,,\ 0.1)$ is selected. This is illustrated in Figure 6. There is more loss in sensitivity and the maximum is not as broad. Also, there is less flexibility on the selection of the observing sequence.

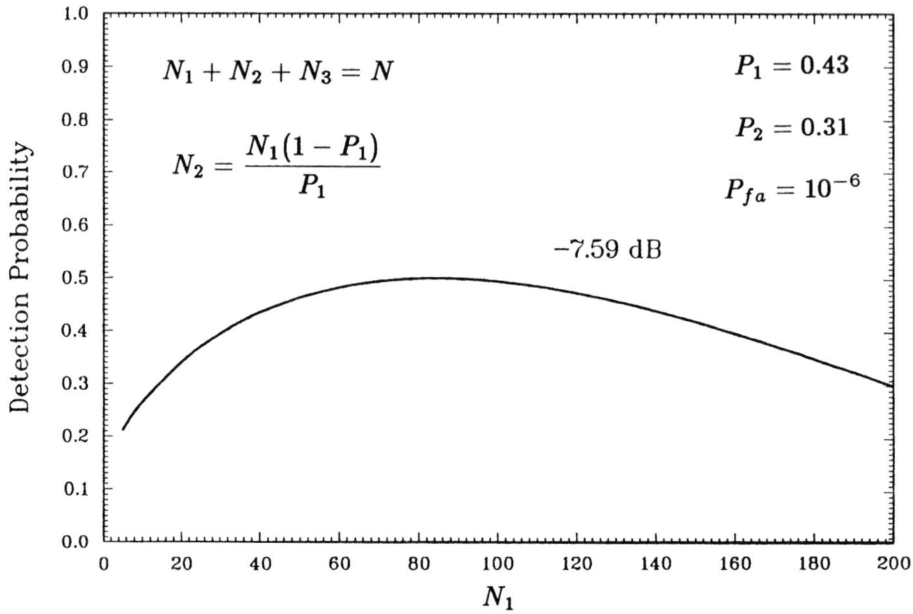

Figure 5. Detection probability as a function of N_1 when the observing sequence is optimized.

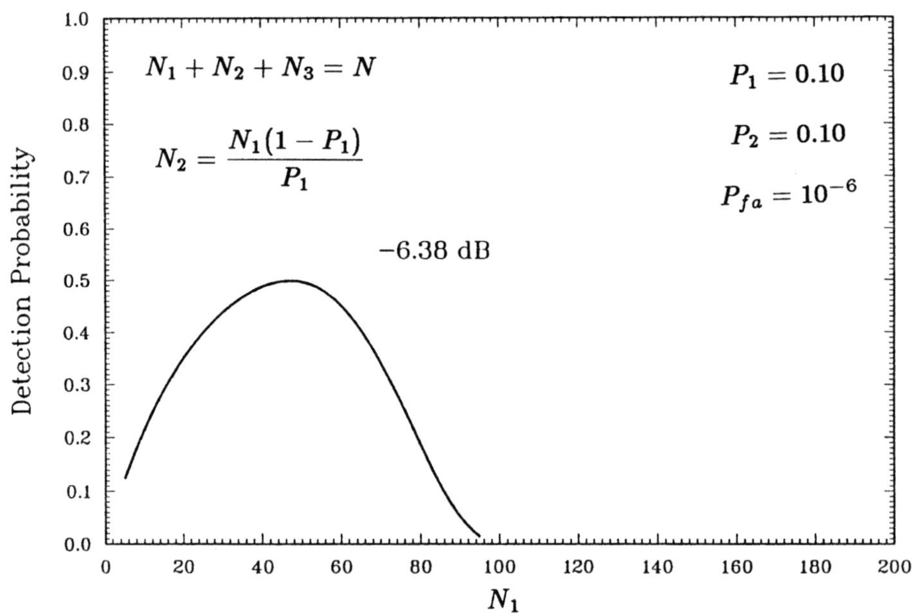

Figure 6. Detection probability as a function of N_1 when the observing sequence is not optimized.

Sensitivity and memory

The optimum r corresponds to a SNR of -7.59 dB. This represents a loss of 0.42 dB when compared with the one-stage procedure. The memory requirement for this case, with $\epsilon = 1$, is $2(91) + 1 = 183$. This is about 9% of the requirement in the one-stage procedure, or a 91% saving. A 91% saving in memory usage costs about 0.42 dB in sensitivity.

8. Effect of changing parameters

In all calculations discussed thus far P_{fa}, N, and P_d have been held fixed at the values given in (1). It is especially interesting to observe what happens if either P_{fa} or N changes.

Certain parameters are almost invariant to changes in N, and others scale in a very nice way. These properties follow from *Appendix B*. For example, in the two-stage procedure if $N = 100$ is the only change made in (1) the optimum observing sequence is $N_1 = 17$, $N_2 = 83$. This is a good example of scaling N_1, N_2, and N by a factor of 1/10. The value for P_1 is almost invariant to changes in N whenever the observing sequence scales as in this case. The corresponding optimum r is 0.597 or -2.24 dB. These and other results for $N = 100$ appear in Table 2.

Table 2. Optimum values for $P_{fa} = 10^{-6}$ and $N = 100$

Procedure	r	SNR(dB)	x	P_1	N_1	N_2	N_3
one-stage	0.554	-2.56		10^{-6}	100		
two-stage	0.597	-2.24		0.17	17	83	
three-stage	0.623	-2.06	0.83	0.41	9	12	79

If the overall false alarm probability is fixed at 10^{-10} with no other changes in (1) there are changes in the optimum values for the parameters. The computational technique is the same as the cases already discussed. The optimum parameters are given in Table 3, and the savings and losses are summarized in Table 4. It is interesting to note from Table 4 that the sensitivity loss scales by a factor of 0.77 in both the two-stage and three-stage procedures.

Table 3. Optimum values for $P_{fa} = 10^{-10}$ and $N = 1000$

Procedure	r	SNR(dB)	x	P_1	N_1	N_2	N_3
one-stage	0.2149	-6.68		10^{-10}	1000		
two-stage	0.2255	-6.47		0.126	126	874	
three-stage	0.2314	-6.36	0.60	0.410	65	93	842

Table 4. Savings and losses, $N = 1000$

procedure	memory saving	sensitivity loss (dB)	
		$P_{fa} = 10^{-6}$	$P_{fa} = 10^{-10}$
two-stage	84%	0.27	0.21
three-stage	91%	0.42	0.32

9. Concluding remarks

There are many ways to save memory at the expense of a loss in sensitivity. An attractive improvement to the method suggested here is by reusing data from the previous stages; that is, do not zero the accumulators at the start of the second and following stages. In this approach the sensitivity calculations are much more difficult since it is necessary to work with truncated (censored) probability densities, and convolutions of these densities with the central and noncentral χ^2 densities. Work on this problem is underway, and preliminary calculations indicate that by reusing data, one can get sensitivity losses as low as 0.1 dB.

Another important problem is to calculate the sensitivities for an observing strategy that involves use of overlapping spectra and Hanning windows [6,7]. This is another interesting challenge because correlations are introduced and the sensitivity calculations become problems of calculating probability distributions of quadratic forms in normal variables [8].

Finally, there is the question of just how should one go about solving the coherent CW detection problem for SETI. (Until now the discussion here has centered on the incoherent CW detection problem using power spectra.) At this point it appears to the author that the "correct" way to do this is by use of the Wigner distribution [9]; however, the constraint of processing data in real time appears to make this approach unrealistic. Various ideas are being pursued by members of the SETI group, including an investigation of how to make the Wigner distribution less computationally intensive.

Acknowledgments

It is a pleasure to thank D. Kent Cullers and other members of the SETI groups at NASA–Ames Research Center and Stanford University for many stimulating discussions about the SETI signal detection problem.

Appendix A

It is assumed that the noise power has a chi-square χ_ν^2 density function with $\nu = 2n$ degrees of freedom [10]. If the average noise power is precisely known, the the noise power can be normalized to unity and the density function can be modified to have mean $\mu = 1$ and variance $\sigma^2 = 1/n$. This (unit mean) version of the density function is given by

$$f_n(x) = \frac{n^n x^{n-1}\, e^{-nx}}{(n-1)!}, \qquad x \geq 0$$

and $f_n(x) = 0$ if $x < 0$. Here n is to be replaced by the appropriate observation period N, N_1, N_2, or N_3. The probability of false alarm P_{fa} is dependent on both the number of terms added n, and the normalized threshold b,

$$P_{fa} = P_{fa}(n,b) = \int_b^\infty f_n(x)\, dx\,.$$

(If the threshold associated with the usual χ^2 density is a, then the normalized threshold is given by $b = a/2n$.)

If a signal is present with the signal power divided by the average noise power equal to r, the density function is a noncentral chi-square [10], usually designated by $\chi_\nu'^2$. In the normalized units the noncentral density is given by

$$f_n'(r,x) = n\left(\frac{x}{r}\right)^{(n-1)/2} e^{-n(r+x)}\, I_{n-1}(2n\sqrt{rx})\,, \qquad x \geq 0$$

and $f_n'(r,x) = 0$ if $x < 0$. The mean is $1 + r$ and the variance is $(1+2r)/n$. Here I_ν is a modified Bessel function of the first kind. For computational purposes it is useful to observe that

$$f_n'(r,x) = f_n(x)\, \mathrm{e}^{-nr}\, {}_0F_1(n; n^2 rx)$$

where ${}_pF_q$ is a generalized hypergeometric function.

The probability of missing a signal characterized by r is given by

$$P_{ms} = P_{ms}(n,r,b) = \int_0^b f_n'(r,x)\, dx\,.$$

The detection probability P_d is simply

$$P_d = 1 - P_{ms}\,.$$

Given a value for P_{fa} (say 10^{-6}) and n (say 1000) it is possible to determine the corresponding threshold b. This value for b is in turn used to determine r when P_d is known.

In a multistage process with m stages, the overall detection probability is given by a product of detection probabilities,

$$P_d = P_d(1)\, P_d(2) \cdots P_d(j) \cdots P_d(m)\,,$$

where $P_d(j)$ is determined for the jth stage as indicated above. Specifically, the replacements are:

$$n \to n_j\,, \qquad b \to b_j\,, \qquad P_{fa}(j) \to P_j\,.$$

Thus

$$P_d(j) = 1 - P_{ms}(n_j, r, b_j)\,.$$

The optimization problem is to fix P_d (say 0.5) and determine the observation times n_j and false alarm probabilities P_j subject to specified constraints such that r is a minimum.

Appendix B

When the equality in (2) holds, the memory requirement in the second stage is the same as in the first stage. For the two-stage procedure this means

$$N_1 + N_2 = N$$

and

$$P_1\left(\frac{N_2}{N_1} + 1\right) = 1\,.$$

If N_2 is eliminated, a simple relation involving P_1, N_1, and N follows,

$$N_1 = NP_1\,.$$

In the three-stage procedure when the equality in (8) holds, the conditions for equal memory in all three stages are

$$N_1 + N_2 + N_3 = N$$

$$N_2 = \frac{N_1(1 - P_1)}{P_1}$$

$$xP_1^2\left(\frac{N_3}{N_2} + 1\right) = 1\,.$$

It is possible to eliminate N_2 and N_3 and obtain a relation involving x, P_1, N_1, and N,

$$N_1 = \frac{NxP_1^3}{1 - P_1 + xP_1^3}\,.$$

References

[1] A good place to start is with the "Selected SETI References and Reading List," with commentary by Charles L. Seeger in *SETI Science Working Group Report*, F. Drake, J. H. Wolfe, and C. L. Seeger, eds., NASA Technical Paper 2244 (1983).*

[2] D. K. Cullers, Ivan R. Linscott, and Bernard M. Oliver, "Signal Processing in SETI," *Comm. ACM,* **28**, 1151–1163 (1985) and *Computer* **18**, No. 11, 37–47 (1985) ACM/IEEE–CS Joint Special Issue.

[3] *The Search for Extraterrestrial Intelligence: SETI*, P. Morrison, J. Billingham, and J. Wolfe, eds., NASA SP–419 (1977). [Reprinted, with trivial deletions by Dover Publications, New York (1979).]

[4] Stanley R. Deans, *The Radon Transform and Some of Its Applications,* John Wiley & Sons, New York (1983).

[5] Stanley R. Deans, "The Radon Transform," in *Mathematical Analysis of Physical Systems,* R. E. Mickens, ed., Van Nostrand Reinhold, New York (1985), pp. 81–133.

[6] P. D. Welch, "The use of fast Fourier transform for the estimation of power spectra: A method based on time averaging over short, modified periodograms," *IEEE Trans. Audio Electroacoust.*, **AU–15**, 70–73 (1967).

[7] F. J. Harris, "On the use of windows for harmonic analysis with the discrete Fourier transform," *Proc. IEEE*, **66**, 51–83 (1978).

[8] Stanley R. Deans, "Distributions of quadratic forms in normal variables," *Bull. Am. Phys. Soc.*, **32**, 49 (1987).

[9] T. A. C. M. Classen and W. F. G. Mecklenbräuker, "The Wigner Distribution — a tool for time-frequency signal analysis, Part II: Discrete-time signals," *Phillips J. Res.* **35**, 276–300, (1980).

[10] N. L. Johnson and Samuel Kotz, *Continuous Univariate Distributions - 2,* John Wiley & Sons, New York (1970).

---

*Anyone who would like to have more details about the SETI program now being planned at NASA (and general SETI literature) can obtain information by writing to the SETI Program Office, NASA–Ames Research Center, Moffett Field CA 94035.

INDUCTIVE INFERENCE BY NEURAL NETWORKS

David Hestenes
Physics Department
Arizona State University
Tempe, AZ 85287

Abstract. Neural networks which emulate many characteristics of human perception have recently been developed. Here we discuss underlying network design principles and mechanisms and explore their relation to theories of signal processing and statistical inference.

We honor Ed Jaynes at this Workshop, because he is the chief architect of a universal theory of statistical inference based on the twin principles of maximum entropy and Bayesian inference. This whole series of Workshops, bringing together scientists and engineers with diverse backgrounds, has served to deepen our understanding of the theory and confirm the claim of universality by documenting the rapidly expanding range of successful applications. Despite the impressive success of the theory, Ed has repeatedly warned us that the theory is incomplete, that there must be additional fundamental principles waiting to be discovered. He has admonished us, moreover, that new principles can only be discovered by grappling with difficult new practical problems. In that spirit, my purpose in this talk is to introduce you to an exciting new direction for research where such discoveries can be expected and, I believe, the greatest examples of statistical inference are yet to be seen.

At this Workshop four years ago I was rash enough to announce the beginning of a new scientific revolution promising to produce a genuine mathematical theory of how brains work. May talk has proved to be prophetic. There has since been an explosive growth of research in the emergent field of neural networks. As an index of that growth I can tell you that in 1983 I helped organize one of the rare conferences on neural networks; about 60 people attended. Last year the American Institute of Physics sponsored a conference on the subject which attracted a turnaway crowd approaching 300. This year the IEEE sponsored the First Annual International Conference on Neural Networks with nearly 2,000 excited participants. The International Neural Network Society was formed this year, and the first issue of its official journal will be published in January. A sense of the great range and richness of research in this field can be gained from the published Proceedings of the two recent conferences ([1],[2]) and several fine volumes of collected papers ([3],[4]).

I am happy to say that the paper I presented at the 1983 Maxent Workshop [5] has helped introduce many people to this new field. From what I have heard, at least a thousand copies of it have been privately distributed throughout the country. As the paper is just now being published after a four year delay, I don't think I can do better today than continue my discussion of the fundamental ideas it dealt with. Happily, the paper has not become outdated, and there is nothing in it I

G. J. Erickson and C. R. Smith (eds.),
Maximum-Entropy and Bayesian Methods in Science and Engineering (Vol. 2), 19–32.

need to take back. Today I intend to supplement it with some recent developments. But my main object is to start building bridges between field represented at this workshop and the field of neural networks.

Let me offer two good reasons for being interested in neural networks, a practical one and a theoretical one. As a matter of utility, neurocomputers can speed up certain kinds of complex computations, especially in the inverse problems so dear to the hearts of many people here. A neurocomputer is a massively interconnected parallel computer. Only this year the first commercial neurocomputers appeared on the marked as peripherals to conventional computers. These are little more than toys for researchers. But I predict that within a year the next wave of neurocomputers will bring us a powerful, practical and affordable computational tool. Moreover, improvements will accrue at a rapid rate for many years.

The theoretical reason for being interested in neural networks is even more exciting than the practical one. To put it briefly, neural networks are capable of inductive inference. The ultimate neural network, the brain, is an inference machine par excellence. The question is, how does the brain do it, and how can we design an artificial network to do as well? I am not referring here to inductive inference with propositions, Bayesian or otherwise. Humans are comparatively poor at that. I wish to refer mainly to the exquisite inference unconsciously performed in human perception. This is the ultimate in complex signal processing. So it invites comparison with the methods of signal processing so often discussed in these workshops.

I THE ADAPTIVE FILTER

I begin by calling your attention to a basic computational structure which plays a fundamental role in conventional signal processing as well as in neural networks. It is also implicit in many computational algorithms which have been discussed at these workshops. It is called an adaptive filter.

A pioneer in applying the adaptive filter to signal processing is Bernie Widrow, who recently published an entire book [6] on the subject. Widrow was also one of the pioneers in neural networks beginning in the fifties. Unfortunately, research in the field collapsed in the sixties when the "meager" results could not match the optimistic predictions or the successes of the serial, digital computers. Widrow jumped ship, but he took the adaptive filter along and made a name for himself by applying it to signal processing. Only last year he discovered that the field of neural networks had revived. He has rejoined it with the optimistic prediction that it will not collapse again, because technology now has the capability to build networks that could only be imagined in the early days.

The design for an adaptive filter is exhibited in Fig. 1. The input variables in the n channels can be regarded collectively as a pattern vector $\mathbf{x} = [x_1, x_2, \ldots, x_n]^T$, where T denotes transpose. Each channel has an adjustable multiplicative weight, and the set of weights can also be represented as a vector $\mathbf{w} = [w_1, w_2, \ldots, w_n]^T$. The filter output y

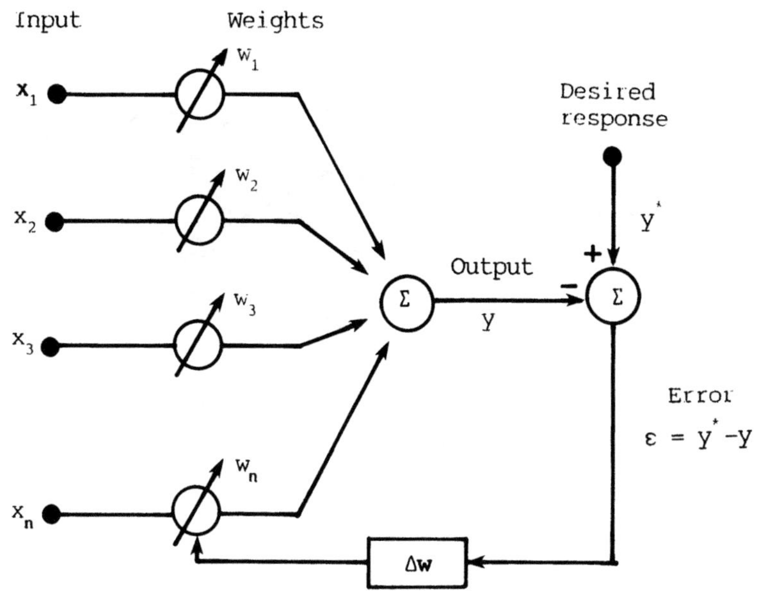

Fig. 1. An adaptive filter

is a linear combination of the inputs as determined by the transfer equation

$$y = \mathbf{w} \cdot \mathbf{x}.$$

This is the basic equation for a linear filter. It becomes an adaptive filter by introducing an adaptive algorithm which specifies an adjustment $\Delta\mathbf{w}$ of the weight vector according some criterion. The simplest and most widely used criterion is to reduce the mean square error $\varepsilon^2 = (\mathbf{y}^* - y)^2$ between the actual output y and some desired output y^*. As Widrow [6] explains in detail, this leads to the least mean square (LMS) adaptive algorithm

$$\Delta\mathbf{w} = \mu(y - y^*)\mathbf{x},$$

where μ is a constant determining the adjustment rate. This is sometimes called the delta rule in neural network applications [3].

Adaptive filters can be classified into two types according to the origins of their inputs. If then inputs x_k all come from different sources, it is of multiple-input type. If the set of inputs $\{x_1, x_2, x_3, \ldots\}$ is a time series from a single source it is of transversal type. For a statistically stationary input, a transversal adaptive filter obeying the delta rule is equivalent to a Wiener filter [6].

To make the connection with neural networks theory, it should be noted that the anatomy of the filter in Fig. 1 is precisely that of an instar

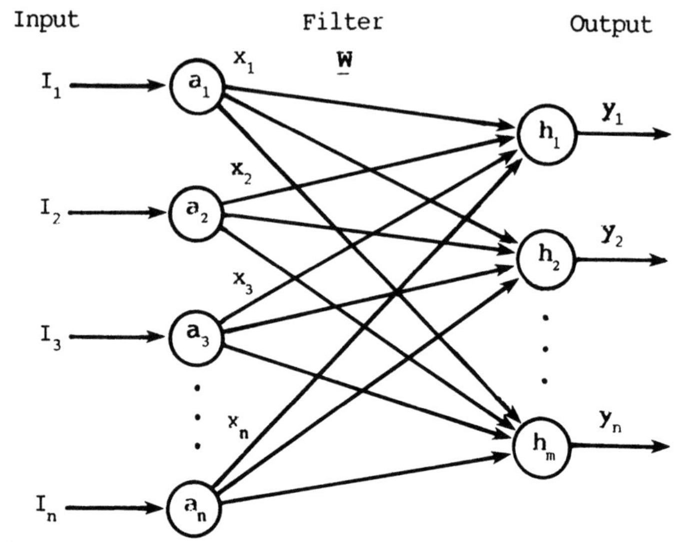

Fig. 2. A linear associator .

in a neural network. The instar was discussed in my previous lecture [5]. It describes multiple axons (or input channels) converging on a single neuron (or processing element). The input signals are multiplicatively "gated" by synaptic weights at synapses and combined additively in the postsynaptic neuron. We turn now to a generalization of the adaptive filter and consider how it functions in more complex networks.

II THE OPTIMAL LINEAR ASSOCIATOR

A generalization of the linear filter is the linear associator diagrammed in Fig. 2. It consists of two layers (or slabs) of processing elements, $\{\mathbf{a}_1, \mathbf{a}_2, \ldots, \mathbf{a}_n\}$ and $\{\mathbf{h}_1, \mathbf{h}_2, \ldots, \mathbf{h}_m\}$. These processing elements are commonly referred to as neurons (or nodes) without necessarily claiming that they represent biological neurons.

The internal state of a neuron is represented by a single variable called its activity. Thus x_i is the activity of neuron $\mathbf{a}_i$ and y_j is the activity of $\mathbf{h}_j$. In this section we deal only with the simplest model of a neuron, for which the activity is equal to the sum of its inputs at every time. Thus, $x_i = I_i$ for the neurons $\mathbf{a}_i$ in Fig. 2. These neurons simply register the inputs and then distribute them to all the neurons on the next level.

Each neuron $\mathbf{h}_j$ in the second level of Fig. 2 is an instar with activity

$$y_j = \mathbf{w}_j \cdot \mathbf{x},$$

where $\mathbf{w}_j$ is its input weight vector. The entire activity pattern $y = [y_1, y_2, \ldots, y_m]^T$ of the output level is thus given by the matrix equation

$$\mathbf{y} = \underline{\mathbf{W}}\mathbf{x}$$

where $\underline{\mathbf{W}} = [\mathbf{w}_1^T, \mathbf{w}_2^T, \ldots, \mathbf{w}_m^T]^T$. This is the equation for a linear filter with transfer function $\underline{\mathbf{W}}$. It maps an input pattern $\mathbf{x}$ into an output pattern $\mathbf{y}$. We turn it into an adaptive filter by introducing an adaptive algorithm. The generalization of the LMS algorithm to this case is

$$\Delta\underline{\mathbf{W}} = \mu(\mathbf{y}^* - \underline{\mathbf{W}}\mathbf{x})\mathbf{x},$$

where $\mathbf{y}^*$ is the desired output pattern.

This adaptive filter can be regarded as a <u>linear associator</u>, because it can be trained to associate a desired output $\mathbf{y}^*$ with a given input $\mathbf{x}$. A "teacher" repeatedly presents the desired input-output pair $\{\mathbf{x}, \mathbf{y}^*\}$ and the weight is adjusted by the LMS rule. It can be proved that the actual outputs will converge to the desired output ([6],[7]). For that reason the associator is said to be optimal.

The associator can learn more than one input-output, or if you will, stimulus-response (S-R) pair. Presented with a teaching input $\{\mathbf{x}_k, \mathbf{y}_k\}$, $k = 1, 2, \ldots, p$, the S-R pairs are <u>stored</u> in the associator by the LMS-rule

$$\Delta\underline{\mathbf{W}} = \mu(\mathbf{y}_k - \mathbf{W}\mathbf{x}_k)\mathbf{x}_k .$$

Then $\mathbf{y}_k$ can be <u>retrieved</u> from $\mathbf{x}_k$ by

$$\mathbf{y}_k = \underline{\mathbf{W}}\mathbf{x}_k .$$

The learning is perfect if the $\{\mathbf{x}_k, \mathbf{y}_k\}$ pairs are orthogonal to one another. If they are not orthogonal, the learning of one pair interferes with the learning of another. This is a serious limitation of the linear associator. Besides restricting the classes of pattern sets that can be learned, it severely limits the storage capacity of the associator. Another limitation of the linear associator is that it cannot be generalized by adding additional processes levels, because the composite of linear transformations is equivalent to a single linear transformation and hence can be computed by the two-level associator.

The linear associator and various nonlinear generalizations are discussed by Kohonen [7]. His book is a good introduction to neural networks for engineers and physicists.

III <u>BACK-PROPAGATION</u>

Now we consider how some of the limitations of the linear associator can be lifted by introducing nonlinearity into the neuron outputs and adding one more processing layer. In my previous lecture [5] I discussed strong theoretical reasons for requiring that the output

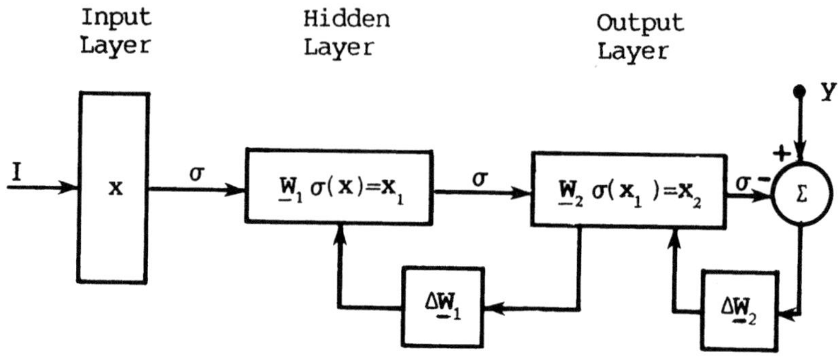

Fig. 3. Backpropagation adaptive algorithm

signal of a neuron be a sigmoid function of its activity. In addition, there is strong experimental evidence that many biological neurons have this property. Thus, for neuron $\mathbf{a}_i$ the output is $s_i = \sigma(x_i)$, where σ is the sigmoid signal function. For the output of an entire slab with activity pattern $\mathbf{x}$ we write $\mathbf{s} = \sigma(\mathbf{x})$.

The specific sigmoid function $\sigma(x) = (1 + e^{-x})^{-1}$ offset by a suitable threshold (or bias) Γ is frequently employed in applications.

The back propagation adaptive algorithm is implemented in a three layer network as shown in Fig. 3. The output $\sigma(\mathbf{x})$ from an activity pattern $\mathbf{x}$ on the input layer is filtered by a transfer function $\underline{\mathbf{W}}_1$ to produce an activity pattern $\mathbf{x}_1 = \underline{\mathbf{W}}_1 \sigma(\mathbf{x})$ on the middle layer. This, in turn, is filtered by W_2 to produce a pattern $\mathbf{x}_2 = \underline{\mathbf{W}}_2 \sigma(\mathbf{x}_1)$ on the output layer. Then, the output $\sigma(\mathbf{x}_2)$ is compared with the desired response $\mathbf{y}$, and the error is fed into the adaptive algorithm to determine an adjustment $\Delta\underline{\mathbf{W}}_2$ in $\underline{\mathbf{W}}_2$. This adjustment implies an error in the output from the hidden layer which, in turn, determines an adjustment $\Delta\underline{\mathbf{W}}_1$, of $\underline{\mathbf{W}}_1$. This completes the computation cycle of the network, so it is ready to adapt to a new input.

The back propagation algorithm is discussed at length in [3]. We only note here that it is an LMS algorithm, differing from Widrow's algorithm only by including the sigmoid function σ. Like Widrow's algorithm it is simple and LMS-optimal, so it has many potential applications. So far its most impressive success is a network called NETtalk developed by Sejnowski and Rosenberg [8]. This network can be trained to read unrestricted written text aloud in English or in any other language, for that matter. It is trained on a sample of text which has been marked to specify the phonological output to be associated with each letter. It is impressive to listen to its output as the network discovers the difference between vowels and consonants and gradually learns to make more difficult phonological distinctions as it progresses from babbling to quite intelligible speech.

The performance of NETtalk is not quite as good that of some commercial products like DECtalk. However, the commercial products are much more complex and took much more effort to develop. The really significant difference is in how the systems work. DECtalk is supplied with a pronunciation dictionary and a system of phonological rules. In contrast, NETtalk is merely supplied with a set of examples from which it extracts a set of features to be associated with each phoneme. It learns to reconcile rules and exceptions by becoming sensitive to the contexts in which they apply. Thus it exhibits some of the characteristics of human learning.

It is important to understand the function of the hidden layer. It develops a recoding of the inputs into classes with invariant features to be associated with the various outputs. For example, in NETtalk after training some neurons in the hidden layer are turned on only by vowels or consonants while other features of speech are related to groups of neurons. The coding is so subtle, however, that it has not proved possible to interpret it completely.

An important justification for the hidden layer comes from a theorem by Kolmogorov. The adaptation of the theorem to neural networks is discussed by Hecht-Nielsen [2]. It asserts that any continuous mapping from the n-dimensional unit cube $[0,1]^n$ to the m-dimensional real space R^m can be implemented exactly by a 3 layer neural network with n, 2n + 1 and m neurons in the respective layers. The back propagation algorithm provides a means for constructing such a mapping from specified input-output pairs.

A defect of the conventional back propagation algorithm is that it is nonlocal in the sense that it uses information available only at the output layer to compute weight adjustments at the hidden layer. For that reason, back propagation is not a plausible computation scheme in biological networks.

IV STATE OF THE ART

The networks we have examined so far are capable of learning only under the supervision of a teacher who tells them the correct responses to their inputs. Now we consider the design of networks capable of unsupervised learning. Design principles for a class of such networks have been developed in the Adaptive Resonance Theory (ART) of Grossberg and Carpenter ([4],[9]). Among many alternative pattern recognition theories in the literature, ART is by far the most advanced. Its great strength is in identifying and resolving fundamental problems which the others have overlooked or ignored. The theory is still under development. It has been implemented and tested in a series of network models, ART-1 in [9] and ART-2 in [2], with successively greater complexity and capabilities. Although the performance of these models is not yet as spectacular as that of NETtalk, considering the theory underlying their designs, one can anticipate impressive results in the future.

My previous lecture [5] included an introduction to ART. Here we will go over some of the same ideas, but from a different angle to help relate them to ideas about statistical inference which often appear in these workshops.

The basic ART model is a two level network like the one in Fig. 2. To build it we start with Fig. 2 and elaborate. To understand what the network does, it is helpful to provide its elements with an interpretation. For the purpose of pattern recognition, we can interpret $\mathbf{a}_i$ as attribute detector and its activity x_i as a measure of the degree to which the attribute is present in the input pattern $\mathbf{I} = [I_1, I_2, \ldots, I_n]^T$. The pattern $\mathbf{x} = [x_1, x_2, \ldots, x_n]^T$ is therefore a profile of the attributes detected in the input $\mathbf{I}$. The kind of attributes detected will depend on the source of the input $\mathbf{I}$ including any preprocessing they have undergone. For example, the input might be the power spectrum of a time series, or the I_i might be pixel intensities from an image detector. In an expert system for medical diagnosis the I_i could be the results of medical tests and observations.

The first processing stage is the registration of an attribute pattern $\mathbf{x}(t)$ computed from the input $\mathbf{I}(t)$. Depending on the application, the attribute slab may be designed so the $\mathbf{a}_i$ interact "laterally" with one another. We need not repeat the discussion in [5] of the lateral interactions and the dynamical equations of the $\mathbf{a}_i$. It will suffice to mention some of their consequences. First of all, the relaxation time in the equations of motion for the x_i is short compared to time variations of the input, so a stable pattern $\mathbf{x}(t)$ tracks the input $\mathbf{I}(t)$. Registration normalizes the input pattern and may contrast enhance it. The normalization condition can be written

$$\sum_i x_i = 1.$$

This makes it possible to interpret the attribute pattern $\mathbf{x}$ as a probability distribution with

$$x_i = P(\mathbf{a}_i \mid \mathbf{I}),$$

which says that x_i represent the probability of attribute $\mathbf{a}_i$ given $\mathbf{I}$.

A pattern classifier places patterns with similar attribute profiles in a single category which may be regarded as an object defined by the pattern profile. Whether the object is to be regarded as real or abstract is another matter. The ART network can be designed so that each neuron $\mathbf{h}_j$ in the second layer represents a distinct object (or category). Thus $\mathbf{h}_j$ can be regarded as an object detector, and its activation signifies "presence" of the object.

The object layer is activated by bottom-up input from the attribute layer. The signal $\mathbf{s} = \sigma(\mathbf{x})$ from the object layer is filtered by a transfer matrix $\underline{\mathbf{W}}$ to produce an input $\mathbf{T}$ to the object layer given by

$$\mathbf{T} = \mathbf{W}\mathbf{s}$$

In component form this can be written

$$T_j = \sum_i w_{ji} \; s_i$$

This equation can be given a probabilistic interpretation as follows. The attribute neurons often operate in the linear range of the signal function, in which case we can write $s_i = \sigma(x_i) = x_i = P(\mathbf{a}_i)$. Ignoring a possible normalization factor, we can interpret w_{ji} as the probability of object $\mathbf{h}_j$ given attribute $\mathbf{a}_i$ and write

$$w_{ji} = P(\mathbf{h}_j \,|\mathbf{a}_i).$$

The transfer equation then gives us

$$T_j = \sum_i P(\mathbf{h}_j \,|\mathbf{a}_i)\; P(\mathbf{a}_i \,|\mathbf{I}) = P(\mathbf{h}_j \,|\mathbf{I}).$$

Thus T_j is the probability that object $\mathbf{h}_j$ is present.

An alternative possibility is to follow Uttley [10] and interpret w_{ji} as the channel entropy

$$w_{ji} = \log P(\mathbf{h}_j \,|\mathbf{a}_i)$$

Then

$$T_j = \sum_i P(\mathbf{a}_i \,|\mathbf{I}) \log P(\mathbf{h}_j \,|\mathbf{a}_i)$$

can be interpreted as the evidence for $\mathbf{h}_j$. This makes sense because evidence from independent sources is additive. However, we will stick with the probability interpretation of the transfer matrix. It should be understood that these interpretations are by no means essential to neural network theory. They are suggested here to facilitate comparison with more conventional approaches to signal processing and pattern recognition.

The bottom-up input pattern **T** activates the object layer. We suppose that the lateral interactions in this layer are designed to strongly contrast enhance the input. Indeed, we suppose the enhancement is so strong that only the neuron $\mathbf{h}_j$ with largest input T_j is activated. Thus, pattern registration in the object layer is a decision as to which category the input belongs. The decision is the conventional "Bayesian choice" of the category which is most probable on the given evidence.

Of course, the accuracy of the classification depends crucially on the filter $\underline{W}$ so we must specify how it is determined. The filter is adaptive with an adaptive algorithm in the form a <u>learning law</u> which, in the simplest case, is a diffential equation of the form

$$\tau \frac{dw_{ji}}{dt} = y_j (s_i - w_{ji})$$

where $s_i = \sigma(x_i)$, as before, and τ is a constant. Such learning laws and their implications were discussed in my previous talk [5]. In the mean time, this specific form of the learning law has received strong experimental support in the work of Levy [11] and other. It also finds theoretical support in ART where it is shown to be necessary for efficient category formation. The term $y_j s_i$ in the learning law is said to be Hebbian while the term $-y_j w_{ji}$ is said to be anti-Hebbian. For attributes $\mathbf{a}_i$ with $s_i > 0$ when $y_j > 0$, the Hebbian term dominates and, as shown in [3], w_{ji} asymptotically becomes essentially the correlation function of y_j and s_i. On the other hand, when $s_i = 0$, the anti-Hebbian term decreases w_{ji} exponentially, which is to say that attribute $\mathbf{a}_i$ is irrelevant to the identification of object $\mathbf{h}_j$.

This process of adaptively forming the object recognition filter **W** was called the Instar coding theorem in [5]. It should be recognized as a kind of inductive inference, because it spontaneously develops a general category from similarities among a few input patterns. The inference process is sharpened by additional structure of an ART network which we discuss next.

We now suppose that the output $\mathbf{U} = \sigma(\mathbf{y})$ of the object layer feeds back through an adaptive filter $\mathbf{Z}$ to produce a top-down input $\mathbf{J} = \mathbf{ZU}$ to the attribute layer, as indicated in Fig. 4. This is like folding the three level network in Fig. 3 at the hidden level and identifying the output layer with the input layer. Our objective in the rest of this talk will be to understand qualitatively the function of the top-down feedback to the attribute layer.

An interpretation of the top-down input **J** is supplied by the outstar learning theorem discussed in the previous lecture [5]. The learning law for the top-down filter **Z** is essentially the same as for the bottom-up filter **W**, but the boundary conditions are different. According to the outstar theorem, while an object detector $\mathbf{h}_j$ is on its component of the filter **Z** samples the activity patterns that appear on the attribute layer and asymptotically learns average of those patterns. This average pattern is called a prototype of the category (or object) $\mathbf{h}_j$. The entire **Z** therefore contains prototypes for all the categories, so we may interpret it as a prototype filter. The prototype of $\mathbf{h}_j$ can be regarded as the most typical pattern in the category. Indeed, its attribute profile can be taken as a definition of the category. Note that prototype learning is a kind of inductive inference complementary to object recognition learning. Surely a complete theory of inductive inference should recognize and reconcile both kinds. Adaptive Resonance Theory aims to do that.

We can assert now that the pattern $\mathbf{J} = \mathbf{Z}\sigma(\mathbf{y})$ is a prototype of the object $\mathbf{h}_j$ which activated it. This prototype is read into the attribute layer where it can be compared with the original input **I**. However, the comparison is indirect and subtle, because it is only incidental to the functioning of the network.

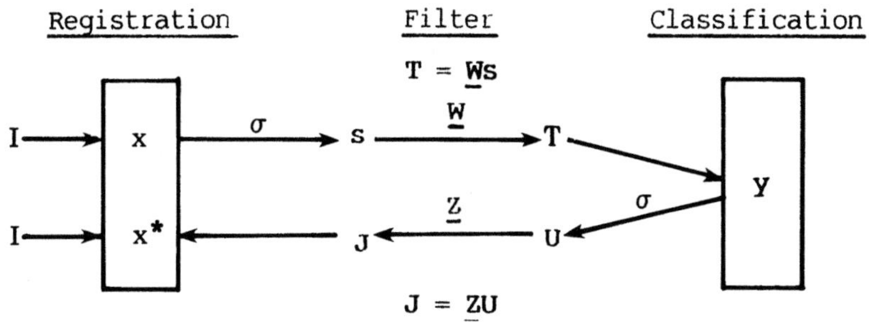

Fig. 4. Activation sequence in an ART network

The bottom-up input **I** and the top-down input **J** combine additively to a resultant input **I+J** which activates a new attribute pattern $\mathbf{x}^*$. Now there are two possibilities, depending on whether or not the output signal $\mathbf{s}^* = \sigma(\mathbf{x}^*)$ maintains the activation of the category detector $\mathbf{h}_j$ which was turned on initially by the bottom-up signal $\mathbf{s} = \sigma(\mathbf{x})$. If it does, we say that the patterns **I** and **J** are <u>matched</u>. Otherwise, we say that **I** and **J** are <u>mismatched</u>. The processing in these two cases is quite different, so we discuss them separately.

The precise condition for a <u>match</u> of **I** and **J** depends on the nonlinear dynamics of the attribute layer. Without getting into mathematical details, it can be said that a match requires significant overlap of the two patterns. The prototype pattern **J** defines <u>features</u> of the object $\mathbf{h}_j$ which are amplified in the attribute pattern $\mathbf{x}^*$ at neurons where they overlap the input pattern **I**. This generates a resonant exchange of signals between attribute and object layers which is maintained until the driving input **I** changes significantly. Such a state is called an <u>adaptive resonance</u>. In the (unlikely) event that the bottom-up input is a prototype, we have **I** = **J** and, for signal functions in the linear range, we have roughly

$$\mathbf{x}^* = \underline{Z}\mathbf{y} = \underline{ZW}\mathbf{x} = \mathbf{x}.$$

Thus, prototypes are eigenvectors of the composite $\underline{\mathbf{ZW}}$ filter. This conclusion is only qualitative, but it is suggestive. More generally, we have a similar eigenvector relation on subspaces where the input **I** overlaps the prototype.

In an adaptive resonance the attribute pattern $\mathbf{x}^*$ can be regarded as a <u>global percept</u> of the network reconciling external evidence **I** with internal evidence **J** derived from past experience. Besides accentuating features of **I** where it overlaps, the prototype **J** performs a kind of image restoration, filling-in pieces of the expected pattern which are missing from **I**. As many psychological experiments show, this so-called Gestalt filling-in is characteristic of human perception.

An adaptive resonance drives changes in the adaptive filters $\mathbf{W}$ and $\underline{\mathbf{Z}}$ to incorporate information in the present input **I**. Two things happen to $\underline{\mathbf{Z}}$. First, features of the prototype which occur most frequently in the external inputs are accentuated by the Hebbian term in the learning law. Thus, the prototype evolves a set of critical features which are most essential to identifying the object. If attribute $\mathbf{a}_i$ is active whenever object $\mathbf{h}_j$ is identified, for linear signaling the Hebbian term in the learning equations implies (except for a possible scale factor) an equivalence of correlation functions so that

$$w_{ji} = z_{ij}$$

In other words, on the subspace of critical features $\underline{\mathbf{Z}}$ is the transpose of $\mathbf{W}$. Consequently, on this subspace $\underline{\mathbf{Z}}\mathbf{W} = \mathbf{W}^T\mathbf{W}$ is a symmetric matrix with real eigenvalues and eigenvectors. This agrees with our previous conclusion that prototypes are eigenvectors of $\underline{\mathbf{Z}}\mathbf{W}$. The second thing that happens to $\underline{\mathbf{Z}}$ in an adaptive resonance is that, for attributes $\mathbf{a}_i$ which are not activated above the threshold of the signal function, $\sigma(x_i) = 0$, so the anti-Hebbian term drives z_{ij} toward zero. Thus, insignificant features are suppressed while critical features in the prototype are enhanced. In this way the prototype evolves a sharper definition of the object.

As noted before, the adaptive process just described is a process of inductive inference. It incorporates new information into more predicative representations of experience. As such, it invites comparisons with Bayes' theorem, which is the formal mechanism for induction in the statistical theory discussed in these workshops. There does not appear to be any exact analog of Bayes' theorem in the ART networks. However, the networks do "strive" for a globally consistent representation of the available information, and Bayes' theorem is a consequence of consistency requirements in statistical inference. One consequence of consistency is the relation $w_{ji} = z_{ij}$ for critical feature filters. With the probalistic interpretation this becomes

$$P(\mathbf{h}_j|\mathbf{a}_i) = P(\mathbf{a}_i|\mathbf{h}_j),$$

a somewhat stronger relation than can be derived from Bayes' theorem alone. Perhaps the network is able to infer more from the consistency requirement than is in Bayes' theorem. This may in part be due to the fact that, in contrast to statistical theory, the network does not separate inferences from decisions. Also, there is another component of network inference to which we now turn.

A mismatch between the external input **I** and the prototype **J** implies that the initial bottom-up identification of category $\mathbf{h}_j$ was mistaken, so $\mathbf{h}_j$ should be shut off and alternative classifications should be considered. That requires additional network mechanisms which are too elaborate to consider here. Details are given in [9]. We must be content here with a few quantitative comments. Mismatch initiates a process of rapid reset and search among the available categories until a satisfactory match is found or a new category is formed. The possibility of mismatch increases as the categories become more sharply defined by sharp prototypes. Sharp prototypes have strong classification power, so they must be protected against adventitious recoding by mismatched patterns. The reset mechanism does this.

To conclude, let met summarize the main characteristics of an ART network. There are three. The network is self-organizing, self-stabilizing and self-scaling in response to pattern complexity.

We have not had the opportunity to discuss the self-scaling property. As explained by Carpenter and Grossberg [9], self-scaling is due to a refinement of the pattern matching mechanism in the attribute layer. The mechanism compares whole patterns instead of comparing the parts of patterns separately. This entails a definition of noise that depends on pattern complexity. More complex patterns can tolerate more noise, because they have more structure as a basis for discrimination. Theoretically, it appears that self-scaling is essential for the network to be capable of developing distinct categories for subset and superset patterns.

Self-organization is the most fundamental property of the ART network. The network learns inductively without a teacher in an arbitarily complex environment (which determines the external inputs to the network). It recognizes similarities among input patterns from which it develops a pattern classification code unique to its experience. The code learning is a progressive refinement of distinctions from which there emerges a set of critical features which may become stabilized as invariants defining abstract categories and real objects. The network continually updates its knowledge base as it recognizes objects by adaptive resonance.

Self-stabilization is essential to protect the evolving object recognition code. It is achieved by a sharpening of prototypes through resonant feedback matching and a reset mechanism which rapidly corrects errors in object identification and initiates a search for an alternative classification. This makes the network sensitive to novelty in its environment. The network analyzes its experience with a kind of open logic that creates new categories as needed.

The exciting thing about ART is that it captures many characteristics of human perception. Of course, it is to understand perception that ART is being developed.

REFERENCES

[1] J. Denker (ed.), 1986, Neural Networks for Computing, American Institute of Physics, NY.

[2] Proceedings of the IEEE First Annual International Conference of Neural Networks, 1987, IEEE.

[3] D. Rumelhart and J. McClelland (eds.), 1986, Parallel Distributed Processing, MIT Press, Cambridge Mass., Vol. I, II.

[4] S. Grossberg, 1987, The Adaptive Brain, I: Cognition, learning, reinforcement and rhythm, II Vision, speech, language and motor control, Amsterdam, Elsevier/North-Holland.

[5] D. Hestenes, 1987, "How the Brain Works, the next great scientific revolution", in C. Ray Smith (ed.) Maximum Entropy and Bayesian Spectral Analysis and Estimation, Reidel, Dordrecht/Boston, p. 173-205.

[6] B. Widrow and S.D. Stearns, 1985, Adaptive Signal Processing, Prentice-Hall Inc.

[7] T. Kohonen, 1984, Self-Organization and Associative Memory, Springer-Verlag, Berlin.

[8] T. Sejnowski and C. Rosenberg, 1986, "NETtalk: A parallel network that learns to read aloud", Technical Report JHU/EECS-86/01 John Hopkins University, Baltimore, MD.

[9] G. Carpenter and S. Grossberg, 1987, "A Massively Parallel Architecture for a Self-Organizing Neural Pattern Recognition Machine", Computer Vision, Graphics and Image Processing, 37, 54-115.

[10] A. Uttley, 1979, Information Transmission in the Nervous System, Academic Press, NY.

[11] W.B. Levy, J.A. Anderson, S. Lehmhuhle, 1985, Synaptic Modification, Neuron Selectivity and Nervous System Organization, Lawrence Erlbaum, Hillsdale, NJ.

ON THE EFFICIENCY OF A CLASS OF MAXIMUM ENTROPY ESTIMATORS

C.C. Rodriguez*
State University of New York at Albany
Department of Mathematics and Statistics
Albany, New York 12222

*This research was supported by PHS grant number 1-R01-CA41171-01A1 awarded by the National Cancer Institute, DHHS.

1 INTRODUCTION

Let $\mathbb{P}$ be a family of probability measures and let $P_0 \in \mathbb{P}$. A classical problem of statistical inference is the estimation of an euclidean parameter $\nu(P_0)$ when n indepedent and identically distributed (iid) observations from P_0 are available.

The parametric case i.e. the case when $\mathbb{P}$ is a regular parametric model has been extensively investigated by LeCam (1956), (1969), (1970) and many others during the past thirty years. The asymptotic theory in nonparametric models has been the subject of comparatively much more recent investigations (an exception being Stein, 1956). In this paper we study asymptotic efficiency in nonparametric models with the approach of Koshevnik and Levit (1976), Pfanzagl (1982) and specially Bickel, Ritov and Wellner (1987). We are concerned with nonparametric models P of the form

$$\mathbb{P} = \{P \text{ on } \Omega : \int v(x)P(dx) \geq 0 \} \tag{1.1}$$

and

$$\mathbb{P} = \{P \text{ on } \Omega : \int v(x)P(dx) = 0\} . \tag{1.2}$$

where Ω is a fixed compact subset of $\mathbb{R}$ and $v:\Omega \to \mathbb{R}$ is a fixed bounded function. The integral sign above (as in the rest of the paper) denotes integration over Ω. Apparently, more general models $\mathbb{P}$ are obtained by allowing Ω and v(x) to be multidimensional and/or unbounded. However, the extension of the results presented in this paper to the case when Ω and v(x) are finite dimensional is trivial, involving mainly notational changes. The infinite dimensional case and some aspects of unboundedness involve further technicalities and they will not be considered here.

G. J. Erickson and C. R. Smith (eds.),
Maximum-Entropy and Bayesian Methods in Science and Engineering (Vol. 2), 33–49.

Models of the type (1.1) and (1.2) naturally appear as a priori descriptions of complex systems where the only information available is about the expectation of a function v. A paradigmatic example of (1.2) occurs in statistical mechanics (see Jaynes 1957). There, it is required to make predictions about the state of the system from macroscopic measurements (e.g. of pressure and temperature) obtained as global interactions (e.g. averages) with the system as a whole.

2 THE MAXIMUM ENTROPY PRINCIPLE

We aim to use the Maximum Entropy Principle to estimate a real parameter $\nu(P_0)$ where $P_0 \in \mathbb{P}$ and data from P_0 are available. The Maximum Entropy Principle can be defined in general terms as follows:

Let P be a probability measure and Q be a sigma-finite measure on the same measurable space Ω and denote by

$$I(P:Q) = \begin{cases} \int \log \frac{dP}{dQ}\, dP & \text{if } P \ll Q \\ +\infty & \text{otherwise}, \end{cases} \tag{2.1}$$

the Kullback number between P and Q. Where $P \ll Q$ denotes absolute continuity of P with respect to Q (i.e. events with positive Q-measure are also possible events for P) and dP/dQ denotes the Radon-Nikodym derivative.

The quantity I(P:Q) measures the mean information per discrimination in favor of P, against Q when sampling from P (see Kullback, 1959 page 5). The measure Q is interpreted as the best current description of the underlying probability measure $P_0 \in \mathbb{P}$. Where the class $\mathbb{P}$ is assumed to contain all the available a priori information about P_0. If $Q \notin \mathbb{P}$ the maximum entropy principle suggests to replace Q (which initially might be an improper σ-finite measure e.g. Lebesgue measure) by its I-projection on $\mathbb{P}$ i.e. the solution, P*, of the variational problem

$$\min_{P \in \mathbb{P}} I(P:Q) \,. \tag{2.2}$$

As stated here this method of inference was first suggested by Jaynes (1978) even though a special case appears in Kullback (1959). General properties of I-projections are investigated in Csiszar (1975) and connections with the general problem of smoothing can be found in Rodriguez (1986).

The solution of (2.2) gives the probability measure $P^* \in \mathcal{P}$ that is most difficult to discriminate from the intial guess Q. If we have reasons to believe that Q is a good initial guess but it does not agree with all the prior information (i.e. $Q \notin \mathcal{P}$) then, the solution P* appears as a compromise between our initial guess and the prior information that $P_0 \in \mathcal{P}$. We can therefore say that P* is the minimum modification of Q (in the sense of being most difficult to discriminate from Q) that agrees with the prior information. This interpretation of the separation between P and Q measured by the Kullback number I(P:Q) may be regarded by many as cogent enough to justify its use in a number of special cases. However, I believe that there are even more compelling reasons to think of (2.2) as a fundamental variational problem of statistical inference. Some of these reasons are: The axiomatic derivation of Shore and Johnson (1980), (1981). (Even though, I think that we should expect to see in the future refinements of these axioms, Shore and Johnson have already shown the way toward the logical necessity of the variational problem (2.2)). Other important facts are the strong connection that exists between the Kullback number and the theory of sufficient statistics (see Kullback and Leibler 1956) and (thus) with exponential families (see below, Lemma (3.4) and Kullback, 1959), the asymptotic agreement with conditionalization as proved in VanCampenhout and Cover (1981) and the striking appearence of (2.2) in thermodynamics (see Jaynes 1957a,b, and Gibbs, 1902). If I were pushed to hastily give an analogy I would say that the Kullback number is to inference as the Hamiltonian is to mechanics.

Notice, that by letting Q be a σ-finite measure the variational setting given in (2.2) can readily handle with mathematical rigour (through the Radon-Nikodym theorem) the use of improper noninformative priors. This is a very desirable feature since improper priors seem to be unavoidable even in simple problems (see Mandelbrot (1967), Rodriguez, 1987) and alternative theories propose to replace Kolmogorov's axioms with an axiomatic system for conditional probability (see Renyi, 1955) which is of limited applicability.

If $P^* \in \mathcal{P}$ is the soltuion of (2.2) we call $\nu(P^*)$ the maximum entropy estimator of $\nu(P_0)$. We show in this paper that when P is as in (1.2) and Q is the empirical measure, P_n, of a random sample of size n from P_0 then, for a large class of functionals $\nu(\cdot)$ the maximum entropy estimators $\nu(P^*)$ are asymptotically efficient.

3 STATEMENT OF THE MAIN RESULT

If E is a Banach space we denote by $\|\cdot\|_E$ its norm. The following four lemmas will be useful in the sequel:

Lemma (3.1):

a) If $P<<Q$ then

$$I(P:Q) \leq \log \left\|\frac{dP}{dQ}\right\|_{L_1(P)} = 2 \log\left\|\frac{dP}{dQ}\right\|_{L_2(Q)}$$

with equality if and only if dP/dQ is constant a.e.-Q.

b) If $P<<Q$ and $Q<<P$ then

$$I(P:Q) \geq \log\{1/Q(\Omega)\}$$

with equality if and only if dP/dQ is constant a.e.-P.

Proof: a) Since P is a probability measure and log is strictly concave the result follows from Jensen's inequality. From the change of variables theorem we have

$$\int\frac{dP}{dQ}\, dP = \int\left(\frac{dP}{dQ}\right)^2 dQ \; .$$

Hence,

$$\left\|\frac{dP}{dQ}\right\|_{L_1(P)} = \left\|\frac{dP}{dQ}\right\|^2_{L_2(Q)} \; .$$

b) The function $\varphi \equiv -\log$ is strictly convex and P is a probability then Jensen's inequality implies

$$I(P:Q) = E_P\left[\varphi\left(\frac{dQ}{dP}\right)\right] \geq \varphi\left(\int dQ\right) = \log\{1|Q(\Omega)\}$$

with equality if and only if $\frac{dQ}{dP}=1/(dP/dQ)$ is constant a.e.-P. ∎

Notice that if $Q(\Omega)=\infty$ the inequality b) reduces to the trivial statement $I(P:Q)\geq-\infty$ and if $Q(\Omega)=1$ we obtain the classical convexity inequality $I(P:Q)\geq 0$ valid for probability measures P and Q.

Before we enunicate the next lemma we need the following:

<u>Definition</u>: The set $\{Q_t : t \in \mathbb{R}\}$ of probability measures on Ω is said to be the exponential family, with sufficient statistic $v(x)$, generated by the measure Q if and only if their Q-densities are given by

$$\frac{dQ_t}{dQ}(x) = \exp\{tv(x) - \beta(t,Q)\} \quad \text{for} \quad x \text{ a.e.-}Q \qquad (3.2)$$

where $\beta(t,Q)$ is the constant such that $Q_t(\Omega)=1$ $\forall t$ i.e.

$$\beta(t,Q) = \log \int \exp\{tv(x)\}\, dQ(x)$$

<u>Lemma</u> (3.3):

If $t^* \in \mathbb{R}$ is such that $\int v(x)\,dQ_{t^*}=0$ then,

$$t^* \int v(x)\,dQ \leq 0 .$$

<u>Proof</u>:

Differentiating with respect to t we obtain

$$\beta'(t,Q) = \int v\,dQ_t$$

which is strictly increasing in t since

$$\beta''(t,Q) = \int v^2 dQ_t - \left[\int v\,dQ_t\right]^2 = \operatorname{var}_t(v(x)) > 0 \quad \forall t .$$

the result follows from this fact by noticing that

$$\beta'(0,Q) = \int v\,dQ \text{ and } \beta'(t^*,Q) = 0 .$$

Thus, if $\int v\,dQ > (<)\, 0$ we must have $t^* < (>)\, 0$ since $\beta'(t,Q)$ is strictly inceasing as a function of t. ∎

<u>Lemma</u> (3.4):

If P is defined as in (1.1) and the measure Q satisfies

$$\int v(x)\,dQ < 0 \;.$$

Then, the solution to the problem (2.2) is given by $P^*=Q_{t*}$.

i.e. the I-projection of the measure $Q\notin \mathcal{P}$ onto $\mathcal{P}$ is the member of the exponential family generated by Q and v that lies on the boundary of $\mathcal{P}$. Notice also that if $Q\in \mathcal{P}$ then $P^*=Q_0=Q$.

Proof:

P and Q_{t*} are probability measures dominated by Q. Then, Lemma (3.1)b) implies

$$0 \leq I(P:Q_{t*}) = \int \log\frac{dP/dQ}{dQ_{t*}/dQ}\, dP \;.$$

Hence, $I(P:Q)\geq\int\log\frac{dQ_{t*}}{dQ}\,dP$. Using (3.2) and Lemma (3.3) we have

$$I(P:Q) \geq -\beta(t^*,Q) \quad \forall P\in \mathcal{P} \tag{3.5}$$

and since $I(Q_{t*}:Q)=-\beta(t^*,Q)$ the result follows from (3.5)(i.e. the lower bound in (3.5) is uniquely achieved for $P=Q_{t*}\in \mathcal{P}$). The same argument (without Lemma (3.3)) shows that $P^*=Q_{t*}$ when $\mathcal{P}$ is (1.2).

An immediate consequence of Lemma (3.4) is

Corollary (3.6):

Let $x_1,x_2,\ldots,x_n$ be n iid observations from a random variable X with probability measure P_0. Then,
a) With the sample as the only information about P_0 the maximum entropy principle generates the empirical measure P_n as the estimate of P_0. i.e. if $A\subset\Omega$ and I_A denotes the indicator function of A,

$$P_n(A) = \frac{1}{n}\sum_{i=1}^{n} I_A(x_i) \;.$$

b) Moreover, if we assume $P_0\in \mathcal{P}$ ($\mathcal{P}$ as in (1.1) or (1.2)) the maximum entropy principle modifies the intial guess P_n to P_n^*

where for i=1,2,...,n

$$P_n^*(\{x_i\}) = \frac{1}{n}\exp\{t^*v(x_i) - \beta(t^*,P_n)\} \tag{3.7}$$

Proof: a) Apply Lemma (3.4) with v=0 in (3.2). b) Apply Lemma (3.4) with $Q=P_n$ and use (3.2). ∎

We can now present the main result of this paper.

Theorem (3.8):
Let $\mathbb{P}$ be defined as in (1.2) and let $T:\mathbb{R}^k \to \mathbb{R}$ be a function satisfying:

i) T is twice continuously differentiable at $\mu \in \mathbb{R}^k$ (i.e. $\dot{T}$ and the Hessian $\ddot{T}$ exist and they are continuous at $\mu \in \mathbb{R}^k$).

ii) $\dot{T}=(\dot{T}_1,\ldots,\dot{T}_k)^t$ is such that $\dot{T}_j(\mu)\neq 0$ $\forall j=1,2,\ldots,k$.

Define the real parameter $\nu:\mathbb{R}\to\mathbb{R}$ by

$$\nu(P) = T(\mu(P)) = T(\mu_1,\mu_2,\ldots,\mu_k)$$

where

$$\mu_j = \mu_j(P) = \int u_j(x)\,dP(x)$$

with the u_j's fixed bounded and real functions defined on Ω.

Given the sample $x_1,x_2,\ldots,x_n$ (iid) from $P_0 \in \mathbb{P}$ the estimator

$$\nu(P^*) = T(\mu(P^*))$$

where $$\mu(P^*) = \{\frac{1}{n}\sum_{i=1}^{n} u_j(x_i)\exp\{t^*v(x_i)-\beta(t^*,P_n)\}\}_{j=1}^{K}$$

(which by Corollary (3.6) is the maximum entropy estimator of $\nu(P)$) is <u>asymptotically efficient</u>.

i.e. $\nu(P^*)$ achieves the Fréchet (Cramer-Rao) lower bound on each one-dimensional regular parametric submodel in $\mathbb{P}$ that passes through P_0. This concept of efficiency in nonparametric models appears informally in Stein (1956) and it is developed and extended in Koshevnik and Levit (1976). The proof presented here is based on the calculation of the efficient influence function. This is obtained by projecting on the tangent space of $\mathbb{P}$ at P_0 as presented in Bickel, Ritov and Wellner (1987) (see also Pfanzagl, 1982).

Some consequences of this theorem partially overlap with those in Haberman (1984) even though the generality, the context in which they appear, the emphasis given and the methods of proof are different.

4 PROOF OF THE MAIN RESULT

If we assume that $\mathbb{P}$ is dominated by a σ-finite measure μ we can imbed $\mathbb{P}$ in the Hilbert space $L_2(\mu)$ through the transformation

$$\mathbb{P} \ni P \mapsto s = \left[\frac{dP}{d\mu}\right]^{1/2} \in L_2(\mu) \ .$$

For fixed $P_0 \in \mathbb{P}$ a useful local imbedding of $\mathbb{P}$ into $L_2(P_0)$ is obtained through the transformation

$$P \mapsto r = 2\left[\frac{s}{s_o} - 1\right] 1\left[s_o > 0\right]$$

where $s_o = \left[\frac{dP_0}{d\mu}\right]^{1/2}$. Therefore, considering P as a subset of $L_2(P_0)$ we obtain that the tangent space at P_0 is given by (see Bickel, Ritov and Wellner (1987) Chapter 3, Example 3 or Pfanzgal (1982) Proposition (4.5.1), p.75)

$$\dot{\mathbb{P}} = \{h \in L_2(P_0) \ : \ E_0 h = 0 \text{ and } < \dot{\gamma}_v(P_0), h>_0 = 0\}$$

where E_0 denotes expectation w/r to P_0, $<\cdot,\cdot>_0$ denotes the inner product in $L_2(P_0)$ and the gradient $\dot{\gamma}_v$ (the Gâteaux derivative at s_o of $\gamma_v(P) = \int v dP$) satisfies

$$<\dot{\gamma}_v(s_o),h> = \frac{\partial}{\partial\eta}\gamma_v(s_o+\eta h)|_{\eta=0} = \frac{\partial}{\partial\eta}\{\int v(s_o+\eta h)^2 d\mu\}|_{\eta=0} =$$

$$= \int 2s_o vh d\mu$$

hence, $$\dot{\tilde{\gamma}}_v = \dot{\gamma}_v(s_o)/(2s_o) = v$$

on the other hand if $\nu(P)=\int u(x)dP(x)$ for u fixed and bounded then

$$\dot{\tilde{\nu}} = \dot{\nu}(s_o)/(2s_o) = u .$$

Using Theorem 1 of Chapter 3, Section 3 in Bickel, Ritov and Wellner (1987) (see also Koshevnit and Levit, 1976) we have that the efficient influence function of ν at P_0 is given by

$$\tilde{\ell} = \Pi_0(\dot{\tilde{\nu}}|\dot{P})$$

where $\Pi_0(\cdot|\dot{P})$ denotes the projection operator in $L_2(P_0)$ onto $\dot{P}$. We have

$$\tilde{\ell} = \Pi_0(u|[1,v]^\perp) = u - \Pi_0(u|[1,v])$$

where [1,v] denotes the closed linear span generated by the functions 1 (constant 1) and v in $L_2(P_0)$. Clearly $1\perp v$ in $L_2(P_0)$ (since $P_0\in P$) and therefore

$$\tilde{\ell}(x) = u(x) - \left[\frac{\int u(y)v(y)dP_0(y)}{\int v^2(u)dP_0(y)}\right]v(x) - \nu . \tag{4.1}$$

We want to show first that

$$\nu_n = \nu(P_n^*) = \sum_{i=1}^{n} u(x_i)q_n(x_i) \tag{4.2}$$

with

$$q_n(x_i) = \frac{1}{n} \exp\{t^* v(x_i) - \beta(t^*, P_n)\}$$

is an asymptotically efficient estimate of $\nu(P)$. It is sufficient to show that ν_n has efficient influence function. To obtain the influence function of ν_n we use Theorem 2 from Section 2.5 in Bickel, Ritov and Wellner (1987). (See also Huber, 1981, p.133 Corollary 3.2). This theorem states that under regular conditions, asymptotic M-estimates are asymptotically linear estimates of $\nu(P)$ with influence function

$$\frac{\psi(\cdot, \nu(P))}{-\int \frac{\partial \psi}{\partial \nu}(x, \nu(P))\, dP(x)} \tag{4.3}$$

where $|\psi(x, \nu(P))|$ considered as a function of x belongs to $L_2(P)$ with

$$\int \psi(x, \nu(P))\, dP(x) = 0 \quad \text{for all } P \in \mathcal{P}$$

and the estimator, ν_n, satisfies

$$\sqrt{n} \int \psi(x, \nu_n)\, dP_n(x) \overset{P}{\to} 0 \text{ as } n \to \infty$$

(i.e. ν_n is an asymptotic M-estimate of ν based on ψ).

Notice first that if we denote by t_n the parameter t^* when $Q=P_n$ then we have

$$\frac{1}{n} \sum_{i=1}^{n} v(x_i) \exp\{t_n v(x_i)\} = 0 .$$

But for x_i fixed

$$\exp\{t_n v(x_i)\} = 1 + t_n v(x_i) + o(t_n v(x_i)) .$$

Hence,

$$\frac{1}{n} \sum_{i=1}^{n} v(x_i) e^{t_n v(x_i)} = \frac{1}{n} \sum_{i=1}^{n} v(x_i) + t_n \frac{1}{n} \sum_{i=1}^{n} v^2(x_i) +$$

$$+ o_P\left[t_n \frac{1}{n} \sum_{i=1}^{n} v^2(x_i)\right] = 0 .$$

Therefore,

$$t_n = \frac{-\sum_{i=1}^{n} v(x_i)}{\sum_{i=1}^{n} v^2(x_i)} + o_P(n^{-1/2}) . \quad (4.4)$$

Since by the central limit theorem (CLT) applied to $\Sigma v(x_i)/n$ we have $o_P(t_n)=o_P(n^{-1/2})$. Hence, if $P \in \mathcal{P}$ we can write

$$\frac{1}{n} \sum_{i=1}^{n} v(x_i) \xrightarrow{P} 0 \quad \text{and} \quad \frac{1}{n} \sum_{i=1}^{n} v^2(x_i) \xrightarrow{P} E_P v^2$$

by the weak law of large numbers (WLLN) (since v is bounded). Thus using (4.4) we obtain

$$t_n \xrightarrow{P} 0 \text{ i.e. } t_n = o_P(n^0) \text{ as } n\to\infty . \quad (4.5)$$

We now check the conditions of the theorem on M-estimates

1) ν_n (given by (4.2)) is a consistent estimate of $\nu(P)$ i.e.

$$\nu_n = E_P u + o_P(n^0) \quad \text{for all } P \in \mathcal{P} . \quad (4.6)$$

To see this consider for x fixed $q_t(x)=r(t)$ as a function of t. Then,

$$r(t) = r(0) + tr'(0) + o(t) \text{ as } t\to 0$$

with $r(0)=1$ and $r'(0)=[v(x)\beta'(t,Q)]\exp\{tv(x)-\beta(t,Q)\}|_{t=0}$ $=v(x)$. Hence $\frac{dQ_t}{dQ}(x)=Q_t(x)=1+tv(x)+o(t)$ as $t\to 0$ replacing in (4.2) and using the fact that $o_P(t_n)=o_P(n^{-1/2})$ (see (4.4)) we have

$$\nu_n = \frac{1}{n} \sum_{i=1}^{n} u(x_i) + t_n\left\{\frac{1}{n} \sum_{i=1}^{n} u(x_i)\right\} + o_P(n^{-1/2}). \quad (4.7)$$

From (4.5) and the consistency of the sample means we obtain the RHS of (4.6).

2) ν_n is an asymptotic M-estimate of $\nu(P)$ with efficient influence function. i.e. for

$$W_n(\nu(P)) = \frac{1}{n}\sum_{i=1}^{n}\tilde{\ell}(x_i,\nu) = \frac{1}{n}\sum_{i=1}^{n}u(x_i) - \nu(P) - \left\{\frac{\int uv\,dP}{\int v^2 dP}\right\}\left[\frac{1}{n}\sum_{i=1}^{n} v(x_i)\right]$$

we only need to show that

$$\sqrt{n}\; W_n(\nu_n) = o_P(n^0) \text{ as } n\to\infty$$

which is clearly true since from (4.7) we can write

$$\sqrt{n}\; W_n(\nu_n) =$$

$$\left\{-t_n \frac{1}{n}\sum_{i=1}^{n} u(x_i)v(x_i) - \frac{\int uv\,dP_n^*}{\int v^2 dP_n^*}\frac{1}{n}\sum_{i=1}^{n} v(x_i)\right\}\sqrt{n} + o_P(n^0)$$

$$= \left\{\frac{\frac{1}{n}\sum_{i=1}^{n} u(x_i)v(x_i)}{\frac{1}{n}\sum_{i=1}^{n} v^2(x_i)} - \frac{\int uv\,dP_n^*}{\int v^2 dP_n^*}\right\}\sqrt{n}\left[\frac{1}{n}\sum_{i=1}^{n} v(x_i)\right] + o_P(n^0)$$

where we have used (4.4) in this last equation. From (4.7) the WLLN and the CLT applied to $\bar{v}_n$ (since $P \in \mathcal{P}$) the last equation is $o_P(n^0)o_P(n^0)+o_P(n^0)=o_P(n^0)$.

3) All the other regularity conditions of the theorem are trivially satisfied. Moreover

$$\dot{W}_n(\nu) = -1 = \dot{W}(\nu) = \int \frac{\partial\tilde{\ell}}{\partial\nu}\,dP .$$

Therefore we conclude that ν_n is an asymptotically linear estimate of $\nu(P)$ with influence function (given by (4.3))

$\tilde{\ell}$, i.e., efficient.

We now show the theorem when $\nu(P)$ is defined as in Theorem (3.8). From the chain rule we can write (imbedding P in $L_2(\mu)$)

$$\nu(s+\eta h) = T(\mu(s+\eta h)) = T(\mu(s)) + \eta<\dot{\mu}\cdot\dot{T},h> + o(\eta)$$

hence,

$$\dot{\nu}(s) = \dot{\mu}(s)\cdot\dot{T}(\mu(s)) .$$

Thus

$$\dot{\tilde{\nu}}(x) = \frac{\dot{\nu}}{2s_o} = \sum_{j=1}^{k} \dot{T}_j \, u_j(x)$$

and projecting $\dot{\tilde{\nu}}$ on $\dot{P}=[1,v]^{\perp}$ we obtain the efficient influence function $\ell^*(x,\nu)$ that can be written as

$$\ell^*(x,\nu(p)) = \sum_{j=1}^{k} \dot{T}_j(P)\left\{u_j(x)-\mu_j(P) - \frac{\int u_j v dP}{\int v^2 dP} v(x)\right\}$$

$$= \sum_{j=1}^{k} \dot{T}_j(P) \, \tilde{\ell}_j(x,\mu_j(P))$$

where $\tilde{\ell}_j$ is the efficient influence function of μ_j (see (4.1) when $u=u_j$). We now check (once more) the hypothesis of the theorem on M-estimates for $\nu_n=T(\mu(P_n^*))$.

1) ν_n is consistent. It follows from the fact that

$$\mu(P_n^*) = \mu(P) + o_P(n^0)$$

which in turn follows by using (4.6) for each of the components of μ and the assumed continuity of T at $\mu(P)$. i.e.

$$\nu_n = T(\mu(P) + o_P(n^0)) = \nu(P) + o_P(u^0) .$$

2) ν_n is an asymptotic M-estimate based on

$$W_n(\nu) = \frac{1}{n} \sum_{i=1}^{n} \ell^*(x_i,\nu) = \sum_{j=1}^{k} \dot{T}_j(P) W_n^{(j)}(\mu_j)$$

where $$W_n^{(j)}(\mu_j) = \frac{1}{n} \sum_{i=1}^{n} \tilde{\ell}_j(x_i,\mu_j(P)) .$$

This follows from the fact that

$$\sqrt{n}\, W_n(\nu_n) = \sum_{j=1}^{k} \dot{T}_j(P_n^*)\left\{\sqrt{n}\, W_n^{(j)}(\mu_{jn})\right\}$$

$$= o_P(n^0) \sum_{j=1}^{k} \dot{T}_j(P_n^*) = o_P(n^0)$$

since $\dot{T}$ is continuous by hypothesis hence,

$$\dot{T}(P_n^*) = \dot{T}(P) + o_P(n^0) .$$

Moreover,

$$\dot{W}_n(\nu) = \sum_{j=1}^{k}\left[\left\{\sum_{\ell=1}^{k} \ddot{T}_{j\ell}(\mu)\frac{\partial \mu_\ell}{\partial \nu}\right\} W_n^{(j)}(\mu_j) - \dot{T}_j(\mu)\frac{\partial \mu_j}{\partial \nu}\right] .$$

Since $\dot{W}_n^{(j)}(\mu_j)=-1$. An application of the inverse function theorem (that holds from (3.8)ii)) assures the continuity of $\frac{\partial \mu_j}{\partial \nu}(\nu)$ for $j=1,2,\ldots,k$. Hence, (3.8)i) and ii) are sufficient for $\dot{W}_n(\nu)$ to exist and to be continuous. Moreover,

$$\dot{W}_n^{(\nu)} = \sum_{\ell=1}^{k} \left\{\sum_{\ell=1}^{k} \ddot{T}_{j\ell}(\mu)\frac{\partial \mu_\ell}{\partial \nu}\right\} W_n^{(j)}(\mu_j) - 1 .$$

Since $$\sum_{j=1}^{k} \dot{T}_j(\mu)\frac{\partial \mu_j}{\partial \nu} = \frac{\partial T}{\partial \nu} = 1 \ .$$

Hence,

$$\dot{W}(\nu) = \int \frac{\partial \ell^*}{\partial \nu}(x,\nu)P(dx) = \sum_{j=1}^{k}\sum_{\ell=1}^{k} \ddot{T}_{j\ell}\frac{\partial \mu_\ell}{\partial \nu}\int \tilde{\ell}_j(x,\mu_j)dP(x)-1$$

and since

$$\int \tilde{\ell}_j(x,\mu_j(P))dP(x) = 0$$

we obtain $\dot{W}(\nu)=-1$. Again, all the hypothesis of the theorem on M-estimates are satisfied and thus, ν_n is an asymptotically linear estimate of $\nu(P)=T(\mu(P))$ with influence function given by (see (4.3))
$-\dot{W}^{-1}(\nu)\ell^*(x,\nu)=\ell^*(x,\nu)$ i.e. efficient. Q.E.D.

Acknowledgement:
I would like to thank Peter Bickel for introducing me to the theory of efficiency in nonparametric models.

REFERENCES

Bickel, P. Ritov, L. and Wellner, J., (1987). Semiparametric Models. Book manuscript (to appear).

Gibbs, J.W. (1902). Elementary Principles in Statistical Mechanics, Yale University Press, New Haven, Conn. Reprinted by Dover Publications, New York, 1960.

Haberman, S.J. (1984). Adjustment by minimum discriminant information. Annals of Statistics, Vol.12, No.3, 971-988.

Huber, P.J. (1981). Robust Statistics, John Wiley & Sons, New York.

Jaynes, E.T. (1957). Information theory and Statistical mechanics. Physics Review, 106, 620-630. Reprinted in Jaynes (1983).

Jaynes, E.T. (1978). Where do we stand on maximum entropy?, The Maximum Entropy Formalism, R.D. Levine and R. Tribus, Editors, M.I.T. Press, Cambridge, Mass. Reprinted in Jaynes (1983).

Jaynes, E.T. (1983). Papers on Probability, Statistics and Statistical Physics, P.D. Rosenkrantz, Editor. D. Reidel Publishing Co.

Koshevnik, Yu. A. and Levit, B. Ya. (1976). On a non-parametric analog of the information matrix. Theor. Prob. Applic., vol. 21, 738-753.

Kullback, S. (1959). Information Theory and Statistics. John Wiley and Sons, New York. Reprinted by Dover Publications, N.Y. 1968.

Kullback, S. and Leibler, R.A. (1951). On information and sufficiency. Ann. Math. Statist., 22, 79-86.

LeCam, L. (1956). On the asymptotic theory of estimation and testing hypotheses. Proc. Third Berkeley Symp. Math. Statist. Prob., vol. 1, 129-156, University of California Press, Berkeley.

LeCam, L. (1969). Theorie Asymptotique de la Decision Statistique, Les Presses de l'Universite de Montreal, Montreal.

LeCam, L. (1970). On the assumptions used to prove asymptotic normality of maximum likelihood estimates. Ann. Math. Statist., vol. 41, 802-828.

Mandelbrot, B.B. (1967). Sporadic random functions and conditional spectral analysis; self-similar examples and limits. Proc. Fifth Berkeley Symp. Math. Statist. Prob., vol. 3, 155-179, University of California Press, Berkeley.

Pfanzagl, J. (1982). Contributions to a general Asymptotic Statistical Theory. Lecture Notes in Statistics, Springer, New York.

Renyi, A. (1955). On a new axiomatic theory of probability. Acta Mathematica Hungarica, 6, 285-335.

Rodriguez, C. (1986). Maximum Entropy Smoothing, Techn. Report No. 30, Dept. of Math. and Stat. SUNY at Albany (to appear).

Rodriguez, C. (1987). Understanding ignorance. Proc. Sixth Workshop on Maximum Entropy and Bayesian Methods in Applied Statistics (to appear).

Shore, J.E. and Johnson, R.W. (1980). Axiomatic derivations of the principle of maximum entropy and the principle of minimum cross-entropy, IEEE Trans. on Information Theory, IT-26, 26-37.

Shore, J.E. and Johnson, R.W. (1981). Properties of cross entropy minimization, IEEE Trans. on Information Theory, IT-27,472-482.

Stein, C. (1956). Efficient nonparametric testing and estimation. Proc. Third Berkeley Symp. Math. Statist. Prob., vol. 1, 187-195.

Van Campenhout, J. and Cover, T.M. (1981). Maximum entropy and conditional probability. IEEE Trans. on Information Theory. IT-27, 483-489.

QUANTUM STATISTICAL MECHANICS IN PHASE SPACE AND THE CLASSICAL LIMIT

Y. Tikochinsky
Racah Institute of Physics, The Hebrew University of Jerusalem, Jerusalem 91904, Israel

D. Shalitin
Racah Institute of Physics, The Hebrew University of Jerusalem, Jerusalem 91904, Israel

Abstract. The Wigner formulation of quantum mechanics in phase space is reviewed. Using this formulation, the classical limit of the quantum mechanical description of a system characterized by a given sharp value A of an observable $\hat{A}$, or by a given expectation value $\langle\hat{A}\rangle$ of the same observable, is discussed. It is shown that, in the limit $\hbar \to 0$, the quantum description of the system reduces to that given by the corresponding classical microcanonical or canonical distributions. In particular, the condition for the classical entropy to coincide with the limiting value of the quantum entropy, is spelled out. The first quantum corrections to the classical description are calculated. It is shown that, for a system of non-interacting identical particles, the first correction due to the Pauli principle is proportional to $\hbar^3$. A simple relation between the normal-antinormal correspondence of operators and the Wigner (inverse Weyl) correspondence is established.

I INTRODUCTION

Among the various equivalent formulations of quantum mechanics the formulation by Wigner (1932), employing phase space, is probably the best starting point for discussing the classical limit. As it turns out, the typical appearance of Planck's constant $\hbar$ in denominators in the usual (configuration space) formulations is replaced by $\hbar$ appearing in numerators in the Wigner formulation. Thus, instead of asymptotic expansions, an easy passage to the limit $\hbar \to 0$ and a straighforward expansion in powers of $\hbar$ are made possible.

The main goal of the Wigner formulation is to express expectation values of dynamical variables as integrals over phase space:

$$\langle\hat{A}\rangle = \mathrm{tr}(\hat{\rho}\hat{A}) = \int \rho(x,p)A(x,p)dxdp/(2\pi\hbar), \qquad (1.1)$$

where $\rho(x,p)$ and $A(x,p)$ are two real functions corresponding to the hermitian operators $\hat{\rho}$ and $\hat{A}$, ($\hat{\rho}$ being the density operator)

$$\hat{\rho} \to \rho(x,p) \qquad , \qquad \hat{A} \longrightarrow A(x,p) \ . \qquad (1.2)$$

G. J. Erickson and C. R. Smith (eds.),
Maximum-Entropy and Bayesian Methods in Science and Engineering (Vol. 2), 51–82.

As indicated by the different arrows, the mapping $\hat{\rho} \to \rho(x,p)$ and $\hat{A} \dashrightarrow A(x,p)$ are in general different (though related). By writing the expectation value $\langle\hat{A}\rangle$ in the classical form above we certainly do not mean that Heisenberg's uncertainty principle regarding the simultaneous determination of x and p, can be bypassed. Yet, the exact classical form above can be achieved, in fact in infinitely many ways, (Agarwal & Wolf 1970) at the price of relaxing our expectations regarding the "classical" density $\rho(x,p)$. The function $\rho(x,p)$ will turn out to have some peculiar properties such as obtaining positive and negative values at different regions of phase space. Thus $\rho(x,p)$ cannot serve as a proper probability density in phase space and is accordingly named quasidensity.

It is perhaps surprising that fifty years or so after its introduction, Wigner's formulation is still a source for basic research (O'Connell & Wigner 1984) and its implications concerning the classical limit of quantum mechanics are still not fully recognized. Thus, in a recent article dealing with a family of mappings of the form (1.2) satisfying (1.1), Wang (1986) proves that if, for a given mapping, the functions $\rho(x,p)$ and $A(x,p)$ approach the limits $\rho_c(x,p)$ and $A_c(x,p)$ respectively as $\hbar \to 0$, then the same limit is attained by all members of the family. He calls the limiting functions "classical" and leaves us wondering just there: Does the limit $\hbar \to 0$ exist? What is the relation between the limiting functions $\rho_c(x,p)$ and $A_c(x,p)$ and the functions $\rho_{cl}(x,p)$ and $A_{cl}(x,p)$ that a classical physicist would have used to describe the same physical system? In order to answer these questions, we must first clarify what we mean by the phrase "the classical limit as $\hbar \to 0$". Imagine two physicists, a classical one and a contemporary one, presented independently with a well defined physical problem: e.g. A system described by a given classical Hamiltonian $H(x,p)$ is known to have sharp energy E. Alternatively, the exact energy E is not known but its expectation value $\langle E\rangle$ is known. Find the density $\rho_{cl}(x,p)$ or the quasidensity $\rho(x,p)$ appropriate for the system, and compare the results for $\rho(x,p)$ and for expectation values of dynamical variables calculated according to (1.1) in the limit $\hbar \to 0$, (that is, when $\hbar$ is small compared to typical quantities of the system), with the corresponding results obtained by the classical physicist. Moreover, express the quantum corrections to the classical results in ascending powers of Planck's constant $\hbar$. In the following we shall follow precisely this route.

Section II reviews Weyl's rule (Weyl (1927, 1931)) for associating a hermitian opererator $\hat{A}$ with a given classical observale $A(x,p)$. The inverse mapping, namely $\hat{A} \to A(x,p)$ and $\hat{\rho} \to \rho(x,p)$ will turn out to satisfy (1.1). As recognized by Moyal (1949), the "classical" function $A(x,p)$ which upon quantization a la Weyl yields the given operator $\hat{A}$, is precisely the Wigner function for the operator $\hat{A}$. To keep the notation simple, we shall treat a single particle in one dimension. The generalization to N(non identical) particles in three dimensions is straightforward. We shall note in passing that, according to Weyl, Dirac's quantization rule, namely, the association of a commutator with a given Poisson bracket, is in general, satisfied only to first order in $\hbar$. It may be a shocking revelation to some readers to find that this result is independent of the Weyl quantization. Dirac's rule for observables $A(x,p)$ more complicated than bilinear forms in x and p, is in general, self contradictory. In Sec. III the quantum mechanical equations are translated to phase space language. As an example, we solve for the eigen values and eigen quasidensities of the harmonic oscillator. Section IV is devoted to the discussion of the classical limit. We show that the quasidensity $\rho(x,p)$ for a system with a given sharp energy E, or a given expectation value $\langle E \rangle$, approaches respectively the classical microcanonical or canonical distribution $\rho_{cl}(x,p)$ as $\hbar \to 0$. Moreover, the quantum expectation values for any observale $\hat{A}$ coincide, in this limit, with the classical ones. By the principle of maximum entropy, the same results are true for a system characterized by a sharp value A or an expectation value $\langle \hat{A} \rangle$ of any observable $\hat{A}$. As is well known, classical entropy is defined only up to an additive constant. The condition for the classical entropy to coincide with the limiting value of the quantum entropy is also spelled out. In Sec. V we treat the quantum corrections to the classical limit. As was shown by Wigner (1932) (and as expected from an expansion of a real quantity $\rho(x,p)$ in powers of the imaginary parameter $i\hbar$), only even powers of $\hbar$ enter the expansion. Finally, in Sec. VI, we take up the hitherto neglected identity of particles. We show that the first correction due to the identity of N non-interacting particles is proportional to $\hbar^3$. This is done both formally, using the language of second quantization, and classically, by counting the number of partitions of N particles with two (or more) particles in a cell. The treatment is incomplete in the sense that spin degrees of freedom were not taken explicitly into account. An example of another correspondence of the type (1.2) satisfying (1.1), namely, the normal-antinormal correspondence, is given in Appendix C. The (little known but simple) relation of this correspondence to the Wigner (inverse Weyl) one is spelled out. For a comprehensive review of the role of space phase quasidensities in physics, the reader is referred to the recent work of Hillery et al (1984), where additional references can be found.

II WEYL QUANTIZATION AND THE WIGNER FUNCTION

A Weyl Quantization and its inverse

Weyl (1927) suggested the following general and simple quantization rule. Let $A(x,p)$ be a classical observable representable as a Fourier integral

$$A(x,p) = \frac{\hbar}{2\pi} \int \bar{A}(u,v) e^{i(ux+vp)} du dv \qquad (2.1)$$

where

$$\bar{A}(u,v) = \frac{1}{2\pi\hbar} \int A(x,p) e^{-i(ux+vp)} dx dp \ . \qquad (2.2)$$

Then the operator $\hat{A}$, corresponding to $A(x,p)$, is obtained by replacing the exponential function in (2.1) by the exponential operator $\exp[i(u\hat{x} + v\hat{p})]$, where $\hat{x}$ and $\hat{p}$ are two hermitian operators satisfying the commutation relation $[\hat{x},\hat{p}] = i\hbar$. (we have inserted

Planck's constant $\hbar$ in Eqs. (2.1) and (2.2) in order to take care of the dimensionality.) Using Eq. (2.2), we have,

$$\hat{A} = \frac{\hbar}{2\pi} \int \bar{A}(u,v) e^{i(u\hat{x}+v\hat{p})} dudv \tag{2.3a}$$

or

$$\hat{A} = \frac{1}{(2\pi)^2} \int A(x,p) e^{-i(ux+vp)} e^{i(u\hat{x}+v\hat{p})} dxdpdudv \ . \tag{2.3b}$$

Some simple properties of the correspondence $A(x,p) \to \hat{A}$ follow immediately from (2.3):

$$A(x,p) \text{ real} \longrightarrow \hat{A} \text{ hermitian} \tag{2.4a}$$

$$A(x) \to \hat{A} = A(\hat{x}) \quad , \quad B(p) \to \hat{B} = B(\hat{p}) \tag{2.4b}$$

$$\alpha A(x,p) + \beta B(x,p) \to \alpha\hat{A} + \beta\hat{B} \quad \text{(linearity)} \tag{2.4c}$$

$$x^2 p \to (\hat{x}^2\hat{p} + \hat{p}\hat{x}^2 + \hat{x}\hat{p}\hat{x})/3 \quad \text{(example of the Weyl symmetrization)} \tag{2.4d}$$

Conversely, given an operator $\hat{A}$ represented by (2.3) and using the orthogonality relation

$$\mathrm{tr}[e^{i(u\hat{x}+v\hat{p})}] = \frac{2\pi}{\hbar} \delta(u)\delta(v) \ , \tag{2.5}$$

we obtain

$$A(x,p) = \frac{\hbar}{2\pi} \int \mathrm{tr}[\hat{A} e^{-i(u\hat{x}+v\hat{p})}] e^{i(ux+vp)} dudv \ . \tag{2.6}$$

In deriving the last two Equations, we have made use of the following regrouping property of exponential operators (see e.g. Messiah 1961)

$$e^{\hat{A}+\hat{B}} = e^{-\frac{1}{2}[\hat{A},\hat{B}]} e^{\hat{A}} e^{\hat{B}} \tag{2.7a}$$

which is valid provided $[\hat{A},[\hat{A},\hat{B}]] = [\hat{B},[\hat{A},\hat{B}]] = 0$.

For example,

$$e^{i(u\hat{x}+v\hat{p})} = e^{\frac{1}{2}\hbar uv} e^{iu\hat{x}} e^{iv\hat{p}} \ . \tag{2.7b}$$

Again, taking the trace of Eq. (2.3) with the aid of Eq. (2.5) we secure the important result

$$\mathrm{tr}(\hat{A}) = \int A(x,p) \frac{dxdp}{2\pi\hbar} \tag{2.8}$$

B <u>The Wigner function</u>

In order to simplify the expression for the function $A(x,p)$ corresponding to the operator $\hat{A}$, we shall employ the x-representation to calculate the trace in Eq. (2.6). A short calculation using Eqs. (2.7) yields

$$A(x,p) = \int < x-\tfrac{1}{2}x'|\hat{A}|x + \tfrac{1}{2}x'> \exp(ipx'/\hbar) dx' \ . \tag{2.9}$$

Alternatively, using the p-representation, we obtain

$$A(x,p) = \int \langle p-\tfrac{1}{2}p'|\hat{A}|p+\tfrac{1}{2}p'\rangle \exp(-ixp'/\hbar)dp \ . \tag{2.10}$$

In particular, for the density matrix $\hat{\rho} = |\psi\rangle\langle\psi|$ representing the pure state $|\psi\rangle$, we have

$$\rho(x,p) = \int \psi(x-\tfrac{1}{2}x')\psi^*(x+\tfrac{1}{2}x')e^{\frac{i}{\hbar}px'} dx'$$

$$= \int \phi(p-\tfrac{1}{2}p')\phi^*(p+\tfrac{1}{2}p')e^{-\frac{i}{\hbar}xp'} dp' \ , \tag{2.11}$$

where $\psi(x) = \langle x|\psi\rangle$ is the wave function and $\phi(p) = \langle p|\psi\rangle$ is its Fourier transform. Equation (2.11) for $\rho(x,p)$ is precisely the one introduced (on different grounds) by Wigner (1932). The identification of the Wigner function with the function $\rho(x,p)$ which upon quantization a la Weyl yields the operator $\hat{\rho}$, is due to Moyal (1949). Correspondingly, the function $A(x,p)$ of Eqs. (2.6), (2.9) or (2.10) is called the Wigner function for the operator $\hat{A}$.

We shall now list some of the properties of the Wigner function. From (2.9) or (2.10) it is clear that

$$\hat{A} \text{ hermitian implies } A(x,p) \text{ real.} \tag{2.12}$$

Upon integrating Eq. (2.11) on p or on x we obtain for the marginals

$$\int \rho(x,p)\frac{dp}{2\pi\hbar} = |\psi(x)|^2 \ , \quad \int \rho(x,p)\frac{dx}{2\pi\hbar} = |\phi(p)|^2 . \tag{2.13a}$$

Finally, if $\psi(x)$ is properly normalized, we obtain, in accordance with Eq. (2.8),

$$\int \rho(x,p) \frac{dxdp}{2\pi\hbar} = 1 \ . \tag{2.13b}$$

Equation (2.12) guarantees the reality of $\rho(x,p)$. On the other hand, it is easy to see (Hillery et al 1984) that the Wigner function cannot be, in general, positive everywhere. To this end, consider the integral of two Wigner functions $\rho_1(x,p)$ and $\rho_2(x,p)$ corresponding to the two pure states $|\psi_1\rangle$ and $|\psi_2\rangle$. A short calculation yields

$$\int \rho_1(x,p)\rho_2(x,p)\frac{dxdp}{2\pi\hbar} = \left|\int \psi_1^*(x)\psi_2(x)dx\right|^2 =$$

$$= |\langle\psi_1|\psi_2\rangle|^2 \geq 0 \ . \tag{2.13c}$$

In particular, if the two states are orthogonal,

$$\int \rho_1(x,p)\rho_2(x,p)\frac{dxdp}{2\pi\hbar} = 0 \quad \text{for} \quad \langle\psi_1|\psi_2\rangle = 0 \ ,$$

Showing that ρ_1 and ρ_2 cannot be both positive everywhere.

C <u>Wigner function for product of operators</u>

Let $C_{\hat{A}\hat{B}}(x,p)$ denote the phase space function which upon Weyl's quantization yields the operator $\hat{C} = \hat{A}\hat{B}$. We shall give three equivalent expressions for this Wigner function. Using Eq. (2.6) for $\hat{C}$, we obtain an integral representation for the function C.

$$C_{\hat{A}\hat{B}}(x,p) = (\pi\hbar)^{-2}\int dx_1 dx_2\, dp_1 dp_2 A(x_1,p_1)B(x_2,p_2)$$
$$\times \exp[2\tfrac{i}{\hbar}x(p_1-p_2)]\exp[-2\tfrac{i}{\hbar}p(x_1-x_2)]\exp[\tfrac{i}{\hbar}(x_1p_2-x_2p_1)] \qquad (2.14)$$

Inserting the last expression in Eq. (2.8), we secure for finite traces

$$\mathrm{tr}(\hat{A}\hat{B})=\mathrm{tr}(\hat{B}\hat{A})=\int C_{\hat{A}\hat{B}}(x,p)\frac{dxdp}{2\pi\hbar} = \int A(x,p)B(x,p)\frac{dxdp}{2\pi\hbar}\ . \qquad (2.15)$$

In particular, for $\hat{B} = \hat{\rho}$, we have,

$$\langle\hat{A}\rangle = \mathrm{tr}(\hat{\rho}\hat{A}) = \int \rho(x,p)A(x,p)dxdp/2\pi\hbar \qquad (2.16)$$

which proves that the mapping (2.6) does indeed fulfill Eq. (1.1). The second expression (operator form) of the Wigner function for the product of two operators is

$$C_{\hat{A}\hat{B}}(x,p) = \exp(i\tfrac{\hbar}{2}P_{ij})A(x_i,p_i)B(x_j,p_j)\Big|_{\substack{x_i=x_j=x\\ p_i=p_j=p}} \qquad (2.17a)$$

where

$$P_{ij} \equiv \frac{\partial}{\partial x_i}\frac{\partial}{\partial p_j} - \frac{\partial}{\partial p_i}\frac{\partial}{\partial x_j} \equiv D_{x_i}D_{p_j} - D_{p_i}D_{x_j} \qquad (2.17b)$$

is the Poisson bracket operator and the limit $x_i=x_j=x$, $p_i=p_j=p$ is taken after performing the differentiations. In Appendix A, the Wigner function is generalized for the product of n operators:

$$C_{\hat{A}_1..\hat{A}_n}(x,p)=\exp(i\tfrac{\hbar}{2}\sum_{i<j}^{n}P_{ij})\prod_{i=1}^{n}A_i(x_i,p_i)\Big|_{\substack{x_1=..=x_n=x\\ p_1=..=p_n=p}} \qquad (2.18)$$

Equivalent results were obtained by Groenwold (1945). Let us apply Eq. (2.17), for the commutator $\hat{C} = [\hat{A},\hat{B}]$:

$$C_{[\hat{A},\hat{B}]}(x,p)=\exp(i\tfrac{\hbar}{2}P_{12})[A(1)B(2)-A(2)B(1)]\Big|$$
$$=[\exp(i\tfrac{\hbar}{2}P_{12})-\exp(-i\tfrac{\hbar}{2}P_{12})]A(1)B(2)\Big|$$
$$=2i\sin(\tfrac{\hbar}{2}P_{12})A(1)B(2)\Big|_{\substack{x_1=...=x_n=x\\ p_1=...=p_n=p}} \qquad (2.19)$$

where the shorthand notation $A(1) = A(x_1,p_1)$ etc. has been used.

Expanding the sine operator into a power series, we have,

$$C_{[\hat{A},\hat{B}]}(x,p) = 2i \sum_{k=0} \frac{(-1)^k}{(2k+1)!} (\frac{\hbar}{2}P_{12})^{2k+1} A(1)B(2)$$

$$= i\hbar\{A(x,p),\ B(x,p)\} + O(\hbar^3) \tag{2.20}$$

where

$$\{A,B\} \equiv \frac{\partial A}{\partial x}\frac{\partial B}{\partial p} - \frac{\partial A}{\partial p}\frac{\partial B}{\partial x} \tag{2.21}$$

is the (classical) Poisson brackets for the observable $A(x,p)$ and $B(x,p)$. (Note that if at least one of the operators $\hat{A},\hat{B}$ is at most bilinear in $\hat{x}$ and $\hat{p}$, $C_{[\hat{A},\hat{B}]} = i\hbar\{A(x,p),\ B(x,p)\}$ is exactly satisfied.) Thus, according to Weyl, the famous Dirac quantization rule

$$\{A,B\} \rightarrow -\frac{i}{\hbar}[\hat{A},\hat{B}] \tag{2.22}$$

is in general obeyed only to first order of $\hbar$. This perturbing result is, in fact, independent of the Weyl quantization. The following example due to G. Rosen (1969) shows that, in general, (that is, for observables more complicated than bilinear forms in x and p), Dirac quantization rule may be self contradictory. Let

$$A(x,p) = x^2p^2 = \frac{1}{9}\{x^3,p^3\} = \frac{1}{24}\{\{x^2,A\}\ ,\ p^2\}\ . \tag{2.23}$$

Applying Eq. (2.22) to the first Poisson bracket, and using the commutation relation $[x,p] = i\hbar$, we have

$$A \rightarrow \hat{A} = -\frac{i}{9\hbar}[\hat{x}^3,\hat{p}^3] = \frac{1}{2}(\hat{p}^2\hat{x}^2 + \hat{x}^2\hat{p}^2) + \frac{1}{3}\hbar^2\ . \tag{2.24a}$$

But now, apply Dirac's rule to the second Poisson bracket in Eq. (2.23) and insert the result (2.24a) to obtain a <u>different</u> result

$$A \rightarrow \hat{A} = -\frac{i}{24\hbar}[-\frac{i}{\hbar}[\hat{x}^2,\hat{A}],\hat{p}^2] = \frac{1}{2}(\hat{p}^2\hat{x}^2+\hat{x}^2\hat{p}^2) + \frac{1}{2}\hbar^2\ . \tag{2.24b}$$

We shall postpone the third expression for the Wigner function $C_{\hat{A}\hat{B}}(x,p)$ to the next Section (see Eq. (3.10)) and turn now to complete our account of Weyl's quantization by introducing dynamics.

D <u>Equations of motion</u>

From Hamilton's equations of motion

$$\dot{x} = \frac{\partial H}{\partial p} = -\{H,x\}\quad ,\quad \dot{p} = -\frac{\partial H}{\partial x} = -\{H,p\} \tag{2.25}$$

we have at time $t = \Delta t$

$$x(\Delta t) = x(0) - \Delta t\{H(x(0),p(0)),x(0)\} + O(\Delta t)^2)$$

$$p(\Delta t) = p(0) - \Delta t\{H(x(0),p(0)),p(0)\} + O((\Delta t)^2). \quad (2.26)$$

We now <u>define</u> operators $\hat{x}(\Delta t)$ and $\hat{p}(\Delta t)$ at time $t = \Delta t$ by

$$\hat{x}(\Delta t) \equiv \widehat{x(\Delta t)} \quad , \quad \hat{p}(\Delta t) \equiv \widehat{p(\Delta t)}$$

where $\widehat{x(\Delta t)}$ and $\widehat{p(\Delta t)}$ are the operators obtained from the right hand side of Eq. (2.26) by applying the quantization rule (2.3) at time $t = 0$. From Eq. (2.20), we have,

$$\widehat{\{H,x\}}_{t=0} = -\frac{i}{\hbar}[\hat{H},\hat{x}] \text{ and } \widehat{\{H,p\}}_{t=0} = -\frac{i}{\hbar}[\hat{H},\hat{p}]. \quad (2.27)$$

Thus, $\hat{x}(t)$ and $\hat{p}(t)$ are <u>defined</u> as the operator solution to the Heisenberg equation of motion

$$\frac{d\hat{x}}{dt} = \frac{i}{\hbar}[\hat{H},\hat{x}] \quad , \quad \frac{d\hat{p}}{dt} = \frac{i}{\hbar}[\hat{H},\hat{p}] \quad (2.28)$$

subject to the initial condition $\hat{x}(0)$ and $\hat{p}(0)$. Note that this definition is <u>invariant</u> to the choice of the quantization time $t = 0$. That is, quantizing at $t = 0$ and propagating with (2.28) is equivalent to propagating with (2.25) to time $t = t_1$ and then quantizing at t_1. Note also that by Eq. (2.28) the commutator $[\hat{x}(t), \hat{p}(t)]$ is time independent and hence satisfies $[\hat{x}(t), \hat{p}(t)] = i\hbar$. We now <u>define</u> a general operator at time t by Eq. (2.3) with $\hat{x}$ and $\hat{p}$ replaced by $\hat{x}(t)$ and $\hat{p}(t)$, that is,

$$\hat{A}(t) \equiv \frac{\hbar}{2\pi} \int \tilde{A}(u,v)e^{i[u\hat{x}(t)t+v\hat{p}(t)]}dudv. \quad (2.29)$$

It remains to show that $\hat{A}(t)$ satisfies the Heisenberg equation of motion

$$i\hbar \frac{d\hat{A}}{dt} = -[\hat{H},\hat{A}]. \quad (2.30)$$

This result is proved in Appendix B. Note that, in general,

$$\widehat{\frac{dA}{dt}} = -\widehat{\{H,A\}} \neq \frac{i}{\hbar}[\hat{H},\hat{A}] = \frac{d\hat{A}}{dt}. \quad (2.31)$$

We note for further reference that, by Schrödinger's equation of motion

$$i\hbar \frac{\partial}{\partial t}|\psi\rangle = H|\psi\rangle , \quad (2.32)$$

the density operator

$$\hat{\rho} = \sum_i W_i|\psi_i\rangle\langle\psi_i| , \quad 0 \le W_i \le 1, \quad \sum_i W_i = 1 \quad (2.33)$$

(with $\{|\psi_i\rangle\}$ a complete orthonormal set), satisfies the quantum mechanical Liouville equation

$$i\hbar \frac{\partial\hat{\rho}}{\partial t} = [\hat{H}, \hat{\rho}]. \quad (2.34)$$

We are now ready to translate the equations of quantum mechanics to the language of phase space.

III TRANSLATION OF THE QUANTUM MECHANICAL EQUATIONS TO PHASE SPACE

Since the inverse Weyl correspondence $\hat{A} \to A(x,p)$ (Eq. (2.6)) is a linear operation, we immediately obtain from the Heisenberg equation of motion (2.30), by use of Eq. (2.19)

$$i\hbar \frac{\partial A}{\partial t}(x,p,t) = -2i \sin(\frac{\hbar}{2}P_{12})H(1)A(2)\Big|_{\substack{x_1=x_2=x \\ p_1=p_2=p}} \tag{3.1}$$

Similarly, Liouville's equation (2.34) yields

$$i\hbar \frac{\partial \rho}{\partial t}(x,p,t)=2i \sin(\frac{\hbar}{2}P_{12})H(1)\rho(2)\Big|_{\substack{x_1=x_2=x \\ p_1=p_2=p}} . \tag{3.2}$$

Let us spell out the last equation for a Hamiltonian of the form

$$H = \frac{p^2}{2m} + V(x) . \tag{3.3}$$

Due to the simple p-dependence of the Hamiltonian, Eq. (3.2) reduces to

$$\frac{\partial \rho}{\partial t} = \{H,\rho\} + \sum_{k=1} \frac{(-1)^k}{(2k+1)!}(\frac{\hbar}{2})^{2k} V^{(2k+1)}(x)\frac{\partial^{2k+1}}{\partial p^{2k+1}}\rho(x,p,t) \tag{3.4}$$

where $V^{(2k+1)}(x)$ denotes the (2k+1)th derivative of the potential. Note that, for a harmonic oscillator, the quantum mechanical Liouville equation (3.4) reduces to the classical Liouville equation

$$\frac{\partial \rho}{\partial t} = \{H,\rho\} , \tag{3.5}$$

and the only difference between the quantum mechanical treatment and the classical one in this case, enters through the initial conditions as indicated by Tatarskii (1983).

It is evident that all of quantum mechanics could be translated into and solved in phase space. Thus, for example, the Schrödinger equation for eigen values and eigen states

$$\hat{H}|\psi\rangle = E|\psi\rangle \quad \text{or} \quad \hat{H}|\psi\rangle\langle\psi| = E|\psi\rangle\langle\psi| \tag{3.6}$$

i.e.

$$\hat{H}\hat{\rho} = E\hat{\rho} \quad \text{or} \quad \hat{\rho}\hat{H} = E\hat{\rho} \tag{3.7}$$

translates into

$$e^{\frac{i}{2}\hbar P_{12}} H(1)\rho(2)\Big|_{\substack{x_1=x_2=x \\ p_1=p_2=p}} = E\rho(x,p) \tag{3.8a}$$

with the boundary condition

$$\rho(x,p) \to 0 \qquad \text{for} \quad |x|\to\infty \qquad \text{or} \qquad |p|\to\infty \quad . \tag{3.8b}$$

So that $\rho(x,p)$ is normalizable. Since,

$$e^{\frac{i}{2}\hbar P_{12}} H(1)\rho(2)\Big| = e^{\frac{i}{\hbar}(D_{x_1}D_{p_2}-D_{p_1}D_{x_2})} H(1)\rho(2)\Big|$$

$$= H(x + \frac{i}{2}\hbar\,\frac{\partial}{\partial p}\ ,\ p - \frac{i}{2}\hbar\,\frac{\partial}{\partial x})\rho(x,p) \tag{3.9}$$

for the Hamiltonian (3.3), we have (the promised third form)

$$H(x+\frac{i}{2}\hbar\,\frac{\partial}{\partial p}\ ,\ p - \frac{i}{2}\hbar\,\frac{\partial}{\partial x})\rho(x,p) = E\rho(x,p) \tag{3.10}$$

with the boundary condition $\rho(x,p) \to 0$ at infinity. (See also Agarwal & Wolf 1970 and Hillery et al 1984.) This symmetrical form of the Schrödinger equation in phase space should be compared with the usual (configuration space) form

$$H(x,-i\hbar\,\frac{\partial}{\partial x})\psi(x) = E\psi(x) \tag{3.11}$$

with the boundary condition $\psi(x) \to 0$ at infinity. Returning to Eq. (3.8), we note, that by the second part of Eq. (3.7) $e^{-\frac{i}{2}\hbar P_{12}} H(1)\rho(2)\Big| = E\rho(x,p)$. Hence all imaginary terms drop (as they should) from Eq. (3.8) and we have

$$\cos(\frac{\hbar}{2}P_{12})H(1)\rho(2)\Big|_{\substack{x_1=x_2=x \\ p_1=p_2=p}} = E\rho(x,p) \ . \tag{3.12}$$

For the simple Hamiltonian (3.3), Eq. (3.12) takes the explicit form

$$H\rho - \frac{\hbar^2}{8}[V''(x)(\partial^2\rho/\partial p^2) + (1/m)(\partial^2\rho/\partial x^2)]$$

$$+ \sum_{k=2} \frac{(-1)^k}{(2k)!}\,(\frac{\hbar}{2})^{2k} V^{(2k)}(x)\,\frac{\partial^{2k}\rho}{\partial p^{2k}} = E\rho \ . \tag{3.13}$$

As an example, consider the harmonic oscillator. Since the potential is quadratic in x the Schrödinger equation (3.13) for the Hamiltonian $H = p^2/(2m) + (m/2)\omega^2 x^2$ reduces to

$$H\rho - \frac{\hbar^2}{8}[m\omega^2(\partial^2\rho/\partial p^2)+(1/m)(\partial^2\rho/\partial x^2)] = E\rho . \tag{3.14}$$

The vanishing of the imaginary terms, namely, $i\sin(\frac{\hbar}{2} P_{12})(H(1)\rho(2)| = 0$ now implies

$$\{H,\rho\} = 0 \ , \quad \text{and hence, } \rho = \rho(H(x,p)) \ . \tag{3.15}$$

Substituting $\rho = f(H)$ in Eq. (3.14), we obtain for $f(H)$ the ordinary differential equation

$$Hf'' + f' + (E-H)/(\hbar\omega/2)^2 f = 0 \ . \tag{3.16}$$

The asymptotic behaviour

$$f'' - (2/\hbar\omega)^2 f \to 0 \quad \text{as} \quad H \to \infty$$

suggests the substitution

$$f = e^{-2H/(\hbar\omega)} g(H) \ , \tag{3.17}$$

whence

$$Hg'' + g'(1 - \frac{4H}{\hbar\omega}) + \frac{2}{\hbar\omega}(\frac{2E}{\hbar\omega} - 1)g = 0 \ . \tag{3.18}$$

Expanding g(H) in a power series $g(H) = \Sigma\, a_n H^n$ and substituting in the last equation, we obtain the recurrence relation

$$(n+1)^2 a_{n+1} + \frac{2}{\hbar\omega}[(\frac{2E}{\hbar\omega} -1)-2n]a_n = 0 \quad , \quad n = 0,1,2,... \tag{3.19}$$

We therefore have

$$\frac{a_{n+1}}{a_n} \underset{n\to\infty}{\sim} \frac{4}{\hbar\omega}\frac{1}{n+1} \ ,$$

and unless g(H) is a polynomial, $g(H) \sim e^{4H/(\hbar\omega)}$ and by (3.17) f(H) is not normalizable. We therefore have

$$E_n = \hbar\omega(n+\tfrac{1}{2}) \quad , \quad n = 0,1,2,... \tag{3.20}$$

for the eigen values, and the normalized eigen quasidensities are given by

$$\rho_n(x,p) = (-1)^n 2e^{-2H/(\hbar\omega)} L_n(\frac{4H}{\hbar\omega}) ; \int \rho_n(x,p)\frac{dxdp}{2\pi\hbar} = 1 \ , \tag{3.21}$$

where $L_n(x)$ is the Laguerre polynomial. Thus, the solution of the harmonic oscillator problem in phase space follows the same lines as the corresponding solution in configuaration space with the amusing result that Hermite polynomials are replaced by Laguerre polynomials. The result (3.21) has been first given by Groenewold (1945). An alternative derivation is presented by Hillery et al (1984).

IV THE CLASSICAL LIMIT

A System with a sharp energy E

From the Schrödinger equation (3.13) we have

$$(H-E)\rho = O(\hbar^2) \ . \tag{4.1}$$

Hence, as $\hbar \to 0$

$$\rho \to 0 \qquad \text{for} \quad H \neq E \tag{4.2}$$

in such a way that the normalization condition $\int \rho(x,p)dxdp/(2\pi\hbar) = 1$ is fulfilled. In order to estimate the rate at which the quasi-distribution $\rho(x,p)$ approaches the microcanonical distribution (4.2), let us use the following relations:

$$\langle\hat{H}\rangle = E = \int \rho H \frac{dxdp}{2\pi\hbar} \ , \tag{4.3a}$$

$$\langle\hat{H}^2\rangle = E^2 = \int \rho \exp\frac{i}{2}\hbar P_{12} H(1)H(2)\frac{dxdp}{2\pi\hbar}$$

$$= \int \rho[H^2 - \frac{\hbar^2}{4m} V''(x)]\frac{dxdp}{2\pi\hbar} \ . \tag{4.3b}$$

Defining the "width" δH by

$$(\delta H)^2 \equiv \int \rho H^2 \frac{dxdp}{2\pi\hbar} - (\int \rho H \frac{dxdp}{2\pi\hbar})^2 \ , \tag{4.4}$$

we have,

$$(\delta H)^2 = \frac{\hbar^2}{4m} \int \rho V''(x) \frac{dxdp}{2\pi\hbar} \ . \tag{4.5}$$

Thus, the "width" δH is proportional to $\hbar$.

As an example, let us examine in detail the case of the harmonic oscillator solved in the previous Section. For a given energy $E_n = \hbar\omega(n+\frac{1}{2})$, $\hbar \to 0$ implies $n \to \infty$. Hence the following asymptotic expressions with $\nu \equiv 2(2n+1)$ apply to Eq. (3.21):

$$\rho(x,p) \sim 4\sin\{[\nu(2\theta-\sin 2\theta)+\pi]/4\}/(\pi\nu\sin 2\theta)^{\frac{1}{2}} + O(\frac{1}{n})$$

$$\text{for } \cos^2\theta \equiv H/E < 1 \ , \tag{4.6a}$$

and

$$\rho(x,p) \sim 2\exp\{-[\nu(\sinh 2\theta - 2\theta)/4]\}/(\pi\nu\sinh 2\theta)^{\frac{1}{2}} + O(\frac{1}{n})$$
$$\text{for } \cosh^2\theta \equiv H/E > 1 \ . \tag{4.6b}$$

(Erdély (1953) also gives a connection formula between the two regions $H/E < 1$ and $H/E > 1$). Thus, for $H < E$, $\rho(H)$ oscillates with an amplitude dying as $n^{-\frac{1}{2}}$ while for $H > E$, $\rho(H)$ decays as $e^{-\alpha n}$. For a much more careful and thorough discussion of the semiclassical behavior of $\rho(x,p)$, the reader is referred to the work of Berry (1977).

B System with a given expectation value <E>

We shall divide our discussion of a system with a given expectation value $\langle E\rangle$ i.e., a system in contact with a heat bath at temperature T, into three parts. We shall first give the quantum mechanical treatment and establish the limit $\hbar \to 0$, then the classical treatment, and finally, by comparing the expression obtained for the quantum entropy in the limit $\hbar \to 0$ with the corresponding classical expression, establish the condition under which all the quantum mechanical results coincide (in the limit $\hbar \to 0$) with the corresponding classical results.

(a) Quantum mechanics. As is well known, the canonical density operator is given by

$$\hat{\rho} = e^{-\beta\hat{H}}/Z(\beta) \qquad \text{where} \qquad Z(\beta) = \mathrm{tr}(e^{-\beta\hat{H}}) \tag{4.7}$$

is the partition function and $\beta = 1/(kT)$ is the inverse temperature. Let

$$R \equiv e^{-\beta\hat{H}} = \sum_n \frac{(-\beta)^n}{n!} \hat{H}^n \tag{4.8}$$

denote the unnormalized density operator. By Eq. (2.18) the corresponding Wigner function satisfies

$$R(x,p) = \sum_n \frac{(-\beta)^n}{n!} e^{\frac{i}{2}\hbar \sum_{i<j}^{n} P_{ij}} H(1)H(2)\ldots H(n)\Big|_{\substack{x_1=\ldots=x_n=x \\ p_1=\ldots=p_n=p}}$$

$$\xrightarrow[\hbar\to 0]{} \sum_n \frac{(-\beta)^n}{n!} H^n(x,p) = e^{-\beta H(x,p)} \ . \tag{4.9}$$

Note that, by the same argument, any operator function $f(A) = \sum_n f_n A^n$ satisfies

$$C_{f(\hat{A})}(x,p) \longrightarrow f(A(x,p)) \text{ as } \hbar \to 0 \ . \tag{4.10}$$

Hence

$$Z(\beta) = \int R(x,p) \frac{dxdp}{2\pi\hbar} \underset{\hbar\to 0}{\sim} \int e^{-\beta H} \frac{dxdp}{2\pi\hbar} \tag{4.11}$$

and

$$\langle A\rangle = \int \rho A \frac{dxdp}{2\pi\hbar} \underset{\hbar\to 0}{\sim} \int A e^{-\beta H} \frac{dxdp}{2\pi\hbar} \Big/ \int e^{-\beta H} \frac{dxdp}{2\pi h} , \tag{4.12}$$

where Eq. (2.16) has been involved.

(b) Classical mechanics. The classical canonical density is

$$\rho_{cl}(x,p) = e^{-\beta H(x,p)}/Z_{cl}(\beta) \ , \ Z_{cl}(\beta) = \int e^{-\beta H} \frac{dxdp}{2\pi\hbar_c} , \tag{4.13}$$

where $\hbar_c$ is a (small) unknown quantity having the dimension of action. Since $\hbar_c$ cancels from all expressions for expectation values

$$\langle A\rangle_{cl} = \int \rho_{cl} A \frac{dxdp}{2\pi\hbar_c} = \int e^{-\beta H} A \frac{dxdp}{2\pi\hbar_c} \Big/ \int e^{-\beta H} \frac{dxdp}{2\pi\hbar_c} , \tag{4.14}$$

it is usually dropped out. It is clear, however, that in order to make Eq. (4.13)dimensionally correct, one must assume the existence of $\hbar_c$. Comparing Eq. (4.12) with Eq. (4.14), we see, that, for any observale,

$$\langle\hat{A}\rangle \underset{\hbar\to 0}{\longrightarrow} \langle A\rangle_{cl} . \tag{4.15}$$

(c) The entropy. The quantum mechanical expression for the entropy is

$$S = - k \ \mathrm{tr}(\hat{\rho} \log \hat{\rho}) . \tag{4.16}$$

Using Eqs. (4.7), (4.11) and (4.15), we have,

$$S = k[\beta\langle E\rangle + \log Z(\beta)] \underset{\hbar\to 0}{\sim} k[\beta\langle E\rangle_{cl} + \log\int e^{-\beta H} \frac{dxdp}{2\pi\hbar}] \tag{4.17}$$

The corresponding classical expression is (see Jaynes 1963):

$$S_{cl} = -k \int P(x,p) \log \frac{P(x,p)}{m(x,p)} dxdp , \tag{4.18a}$$

where

$$P(x,p) \equiv \frac{1}{2\pi\hbar_c} \rho_{cl}(x,p) \ , \quad \int P(x,p) dxdp = 1 , \tag{4.18b}$$

and

$$m(x,p) = 1/(2\pi\hbar_c) \tag{4.18c}$$

is the prior or density of states in phase space.

Hence, using Eq. (4.13),

$$S_{cl} = k[\beta\langle E\rangle_{cl} + \log \int e^{-\beta H} \frac{dxdp}{2\pi\hbar_c}] \ . \tag{4.19}$$

Comparing Eq. (4.17) with Eq. (4.19), we see, that the limiting expression for the quantum entropy coincides with the classical one, if and only if the unknown quantity, $\hbar_c$ is chosen as

$$\hbar_c = \hbar \ . \tag{4.20}$$

Thus the limiting expression for the quantum entropy determines uniquely the hitherto undetermined expression (to within a constant) for the classical entropy. By the same token, we have,

$$\rho(x,p) \underset{\hbar\to 0}{\sim} \rho_{cl}(x,p) \ . \tag{4.21}$$

We end this Section by noting that, according to the principle of maximum entropy, all the results obtained for the observable $\hat{H}$ (given E or $\langle E\rangle$), are valid for any observable $\hat{A}$ (given a sharp value A or an expectation value $\langle\hat{A}\rangle$). For further information on the principle of maximum entropy, consult Jaynes (1957), Shore & Johnson (1980, 1983) and Tikochinsky et al (1984, 1985).

V QUANTUM CORRECTIONS TO THE CLASSICAL CANONICAL DISTRIBUTION

Let $\hat{R} = e^{-\beta\hat{H}}$ denote the unnormalized canonical density operator. From the explicit expression (4.9) one can derive an expansion in powers of $\hbar$ for the Wigner function R(x,p). First note that only even powers of $\hbar$ enter. Since the permutation $\begin{pmatrix}1 & 2\ldots & n\\ n & n-1\ldots & 1\end{pmatrix}$ carries Eq. (4.9) into

$$R(x,p) = \sum_n \frac{(-\beta)^n}{n!} e^{-\frac{i}{2}\hbar \sum_{i<j}^{n} P_{ij}} H(1)H(2)\ldots H(n) \Big|_{\substack{x_1=\ldots=x_n=x \\ p_1=\ldots=p_n=p}} \tag{5.1}$$

we have,

$$R(x,p) = \sum \frac{(-\beta)^n}{n!} \cos\left(\frac{\hbar}{2} \sum_{i<j}^{n} P_{ij}\right) H(1)H(2)\ldots H(n) \Big|_{\substack{x_1=\ldots=x_n=x \\ p_1=\ldots=p_n=p}} \ . \tag{5.2}$$

The first correction, proportional to $\hbar^2$, using this explicit expression can be carried out by direct counting. However direct counting becomes rapidly intricate. An efficient (indirect) way to calculate the first quantum corrections to $R(x,p) \simeq R_{cl}(x,p) = \exp(-\beta H(x,p))$, has been given by Wigner (1932). This way which consists of approximate successive solutions to the Bloch equation, will now be reviewed. From the definition $\hat{R} = \exp(-\beta\hat{H})$ we see that the (unnormalized) canonical operator satisfies the Bloch equation

$$\frac{\partial \hat{R}}{\partial \beta} = - \hat{H}\hat{R} = -\hat{R}\hat{H} \tag{5.3}$$

with the initial condition $\hat{R}(\beta = 0) = 1$.
Translating the Bloch equation into phase space, we obtain,

$$\frac{\partial R}{\partial \beta}(x,p,\beta) = -\cos(\frac{\hbar}{2} P_{12})H(1)R(2)\Big|_{\substack{x_1=x_2=x \\ p_1=p_2=p}} \tag{5.4}$$

with $R(x,p,\beta=0) = 1$.
Explicitly, for the simple Hamiltonian (3.3), we have,

$$\begin{aligned}\frac{\partial R}{\partial \beta} &= - HR + \frac{\hbar^2}{8}(V'' \frac{\partial^2 R}{\partial p^2} + \frac{1}{m} \frac{\partial^2 R}{\partial x^2}) \\ &- \sum_{k=2} \frac{(-1)^k}{(2k)!} (\frac{\hbar}{2})^{2k} V^{(2k)}(x) \frac{\partial^2 k}{\partial p^{2k}} R\end{aligned} \tag{5.5}$$

Subject to the initial condition $R(\beta=0) = 1$.
To obtain the first quantum correction to $R_{cl}(x,p) = e^{-\beta H(x,p)}$ we insert

$$R \simeq e^{-\beta H}[1-\hbar^2\bar{f}_2] \simeq e^{-\beta H[1+\hbar^2 f_2]} ; \bar{f}_2=\beta H f_2 \quad . \tag{5.6}$$

into Eq. (5.5), and by keeping only terms up to order $\hbar^2$, we secure for f_2 ,

$$\frac{\partial}{\partial \beta}(\beta f_2) = \frac{\beta}{8H}[\frac{2}{m}V'' - \frac{\beta}{m}(\frac{p^2}{m}V'' + V'^2)] \tag{5.7a}$$

$$f_2 = \frac{\beta}{8mH}[V'' - \frac{\beta}{3}(\frac{p^2}{m} V'' + V'^2)] \ . \tag{5.7b}$$

Hence

$$R=e^{-\beta H}\{1- \frac{\hbar^2\beta^2}{8m}[V''- \frac{\beta}{3}(\frac{p^2}{m}V'' + V'^2)]\} + O(\hbar^4) \ . \tag{5.8}$$

Correspondingly, we obtain for the partition function $Z(\beta) = \int R \, dxdp/(2\pi\hbar)$ and for the free energy $F = - kT \log Z(\beta)$

$$Z = Z_{cl}[1 - \frac{\hbar^2\beta^3}{24m} \langle V'^2\rangle_{cl} + O(\hbar^4)] \ , \tag{5.9}$$

$$F = F_{cl} + \frac{\hbar^2\beta^2}{24m} \langle V'^2\rangle_{cl} + O(\hbar^4) \ , \tag{5.10}$$

where the partial integration

$$\int e^{-\beta H}V''dx = \beta \int e^{-\beta H}V'^2 dx$$

has been utilized. From Eq. (5.10) we see that the first quantum correction to the free energy is positive, proportional to the average of the squared force and diminishes as the temperature and/or the mass increase. It is useful to express the first quantum correction to the Maxwellian distribution in terms of effective temperature and effective force (see Landau & Lifshitz 1968):

$$W(p) \propto \int R(x,p)\,dx \propto \exp(-\beta_{eff}p^2/2m)\,, \tag{5.11a}$$

$$1/\beta_{eff} = T_{eff} \equiv T + \frac{\hbar^2 \langle V'^2 \rangle_{cl}}{12mk^3T^2}\;; \tag{5.11b}$$

$$W(x) \propto \int R(x,p)\,dp \propto \exp(-\beta V_{eff})\,, \tag{5.12a}$$

$$V_{eff} \equiv V - \frac{\hbar^2\beta^2}{24m} V'^2 + \frac{\hbar^2\beta}{12m} V''\,. \tag{5.12b}$$

As expected, quantum "smearing" leads to an effective temperature higher than the classical temperature. The next quantum correction and a generalization to N particles (neglecting the possible identity of the particles) can be found in the original work of Wigner (1932).

We end this Section by solving exactly the Bloch equation for the harmonic oscillator (See Imre et al 1967 and Hillery et al 1984). An alternative direct calculation, using the inverse Weyl mapping (2.6), is given in Appendix C. The Bloch equation for the harmonic oscillator reads

$$\frac{\partial R}{\partial \beta} = -HR + \frac{\hbar^2}{8}\left[m\omega^2 \frac{\partial^2 R}{\partial p^2} + \frac{1}{m}\frac{\partial^2 R}{\partial x^2}\right], \qquad R(\beta=0) = 1. \tag{5.13}$$

The ansatz

$$R = e^{-A(\beta)H - B(\beta)}\,, \quad A(0) = B(0) = 0 \tag{5.14}$$

leads, by comparing the coefficients of equal powers of x and p, to the ordinary differential equations

$$A' = 1 - \frac{\hbar^2\omega^2}{4} A^2\,, \quad A(0)=0; \quad B' = \frac{\hbar^2\omega^2}{4}\,, \quad B(0) = 0. \tag{5.15}$$

Hence

$$A = \frac{2}{\hbar\omega}\,\mathrm{tgh}\left(\frac{\beta\hbar\omega}{2}\right), \quad B = \log\,\cosh\left(\frac{\beta\hbar\omega}{2}\right), \tag{5.16}$$

$$\begin{aligned} R &= \exp\{-\beta H[\mathrm{tgh}(\tfrac{\beta\hbar\omega}{2})/(\tfrac{\beta\hbar\omega}{2})]\}/\cosh(\tfrac{\beta\hbar\omega}{2}) \\ &= e^{-\beta H}[1 + \tfrac{1}{24}(\beta\hbar\omega)^2(-3+2\beta H) + \ldots]\;; \end{aligned} \tag{5.17a}$$

$$Z(\beta) = [2\ \sinh(\frac{\beta\hbar\omega}{2})]^{-1} = (\beta\hbar\omega)^{-1} - (\beta\hbar\omega)/24 + \dots\ ; \qquad (5.17b)$$

$$\rho(x,p) = 2\text{tgh}(\frac{\beta\hbar\omega}{2})\exp\{-\beta H[\text{tgh}(\frac{\beta\hbar\omega}{2})/(\frac{\beta\hbar\omega}{2})]\}. \qquad (5.17c)$$

Note, that from the calculated partition function, one can read the energy levels and the degeneracies:

$$Z = \sum_n e^{-\beta E_n} = [2\ \sinh(\frac{\beta\hbar\omega}{2})]^{-1} = \exp(-\beta\hbar\omega/2)[1-\exp(-\beta\hbar\omega)]^{-1}$$

$$= e^{-\beta\hbar\omega/2} \sum_n e^{-\beta\hbar\omega n} \quad . \qquad (5.18)$$

Hence, $E_n = \hbar\omega(n+\frac{1}{2})$ and $g_n = 1$.

VI N IDENTICAL PARTICLES - CORRECTIONS DUE TO THE PAULI PRINCIPLE

Up to now we have dealt with the Wigner function for a single particle in one dimension. The generalization to N particles in three dimensions is straightforward. Indeed, using the notation

$$ux = \sum_{i=1}^{N} \vec{u}_i \cdot \vec{x}_i \quad , \quad dx = d^3x_1 d^3x_2 \dots d^3x_N \qquad (6.1)$$

etc., we can write the generalized version of Eqs. (2.3) and (2.6) as

$$\hat{A} = \frac{1}{(2\pi)^{6N}} \int A(x,p) e^{-i(ux+vp)} e^{i(u\hat{x}+v\hat{p})} dx dp du dv \qquad (6.2)$$

and conversely

$$A(x,p) = (\frac{\hbar}{2\pi})^{3N} \int \text{tr}[\hat{A}\ e^{-i(u\hat{x}+v\hat{p})}] e^{i(ux+vp)} du dv \ . \qquad (6.3)$$

The formulas above are valid for N distinguishable particles. If the particles are identical, we must take the Pauli principle into consideration. That is, we must use symmetric or antisymmetric wave functions according to the particles being bosons or fermions. In order to facilitate the discussion, we shall use the language of second quantization. Our formal derivation will show that the first correction due to the identity of N non-interacting (3-dimensional) particles is proportional to $\hbar^3$. We shall also argue that this correction is essentially a "classical" one and could have been found by Gibbs, had he been told that $\hbar_c$ (whatever its value might be) is finite. It must be admitted, however, that our treatment is incomplete inasmuch as spin variables are not explicitly taken into account.

A Notation of second quantization

Let $|\psi_N\rangle$ denote an N-particle state with definite symmetry (describing bosons or fermions), and let $\psi_N(\vec{x}_1,\vec{x}_2,\ldots,\vec{x}_N) = \langle\vec{x}_1,\vec{x}_2,\ldots,\vec{x}_N|\psi_N\rangle$ denote the corresponding wave function. Then (see e.g. Schweber 1961)

$$\psi_N(\vec{x}_1,\vec{x}_2,\ldots,\vec{x}_N) = \langle 0|\psi(\vec{x}_N)\ldots\psi(\vec{x}_1)|\psi_N\rangle/(N!)^{\frac{1}{2}} \tag{6.4}$$

where
$$\psi(x) = \sum_i \psi_i(\vec{x})a_i \tag{6.5}$$

is the destruction operator for a single particle at the location $\vec{x}$, and $\{\psi_i(\vec{x})\}$ is a complete orthonormal set of single particle wave functions. Here $|0\rangle$ denotes the state with no particles (the vacuum state). The field operators $\psi(\vec{x})$ and $\psi^+(\vec{x}) = \sum_i \psi_i{}^*(\vec{x})a_i{}^+$ satisfy the commutation (-), or anticommutation (+) relations:

$$[\psi(\vec{x}),\ \psi^+(\vec{x}')]_{\pm} = \delta(\vec{x}-\vec{x}')\quad,\ [\psi(\vec{x}),\ \psi(\vec{x}')]_{\pm} = 0\ . \tag{6.6}$$

Equation (6.4) together with the commutation (anticommutation) relations (6.6) contain in a nutshell all the results needed to take care of the Pauli principle.

B Wigner functions with symmetrization

Let $\{|\psi_N{}^{(\alpha)}\rangle\}$ be a complete orthonormal set of N-particle symmetrized states. We shall use the verb to symmetrize both ways, meaning to make symmetric or antisymmetric. Introduce the notation

$$\mathrm{tr}^\theta(\hat{A}) \equiv \sum_\alpha \langle\psi_N{}^{(\alpha)}|\hat{A}|\psi_N{}^{(\alpha)}\rangle \tag{6.7}$$

for the trace taken with states of definite symmetry θ . In terms of this notation, define the Wigner function with symmetrization for the operator $\hat{A}$ by

$$A_\theta(x,p) = (\hbar/2\pi)^{3N} \int \mathrm{tr}^\theta[\hat{A}e^{-i(u\hat{x}+v\hat{p})}]e^{i(ux+vp)}du dv. \tag{6.8}$$

As will be shown later, the last definition is the proper modification of Eq. (6.3) for the case of N identical particles. Using Eq. (6.4) and the fact that for a symmetrical N-particle operator $\hat{A}$, the state $\hat{A}|\psi_N{}^{(\alpha)}\rangle$ has the same symmetry as the state $|\psi_N{}^{(\alpha)}\rangle$, it is not difficult to show that

$$\mathrm{tr}^\theta[\hat{A}\ \exp(-i(u\hat{x}+v\hat{p}))] = (1/N!)\exp(\tfrac{1}{2}i\hbar uv)\int dy\ \exp(-iuy)$$
$$\times \langle 0|\psi(\vec{y}_N-\hbar\vec{v}_N)\ldots\psi(\vec{y}_1-\hbar\vec{v}_1)\hat{A}\psi^+(\vec{y}_1)\ldots\psi^+(y_N)|\rangle \tag{6.9}$$

Inserting this result into the definition (6.8) we have explicitly

$$A_\theta(x,p) = (1/N!) \int dx'\ \exp(ix'p/\hbar)\ \times$$

$$\times \langle 0|\psi(\vec{x}_N-\tfrac{1}{2}\vec{x}_N')\ldots\psi(\vec{x}_1-\tfrac{1}{2}\vec{x}_1')\hat{A}\psi^+(\vec{x}_1+\tfrac{1}{2}\vec{x}_1')\ldots\psi^+(\vec{x}_N+\tfrac{1}{2}\vec{x}_N')|0\rangle \ . \qquad (6.10)$$

Another form (Hillery et al 1984) for the Wigner function with symmetrization can be obtained from Eq. (6.10) by noting that the matrix element $\langle 0|\ldots|0\rangle$ can be replaced without change by the sum

$$\sum_\alpha \langle\psi_N^{(\alpha)}|\ldots|\psi_N^{(\alpha)}\rangle = tr^\theta(\ldots) \ .$$

Using the cyclic property of the trace, Eq. (6.10) can then be rewritten as

$$A_\theta(x,p) = (1/N!)\int dx' \exp(ix'p/\hbar)$$

$$\times\ tr^\theta[\hat{A}\psi^+(\vec{x}_1+\tfrac{1}{2}\vec{x}_1')\ldots\psi^+(\vec{x}_N+\tfrac{1}{2}\vec{x}_N')\psi(\vec{x}_N-\tfrac{1}{2}\vec{x}_N')\ldots(\vec{x}_1-\tfrac{1}{2}\vec{x}_1')] \ . \qquad (6.11)$$

However, we shall make no use of the form (6.11) in the present work.

C <u>Passage from the "Schrödinger picture" to the "Heisenberg picture"</u>

Instead of working in the subspace of physical states spanned by $\{|\psi_N^{(\alpha)}\rangle\}$, we could use the identity

$$tr^\theta(\hat{A}) = \sum_\alpha \langle\psi_N^{(\alpha)}|\hat{A}|\psi_N^{(\alpha)}\rangle = tr \sum_\alpha |\psi_N^{(\alpha)}\rangle\langle\psi_N^{(\alpha)}|\,\hat{A} \qquad (6.12)$$

to rewrite $tr^\theta(\hat{A})$ as

$$tr^\theta(\hat{A}) = tr(\hat{P}_\theta\hat{A}) = tr(\hat{A}\hat{P}_\theta) \ , \qquad (6.13)$$

where

$$\hat{P}_\theta \equiv \sum_\alpha |\psi_N^{(\alpha)}\rangle\langle\psi_N^{(\alpha)}| \qquad (6.14)$$

is the projection operator onto the space of symmetrized states. Note that in the right hand side of Eq. (6.13) tr stands for the regular trace and the burden of the Pauli principle is shifted from the state $|\psi_N^{(\alpha)}\rangle$ to the projection operator $\hat{P}_\theta$. This transition amounts to a passage from the "Schrödinger picture" to the "Heisenberg picture". Note also that if $\hat{A}$ is an N-particle symmetric and hermitian operator, then (Imre et al 1967):

$$\hat{P}_\theta\hat{A} = \hat{A}\hat{P}_\theta \ . \qquad (6.15)$$

Indeed, decompose a general N-particle state $|\psi\rangle$ into two orthogonal components

$$|\psi\rangle = |\psi_\theta\rangle + |\psi_{\bar{\theta}}\rangle \ , \qquad (6.16)$$

where $|\psi_\theta\rangle = \hat{P}_\theta|\psi\rangle$. Then

$$\hat{A}\hat{P}_\theta|\psi\rangle = \hat{A}|\psi_\theta\rangle \ ; \qquad (6.17)$$

$$\hat{P}_\theta\hat{A}|\psi\rangle = \hat{P}_\theta\hat{A}|\psi_\theta\rangle + \hat{P}_\theta\hat{A}|\psi_{\bar{\theta}}\rangle = \hat{P}_\theta\hat{A}|\psi_\theta\rangle = \hat{A}|\psi_\theta\rangle \ . \tag{6.18}$$

$\hat{P}_\theta\hat{A}|\psi_{\bar{\theta}}\rangle$ above drops, since $\hat{A}|\psi_{\bar{\theta}}\rangle$ has a vanishing component in the subspace spanned by the symmetrized states:

$$\langle\phi\hat{A}_\theta|\psi_\theta\rangle = \langle\hat{A}\phi_\theta|\psi_\theta\rangle = 0 \ . \tag{6.19}$$

We shall now work out the connection between the Wigner function with symmetrization and the regular Wigner function. Using Eq. (6.13) we can rewrite the definition (6.8) as

$$A_\theta(x,p)=(\hbar/2\pi)^{3N}\int \mathrm{tr}[\hat{P}_\theta\hat{A}\exp(-i(u\hat{x}+v\hat{p}))] \times$$
$$\times \exp[i(ux+vp)]dudv = C_{\hat{P}_\theta\hat{A}}(x,p) \ , \tag{6.20}$$

where $C_{\hat{P}_\theta\hat{A}}(x,p)$ is the regular Wigner function for the product $\hat{C}=\hat{P}_\theta\hat{A}$ (see Sec. II.C.). In order to make use of the last relation, we shall need an expression for the (regular) Wigner function of the operator $\hat{P}_\theta$. By Eqs. (6.3), (6.13) and (6.10) this function is given by

$$C_{\hat{P}_\theta}(x,p) = I_\theta(x,p) \ , \tag{6.21}$$

where

$$I_\theta(x,p) \equiv (1/N!) \int dx'\exp(ix'p/\hbar) \times$$
$$\langle 0|\psi(\vec{x}_N-\tfrac{1}{2}\vec{x}_N')\ldots\psi(\vec{x}_1-\tfrac{1}{2}\vec{x}_1')\psi^+(\vec{x}_1+\tfrac{1}{2}\vec{x}_1')\ldots\psi^+(\vec{x}_N+\tfrac{1}{2}\vec{x}_{N'})|0\rangle \tag{6.22}$$

is the Wigner function with symmetrization for the identity operator. Inserting this result into Eq. (6.20) and using expression (2.17) for the function $C_{\hat{P}_\theta\hat{A}}(x,p)$, we secure (Imre et al 1967)

$$A_\theta(x,p)=\exp[\tfrac{1}{2}i\hbar P_{12}]I_\theta(x^{(1)},p^{(1)})A(x^{(2)},p^{(2)})\Big|_{\substack{x^{(1)}=x^{(2)}=x\\ p^{(1)}=p^{(2)}=p}}$$
$$= \exp[\tfrac{1}{2}i\hbar P_{12}]A(x^{(1)},p^{(1)})I_\theta(x^{(2)},p^{(2)})\Big|_{\substack{x^{(1)}=x^{(2)}=x\\ p^{(1)}=p^{(2)}=p}} \tag{6.23}$$

This is the desired relation between the Wigner function with symmetrization and the regular Wigner function. The second line of Eq. (6.23) is obtained from the first line by utilizing the property (6.15). Since $\mathrm{tr}^\theta(\hat{A}\hat{B}) = \mathrm{tr}(\hat{P}_\theta\hat{A}\hat{B})$ we have, with the aid of Eqs. (6.20), (2.15) and (6.15),

$$\mathrm{tr}^\theta(\hat{A}\hat{B}) = \int A_\theta(x,p)B(x,p)dxdp/(2\pi\hbar)^{3N}$$
$$= \int A(x,p)B_\theta(x,p)dxdp/(2\pi\hbar)^{3N} \tag{6.24}$$

In particular, for $\hat{B} = 1$,

$$\mathrm{tr}^{\theta}(\hat{A}) = \int A(x,p)I_{\theta}(x,p)dxdp/(2\pi\hbar)^{3N} . \tag{6.25}$$

Finally, the Wigner function with symmetrization for the product of two operators $\hat{C} = \hat{A}\hat{B}$ is (Imre et al 1967)

$$C^{\theta}_{\hat{A}\hat{B}}(x,p) \overset{(6.20)}{=\!=} C_{\hat{P}_{\theta}\hat{A}\hat{B}}(x,p) \overset{(2.17)}{=\!=}$$

$$(\exp[\tfrac{1}{2}i\hbar P_{12}]C_{\hat{P}_{\theta}\hat{A}}(x^{(1)},p^{(1)})B(x^{(2)},p^{(2)})|$$

$$\overset{(6.20)}{=\!=} (\exp[\tfrac{1}{2}i\hbar P_{12}]A_{\theta}(x^{(1)},p^{(1)})B(x^{(2)},p^{(2)})\Big|_{\substack{x^{(1)}=x^{(2)}=x \\ p^{(1)}=p^{(2)}=p}}$$

$$\overset{(6.15)}{=\!=} (\exp[\tfrac{1}{2}i\hbar P_{12}]A(x^{(1)},p^{(1)})B_{\theta}(x^{(2)},p^{(2)})\Big|_{\substack{x^{(1)}=x^{(2)}=x \\ p^{(1)}=p^{(2)}=p}} \tag{6.26}$$

Now, that we have all the machinery of the former Sections at our disposal, we can repeat the game. For example, the Bloch equation for $\hat{R} = e^{-\beta\hat{H}}$ is

$$\partial\hat{R}/\partial\beta = -\hat{H}\hat{R} = -\hat{R}\hat{H} \; ; \quad \hat{R}(\beta=0) = 1 \tag{6.27}$$

translates, in the case of N identical particles, into

$$\partial R_{\theta}(x,p,\beta)/\partial\beta = -\cos(\tfrac{1}{2}\hbar P_{12})H^{(1)}R_{\theta}^{(2)}\Big|_{\substack{x^{(1)}=x^{(2)}=x \\ p^{(1)}=p^{(2)}=p}} \tag{6.28a}$$

with the initial condition

$$R_{\theta}(\beta=0) = I_{\theta}(x,p) . \tag{6.28b}$$

Note that the differential equation remains unchanged (see Eq. (5.4)) but the initial condition changes.

D Quantum corrections due to the identity of particles - the case of ideal gas

By Eqs. (6.25) and (6.22) the partition function for N non-interacting identical particles is

$$Z = \mathrm{tr}^{\theta}(\hat{R}) = \int R(x,p)I_{\theta}(x,p)dxdp/(2\pi\hbar)^{3N} =$$

$$= \int dxdp(2\pi\hbar)^{-3N}R(x,p) \times$$

$$\frac{1}{N!}\int dy e^{\frac{i}{\hbar}py} <0|\psi(\vec{x}_N-\tfrac{1}{2}\vec{y}_N)\ldots\psi(\vec{x}_1-\tfrac{1}{2}\vec{y}_1)\psi^+(\vec{x}_1+\tfrac{1}{2}\vec{y}_1)\ldots\psi^+(\vec{x}_N+\tfrac{1}{2}\vec{y}_N)|0> \quad (6.29)$$

where

$$\hat{R} = \exp(-\beta\hat{H}) \quad \text{and} \quad \hat{H} = \sum_{i=1}^{N} \hat{\vec{p}}_i^{\,2}/2m = \hat{p}\hat{p}/2m \quad . \quad (6.30)$$

In the zeroth (classical) approximation, the Wigner function for $\hat{R}$ is simply $\exp(-\beta H) = \exp(-\beta pp/2m)$ (see Eq. (5.8)). We wish to calculate the first quantum correction to the classical partition function

$$Z_0 = (1/N!)\int \exp(-\beta\Sigma\vec{p}_i^{\,2}/2m)\,dxdp/(2\pi\hbar)^{3N} \quad (6.31)$$

due to the identity of particles. To this end, we shall first perform the p-integration in Eq. (6.29). Let

$$J(x) \equiv (1/N!)\int dp \exp(\beta pp/2m)\int dy \exp(ipy/\hbar) I(x,y) \quad (6.32)$$

where

$$I(x,y) \equiv <0|\psi(\vec{x}_N-\tfrac{1}{2}\vec{y}_N)\ldots\psi(\vec{x}_1-\tfrac{1}{2}\vec{y}_1)\psi^+(\vec{x}_1+\tfrac{1}{2}\vec{y}_1)\ldots\psi^+(\vec{x}_N+\tfrac{1}{2}\vec{y}_N)|0> \quad . \quad (6.33)$$

Performing the p-integration, we have,

$$J(x) = (1/N!)(2\pi m/\beta)^{3N/2}\int dy \exp[-(\vec{y}_1^{\,2}+\ldots+\vec{y}_N^{\,2})]/(2\hbar^2\beta/m) I(x,y) . \quad (6.34)$$

We now use the commutation (anticommutation) relations (6.6) to bring all the creation operators in $I(x,y)$ to the left of the destruction operators (i.e. bring to normal form). For example, in the case of bosons, the contraction of particle i with particle j gives

$$\psi(\vec{x}_i-\tfrac{1}{2}\vec{y}_i)\psi^+(\vec{x}_j+\tfrac{1}{2}\vec{y}_j) = \psi^+(\vec{x}_j+\tfrac{1}{2}\vec{y}_j)\psi(\vec{x}_i-\tfrac{1}{2}\vec{y}_i)+\delta(\vec{x}_j-\vec{x}_i+\tfrac{1}{2}(\vec{y}_i+\vec{y}_j)) \quad . \quad (6.35)$$

Since operators in the normal form do not contribute when sandwiched between vacuum states, we end up with a sum of N! terms each of which is a product of N delta functions. Thus, assuming for example N = 4 bosons, the following terms appear (among others)

$$(1\text{-}1 \quad 2\text{-}2 \quad 3\text{-}3 \quad 4\text{-}4) = \delta(\vec{y}_1)\delta(\vec{y}_2)\delta(\vec{y}_3)\delta(\vec{y}_4) \quad (6.36a)$$

$$(1\text{-}1 \quad 2\text{-}2 \quad 3\text{-}4 \quad 4\text{-}3) = \delta(\vec{y}_1)\delta(\vec{y}_2)\delta(\vec{x}_3-\vec{x}_4+\tfrac{1}{2}(\vec{y}_3+\vec{y}_4)) \times \delta(\vec{x}_4-\vec{x}_3+\tfrac{1}{2}(\vec{y}_3+\vec{y}_4)) \quad (6.36b)$$

$$(1\text{-}1 \quad 2\text{-}3 \quad 3\text{-}4 \quad 4\text{-}2) = \delta(\vec{y}_1)\delta(\vec{x}_3-\vec{x}_2+\tfrac{1}{2}(\vec{y}_2+\vec{y}_3)) \times\delta(\vec{x}_4-\vec{y}_3+\tfrac{1}{2}(\vec{y}_3+\vec{y}_4))\delta(\vec{x}_2-\vec{x}_4+\tfrac{1}{2}(\vec{y}_2+\vec{y}_4)) \quad . \quad (3.36c)$$

Here (1-1 2-2 3-3 4-4) denotes the terms arising from the contraction of particle 1 with particle 1, particle 2 with particle 2, etc. Returning to Eq. (6.34), we note, that the contribution to the integration over $\vec{y}_i$ comes mainly from the region $|\vec{y}_i| \lesssim (\hbar^2\beta/m)^{\frac{1}{2}}$. Hence the term (6.36b) describes two particles in a cell of volume $v \simeq (\hbar^2\beta/m)^{3/2}$, (3.36c) describes three particles in a cell of volume $v \simeq (\hbar^2\beta/m)^{3/2}$ etc. Let k=N, N-1,...,1 denote the number of different cells into which the N particles are distributed, and let $N_i^{(k)}$ denote the number of particles in cell i $(\Sigma N_i^{(k)} = N)$. Then, integrating J(x) over x (see Eq. (6.32)), the partition into k cells contributes to Z (β) in proportion to

$$J_k \equiv V^k (\hbar^2\beta/m)^{3[(N_1^{(k)}-1)+\ldots+(N_k^{(k)}-1)]/2} = V^k (\hbar^2\beta/m)^{3(N-k)/2} \quad (6.37)$$

where V is the total volume of the system. Thus, when all N particles are distributed into N distinct cell $J_N = V^N$, when exactly two of the particles shares the same cell $J_{N-1} = V^{N-1}\hbar^3\beta^{3/2}/m^{3/2}$ etc. Equation (6.37) thus shows, that the quantum corrections to the classical partition function $Z_o(\beta)$ are proportional to $\hbar^3$, $\hbar^6$,etc.

But the same corrections are also needed in classical physics, and could have been calculated by Gibbs, had he known that $\hbar_c$ is finite! Indeed, Gibbs realized that for N identical particles, in order to have entropy as an extensive quantity, all N! arrangements of the particles in N different cells should be counted as one state, and the partition function (for non interacting particles) should read

$$\begin{aligned} Z_{cl} &= (1/N!)\int dxdp\,(2\pi\hbar_c^2)^{-3N}\exp(-\beta\Sigma\vec{p}_i^{\,2}/2m) \\ &= (1/N!)\,[V(m/2\pi\hbar_c^2\beta)^{3/2}]^N \quad . \end{aligned} \quad (6.38)$$

But, if $\hbar_c$ is finite, the number of available cells is also finite and corrections due to distributions with two or more particles in a cell should be made. Thus, if N particles occupy $k < N$ cells each with volume $v = (2\pi\hbar_c^2\beta/m)^{3/2}$, one should increase Z_{cl} by

$$J_k = V^k (2\pi\hbar_c^2\beta/m)^{3[(N_1^{(k)}-1)+\ldots+(N_k^{(k)}-1)]/2} = V^k (2\pi\hbar_c^2\beta/m)^{3(N-K)/2} \quad , \quad (6.39)$$

that is, correct Z_{cl} by terms proportional to $\hbar_c^3, \hbar_c^6, \hbar_c^9$ etc.

Let us return to Eq. (6.29) to calculate the dependence of the first quantum correction on the number of particles. Since the number of arrangements with exactly two particles in a cell is N(N-1)/2, we have, with the aid of Eqs. (6.32) to (6.37)

$$Z = \int dx\,(2\pi\hbar)^{-3N} J(x)$$

$$\simeq \int \frac{dx}{(2\pi\hbar)^{3N}} (1/N!)(2\pi m/\beta)^{3N/2}$$

$$\times \ [\int dy \exp[-(\vec{y}_1^2+\ldots+\vec{y}_N^2)/(2\hbar^2\beta/m)]\,\delta(\vec{y}_1)\ldots\delta(\vec{y}_N)$$

$$\pm \ \binom{N}{2}\int d^3y_1 d^3y_2 \times \exp\Big(-\frac{\vec{y}_1^2+\vec{y}_2^2}{2\hbar^2\beta/m}\Big)\ \delta\Big(\vec{x}_2-\vec{x}_1+\frac{\vec{y}_1+\vec{y}_2}{2}\Big)\delta\Big(\vec{x}_1-\vec{x}_2+\frac{\vec{y}_1+\vec{y}_2}{2}\Big)]$$

$$= \frac{1}{N!}[V(\frac{m}{2\pi\hbar^2\beta})^{3/2}]^N \pm \frac{1}{N!}\,\frac{N(N-1)}{2^{5/2}}[V\frac{m}{2\pi\beta\hbar^2})^{3/2}]^{N-1} \equiv Z_o + \Delta Z \qquad (6.40)$$

That is, the first correction to $\log Z = \log Z_o[1+\Delta Z/Z_o] \simeq \log Z_o + \Delta Z/Z_o$ is

$$\Delta Z/Z_o = \pm \tfrac{1}{2} N(N-1)(\pi^{3/2}\beta^{3/2}/Vm^{3/2})\hbar^3 \qquad (6.41)$$

where the + and - signs refer to bosons and fermions respectively. In the derivation of the last result we have neglected the spin degrees of freedom. Since the number of states available to a particle with spin s confined to the volume V, is proportional to gV (with $g = 2s+1$), a plausible correction to Eq. (6.41) is

$$\Delta Z/Z_o = \pm \tfrac{1}{2}N(N-1)(\pi^{3/2}\beta^{3/2}/gVm^{3/2})\hbar^3 \ . \qquad (6.42)$$

As a comparison to the result

$$\Delta Z/Z_o = \pm \tfrac{1}{2}N^2(\pi^{3/2}\beta^{3/2}/gV\,m^{3/2})\hbar^3 \qquad (6.43)$$

given by Landau and Lifshitz for $N \to \infty$ shows, Eq. (6.42) is indeed the appropriate expression for finite N.

APPENDIX A: THE WIGNER FUNCTION FOR A PRODUCT OF n OPERATORS

By Eqs. (2.6) and (2.3) the Wigner function for the operator $\hat{C} = \hat{A}_1\hat{A}_2\ldots\hat{A}_n$ is

$$C_{\hat{A}_1\ldots\hat{A}_n}(x,p) = \frac{\hbar}{2\pi}\int tr[\hat{A}_1\ldots\hat{A}_n e^{-i(u\hat{x}+v\hat{p})}]e^{i(ux+vp)}dudv$$

$$= \frac{\hbar}{2\pi}\int dudv e^{i(ux+vp)} \prod_{i=1}^{n} \frac{1}{(2\pi)^2}\int dx_i dp_i du_i dv_i \ A_i(x_i,p_i)$$

$$\times \ \exp[-i(u_i x_i + v_i p_i)]\,(tr\hat{F}(u_1,\ldots,u_n,v_1,\ldots,v_n,u,v); \qquad (A.1)$$

$$\hat{F} = \exp[i(u_1\hat{x}+v_1\hat{p})]\ldots\exp[i(u_n\hat{x}+v_n\hat{p})]\exp[-i(ux+vp)]. \quad (A.2)$$

Using the grouping property (2.7), we can rewrite $\hat{F}$ as

$$\hat{F} = \exp[-\tfrac{i}{2}\hbar\sum_{i<j}(u_iv_j-v_iu_j)]\exp[\tfrac{i}{2}\hbar\Sigma(u_iv-v_iu)]$$

$$\times \exp(i[(\Sigma u_i-u)\hat{x} + (\Sigma v_i-v)\hat{p}]). \quad (A.3)$$

Hence, by Eq. (2.5)

$$\mathrm{tr}\hat{F} = \frac{2\pi}{\hbar}\delta(\Sigma u_i-u)\delta(\Sigma v_i-v)$$

$$\times \exp[-\tfrac{i}{2}\hbar\sum_{i<j}(u_iv_j-v_iu_j)] \ . \quad (A.4)$$

Substituting the last result in Eq. (A.1), we have,

$$C_{\hat{A}_1\ldots\hat{A}_n}(x,p) = (2\pi)^{-2n}\int\exp[-\tfrac{i}{2}\hbar\sum_{i<j}(u_iv_j-v_iu_j)]$$

$$\times \exp\{i\Sigma[u_i(x-x_i) + v_i(p-p_i)]\}$$

$$\times A_1(x_1,p_1)\ldots A_n(x_n,p_n)d^nxd^npd^nud^nv \ . \quad (A.5)$$

Now, by using the identity

$$\exp[-\tfrac{i}{2}\hbar\sum_{i<j}(u_iv_j-v_iu_j)]\exp(i\Sigma[u_i(x-x_i)+v_i(p-p_i)])$$

$$= \exp(\tfrac{i}{2}\hbar\sum_{i<j}P_{ij})\exp(i\Sigma[u_i(x-x_i)+v_i(p-p_i)]) \ , \quad (A.6)$$

where $P_{ij} = \frac{\partial}{\partial x_i}\frac{\partial}{\partial p_j} - \frac{\partial}{\partial p_i}\frac{\partial}{\partial x_j}$, we can rewrite Eq. (A.5) as

$$C_{\hat{A}_1\ldots\hat{A}_n}(x,p) = (2\pi)^{-2n}\int d^nxd^npA_1(x_1,p_1)\ldots A_n(x_n,p_n)$$

$$\times \exp[\tfrac{i}{2}\hbar\sum_{i<j}P_{ij}]\int d^nud^nv\ \exp(i\Sigma[u_i(x-x_i)+v_i(p-p_i)])$$

$$= \int d^nxd^npA_1(x_1,p_1)\ldots A_n(x_n,p_n)\exp[\tfrac{i}{2}\hbar\sum_{i<j}P_{ij}]\prod_{i=1}^{n}\delta(x-x_i)\delta(p-p_i). \quad (A.7)$$

Finally, using the partial integration formula

$$\int(\partial^k/\partial x_1^k)\delta(x-x_1)f(x_1)dx_1 = (-1)^k\int\delta(x-x_1)(\partial^k/\partial x_1^k)f(x_1)dx_1 \ , \quad (A.8)$$

and noting that the integrations in (A.7) come in pairs $dx_i dp_i$, we secure,

$$C_{\hat{A}_1 \ldots \hat{A}_n}(x,p)=\exp\left(\frac{i\hbar}{2}\sum_{i<j}^{n} P_{ij}\right)A_1(x_1,p_1)\ldots A_n(x_n,p_n)\Big|_{\substack{x_1=..=x_n=x \\ p_1=..=p_n=p}} \quad (A.9)$$

This result coincides with Eq. (2.18).

APPENDIX B: DERIVATION OF THE HEISENBERG EQUATION OF MOTION

Let $\hat{B}(t) \equiv u\hat{x}(t) + v\hat{p}(t)$. Then, by Eq. (2.28),

$$\frac{d\hat{B}}{dt} = \frac{i}{\hbar}(u[\hat{H},\hat{x}]+v[\hat{H},\hat{p}]) = \frac{i}{\hbar}[\hat{H},\hat{B}] \ . \quad (B.1)$$

Assuming

$$\frac{d}{dt}\hat{B}^n = \frac{i}{\hbar}[\hat{H},\hat{B}^n] \quad (B.2)$$

we have

$$\frac{d}{dt}\hat{B}^{n+1}=\frac{d}{dt}(\hat{B}\hat{B}^n)=\frac{d\hat{B}}{dt}\hat{B}^n + \hat{B}\frac{d}{dt}\hat{B}^n = \frac{i}{\hbar}([\hat{H},\hat{B}]\hat{B}^n+\hat{B}[\hat{H},\hat{B}^n])$$

$$= \frac{i}{\hbar}[\hat{H},\hat{B}\hat{B}^n] = \frac{i}{\hbar}[\hat{H},\hat{B}^{n+1}] \quad , \quad (B.3)$$

which proves the relation (B.2). How, by Eq. (2.29), a general operator at time t is defined by

$$\hat{A}(t)=\frac{\hbar}{2\pi}\int\bar{A}(u,v)e^{i\hat{B}(t)}dudv=\frac{\hbar}{2\pi}\int\bar{A}(u,v)\Sigma\frac{i^n}{n!}\hat{B}^n(t)dudv. \quad (B.4)$$

Using the result (B.2), we secure the Heisenberg equation of motion

$$\frac{d\hat{A}}{dt} = \frac{\hbar}{2\pi}\int\bar{A}(u,v)\sum\frac{i^n}{n!}\ \frac{i}{\hbar}[\hat{H},\hat{B}^n]dudv = \frac{i}{\hbar}[\hat{H},\hat{A}]. \quad (B.5)$$

APPENDIX C: THE NORMAL-ANTINORMAL MAPPING

The purpose of this Appendix is to review briefly another mapping of the form (1.2) satisfying Eq. (1.1), namely, the normal-antinormal mapping, and to establish its (little known) connection with the Wigner (inverse Weyl) mapping. A more detailed treatment of the normal and antinormal mappings and of coherent states may be found in textbooks, e.g. Louisell (1973). The following results have already been derived, although differently, by Agarwal & Wolf (1970). As an application, we shall use this mapping to calculate directly from Eq. (2.6) the exact Wigner function (5.17) for the harmonic oscillator. Let us define the distruction and creation operators for a (one dimensional) particle of mass m:

$$a = 2^{-\frac{1}{2}}(\hat{x}/x_o + i\hat{p}/p_o) \quad \text{and} \quad a^+ = 2^{-\frac{1}{2}}(\hat{x}/x_o - i\hat{p}/p_o) \qquad (C.1)$$

where $x_o = [\hbar/(m\omega)]^{\frac{1}{2}}$ and $p_o = (m\hbar\omega)^{\frac{1}{2}}$. Let $\hat{A}$ be a general (analytic) operator of the form

$$\hat{A} = \sum C_{\ldots ijk\ell\ldots} \ldots a^i a^{+j} a^k a^{+l} \ldots \quad . \qquad (C.2)$$

By using the commutation relation $[a,a^+] = 1$, one can bring all the creation operators to the left of the destruction operators. This results in the unique normal expansion

$$\hat{A} = \tilde{A}^{(n)}(a,a^+) = \sum_{r,s} A_{rs}^{(n)} a^{+r} a^s \quad . \qquad (C.3)$$

Similarly, by bringing all the destruction operators to the left of the creation operators, one obtains the unique antinormal expansion

$$\hat{A} = \tilde{A}^{(a)}(a,a^+) = \sum_{r,s} A_{rs}^{(a)} a^r a^{+s} \quad . \qquad (C.4)$$

These two forms arise naturally in the coherent state representation defined by

$$a|\alpha\rangle = \alpha|\alpha\rangle \quad , \quad \langle\alpha|\alpha\rangle = 1 \quad , \qquad (C.5)$$

where $\alpha = (\alpha_R + i\alpha_I)$ is a complex number. Thus, the diagonal elements of $\hat{A}$ in this representation are given by the normal form

$$\langle\alpha|\hat{A}|\alpha\rangle = \tilde{A}^{(n)}(\alpha,\alpha^*) \quad , \qquad (C.6)$$

and its projection-operator expansion by the antinormal form

$$\hat{A} = \int \frac{d^2\alpha}{\pi} \tilde{A}^{(a)}(\alpha,\alpha^*)|\alpha\rangle\langle\alpha| \quad . \qquad (C.7)$$

Here $d^2\alpha = d\alpha_r d\alpha_I$. In terms of the normal and antinormal functions defined in Eqs. (C.3) and (C.4), the expectation value of an operator $\hat{A}$ in a state characterized by a density operator $\hat{\rho}$, is

$$\begin{aligned}\langle\hat{A}\rangle = \mathrm{tr}(\hat{\rho}\hat{A}) &= \int \frac{d^2\alpha}{\pi} \rho^{(a)}(\alpha,\alpha^*)\tilde{A}^{(n)}(\alpha,\alpha^*) \\ &= \int \frac{d^2\alpha}{\pi} \tilde{\rho}^{(n)}(\alpha,\alpha^*)\tilde{A}^{(a)}(\alpha,\alpha^*) \\ &= \int \frac{d^2\alpha}{\pi} \langle\alpha|\hat{\rho}|\alpha\rangle \tilde{A}^{(a)}(\alpha,\alpha^*) \quad . \qquad (C.8)\end{aligned}$$

To obtain the last equality, we have used Eq. (C.6). In particular, we obtain for the trace of an operator $\hat{A}$,

$$\mathrm{tr}(\hat{A}) = \int \frac{d^2\alpha}{\pi} \langle\alpha|\hat{A}|\alpha\rangle \quad . \qquad (C.9)$$

Let

$$\alpha = 2^{-\frac{1}{2}}(x/x_o+ip/p_o) \quad , \quad d^2\alpha/\pi = dxdp/2\pi\hbar \ . \tag{C.10}$$

Define two new functions by

$$A^{(n)}(x,p) \equiv \tilde{A}^{(n)}(\alpha,\alpha^*) \quad \text{and} \quad \tilde{A}^{(a)}(x,p) \equiv \tilde{A}^{(a)}(\alpha,\alpha^*). \tag{C.11}$$

In terms of the new functions Eq. (C.8) reads, in accordance with Eq. (1.1).

$$\langle\hat{A}\rangle = \mathrm{tr}(\hat{\rho}\hat{A}) = \int\rho^{(a)}(x,p)A^{(n)}(x,p)dxdp/2\pi\hbar$$

$$= \int\rho^{(n)}(x,p)A^{(a)}(x,p)dxdp/2\pi\hbar \quad . \tag{C.12}$$

Our first goal is to relate the functions $A^{(n)}(x,p)$ and $A^{(a)}(x,p)$ to the Wigner function $A(x,p)$, which upon quantization according to Eq. (2.3), yields the operator $\hat{A}$. To this end, rewrite the exponential operator in Eq. (2.3) in terms of the operators a and a^+. Using Eq. (C.1), we have

$$\exp[i(u\hat{x}+v\hat{p})]=\exp(i[2^{-\frac{1}{2}}(\tilde{u}+i\tilde{v})a^++2^{-\frac{1}{2}}(\tilde{u}-i\tilde{v})a]) \ ; \tag{C.13}$$

$$\tilde{u} = ux_o \qquad \text{and} \qquad \tilde{v} = vp_o \quad . \tag{C.14}$$

Inserting this expression into Eq. (2.3) and utilizing the grouping property (2.7), we secure,

$$\hat{A} = \tilde{A}^{(n)}(a,a^+)= \frac{\hbar}{2\pi}\int\ddot{A}(u,v)\exp[-\tfrac{1}{4}(\tilde{u}^2+\tilde{v}^2)]\exp[i2^{-\frac{1}{2}}(\tilde{u}+i\tilde{v})a^+]$$

$$\star\ \exp[i2^{-\frac{1}{2}}(\tilde{u}-i\tilde{v})a]dudv$$

$$= A^{(a)}(a,a^+)=\frac{\hbar}{2\pi}\int\ddot{A}(u,v)\exp[\ \tfrac{1}{4}(\tilde{u}^2+\tilde{v}^2)\}\exp[i2^{-\frac{1}{2}}(\tilde{u}-i\tilde{v})a]$$

$$\star\ \exp[i2^{-\frac{1}{2}}(\tilde{u}+i\tilde{v})a^+]dudv \ . \tag{C.15}$$

The functions $\tilde{A}^{(n)}(\alpha,\alpha^*)$ and $\tilde{A}^{(a)}(\alpha,\alpha^*)$ can now be read off Eq. (C.15). Inserting the relation (C.10) and using definition (C.11), we obtain

$$A^{\binom{(n)}{(a)}}(x,p)=\frac{\hbar}{2\pi}\int\ddot{A}(u,v)\exp[\mp\ \tfrac{1}{4}(\tilde{u}^2+\tilde{v}^2)]\exp[i(ux+vp)]dudv \quad . \tag{C.16}$$

Finally, the identity

$$\exp[\mp\ \tfrac{1}{4}\ (u^2x_o^2+v^2p_o^2)]\exp[i(ux+vp)] \quad =$$

$$\exp[\pm\ \tfrac{1}{4}(x_o^2D_x^2+p_o^2D_p^2)]\exp[i(ux+vp)] \quad , \tag{C.17}$$

can be used to rewrite Eq. (C.16) in the form

$$A^{\binom{(n)}{(a)}}(x,p) = \exp[\pm \frac{\hbar}{4}Q]\ \frac{\hbar}{2\pi}\int \tilde{A}(u,v)e^{i(ux+vp)}dudv, \tag{C.18}$$

where

$$Q \equiv ((m\omega)^{-1}\partial^2/\partial x^2 + m\omega\partial^2/\partial p^2) \quad . \tag{C.19}$$

Note that Q is essentially the Laplacian operator in phase space. (The quantity m is arbitrary and can be chosen to have the numerical value $m = 1.$) Equation (C.18) is the desired relation between the functions $A^{(n)}(x,p)$, $A^{(a)}(x,p)$ and $A(x,p)$. It reads

$$A^{(n)}(x,p) = \exp(\frac{\hbar}{4}Q)A(x,p) \quad A^{(a)}(x,p)=\exp(-\frac{\hbar}{4}Q)A(x,p) . \tag{C.20}$$

Note that, in the limit $\hbar \to 0$, $A^{(n)}(x,p) = A^{(a)}(x,p) = A(x,p)$. Another useful relation can be obtained from Eq. (C.20) by invoking the representation (Erdély 1953).

$$H_n(\xi) = 2^n\exp(-D_\xi^2/4)\xi^n \tag{C.21}$$

for the Hermite polynomials. Let

$$A(x,p) = \sum A_{mn}x^m p^n = \sum A_{mn}x_o^m p_o^n (\frac{x}{x_o})^m (\frac{p}{p_o})^n \quad , \tag{C.22}$$

where x_o and p_o are given as in Eq. (C.2). Then

$$A^{(n)}(x,p) = \sum A_{mn}(\frac{x_o}{2i})^m H_m(i\frac{x}{x_o})\cdot(\frac{p_o}{2i})^n H_n(i\frac{p}{p_o}) \quad , \tag{C.23a}$$

$$A^{(a)}(x,p) = \sum A_{mn}(\frac{x_o}{2})^m H_m(\frac{x}{x_o})\cdot(\frac{p_o}{2})^n H_n(\frac{p}{p_o}) \quad . \tag{C.23b}$$

We now apply the representation (C.12) to calculate the Wigner function $R(x,p)$ for the harmonic osicllator (see Eq. (5.17)) directly from Eq. (2.6). The Hamiltonian $\hat{H}$ and $\hat{R}$ in this case are

$$\hat{H} = p^2/(2m) + m\omega^2\hat{x}^2/2 = \hbar\omega(a^+a+\tfrac{1}{2}) \quad , \tag{C.24}$$

$$\hat{R} = \exp(-\beta\hat{H}) = \exp(-\tfrac{1}{2}\lambda)\exp(-\lambda a^+a), \ \lambda = \beta\hbar\omega \quad . \tag{C.25}$$

Using the expansion coefficients

$$\langle\alpha|n\rangle = \exp[-\tfrac{1}{2}|\alpha|^2]\alpha^n/(n!)^{\frac{1}{2}} \quad , \tag{C.26}$$

where $a^+a|n\rangle = n|n\rangle$, we obtain for the normal function $\tilde{R}^{(n)}(\alpha,\alpha^*)$:

$$\tilde{R}^{(n)}(\alpha,\alpha^*) = e^{-\frac{1}{2}\lambda}\langle\alpha|\exp[-\lambda a^+a]|\alpha\rangle$$

$$= e^{-\frac{1}{2}\lambda}\Sigma\langle\alpha|\exp[-\lambda a^+a]|n\rangle\langle n|\alpha\rangle$$

$$= e^{-\frac{1}{2}\lambda}\Sigma\exp(-\lambda n)|\langle n|\alpha\rangle|^2$$

$$= e^{-\frac{1}{2}\lambda}\Sigma(-1)^{\ell}(1-e^{-\lambda})^{\ell}\alpha^{*\ell}\alpha^{\ell}/\ell! \tag{C.27}$$

The antinormal function $[\widetilde{\exp(-i(u\hat{x}+v\hat{p}))}]^{(a)}(\alpha,\alpha^*)$ is given by (see Eqs. (C.13)-(C.15))

$$[\widetilde{\exp(-i(u\hat{x}+v\hat{p}))}]^{(a)}(\alpha,\alpha^*)=e^{\frac{1}{4}(\tilde{u}^2+\tilde{v}^2)}\exp(-i2^{-\frac{1}{2}}[(\tilde{u}-iv)\alpha+(u+iv)\alpha^*]) . \tag{C.28}$$

Hence, by Eqs. (C.12), (C.11) and (C.10),

$$\mathrm{tr}[\hat{R}e^{-i(u\hat{x}+v\hat{p})}]=e^{-\frac{1}{2}\lambda}\exp[\tfrac{1}{4}(u^2x_o^2+v^2p_o^2)]$$

$$\times\int\frac{dxdp}{2\pi\hbar}\sum\frac{(-1)^{\ell}(1-e^{-\lambda})^{\ell}}{\ell!}(\tfrac{1}{2}[(x/x_o)^2+(p/p_o)^2])^{\ell}e^{-i(ux+vp)}$$

$$= e^{-\frac{1}{2}\lambda}e^{\frac{1}{4}(u^2x_o^2+v^2p_o^2)}\int\frac{dxdp}{2\pi\hbar}\exp[-\tfrac{1}{2}\mu((x/x_o)^2+(p/p_o)^2)]e^{-i(ux+vp)} , \tag{C.29}$$

where

$$\mu = 1-e^{-\lambda} . \tag{C.30}$$

Performing the integrations, we obtain,

$$\mathrm{tr}\ [\hat{R}\exp(-i(ux+vp))]$$

$$= \exp\{-\tfrac{1}{4}\coth(\tfrac{\lambda}{2})[(ux_o)^2+(vp_o)^2]\}/(2\sinh(\tfrac{\lambda}{2})) . \tag{C.31}$$

Inserting this result in Eq. (2.6) and integrating over u and v, we finally secure, in accordance with Eq. (5.17),

$$R(x,p)=\exp[-\beta H(x,p)\,\mathrm{tgh}(\lambda/2)/(\lambda/2)]/\cosh(\lambda/2). \tag{C.32}$$

References

Agarwal, G.S. & Wolf, E. (1970). Phys. Rev., D2,2161,2187,2206.
Berry, M.V. (1977). Phil. Trans. Roy. Soc. London, 287.
Erdély, A. (1953). Higher Transcendental Functions Vol. 2, New-York: McGraw-Hill
Groenwold, H.J. (1945). Physica, 12, 405.
Hillery, M., O'Connell, R.F., Scully, M.O.O. & Wigner, E.P. (1984). Phys. Rep., 106, 122.
Imre, K., Özizmir, E., Rosenbaum, M.& Zweifel, P.F. (1967). J. Math. Phys., 8, 1097.
Jaynes, E.T. (1957). Phys. Rev., 106, 620; 108, 171.

Jaynes, E.T. (1963). In Statistical Physics, Brandeis Summer Institute 1962. ed. K.W. Ford, New-York: Benjamin.
Landau, L.D. & Lifshitz, E.M. (1968). Statistical Physics, Oxford: Pergamon.
Louisell, W.H. (1973). Quantum Statistical Properties of Radiation. New-York: Wiley.
Messiah, A. (1961). Quantum Mechanics. Amsterdam: North-Holland.
Moyal, J.E. (1949). Proc. Cambridge Phil. Soc. 45, 99.
O'Connell,R.F. & Wigner, E.P. (1984). Phys. Rev. A30, 2613.
Rosen, G. (1969). Formulations of Classical and Quantum Dynamical Theory. New-York: Academic.
Schweber, S.S. (1961). An Introduction to Relativistic Quantum Field Theory. New-York: Harper and Row.
Shore, J. & Johnson, R. (1980). IEEE Trans. Infor. Theory IT-26, 26.
\- (1983). IEEE Trans. Infor. Theory IT-29, 942.
Tatarskii, V.I. (1983). Sov. Phys. Usp. 26, 311.
Tikochinsky, Y., Tishby, N. & Levine, R. (1984). Phys. Rev. Lett. 52, 1357; Phys. Rev. A30, 2638.
\- (1985). Phys. Rev. Lett. 55, 337.
Wang, L. (1986). J. Math. Phys. 27, 483.
Weyl, H. (1927). Z. Phys. 46, 1.
\- (1931). The Theory of Groups and Quantum Mechanics. London: Dover.
Wigner, E.P. (1932). Phys. Rev. 40, 749.

SUPERPOSITION EFFECTS IN DIFFERENTIAL ENTROPY AND KULLBACK-LEIBLER INFORMATION

A.K. Rajagopal*, P.J. Lin-Chung and S. Teitler
Naval Research Laboratory, Washington, DC 20735-5000

Abstract. The use of amplitude-based probability densities as distinct from ordinary probability densities means linear combinations of amplitudes introduce both weighting and interference effects. Some inequalities arising from both these effects in differential entropy and Kullback-Leibler information are established.

1 INTRODUCTION

We consider here superposition of probability amplitudes whose squared magnitude forms a probability density. Our particular interest is in characterization of weighting and interference effects as revealed by the properties of the corresponding differential entropy and Kullback-Leibler information.[1,2] In contrast to a differential entropy, a Kullback-Leibler information (also frequently called cross-entropy or relative entropy) is invariant under scale transformations and independent of the choice of the physical dimensions of the variables entering into the definition of the probability density. This invariance is accomplished by requiring the probability density of interest to be compared to a reference probability density. Several inequalities for the differential entropy and Kullback-Leibler information arising when amplitudes in amplitude-based probability densities are superposed will be established in Section 2.

We use a bra-ket notation familiar in quantum mechanics but also applicable in the description of any wave phenomena with complex amplitudes. Consider then a quantity that may be represented by a dimensionless ket vector $|\zeta\rangle$ which is normalized in the sense that $\langle\zeta|\zeta\rangle=1$. The label ζ may correspond to a single eigenket or a superposition of eigenkets in a space with discrete or continuous spectrum. Then if x is the continuous variable of the probability space of interest and $|x\rangle$ is its corresponding eigenket, we are interested in a probability density arising from the complex amplitude $\langle x|\zeta\rangle$. Thus we require $\langle x|\zeta\rangle$ to be L^2 and normalized. We have then

$$\int P_\zeta(x)\, dx = 1 \tag{1}$$

where

$$P_\zeta(x)=|\langle x|\zeta\rangle|^2 \tag{2}$$

*Supported in part by ONR Contract N0001487WX84028

G. J. Erickson and C. R. Smith (eds.),
Maximum-Entropy and Bayesian Methods in Science and Engineering (Vol. 2), 83–88.

and dx represents $d^n x$ if x is a vector variable in an n-dimensional space. The differential entropy is written as

$$S_\zeta(X) = -\int dx\, P_\zeta(x) \log P_\zeta(x) \tag{3}$$

However, S_ζ is properly defined in this way only if $P_\zeta(x)$ (and hence also x) is dimensionless. Even so, S_ζ is usually neither invariant nor necessarily positive. The lack of invariance follows immediately from a consideration of a change of coordinates, say, from x to y with Jacobian $|dx/dy|$. Then

$$S_\zeta(Y) = S_\zeta(X) + \int dx\, P_\zeta(x) \log|dx/dy| \tag{4}$$

In general, the Jacobian is not unity so the second term on the right-hand side does not vanish and $S_\zeta(Y) \neq S_\zeta(X)$. The possible lack of positivity for S_ζ follows from the fact that $P_\zeta(x)$ may have values greater than unity because of a probability density rather than a discrete probability.

To circumvent these problems of dimension, invariance and sign, one may use the Kullback-Leibler information. As an introduction to our consideration of the Kullback-Leibler information, we note that $|\langle x|\zeta\rangle|^2 dx$ can be viewed as a dimensionless element of probability measure

$$d\mu_\zeta \equiv |\langle x|\zeta\rangle|^2 dx = P_\zeta(x) dx \tag{5}$$

Here the physical dimensions (if any) of $|\langle x|\zeta\rangle|^2$ are compensated by those of dx so $d\mu_\zeta$ is dimensionless by definition. In general μ_ζ belongs to a family of measures whose elements can be expressed in terms of a product of amplitude-based probability densities and dx. Of this family, we choose one to be a reference measure. We denote this probability density by

$$P_g(x) \equiv |\langle x|g\rangle|^2 \tag{6}$$

Then $d\mu_\zeta$ can be re-expressed as

$$d\mu_\zeta = \left[P_\zeta(x)/P_g(x)\right] d\mu_g \tag{7}$$

The dimensionless, scale-invariant ratio $P_\zeta(x)/P_g(x)$ can be viewed as a probability density for the measure $d\mu_g$.

The Kullback-Leibler information $I(\zeta,g)$ is an entropy-like functional defined in terms of such a probability density.

$$\begin{aligned} I(\zeta,g) &= \int \left[P_\zeta(x)/P_g(x)\right] \log\left[P_\zeta(x)/P_g(x)\right] d\mu_g \\ &= \int dx\, P_\zeta(x) \log\left[P_\zeta(x)/P_g(x)\right] \end{aligned} \tag{8}$$

An important theorem due to Kullback and Leibler[1,2] provides a constraint on the sign of $I(\zeta,g)$

$$I(\zeta,g) \geq 0 \tag{9}$$

with equality if and only if $P_\zeta(x)=P_g(x)$. We shall refer to inequality (9) and its attendant equality statement as the K-L Theorem. This theorem is a consequence of the convexity of P log P. We see then that $I(\zeta,g)$ addresses all the problems of dimension, invariance and sign that arise in the use of the differential entropy S_ζ.

We are now almost ready to apply the Kullback-Leibler information in a discussion of effects arising from superposition of amplitudes. However, some prior remarks concerning the choice and meaning of $P_g(x)$ seems appropriate. In quantum mechanics, the concept of a reference probability density has already been invoked in the rigged Hilbert space formulation.[3] There the choice for $P_g(x)$ is the ground state probability density. For other phenomena, a similar canonical choice would be the probability density for the lowest mode. More generally, one can make other choices for both quantum mechanics and other wave phenomena depending on the physical situation. (Values of x at which the chosen P_g is zero should be excluded from the domain of x.) Whatever the choice, as Kullback himself has pointed out,[4] $I(\zeta,g)$ provides a hypothesis test for the coincidence of $P_\zeta(x)$ and $P_g(x)$. With this context in hand, we can now turn to the

development of inequalities for the dimensionless differential entropy and the Kullback-Leibler information when superposition is introduced.

2 SOME INEQUALITIES FOR DIFFERENTIAL ENTROPY AND KULLBACK-LEIBLER INFORMATION

We deal with a set of orthonormal functions $\{\langle\zeta_\iota|x\rangle\}$; $\iota=1,2,\ldots,M$ in x-space

$$\int dx \langle\zeta_\iota|x\rangle\langle x|\zeta_j\rangle = \delta_{\iota j} \tag{10}$$

Then each

$$P_{\zeta\iota}(x) = |\langle\zeta_\iota|x\rangle|^2 \tag{11}$$

is a normalized probability density. Consider now that $\langle\zeta|x\rangle$ is a superposition of a few of the $\langle\zeta_\iota|x\rangle$, say M of them.

$$\langle\zeta|x\rangle = \sum_{\iota=1}^{M} a_\iota \langle\zeta_\iota|x\rangle \tag{12}$$

where the a_ι are complex coefficients in general. Now $P_\zeta(x)=|\langle\zeta|x\rangle|^2$ takes the form

$$P_\zeta(x) = P_c(x) + P_{int}(x) \tag{13}$$

where

$$P_c(x) = \sum_{\iota=1}^{M} |a_\iota|^2 P_{\zeta_\iota}(x) \tag{14}$$

$$P_{int}(x) = \mathrm{Re}\Big[\sum_{\iota\neq j} a_\iota a_j^* \langle\zeta_\iota|x\rangle\langle x|\zeta_j\rangle\Big] \tag{15}$$

Also from the normalization of $P_\zeta(x)$, it follows that

$$\sum_{\iota=1}^{M} |a_\iota|^2 = 1 \tag{16}$$

Thus the $|a_\iota|^2$ provide weight factors for the individually normalized $P_\zeta(x)$. In this way, the $|a_\iota|^2$ are responsible for the weighting effects that arise from superposition. Interference effects arise from cross terms in $P_{int}(x)$. From Eq. (10)

$$\int dx\, P_{int}(x) = 0 \tag{17}$$

However, it is also possible that there is an ensemble of superpositions with coefficients having random phases. Then one may consider a random phase approximation in which the ensemble average of the local $P_{int}(x)$ vanishes independently of the integration over x.

$$\langle P_{int}\rangle_{RPA} = 0 \tag{18}$$

Here < > represents an average over random phases. When Eq. (18) is valid, $\langle P_\zeta(x)\rangle_{RPA}$ coincides with $P_c(x)$. However, we note that the random phase approximation for the differential entropy or Kullback-Leibler information does not in general coincide with the corresponding respective quantity defined in terms of $P_c(x)$. Below we provide inequalities for weighting effects.

$$\sum_{\iota=1}^{M} |a_\iota|^2 S_{\zeta_\iota} \leq S_c \leq \sum_{\iota=1}^{M} |a_\iota|^2 S_{\zeta_\iota} - \sum_{\iota=1}^{M} |a_\iota|^2 \log|a_\iota|^2 \tag{19a}$$

$$\sum_{\iota=1}^{M} |a_\iota|^2 I(\zeta_\iota,g) \geq I(c,g) \geq \sum_{\iota=1}^{M} |a_\iota|^2 I(\zeta_\iota,g) + \Sigma|a_\iota|^2 \log|a_\iota|^2 \tag{19b}$$

Here we use the expressions for differential entropy and Kullback-Leibler information given respectively in Eq. (3) and Eq. (8) but with subscripts and arguments labelled by the subscripts of the incorporated probability densities. The inequalities may be established as follows. First (for the lefthand inequality in (19a), apply the K-L theorem (see inequality (9) above) to $-I(\zeta_\iota,c)$ to obtain

$$S_{\zeta\iota} + \int dx\, P_{\zeta\iota}(x) \log P_c(x) \leq 0, \text{ (all } \iota) \tag{20}$$

Multiplying (20) by $|a_\iota|^2$ and summing over $\iota=1,\ldots,M$ gives the lefthand inequality quoted in (19a). Similarly the lefthand inequality in

(19b) can be obtained by starting from (20) with reversed sign and adding and subtracting

$$\int dx\ P_{\zeta\iota}(x)\ \log\ P_g(x)$$

to obtain

$$I(\zeta_\iota,g)\ -\int dx\ P_{\zeta\iota}(x)\ \log\left[P_g(x)/P_c(x)\right] \geq 0 \tag{21}$$

Multiplying (21) by $|a_\iota|^2$ and summing over $\iota=1,\ldots,M$ yields the left-hand inequality quoted in (19b).

To derive the righthand inequalities we note that

$$(\alpha+\beta)\ \log(\alpha+\beta)\ -\ \alpha\ \log\ \alpha\ -\ \beta\ \log\ \beta \geq 0,\ \alpha,\ \beta \geq 0 \tag{22a}$$

Then for $\gamma\geq 0$, apply (22a) for $\beta'=\beta+\gamma$.

$$(\alpha+\beta+\gamma)\ \log(\alpha+\beta+\gamma)\ -\ \alpha\ \log\ \alpha\ -(\beta+\gamma)\ \log(\beta+\gamma) \geq 0 \tag{22b}$$

But (22a) can be applied individually to the last term on the right-hand side of (22b).

$$(\alpha+\beta+\gamma)\ \log(\alpha+\beta+\gamma)\ -\ \alpha\ \log\ \alpha\ -\ \beta\ \log\ \beta\ -\ \gamma\ \log\ \gamma \geq 0 \tag{22c}$$

Hence for a set of $\alpha_\iota \geq 0$, we have

$$\left[\sum_{\iota=1}^{M} \alpha_\iota\right] \log \left[\sum_{\iota=1}^{M} \alpha_\iota\right] - \sum_{\iota=1}^{M} \alpha_\iota\ \log\ \alpha_\iota \geq 0 \tag{23}$$

Taking $\alpha = |a_\iota|^2 P_\zeta(x) \geq 0$ for all x and integrating over x, we obtain the righthand inequality in (19a). The righthand inequality in (19b) follows from (19a) because of Eq. (16) and the definitions of $P_c, I(\zeta_\iota,g)$ and $I(c,g)$.

We have already identified the $|a_\iota|^2$ as weight factors so

$$S_W \equiv -\sum_{\iota=1}^{M} |a_\iota|^2\ \log\ |a_\iota|^2 \tag{24}$$

may be identified as a weighting or mixing entropy. S_W is clearly positive for M>1. Thus the inequalities in (19a) show that the classical differential entropy lies above the weighted sum of component state entropies and below that latter plus the weighting entropy. Similarly (19b) shows the Kullback-Leibler information lies below the weighted sum of component state informations but above the latter sum reduced by S_W.

We turn now to the effect of interference and establish the following inequalities.

$$S_\zeta \leq S_c - \int dx\, P_{int}(x) \log P_c(x) \tag{25a}$$

$$I(\zeta,g) \geq I(c,g) + \int dx\, P_{int} \log \left[P_c(x)/P_g(x)\right] \tag{25b}$$

Just as in the proof of the lefthand inequality of (19a), the inequality (25a) follows a decomposition of the K-L theorem but here for $-I(\zeta,c)$ rather than for $-I(\zeta_1,c)$. Then

$$S_\zeta + \int dx\, P_\zeta(x) \log P_c(x) \leq 0 \tag{26}$$

Inequality (25a) follows directly from the definition of $P_\zeta(x)$ given in Eq. (13). Similarly inequality (25b) follows from the K-L theorem and the definitions of $I(\zeta,g)$, $I(c,g)$ and $P_\zeta(x)$. An interesting point about inequalities (25a,b) is that the respective interference terms may be of either sign so the upper (lower) bound on the full differential entropy (full Kullback-Leibler information) can be larger or smaller than its classical counterpart.

Also it follows from inequalities (25a,b) and Eq. (18) that in the random phase approximation

$$\langle S_\zeta \rangle_{RPA} \leq S_c \tag{27a}$$

$$\langle I(\zeta,g) \rangle_{RPA} \geq I(c,g) \tag{27b}$$

Again it should be emphasized that the random phase approximation that brings coincidence of the phase averaged $P_\zeta(x)$ with $P_c(x)$ does not necessarily bring similar coincidence for the phase averages of S_ζ and $I(\zeta,g)$ with their classical counterparts.

We see then that superposition of amplitudes in amplitude-based probabilities introduce both weighting and interference effects. We have established inequalities for the differential entropy and Kullback-Leibler information to give insight into the nature of, and distinction between these effects. Elsewhere we will apply these inequalities to study superposition effects in some examples from quantum mechanics and other wave phenomena.

1. S. Kullback and R.A. Leibler, Ann Math Stat 22, 79 (1951).
2. S. Kullback, Information Theory and Statistics, (John Wiley & Sons, Inc., New York) (1959).
3. A. Bohm, The Rigged Hilbert Space and Quantum Mechanics, Springer Lecture Notes in Physics, Vol. 78 (Springer-Verlag, New York) (1978).
4. Ref. 2, pp. 4-5.

SUPER VARIATIONAL PRINCIPLES

L. H. Schick

Department of Physics and Astronomy, Univ. of Wyoming, Laramie, WY,82070

ABSTRACT

The principle of maximum entropy is combined with the usual form of variational principle found in quantum mechanics to obtain super variational principles, which may be used when the form of the potential energy operator is partially, or wholly unknown. Details are worked out for a few simple bound-state problems. Possible generalizations are discussed.

I. Introduction

For the past several years, numerous authors[1],including presenters at the annual sessions of this group, have applied the principle of maximum entropy (PME) to a variety of inverse scattering problems.In particular, some of us here at Wyoming[2-4], as well as others elsewhere[5], looked at some quantum mechanical inverse scattering problems. While I was engaged in this work, two points drew my attention. First, except for the application to noisy data which

---

Presented at the Fifth Workshop on Maximum Entropy and Bayesian Methods in Applied Statistics, Laramie, Wyoming, August 5-8, 1985.

G. J. Erickson and C. R. Smith (eds.),
Maximum-Entropy and Bayesian Methods in Science and Engineering (Vol. 2), 89–103.

demanded a χ^2 constraint, the constraints used in conjunction with the PME were generated perturbatively, and hence were linear constraints. Second, although the "exact" theory of quantum inverse problems[6] allowed for the inclusion of bound-state information, the entropy methods that had been applied to quantum problems used only scattering-state data. Working with bound states, I was led naturally to a method which addressed both these points, namely,the variational principle method. It is the incorporation of this method into the PME that is my subject today.

II. The Usual Variational Principle: Bound-State Application

The usual mathematical statement of a variational principle[7] may be expressed as

$$E = E[X], \qquad E_o = E[X_o], \qquad \delta E[X]|_o = 0 \; . \tag{1}$$

That is, E is a functional of the space of functions X, such that when X is a particular function X_o, E takes on the value E_o and E is stationary about E_o. Roughly speaking, for e<<1, if X is within e of X_o then E is within e^2 of E_o. The physical significance of E_o, and of X, as well as the specific form of E, depends on the particular problem to which the principle is applied. The point is that we are interested in calculating E_o, or X_o, or both. Our problem is too complicated to do this directly. Eq.(1) gives us a useful method for estimating these variables. In some cases we obtain the bonus that E is either an upper bound[8], or a lower bound[9] on E_o.

For example, consider the quantum system whose Hamiltonian H is the sum of two known operators, H_o (e.g., the kinetic energy operator) and V (e.g., the potential energy operator). The time-independent Schroedinger equation for this system may be written

$$H X_n = (H_o + V) X_n = E_n X_n , \quad (2)$$

where

$$E_o < E_1 < E_2 < \ldots < E_n < 0 \quad (3)$$

are the bound-state energies of the system. Then[10]

$$E = \langle X | H | X \rangle / \langle X | X \rangle \quad (4)$$

is stationary about each of the X_j, with corresponding stationary value E_j, for j = 0,1, ... ,n. To carry out the variation, a trial function X(a), which depends on a set of parameters, "a", is used in the right side of Eq. (4), to obtain a function E(a), and the partial derivatives of this function with respect to each of the a's are set equal to zero. For an estimate of any single E_j, the values for the set "a" obtained in this way are substituted back into E(a). For an estimate of more than one E_j, the process is repeated with a different trial function used for each energy estimated and with the additional requirements that the trial functions be made mutually orthogonal. If carried out sequentially for E_o, E_1, ..., this process yields an upper bound on each of the energies so estimated.

There are, of course, scattering-state variation principles[11], as well as other types of bound-state variation principles[9]. I shall return to these briefly in my concluding remarks. For the sake of clarity and to keep the number of subscripts from getting out of hand, I shall take as my archetype variational problem the bound state problem just discussed with the further restriction that it is just one binding energy, say E_0, that enters the problem.

III. The Super Variational Principle

Our archetype variational principle may be stated as follows: Given $H = H_0 + V$, estimate E_0 using the method described above. This is a one-for-one proposition; i.e., put in a trial function X, get out an estimate of E_0.

A <u>super</u> variational principle is designed to be used when $H = H_0 + V$ and V is not known. It is obtained by relating V to a probability distribution and using a PME calculation on this distribution with the usual form of variational functional as a constraint. It is a two-for-one proposition. When a trial function X is put in, both an estimate of V, say U, and an estimate of E_0, say E, are returned. The estimate E is a variational estimate of E_0 for the Hamiltonian $H' = H_0 + U$ with the same trial function X. The estimate U is a maximum entropy estimate of V in that the realized probabilty distribution that describes U is as close as possible to whatever prior distribution was assumed, while yielding exactly E as the expectation value of H' in the varied state X. Thus, if E_0 is actually known, by setting $E = E_0$ we achieve the solution to a bound-state inverse problem. If E_0 is an eigenvalue of the prior Hamiltonian, H_o in this case, and X is the corresponding eigenfunction, then U will turn out to be zero; i.e., no new information is added to our prior estimate of the potential by knowledge of E_o, but confirming evidence for our prior is.

For a specific type of formulation of a super variation principle, I shall consider the simple case of a nonrelativistic, spinless particle of mass m under the influence of a central potential V(r), which is to be estimated, and for which we know that an s-wave bound-state with binding energy E exists. (We must assume the bound state has some definite angular momentum. The assumption of a non-s-wave angular momentum needlessly complicates the discussion.)

To begin with we assume no prior information about V was available, so that without the bound-state information the unbiased estimate for the potential is obviously $V(r) = 0$, for all $0 \leqq r \leqq \infty$. Next, as a calculational convenience, we shall work with a discretized form of the problem. Thus, $r \dashrightarrow r_i$ and $V(r) \dashrightarrow V(r_i) = V_i$. At the ith radial point the potential will be written as[4]

$$V_i = V_o \sum_s s\, p_{si} \quad , \quad s = \pm 1, \pm 2, \ldots, \pm N \ , \tag{5}$$

where p_{si} is the probability that $V_i = s\, V_o$. To keep things simple let's confine the discussion to the case N = 1. Then with $p_{1i} = p_i$ and $p_{-1i} = 1 - p_i$,

$$V_i = V_o\, (2\, p_i - 1). \tag{6}$$

The unconstrained entropy, S_o, of the set of p_i is given by

$$S_o = - \sum_i p_i \log p_i - \sum_i (1 - p_i) \log (1 - p_i) \quad . \tag{7}$$

The constraint imposed by the variation equation is incorporated into our formalism by adding to S_o the additional term

$$\Delta S = \lambda [< X_b | H | X_b > - E] = 0 \tag{8}$$

where λ is a Lagrange multiplier and

$$< X_b | H | X_b > = < H_o > + 4 \pi \Delta r \sum_i | X_b(r_i)|^2 V_i r_i^2 \ . \tag{9}$$

Here H_o is the kinetic energy operator of the particle and the subscript b on the trial function $X_b(r_i)$ stands for the set of variation parameters contained in this function.

We may now add Eqs.(7) and (8), set the derivative of the result with respect to each p_j equal to zero, and solve for V_j. Following ref. (4), we may then take the limit V_o ---> ∞ with λV_o^2 --->U_o = finite and use Eq.(8) to eliminate U_o from our result for the potential. Expressing this result in terms of the continuous variable r, we have

$$V_b(r) = R_b(r) [E - < X_b | H_o | X_b >] / < X_b | R_b | X_b >, \tag{10}$$

where

$$R_b(r) = 4 \pi r^2 | X_b(r)|^2 \quad . \tag{11}$$

We have used the subscript b in our result to indicate its dependence on the variation parameters which appear on the right side of this equation.

We now use the potential just obtained as the potential in the variational equation

$$\delta[\langle X_a | H_b | X_a \rangle / \langle X_a | X_a \rangle] = \delta[\langle X_a | H_o + V_b | X_a \rangle / \langle X_a | X_a \rangle] = 0 \ , \quad (12)$$

where X_a has the same functional form as the X_b used to calculate V_b, but we have given the variation parameters in this function a new label, namely "a". This is necessary since what is to be varied in Eq.(12) is the trial function X which is shown explicitly in this equation and not the trial function X which is implicitly contained in this equation via Eq.(10). After the variation in Eq.(12) is carried out, in order to stay consistent with the PME part of our calculation, the parameter set b may be put equal (on a one-to-one basis) with the set a and the values of these parameters (which we denote "c") found from the resulting equations.

Equation (10) with the parameter set b replaced by the parameter set c is the result of our super variation principle. It is clear, since the expectation value of the kinetic energy operator must be positive, that for a negative-energy bound state $V_c(r) < 0$ for all r.

The generalization to cases in which we know more than one bound-state energy may be easily accomplished, so we need not dwell on it here.

If, before we obtain any sort of bound-state information, we already have an estimate of how the potential might differ from zero as a function of r (from scattering experiments, for example), we may incorporate this prior information into our formalism in the usual way[12] via a prior distribution of probabilities, q_i. The prior potential $V^p{}_i$ would be related to the q_i via a relation analagous to Eq.(6), while the form for the unconstrained entropy S_o of Eq.(7) would be replaced by

$$S_o = - \sum_i p_i \log p_i/q_i - \sum_i (1 - p_i) \log (1-p_i)/(1-q_i). \qquad (13)$$

Proceeding as in the case directly above, we would obtain

$$V_b(r) = V^p(r) + R_b(r) [E - \langle X_b | H^p | X_b \rangle] / \langle X_b | R^b | X_b \rangle, \qquad (14)$$

where $H^p = H_o + V^p$ is the prior Hamiltonian, and the rest of the notation is that used in Eq.(10).

Before moving on, we note that if we know the eigenfunctions X^p and corresponding eigenvalues E^p of the prior Hamiltonian, then we may use these eigenfunctions as our trial functions, and obtain

$$V_b(r) - V^p(r) = R^p(r) [E - E^p] / \langle X^p | R^p | X^p \rangle . \qquad (15)$$

This is just the first-order perturbation theory result of ref.(4). However, because of the variation to be carried out in the more general case described here, even though it appears we are dealing with a linear constraint, the super variation principle method is a nonlinear, non-perturbative method.

IV. Examples

We shall now look at some simple examples for each of which we know only one bound-state energy, E_o, and for each of which we use a real trial function with only one variation parameter, b. For such cases we may write

$$V_b(r) = V^P(r) + R_b(r)\,[\,E_o - \langle X_b | H_o | X_b \rangle\,] \,/ \int_o^\infty R_b^2(r)\,dr\,, \quad (16)$$

where $R_b(r)$ may be found from Eq.(11).

A. Coulomb 1s trial function with $V^P(r) = 0$.

$$X_b(r) = (1/\pi\, b^3)^{\frac{1}{2}} \exp(-r/b)\,.$$

Then from Eq.(16)

$$V_b(r) = -(16\hbar^2/3m)(b^{-2} + a_o^{-2})(r/b)^2 \exp(-2r/b)\,,$$

where we have set $E_o = -(\hbar^2/2ma_o^2)$. Setting equal to zero the derivative of $\langle X_a | H_o + V_b | X_a \rangle$ evaluated at $b = a$, we obtain $a = 3^{\frac{1}{2}}a_o$, and hence our super variational result

$$V(r) = -(64\hbar^2/27ma_o^2)\,(r/a_o)^2 \exp(-2r/3^{\frac{1}{2}}a_o)\,.$$

If, in fact, we are dealing with a hydrogen atom in its ground state, so that a_o is the Bohr radius, then the "correct" answer for V(r) is, of course, a Coulomb potential proportional to 1/r. Our result, as is to be expected, doesn't have the wide variation in value of the Coulomb potential and some might say it looks nothing like a Coulomb potential. Predicting a continum of numbers, V(r), from one number, E_o, is a tough business. Within the limits of our trial function, our result is the best we can do. After all, our result is well-behaved, it has the correct properties as required by general physical principles[13] as $r \to 0$ and as $r \to \infty$, and it gives correctly, in a variational sense, the one datum we have. It's a case of "For two cents, plain"[14].

B. Coulomb 1s trial function with a Yukawa prior potential

As another example with a 1s Coulomb trial function, we look at the case where we have a Yukawa form of prior potential; i.e.

$$V^P(r) = V_o^P (r/u)^{-1} \exp(-r/u) \quad ,$$

where both the strength V_o^P and the range u are known. We obtain

$$V(r) = \frac{V_o^P}{(r/u)} \exp(-r/u) - \frac{32\hbar^2}{6m} V_1 (r/a)^2 \exp(-2r/a) \quad ,$$

with

$$V_1 = \frac{1}{a_o^2} + \frac{1}{a^2} + \frac{4 U_o^P u^3}{a (2u + a)^2} \quad .$$

and $U_o^P = (\hbar^2/2m) V_o^P$. The range parameter "a" is found by solving

$$\frac{3}{a^2} - \frac{1}{a_o^2} + 4 U_o^P \frac{u^3 (2u + 5a)}{a (2u + a)^3} = 0 \quad .$$

The limit $u \to 0$ gives back $a^2 = 3a_o^2$, as it should, while for $u \to \infty$, with $U_o^P u \to W_o$ = finite, we obtain

$$(a / a_o)^2 + W_o a - 3 = 0 \quad ,$$

For a Coulomb prior such that $W_o = -2/a_o$, this last yields $a = a_o$ as it should, and, of course, V_1 vanishes.

C. Oscillator ground state trial function with $V^P(r) = 0$.

For the oscillator potential $(1/2)kr^2$, we have a $E_o > 0$, which was not true above. Here our method breaks down. We obtain no real, positive solution for the varied range parameter in our trial function. If $E_o < 0$, then the solution for the potential is perfectly well behaved, but it is negative for all r. In neither case do we obtain a potential which is oscillator-like by being positive for a range of r, especially the region r ---> ∞. This can be achieved, however by choosing a prior potential with these properties.

D. Oscillator ground-state trial function with $V^P(r) = V_o{}^P r^2$

We consider this case as our final example. Here

$$X_b(r) = [\pi b^2]^{3/4} \exp(-r^2/2b^2) ,$$

with $U_o{}^P = (2m/\hbar^2)V_o{}^P$, $E > 0$, and $a_o{}^2 = (3\hbar^2/2mE_o)$. It is quite simple to calculate the result for the potential as a function of $U_o{}^P$, a_o, and the variation parameter "a". Here this parameter must satisfy

$$a^{-2} = - a_o{}^{-2}/3 + [a_o{}^{-4}/9 + 5U_o{}^P/3]^{\frac{1}{2}} .$$

If our prior Hamiltonian has for its lowest state an eigenvalue that reproduces the experimental value a_o; i.e., $U_o{}^P = a_o{}^{-4}$, then this last equation yields $a = a_o$ and our prior potential is our result, as it should be. Otherwise, our result adds a term proportional to $X_a{}^2$ to the prior potential.

V. Possible Generalizations

There are several kinds of generalization that might be applied to the above form of super variation principle. The first kind has to do with the probability distribution used to represent the potential energy. A different, not as transparent, result would be obtained if we did not take V_0 ----> ∞. Further, if we do not take this limit we can use a multi-element distribution to represent each V_i instead of just a two-element distribution[15]. We might even use a continuous distribution as was done by Inguva and Baker-Jarvis[3,4]. With this last approach we would also have a way of representing a potential that depended on more than one variable.

Another possible generalization is to the realm of scattering problems. Any one of a number of well known scattering-state variation principles[11] for various phase shifts could be used as constraints on the entropy of the probability distributon that describes the unknown potential. This sort of application would give us a non-perturbative maximum-entropy approach to the solution of inverse-scattering problems. We could attempt to carry out such a calculation using a series of steps similar to those we used for our bound-state problem. Scattering-state variation principles, however, depend on the potential in a more complicated fashion than Eq.(4) does; e.g., the Schwinger[11] method contains double integrals and an integrand quadratic in the potential, while the Kohn[16] method has an integrand containing derivatives of the wavefunction. Incorporation of these into a super variation principle may lead to very messy equations. These equations may, or may not have a solution. This solution may, or may not be unique. (If it's not unique the one that has the largest entropy is the solution that should be used). We may try to find a solution by an iterative proceedure, but even if a solution exists, this attempt may fail. Clearly, this is only a _possible_ generalization that needs to be investigated further.

More far afield would be the application of this technique to the "blind deconvolution" problem; i. e., the image processing problem in which the point-spread function, as well as the "true" image is unknown. If we could set up a variation functional for the point-spread function we would have turned the problem into that of a "one-eyed deconvolution" problem. (This at least has an acronym which should please our English constituents).

As a final suggestion for an extension of the method, we could consider also relating our trial functions to an independent probability distribution and maximizing the total entropy. In addition to constraints on the entropy due to variational estimates of (say) the binding energies of the system, we could also impose other constraints arising out of other physical conditions that the wave function for the system must satify[17,18].

Clearly their is a rich variety of problems of this type yet to be explored.

References

1. See the list of references given by C. Ray Smith, Ramarao Inguva, and R. L. Morgan, SIAM-AMS Proceedings 14, 1984.
2. John P. Hallorasn and Lee H. Schick, Nucl. Phys. A431, (1984), 189.
3. R. Inguva and J. Baker-Jarvis, Proceedings of the Fourth Workshop on Maximum Entropy and Bayesian Methods in Applied Statistics, Calgary, 1984.
4. L. H. Schick, submitted to Phys. Rev. A, May 1985.
5. Y.M. Engel, E. Friedman, and R. D. Levine, Phys. Lett. 111B, (1982), 279.
6. K. Chadan and P. C. Sabatier, Inverse Problems in Quantum Scattering Theory, Springer-Verlag, New York, 1977, Ch. III.
7. Albert Messiah, Quantum Mechanics, John Wiley & Sons, New York, 1961, Vol. II., Ch. XVIII., 762.
8. Ibid, 765.
9. Tosio Kato, Phys. Rev. 77, (1950), 413.
10. Messiah, loc. cit.
11. John R. Taylor, Scattering Theory: The Quantum Theory on Non--relativistic Collisions, John Wiley & Sons, New York, 1972, 274.
12. See, for example, R. W. Johnson and J. E. Shore, NRL Memorandum Report 547, 1984.
13. Taylor, ibid, 27.
14. Harry Golden, For 2 Cents Plain, The World Publishing Co., New York, 1959

15. L. H. Schick, Proceedings of the Third Workshop on Maximum Entropy and Bayesian Methods in Applied Statistics, Laramie, 1983.

16. Taylor, 277.

17. C. Ray Smith has been working on just this aspect of the problem (private communication).

18. Rabinder N. Madan and Richard Blankenbecler, Phys. Rev. D 17, (1978), 1155. I'm grateful to Dr, Madan for bringing this work to my attention.

Einstein's Reversal of the Boltzmann Principle and Particle Statistics

A.K. Rajagopal

S. Teitler
Naval Research Laboratory
Washington D.C. 20375-5000

Instead of defining entropy as proportional to the logarithm of complexion function as in the Boltzmann principle, Einstein insisted that a complexion function should be defined in terms of the entropy. It is shown explicitly for the case of particle statistics that the Boltzmann principle is an approximation to Einstein's reversal when the entropy is the Shannon-Jaynes entropy, i.e. a Shannon entropy in an appropriate physics context.

According to Boltzmann's principle, the physical entropy ΔS_B associated with configurations of particles may be written in the form (Sommerfeld (1956))

$$\Delta S_B = k \log W_B \tag{1}$$

Here W_B, the Boltzmann complexion function or thermodynamic probability, is the number of microstates or complexions compatible with the contribution ΔS_B to the entropy. As occasionally noted (see for example, Sommerfeld (1956)) and recently thoroughly reviewed by Pais (1982), Einstein objected to the Boltzmann principle on the grounds that W_B was an ad hoc function without a satisfactory theoretical basis. He suggested that a more appropriate approach would be to define a complexion function in terms of the entropy, i.e.

$$W_E = \exp \Delta S/k \tag{2}$$

However Einstein did not have at his disposal a definition of physical entropy that could be used to obtain results such as particle occupation factors.

Indeed the Boltzmann principle has had great success in providing the basis for determination of the latter. However these occupation factors are for most probable values rather than mean values, and their derivation requires use of large numbers' approximations. Alternative derivations of occupation factors such as by the Darwin-Fowler method described for example by Fowler (1966) that involves the method of steepest descents, or by Khinchin (1960) who used limit theorems of probability do apply to mean values but also require use

G. J. Erickson and C. R. Smith (eds.),
Maximum-Entropy and Bayesian Methods in Science and Engineering (Vol. 2), 105–109.

of large numbers' approximations. Recently Rajagopal and Teitler (1987) have shown how one can use a definition of physical entropy based on microscopic properties to obtain particle occupation factors for mean values without any use of large numbers' approximations. The major point of the present note is to show explicitly that Eq.(1) may be derived as an approximation to Eq.(2) by utilizing this same entropy. In other words, the Boltzmann principle is subsidiary to its Einstein reversal with an appropriate definition of entropy.

The latter definition is the Shannon (1948) entropy placed in a physical context. In recognition of E.T. Jaynes' innovative contributions (Jaynes(1957a,b, 1983)) in the application of Shannon entropy in physics, we call this entropy the Shannon-Jaynes entropy.

$$\Delta S_J/k = -\sum_i p_i \log p_i \quad , \quad \sum_i p_i = 1 \tag{3}$$

Elsewhere in these proceedings Tribus (1987) points out that Shannon named the quantity $\Delta S_J/k$ entropy on advice from von Neumann. von Neumann had already introduced a similar quantity in a quantum mechanical context. Von Neumann entropy has the form

$$\Delta S_V/k = -\mathrm{Tr}\hat{\rho}\log\hat{\rho} \quad , \quad \mathrm{Tr}\hat{\rho}=1 \tag{4}$$

where $\hat{\rho}$ is a quantum mechanical density matrix. The evaluation of $\Delta S_V/k$ involves the determination of the eigenvalues of $\hat{\rho}$ which sum to 1. Thus, these eigenvalues may be taken to correspond to the probabilities in the definition of the Shannon-Jaynes entropy. However, the von Neumann quantum mechanical density matrix can be generalized (See for example Gamo (1964) and Klauder and Sudarshan (1968)), to any unit trace-class non-negative hermitian operator including those outside the context of quantum mechanics. We call the generalization a (general) density matrix.

Our viewpoint is that in applications of entropy in physics, i.e. for the Shannon-Jaynes entropy, the p_i in Eq.(4) can be taken to be eigenvalues of a density matrix. As indicated above, we have derived expressions for the Shannon-Jaynes entropy in terms of mean value occupation factors associated respectively with distinguishable and indistinguishable particles where the latter may be in either symmetric or antisymmetric states. The principal assumption used was that the particles could be viewed as individual entities so that eigenvalue probabilities are associated with multiparticle states. The degeneracies of the eigenvalues turn out to be just the respective Boltzmann enumeration factors that are usually identified as the W_B. However as degeneracy factors they only enter in the evaluation of the respective entropies and are not subject to a variational principle in determining occupation factors in equilibrium. Maximizing the respective entropies expressed in terms of mean value occupation factors subject to constraints on total energy and total particle number leads to the usual Boltzmann, Bose-Einstein and Fermi-Dirac occupation factors without any use of large numbers approximation.

The results for the respective entropies prior to maximization are

$$\Delta S_B/k = -N\sum_i p_i \log p_i \tag{5}$$

$$\Delta S_{BE}/k = -\sum_s R_s[\bar{p}_s \log \bar{p}_s - (1+\bar{p}_s)\log(1+\bar{p}_s)] \tag{6}$$

$$\Delta S_{FD}/k = -\sum_s R_s[p_s \log p_s + (1-p_s)\log(1-p_s)] \tag{7}$$

In Eq.(5), N is the total number of distinguishable particles and p_i are the mean value occupation factors for the ith cell in classical phase space characterized by single particle energy ε_i. In Eq.(6), R_s is the number of symmetric states in the sth shell characterized by single particle energy ε_s in which the indistinguishable particles are distributed with mean value occupation factor $\bar{p}_s$. In Eq.(7), R_s is the number of antisymmetric states in the sth shell characterized by single particle energy ε_s in which the indistinguishable particles are distributed with mean value occupation factor p_s.

Given Eqs.(5-7) we can evaluate the corresponding complexion functions W_E from the Einstein reversal of the Boltzmann principle as given in Eq.(2). This evaluation follows immediately if one makes two approximations: (1) replace the abstract definition of occupation factors by a frequency definition; (2) use large numbers' approximations equivalent to repeated use of the first Stirling's approximation

$$N\log N - N \cong \log N! \tag{8}$$

Thus if, in Eq.(5), one defines $p_i = n_i/N$ where n_i is the number of particles in the ith cell, and uses Stirling's approximation, one obtains

$$\log W_{MB} \cong \log \begin{pmatrix} & N & \\ n_1 & n_2 & \dots & n_m \end{pmatrix} \tag{9}$$

Similarly by inserting $\bar{p}_s = n_s/R_s$ in Eq.(6) and again using the Stirling's approximation, one obtains

$$\log W_{BE} \cong \log \prod_s \binom{R_s + n_s}{n_s} \cong \log \prod_s \binom{R_s + n_s - 1}{n_s - 1} \tag{10}$$

Finally by inserting $p_s = n_s/R_s$ in Eq.(7) and using the Stirling's approximation, one obtains

$$\log W_{FD} \equiv \log \prod_s \binom{R_s}{n_s} \tag{11}$$

Equations (9-11) provide the Boltzmann complexion functions as usually expressed respectively for Maxwell-Boltzmann, Bose-Einstein and Fermi-Dirac statistics. With these complexion functions in hand, one could use the Boltzmann principle and the principle of maximum entropy to obtain occupation factors for most probable values with large numbers' approximations in the traditional way. Thus the Boltzmann approach works to undo the approximations made in formulating the Boltzmann principle. It is clear that the preferred path is not to make these approximations in the first place. As general conclusions, we may say that the Boltzmann principle is subsidiary to the Einstein reversal expressed in terms of the Shannon-Jaynes entropy, and it is the Shannon-Jaynes entropy that should be used in obtaining particle occupation factors.

Acknowledgment: AKR is partially approved by ONR Contract N0001987WX24028

References

Fowler, R.H., (1966). Statistical Mechanics, 2nd Ed. (1936), Cambridge University Press.

Gamo, H., (1964). Thermodynamic Entropy of Partially Coherent Beams, J. Phys. Soc. Japan, 19, 1955-1961.

Jaynes, E.T., (1957a). Information Theory and Statistical Mechanics I, Phys. Rev. 106, 620-630.

Jaynes, E.T., (1957b). Information Theory and Statistical Mechanics II, Phys. Rev. 108, 171-190.

Jaynes, E.T., (1983). Papers on Probability, Statistics and Statistical Physics, edited by R.D. Rosenkrantz, D. Reidl Publ. Co.

Khinchin, A.Y., (1960). Mathematical Foundations of Quantum Statistics, Greylock Press.

Klauder, J.R. and Sudarshan, E.C.G., (1968). Fundamentals of Quantum Optics, W.A. Benjamin, Inc. (N.Y.), ch.5.

Pais, A., (1982). 'Subtle is the Lord...', Oxford University Press, ch.4.

Rajagopal, A.K. and Teitler, S., (1987). Particle occupation factors without large numbers' approximation, Physica in press.

Shannon, C.E., (1948). A Mathematical Theory of Communication, Bell Syst. Tech. J. 27, 379-423.

Sommerfeld, A., (1956). Thermodynamics and Statistical Mechanics, Academic Press, N.Y.

Tribus, M., (1987). An Engineer Looks at Bayes, these proceedings.

CLASSICAL ENTROPY OF A COHERENT SPIN STATE: A LOCAL MINIMUM

C. T. Lee
Department of Physics, Alabama A & M University,
Normal, AL 35762, U.S.A.

Abstract. The classical entropy of a coherent state of an N-spin system, which is $N/(N+1)$, was conjectured by Lieb to be the minimum. To prove it was our original motive. But we can only prove that this entropy is a local minimum. An arbitrary N-spin state is represented by N points on the surface of the unit sphere. For a coherent state, these N points condense into a single point. This is the basis for the proof. It is also conjectured here that the maximum entropy is attained when the N points form a possible regular polyhedron.

1 INTRODUCTION

Two kinds of coherent states are widely used in quantum optics to describe the radiation-matter interaction; namely, the Glauber coherent states for the harmonic oscillator (Schrödinger 1926; Glauber 1963; Sudarshan 1963) and the coherent spin states (also called the atomic coherent states) for a system of spins or two-level atoms (Radcliffe 1971; Arecchi *et al*. 1972). Intuitively, coherence can be considered as the opposite of chaos; so one impotant characteristic of coherent states is that they have minimum uncertainties measured by the standard deviation. However, some serious defects in using the standard deviation as a measure of uncertainties have been discussed by Uffink and Hilgevoord (1985) and, on the other hand, Deutsch (1983) has proposed to use entropy as an alternative measure of uncertainties.

Quantum mechanical entropy has been defined by von Neumann as $-\operatorname{tr}(\hat{\rho}\,\ell n\,\hat{\rho})$, where $\hat{\rho}$ is the density matrix or density operator. The trouble with this definition is that any pure quantum state, coherent or not, always has a minimum entropy of 0. This certainly cannot display the unique character of coherent states. For this purpose, a new definition of the entropy of a quantum state, called "classical" entropy, has been introduced by Wehrl (1979) in terms of the Glauber coherent states $|z\rangle$ as

$$S \equiv -\int \frac{dz}{\pi}\,\rho(z)\;\ell n\,\rho(z)\;, \tag{1}$$

G. J. Erickson and C. R. Smith (eds.),
Maximum-Entropy and Bayesian Methods in Science and Engineering (Vol. 2), 111–119.

where

$$\rho(z) \equiv \langle z| \hat{\rho} |z\rangle \tag{2}$$

is the diagonal element of the density matrix and, hence, must be nonnegative. We recognize that the $\rho(z)$ defined by (2) is exactly the probability density in phase space corresponding to antinormal ordering of operators which is called Q-representation in the terminology of coherent state representation (Haken 1970).

It was conjectured by Wehrl (1979) and proved by Lieb (1978) that Glauber coherent states have the minimum entropy 1 as defined by (1).

In the same paper by Lieb (1978), it was also conjectured that a similar definition of entropy for spin states will give the minimum entropy $N/(N+1)$ for the coherent states of an N-spin system.

To prove Lieb's conjecture was the original motive of this work. However, at present time, we are only able to prove that this entropy is a local minimum.

2 GEOMETRIC REPRESENTATION OF PURE SPIN STATES

It is well known that, as far as mathematical formulation is concerned, a system of N two-level atoms is exactly the same as a system of N spins. This equivalence was first used by Dicke (1954) to develop his theory of superradiance. The basic quantum states are denoted by $|J,J_z\rangle$, where J and J_z are, respectively, the eigenvalue of the total and the z-component of the angular momentum. In this paper J will be fixed to be $N/2$; then the eigenstates can be identified by one eigen value only; namely, $|n\rangle$ with $n \equiv N/2 + J_z$ which runs from 0 to N. In terms of these eigenstates, the density matrix of an arbitrary pure state can be written as

$$\hat{\rho}_N \equiv \sum_{n=0}^{N} \sum_{m=0}^{N} C_m^* C_n |n\rangle \langle m| , \tag{3}$$

with the normalization condition

$$\sum_{n=0}^{N} C_n^* C_n = 1 ; \tag{4}$$

and a coherent spin state is defined as

$$|\theta,\phi\rangle_N \equiv \sum_{n=0}^{N} |n\rangle \binom{N}{n}^{1/2} [\cos(\theta/2)]^{N-n} [e^{i\phi} \sin(\theta/2)]^n , \tag{5}$$

where θ and ϕ are the two angles in the standard spherical coordinate system. In the Q-representation, the probability density over the spherical surface for the state defined by (3) can be written as

$$Q_N(\theta,\phi) \equiv \langle\theta,\phi|\hat{\rho}|\theta,\phi\rangle_N = P_N^*(\theta,\phi)\ P_N(\theta,\phi)\ , \tag{6}$$

where

$$P_N(\theta,\phi) \equiv \sum_{n=0}^{N} C_n \begin{bmatrix} N \\ n \end{bmatrix}^{1/2} [\cos(\theta/2)]^{N-n}\ [e^{i\phi}\ \sin(\theta/2)]^n, \tag{7}$$

is a polynomial of degree N. And the classical entropy of the spin state is defined as

$$S_N \equiv -\frac{N+1}{4\pi} \int_0^{\pi} d\theta\ \sin\theta \int_0^{2\pi} d\phi\ Q_N(\theta,\phi)\ \ln Q_N(\theta,\phi)\ . \tag{8}$$

To carry out the integration in (8), it is convenient to reduce $P_N(\theta,\phi)$ of (7) to its N linear factors. It is always possible to express it as the following product

$$P_N(\theta,\phi) \equiv K_N^{1/2} \prod_{i=1}^{N} p(\theta,\phi;\theta_i,\phi_i)\ , \tag{9}$$

where K_N is the normalization constant and

$$p(\theta,\phi;\theta_i,\phi_i)$$

$$\equiv \cos(\theta_i/2)\ \cos(\theta/2) + \sin(\theta_i/2)\ \sin(\theta/2)\ e^{i(\phi-\phi_i)}\ . \tag{10}$$

Substitution of (9) into (6) gives

$$Q_N(\theta,\phi) = K_N \prod_{i=1}^{N} q(\theta,\phi;\theta_i,\phi_i)\ , \tag{11}$$

with

$$q(\theta,\phi;\theta_i,\phi_i) \equiv |p(\theta,\phi;\theta_i,\phi_i)|^2$$

$$= [1 + \cos\theta\ \cos\theta_i + \sin\theta\ \sin\theta_i\ \cos(\phi-\phi_i)]/2\ . \tag{12}$$

From (11) we can see that the probability density function $Q_N(\theta,\phi)$ of an arbitrary pure state of a system of N spins can be identified by N points on the surface of the unit sphere with the coordinates (θ_i,ϕ_i), where i runs from 1 to N. We call this the geometric representation of pure spin states.

3 COHERENT SPIN STATES

In the formulation described in the previous section, the probability density of a coherent spin state $|\theta',\phi'\rangle_N$ can be written as

$$Q_N(\theta,\phi;\theta',\phi') \equiv \langle\theta,\phi|\hat{\rho}(\theta',\phi')|\theta',\phi'\rangle_N = |\langle\theta,\phi|\theta',\phi'\rangle_N|^2$$

$$= [q(\theta,\phi;\theta',\phi')]^N \tag{13}$$

with the normalization constant $K_N = 1$. The implication of (13) is that, for a coherent state, the N points in its geometric representation reduce to a single point (θ',ϕ'). The formula for calculating the entropy of a coherent state is

$$S_N = -\frac{N+1}{4\pi}\int_0^{\pi} d\theta \sin\theta \int_0^{2\pi} d\phi\, [q(\theta,\phi;\theta',\phi')]^N$$

$$\times N \ln q(\theta,\phi;\theta',\phi') . \tag{14}$$

The integration in (14) can be simplified tremendously by using the fact that the it is over the whole surface of the unit sphere and, hence, the result is an invariant under any rotation of the sphere. We can rotate the sphere so that the point (θ',ϕ') coincides with the north pole. Then we have

$$q(\theta,\phi;\theta',\phi') \longrightarrow q(\theta,\phi:0,0) = \cos^2(\theta/2) . \tag{15}$$

Using (15) in (14) we obtain

$$S_N = N/(N+1) . \tag{16}$$

4 TAYLOR SERIES EXPANSION FOR THE ENTROPY FUNCTION

In the neighborhood of a coherent state, a spin state is represented by N points very close to one another. We will determine the Taylor series expansion of the entropy function of a state in this neighborhood. We can again rotate the sphere so that the point (θ',ϕ')

of the coherent state coincides with the north pole. Then, all the θ_i's in (11) are small quantities and $q(\theta,\phi;\theta_i,\phi_i)$ of (12) can be expressed as

$$q(\phi,\phi;\theta_i,\phi_i)$$

$$\approx \cos^2\frac{\theta}{2}\left\{1 + \theta_i \tan\frac{\theta}{2}\cos(\phi-\phi_i) - \frac{1}{4}\theta_i^2\,[1 - \tan^2\frac{\theta}{2}] - \frac{1}{6}\theta_i^3 \tan\frac{\theta}{2}\cos(\phi-\phi_i) + \frac{1}{48}\theta_i^4\,[1 - \tan^2\frac{\theta}{2}]\right\}, \quad (17)$$

where, as well as in the following, the series expansion has been carried to the fourth order of θ_i because, as will be seen later, all the first, second and third order terms in the expansion for the entropy function vanish identically.

4.1 Series expansion for K_N

The normalization constant is to be determined by the condition

$$K_N \frac{N+1}{4\pi}\int_0^{\pi} d\theta \sin\theta \int_0^{2\pi} d\phi \prod_{i=1}^{N} q(\theta,\phi;\theta_i,\phi_i) = 1 . \quad (18)$$

Using (17) in (18) and after some lengthy algebras, we obtain the following series expansion for $1/K_N$

$$K_N \approx 1 - \frac{1}{4}C_1 + \frac{1}{4N}C_2 - \frac{1}{96}D_1 + \frac{N^2-N-1}{32N(N-1)}D_2 + \frac{1}{24N}D_3 + \frac{1}{16N(N-1)}D_4 - \frac{N-2}{16N(N-1)}D_5 - \frac{1}{8N(N-1)}D_6 + \frac{1}{32N(N-1)}D_7 , \quad (19)$$

where the C's and the D's are, respectively, some second order and fourth order expressions to be defined as follows:

$$C_1 \equiv \sum_i \theta_i^{\,2} , \quad (20)$$

$$C_2 \equiv \sum_i \sum_j \theta_i\,\theta_j \cos(\phi_i-\phi_j) , \quad (21)$$

$$D_1 \equiv \sum_i \theta_i^{\,4} , \tag{22}$$

$$D_2 \equiv C_1^{\,2} , \tag{23}$$

$$D_3 \equiv \sum_i \sum_j \theta_i \ \theta_j^{\,3} \cos(\phi_i - \phi_j) , \tag{24}$$

$$D_4 \equiv \sum_i \sum_j \theta_i^{\,2} \ \theta_j^{\,2} \cos^2(\phi_i - \phi_j) , \tag{25}$$

$$D_5 \equiv C_1 \times C_2 , \tag{26}$$

$$D_6 \equiv \sum_i \sum_j \sum_k \theta_i \ \theta_j \ \theta_k^{\,2} \cos(\phi_i - \phi_k) \cos(\phi_j - \phi_k) , \tag{27}$$

$$D_7 \equiv C_2^{\,2} . \tag{28}$$

4.2 Series expansion for S_N

The formula for calculating the entropy of an arbitrary pure state of an N-spin system is

$$S_N = - K_N \frac{N+1}{4\pi} \int_0^{\pi} d\theta \sin\theta \int_0^{2\pi} d\phi \prod_{i=1}^{N} q(\theta,\phi;\theta_i,\phi_i) \times \sum_{\ell=1}^{N} \ell n\, q(\theta,\phi;\theta_\ell,\phi_\ell) - \ell n\, K_N . \tag{29}$$

We can use (17) in (29) and break up the integral into two parts as follows:

$$S_N = K_N \, [I_1 + I_2] - \ell n\, K_N , \tag{30}$$

where we have defined

$$I_1 \equiv - \frac{N+1}{4\pi} \int_0^{\pi} d\theta \sin\theta \int_0^{2\pi} d\phi \prod_{i=1}^{N} q(\theta,\phi;\theta_i,\phi_i) \ N \ \ell n \cos^2\frac{\theta}{2} \tag{31}$$

and

$$I_2 \equiv -\frac{N+1}{4\pi}\int_0^{\pi} d\theta\ \sin\theta \int_0^{2\pi} d\phi \prod_{i=1}^{N} q(\theta,\phi;\theta_i,\phi_i)$$

$$\times \sum_{\ell=1}^{N} \ell n\left[q(\theta,\phi;\theta_\ell,\phi_\ell)/\cos^2(\theta/2)\right] \ . \quad (32)$$

Carrying out the integrations in (31) and (32) we obtain

$$I_1 \approx \frac{N}{N+1}\left\{1 - \frac{1}{4}C_1 + \frac{2N+1}{4N^2}C_2 - \frac{1}{96}D_1 + \frac{N^4-2N^3-2N^2+1}{32(N-1)^2N^2}D_2 + \frac{2N+1}{24N^2}D_3 + \frac{3N^2-1}{16(N-1)^2N^2}D_4 - \frac{N^3-3N^2+1}{8(N-1)^2N^2}D_5 - \frac{3N^2-1}{8(N-1)^2N^2}D_6 + \frac{3N^2-1}{8(N-1)^2N^2}D_7\right\} \quad (33)$$

and

$$I_2 \approx \frac{1}{4}C_1 - \frac{1}{2N}C_2 + \frac{1}{96}D_1 - \frac{N^2-N-1}{16N(N-1)}D_2 - \frac{1}{12N}D_3 - \frac{1}{8N(N-1)}D_4 + \frac{3N-6}{16N(N-1)}D_5 + \frac{3}{8N(N-1)}D_6 - \frac{1}{8N(N-1)}D_7 \quad (34)$$

On the other hand, from (19) we have

$$K_N \approx 1 + \frac{1}{4}C_1 - \frac{1}{4N}C_2 + \frac{1}{96}D_1 + \frac{N^2-N+1}{32N(N-1)}D_2 - \frac{1}{24N}D_3 - \frac{1}{16N(N-1)}D_4 - \frac{1}{16(N-1)}D_5 + \frac{1}{8N(N-1)}D_6 + \frac{N-2}{32N^2(N-1)}D_7 \quad (35)$$

and

$$\ln K_N \approx \frac{1}{4} C_1 - \frac{1}{4N} C_2 + \frac{1}{96} D_1 + \frac{1}{32N(N-1)} D_2 - \frac{1}{24N} D_3 - \frac{1}{16N(N-1)} D_4 + \frac{1}{8N(N-1)} D_5 - \frac{1}{32N^2(N-1)} D_6 - \frac{1}{16N(N-1)} D_7 . \qquad (36)$$

Substitution of (33) - (36) into (31) yields

$$S_N \approx \frac{N}{N+1} + \frac{6N^2-13N+5}{32(N-1)^2(N+1)} \left[D_2 - \frac{2}{N} D_5 + \frac{1}{N^2} D_7 \right] + \frac{1}{16(N-1)^2} \left[D_4 - \frac{2}{N} D_6 + \frac{1}{N^2} D_7 \right] , \qquad (37)$$

where we notice that the coefficients of the two second order terms, C_1 and C_2, are identically zero.

4.3 Conclusion

Using the expressions for the D's given in (23) - (28) in (37), we can rewrite (37) in a simpler and more revealing form; *i.e.*,

$$S_N \approx \frac{N}{N+1} + \frac{3}{16(N+1)} \left[\sum_i \theta_i^2 - \frac{1}{N} \sum_i \sum_j \theta_i \theta_j \cos(\phi_i - \phi_j) \right]^2 + \frac{1}{32(N-1)^2} \left| \sum_i \theta_i^2 e^{2i\phi_i} - \frac{1}{N} \left[\sum_i \theta_i e^{i\phi_i} \right]^2 \right|^2 . \qquad (38)$$

The first term on the right of (38) is the entropy of the coherent state. It is obvious that the rest of the expression is positive semidefinite. This proves that the classical entropy of a coherent spin state is a local minimum. This minimum should be a very flat one because the first nonvanishing higher order terms are of fourth order.

If we set $\theta_i = \theta'$ and $\phi_i = \phi'$ for all i's, the second and the third terms on the right of (38) both vanish; and the entropy is still $N/(N + 1)$. This is as it should be; since in this particular way of variation, all the N points in the geometric representation still stick together and, hence, still represent a coherent state.

5 CONJECTURE ON MAXIMUM ENTROPY STATES

We have seen in this paper that the classical entropy of a spin state attains a minimum when all the N roots of its probability density function represented by N points on the surface of the unit sphere stick together as a single point. Equation (38) indicates that as these points spread out a little, the entropy increases. It is natural to speculate that as these points spread further apart from one another, the entropy will continue to increase until when these points are as far from one another as possibale on the sphere. Therefore we conjecture that the maximum entropy is attained by a pure spin state when the N points in its geometric representation are located as follows: For $N = 2$, the two points are diametrically opposite to each other; for $N = 3$, the three points form an equilateral triangle on a large circle; and for $N = 4, 6, 8, 12, 20$, etc., the N points form regular polyhedrons.

ACKNOWLEDGEMENT

This work was supported by the U. S. Navy Office of Naval Research.

REFERENCES

Arecchi, F.T., Courtens, E., Gilmore, R. & Thomas, H. (1972). Phys. Rev. **A6**, 2211.
Deutsch, D. (1983). Phys. Rev. Lett. **50**, 631.
Dicke, R.H. (1954). Phys. Rev. **93**, 99.
Glauber, R.J. (1963). Phys. Rev. Lett. **10**, 84.
Haken, H. (1970). Laser Theory. In Handbuch der Physik, **XXV/2c**, 62.
Lieb, E.H. (1978). Commun. Math. Phys. **62**, 35.
Radcliffe, J.M. (1971). J. Phys. A: Gen. Phys. **4**, 313
Schrödinger, E. (1926). Naturwissenschaften **14**, 664.
Sudarshan, E.C.G. (1963). Phys. Rev. Lett. **12**, 210.
Uffink, J.B.M. & Hilgevoord, J. (1985). Found. of Phys. **15**, 925.
Wehrl, A. (1979). Rep. Math. Phys. **16**, 353

LEAST MAXIMUM ENTROPY AND MINIMUM UNCERTAINTY COHERENT STATES

A.K. Rajagopal

S. Teitler
Naval Research Laboratory, Washington D.C. 20357-5000

A density matrix form of the Heisenberg uncertainty relation is obtained by means of the principle of maximum entropy (PME) subject to appropriate constraints. The least maximum entropy of zero is attained for the Heisenberg equality in concordance with the known property that the equality holds for appropriate pure states and their unitary equivalents. The case when the constraints involve expectations of a quadratic form in position and momentum operators is considered in detail. A few unitary transformations of importance, namely those corresponding to rotation in phase space, squeezing, and time evolution which leave the Heisenberg equality intact are discussed. The relation of these to the appropriately defined "correlated minimum uncertainty coherent states" are also discussed.

The traditional discussion of the Heisenberg inequality for the dispersions of two arbitrary hermitian operators with dispersions defined with respect to pure states is based upon the Schwartz inequality. This was generalized by Dodonov et.al. (1980) for arbitrary operators (non-hermitian) and density matrix (mixed states). An alternative approach to a density matrix form of the Heisenberg uncertainty relation is to apply the principle of maximum entropy (PME) for the density matrix subject to the given dispersions as constraints. Such a calculation was published by Titus (1979) and Wichmann (1963). In the present paper we examine this latter approach from a somewhat different viewpoint to show several novel consequences of this formulation. It should be pointed out that the Schwartz inequality argument is a more general one than the PME procedure in that the latter holds for the PME density matrix while the former is valid for any general density matrix. In the limit of Heisenberg equality, both approaches lead to pure states which are unitarily related.

Our principal interest is in the discussion of minimum uncertainty coherent states. For this purpose it is sufficient to consider the arbitrary hermitian operators to be dimensionless conjugate position, $\hat{q}$, and momentum, $\hat{p}$, operators. We use units with the Planck constant $\hbar = h/2\pi$ equal to unity. These operators obey the standard canonical commutation rules:

G. J. Erickson and C. R. Smith (eds.),
Maximum-Entropy and Bayesian Methods in Science and Engineering (Vol. 2), 121–126.

$$[\hat{q},\hat{p}]=i, \quad [\hat{q},\hat{q}]=o, \quad [\hat{p},\hat{p}]=o \tag{1}$$

We consider operators relative to their expectations;

$$\hat{P} = \hat{p}-\langle\hat{p}\rangle, \quad \hat{Q}=\hat{q}-\langle\hat{q}\rangle \tag{2}$$

where $\langle X\rangle=\mathrm{Tr}(\hat{\rho}\hat{X})$, $\hat{\rho}$ is the density matrix.

Titus (1979) and Wichmann (1963) have discussed the maximum of the von Neumann entropy

$$S=-\mathrm{Tr}\hat{\rho}\log\hat{\rho} \tag{3}$$

subject to the constraints of given values of

$$\langle\hat{q}\rangle,\langle\hat{p}\rangle,\langle\hat{Q}^2\rangle,\langle\hat{P}^2\rangle, \text{ and } \langle\hat{Q}\hat{P}+\hat{P}\hat{Q}\rangle\equiv 2\langle(\hat{Q}\hat{P})_s\rangle \tag{4}$$

They obtain

$$\hat{\rho}_u^M = Z^{-1} \,\mathrm{Exp}-\kappa_u\hat{K}_u \tag{5}$$

where

$$\hat{K}_u=\langle\hat{P}^2\rangle\hat{Q}^2+\langle\hat{Q}^2\rangle\hat{P}^2-2\langle(\hat{Q}\hat{P})_s\rangle(\hat{Q}\hat{P})_s \tag{6}$$

$$Z = \tfrac{1}{2}(\Omega^2-1)^{\frac{1}{2}} \tag{7}$$

$$\Omega = 2\left[\langle\hat{Q}^2\rangle\langle\hat{P}^2\rangle-\langle(\hat{Q}\hat{P})_s\rangle^2\right]^{\frac{1}{2}} \tag{8}$$

and

$$\kappa_u = \frac{1}{\Omega}\log\left(\frac{\Omega+1}{\Omega-1}\right) \tag{9}$$

With this as a basis we can obtain the following results.

(i) By a straightforward computation, we find the value of the maximum entropy associated with Eq.(5) to be

$$S_u^M= \left(\frac{\Omega+1}{2}\right)\log\left(\frac{\Omega+1}{2}\right) - \left(\frac{\Omega-1}{2}\right)\log\left(\frac{\Omega-1}{2}\right) \tag{10}$$

(ii) The positive definiteness and the trace-class nature of a general density matrix requires that

$$\Omega \geq 1 \tag{11}$$

This is just the generalized Heisenberg inequality given by Dodonov et.al. (1980). Note that this is consistent with the Schwartz inequality as it should be. This then is the entropic derivation of the Heisenberg relation.

(iii) The operator $\hat{K}_u$ has the eigen-expansion:

$$\hat{K}_u\,|m\rangle = (m+\tfrac{1}{2})\Omega\,|m\rangle,\ m=0,1,2,\ldots. \qquad (12)$$

This implies then that

$$\hat{\rho}_u^{\,M} = \sum_{m=o}^{\infty} w_m\,|m\rangle\langle m| \qquad (13)$$

where

$$w_m = \left(\frac{2}{\Omega+1}\right)\left(\frac{\Omega-1}{\Omega+1}\right)^m, \quad o\leq w_m\leq 1$$

$$\sum_{m=o}^{\infty} w_m=1, \quad S_u^M = -\sum_{m=o}^{\infty} w_m \log w_m \geq o \qquad (14)$$

(iv) Allow Ω to vary and in particular consider the value $\Omega=1$. Then from Eqs. (14) and (13) it is clear that

$$\hat{\rho}_u^{\,M}\rightarrow|o\rangle\langle o| \quad \text{and} \quad S_u^M\rightarrow o \qquad (15)$$

Also from Eq. (9), $\kappa_u\rightarrow\infty$ when $\Omega\rightarrow1$.

Thus the minimum Heisenberg uncertainty occurs for the least maximum entropy with concomitant pure state density matrix. In fact, it is the lowest state of the operator $\hat{K}_u$ and all its unitary equivalents.

(v) There is considerable interest in the recent literature (See for example, Schumaker (1986)) concerning the concept of squeezing or spreading of minimum uncertainty states. Squeezing is said to occur if $\langle\hat{Q}^2\rangle<\tfrac{1}{2}$ and spreading if $\langle\hat{Q}^2\rangle>\tfrac{1}{2}$. In much of the literature $\langle(\hat{Q}\hat{P})_s\rangle$ is taken to be zero. We find it convenient to consider such systems as Type I or uncorrelated, and others with $\langle(\hat{Q}\hat{P})_s\rangle\neq o$ as Type II. In both cases, we consider pure states corresponding to the least maximum entropy for which $\Omega=1$. Note that Ω and the entropy S are invarriant under unitary transformations. The reference configuration for $\Omega=1$ is taken to be Type IE which is a Type I system with equal variances. The corresponding state is sometimes called a minimum uncertainty coherent state (Schumaker (1986)) which will be discussed subsequently. We can generate a wide range of Type I and Type II cases from Type IE by means of a two parameter unitary transformation, $\hat{S}(r,\phi)$, based on a squeezing parameter, r, ($o\leq r<\infty$) and a rotation angle ϕ, $-\pi/2<\phi\leq\pi/2$ (Schumaker (1986)).

$$\hat{S}(r,\phi) = \mathrm{Exp}\ \tfrac{1}{2}r(e^{-2i\phi}\,\hat{a}^2 - e^{2i\phi}\,\hat{a}^{\dagger 2}) \qquad (16)$$

where $\hat{a}=(\hat{q}+i\hat{p})/\sqrt{2}$, $\hat{a}^{\dagger} = (\hat{q}-i\hat{p})/\sqrt{2}$

$r=o$ corresponds to Type IE; $r\neq o,\phi=o$ corresponds to Type I; and $r\neq o,\phi\neq o$ corresponds to Type II. Squeezing occurs where $\langle\hat{Q}^2(r,\phi)\rangle=\tfrac{1}{2}(\mathrm{Cosh}2r-\mathrm{Sinh}2r\ \mathrm{Cos}2\phi)<\tfrac{1}{2}$ and spreading occurs when the inquality is reversed.

Type I <u>excluding</u> Type IE are sometimes called minimum uncertainty twisted coherent states. In all cases, the minimum uncertainty condition $\Omega=1$ is maintained. However, not all possibilities for $\langle\hat{Q}^2\rangle$, $\langle\hat{P}^2\rangle$, and $\langle(\hat{Q}\hat{P})_s\rangle$ consistent with $\Omega=1$ are exhausted by the transformation given by Eq.(16).

A more traditional special case is an arbitrary unitary rotation in (Q,P) space through an angle θ represented by (Schumaker (1986)):

$$\hat{R}(\theta)=\mathrm{Exp}\, i\theta\left[(\hat{Q}^2+\hat{P}^2-2)/2\right] \tag{17}$$

It may be verified that $\langle\hat{Q}^2(\theta)\rangle$, $\langle\hat{P}^2(\theta)\rangle$, $\langle(\hat{Q}\hat{P})_s(\theta)\rangle$ are all different from their $\theta=o$ counterparts and yet $\Omega(\theta)=\Omega(o)=1$. Also, type IE remains unaltered under such rotations whereas other type I and type II have different dispersions from those corresponding to their $\theta=o$ values.

Another physically interesting unitary transformation is time evolution. For example, consider the free mass motion (Yuen (1983)):

$$\hat{U}(t) = \exp(-it\hat{P}^2/2) \tag{18}$$

It is easily verified that $\Omega(t)=\Omega(o)=1$ in general. Also, for Type IE, the spreading in time of $\langle\hat{Q}^2(t)\rangle$ is compensated by the build-up of correlation $\langle(\hat{Q}\hat{P})_s(t)\rangle$ for t=o.

(vi) To relate these observations to the concepts of generalized minimum uncertainty coherent states similar to those discussed by Dodonov et.al.(1980) we observe that the operator $\hat{K}_u$ in Eq.(6) can be recast in terms of new non-hermitian operators $\hat{A}^\dagger$ and $\hat{A}$ such that

$$\hat{K}_u=\Omega(\hat{A}^\dagger\hat{A}+\tfrac{1}{2}) \tag{19}$$

$\hat{A}$, $\hat{A}^\dagger$ obey the commutation rules

$$[\hat{A},\hat{A}]=o=[\hat{A}^\dagger,\hat{A}^\dagger],\ [\hat{A},\hat{A}^\dagger]=1 \tag{20}$$

This can always be done because $\hat{K}_u$ is a positive-definite hermitian operator with its spectrum bounded from below. In fact, a straightforward calculation shows that

$$\hat{A} = i\alpha\hat{P}+\beta\,\hat{Q} \tag{21}$$

with α,β complex such that

$$|\beta|^2=\langle\hat{P}^2\rangle/\Omega,\quad |\alpha|^2=\langle\hat{Q}^2\rangle/\Omega$$
$$\mathrm{Cos}\,\theta_{12} = \Omega/2(\langle\hat{P}^2\rangle\langle\hat{Q}^2\rangle)^{\frac{1}{2}} \tag{22}$$

where θ_{12} is the phase difference between the complex numbers α,β. These follow by comparing Eq.(19) with Eq.(6) rexpressed by use of

Eq.(21). We may note that this is a special case given by Eq.(16) when specialized to Type IE. The "generalized minimum uncertainty coherent states" are states defined by

$$\hat{A}|\mu\rangle=\mu|\mu\rangle \tag{23}$$

Here μ is a complex number so that the states $\{|n\rangle\}$ of $\hat{K}_u$ and these are related by the unitary transformation

$$\hat{D}(\mu) = \exp(\mu\hat{A}^{\dagger} - \mu^{*}\hat{A}) \tag{24}$$

with

$$|\mu\rangle=\hat{D}(\mu)|o\rangle = \exp(-\tfrac{1}{2}|\mu|^2) \sum_{n=o}^{\infty} \frac{\mu^n}{(n!)^{1/2}}|n\rangle \tag{25}$$

The completeness of the states $\{|n\rangle\}$ then imply that

$$\int \frac{d^2\mu}{\pi} |\mu\rangle\langle\mu| = \hat{1} = \sum_{n=o}^{\infty} |n\rangle\langle n| \tag{26}$$

and we can show that $\hat{\rho}_u^M$ has matrix elements

$$\langle\mu|\hat{\rho}_u^M|\mu'\rangle = \frac{2}{\Omega+1} \exp-\tfrac{1}{2}\left[|\mu|^2+|\mu'|^2-2\left(\frac{\Omega-1}{\Omega+1}\right)\mu'\mu^*\right] \tag{27}$$

To find S_u^M, the entropy, in the space of $\{|\mu\rangle\}$ one must diagonalize Eq.(27). In the general case, this will lead back to Eq.(14). In the special case of $\Omega=1$, this diagonalization is simple, leading to a single eigenvalue unity and $S_u^{M1}=o$ is obtained. These are expected results because $D(\mu)$ is a unitary operator.

Thus it becomes clear that the states of $\hat{\rho}_u^M$ corresponding to $\Omega=1$ of the least maximum entropy for given dispersions contain a richness buried in a variety of unitary transformations which are different manifestations of the lowest state of the operator $\hat{K}_u$.

Acknowledgement: A.K. Rajagopal was supported in part by ONR contract number N0001487WX24028

References

Dodonov, V.V., Kurmyshev, E.V., and Man'ko, V.I. (1980) Generalized uncertainty relation and correlated coherent states, Phys. Letts 79A, 150-152.

Schumaker, B.L. (1986) Quantum Mechanical pure states with gaussian wave functions, Phys. Repts. 135, 317-408.

Titus, W.J. (1979) Information theory density matrix for a simple quantum system, Am.J.Phys. 47, 357-361.

Wichmann, E.H. (1963) Density matrices arising from incomplete measurements, J. Math. Phys. 4, 886-896.

Yuen, H.P. (1983) Contractive states and the standard quantum limit for monitoring free-mass positions, Phys. Rev. Letts 51, 719-722.

MAXIMUM ENTROPY SPECTROSCOPY - DIMES and MESA

John Skilling
Department of Applied Mathematics
and Theoretical Physics
Silver Street
Cambridge CB3 9EW, England

Abstract. The direct "- f log f" and indirect "log f" entropy formulae have both been used in maximum entropy spectroscopy. The direct form is shown to be appropriate for finding the single "best" spectrum from incomplete data. The indirect form should be used to find an underlying probability distribution function, but not the spectrum itself. Examples show how and why the indirect form is liable to give misleadingly sharp spectra. When the true spectrum is unusually sharp and sparse, as in a simulation due to Fougere, the indirect form can reconstruct it more accurately than the conservative direct form, but other methods which deliberately seek sharp lines are then likely to do even better

Introduction

The idealised classic problem of spectral analysis is to estimate a positive spectrum f(x) from the values of a subset of its Fourier components

$$A_t = \int dx\ f(x)\ e^{2\pi ixt} \tag{1}$$

A related example of such a problem is to infer the power spectrum

$$f_x = \int d^N u\ p(\underline{u})\ N^{-1} \left| \sum_t u_t\ e^{-2\pi ixt/N} \right|^2 \tag{2}$$

$(x=0,1,\ldots,N-1)$ of a N-periodic real time-series $u_0, u_1, \ldots, u_{N-1}$ with unknown probability distribution function $p(\underline{u})$ from incomplete knowledge of the autocorrelation components

$$A_t = \int d^N u\ p(\underline{u})\ N^{-1} \sum_j u_j\ u_{j+t} = \sum_x f_x\ e^{2\pi ixt/N} \tag{3}$$

Maximum entropy (MaxEnt) is the preferred method of assigning any positive distribution, given incomplete data, and it has been applied to the above problem in two different ways (Jaynes 1982). Confusingly, both have at times been called "the maximum entropy method", with the definite article as if the other method did not exist.

The direct method (Shore & Johnson 1980; Skilling 1986), to

G. J. Erickson and C. R. Smith (eds.),
Maximum-Entropy and Bayesian Methods in Science and Engineering (Vol. 2), 127–145.

which we may assign the acronym DIMES (DIrect Maximum Entropy Spectroscopy), chooses that spectrum f(x) which has greatest (generalised) entropy

$$S(f) = \int dx \; [\; f(x) - m(x) - \log(f(x)/m(x)) \;] \qquad (4)$$

subject to the given constraints. Here m(x) is a prior model, which is often taken to be a constant. This method has had practical success in interferometry, deconvolution, tomography, spectroscopy and elsewhere (e.g. Gull & Skilling 1984a).

The indirect method (Burg 1967, 1972; Ulrych and Bishop 1975) has the acronym MESA (Maximum Entropy Spectral Analysis). It selects the underlying probability distribution by maximising its entropy

$$S(p) = - \int d^N u \; p(\underline{u}) \log p(\underline{u}) \qquad (5)$$

or, occasionally (Navaza 1985), the corresponding cross-entropy with non-uniform model $m(\underline{u})$. Then the corresponding power spectrum (2) is displayed as the result of the calculation. The MESA approach is adopted by about 80% of the maximum entropy papers on spectral analysis. However, as has been noted (Fougere et al. 1976), it can be subject to infelicities such as erroneous line-splitting and frequency shifts. Although there are ways of forcing these to be reduced (Fougere 1977), they are not entirely consonant with the maximum entropy formalism. The excellent and widely read book "Numerical Recipes" (Press et al. 1986) gives an example and describes maximum entropy as having "a certain cult popularity", the results as "quirky", and instructs its readers not to believe "that it gives an intrinsically better estimate than is given by other methods". MESA, the book tells us, "has the very cute property of being able to fit sharp spectral features, but there is nothing else magical about its power spectrum estimates".

At the 1986 MaxEnt conference, Fougere (1986) defended the MESA approach by presenting a highly accurate MESA reconstruction from error-free autocorrelation data, and ended with the challenge "Can any other method do better?". This paper takes up the challenge and compares MESA with DIMES and with the geometrical MUSIC method (Schmidt, 1986).

Derivation of DIMES

MaxEnt is intrinsically a selection procedure (Skilling 1986) which produces the single "best" positive distribution f(x) from incomplete data. For the "best" distribution to be properly defined, there must be a transitive "better than" relationship between distributions. Any transitive ranking can be mapped onto a "greater than" relationship between real numbers. Hence we seek a regularisation procedure, in which f is chosen by maximising a functional S(f).

One property which a useful reconstruction algorithm ought to have is "subset independence" (Shore & Johnson, 1980). This means that if two datasets pertain to two separate distributions, then the same distribution should be recovered either by processing the datasets independently or by processing them together as one joint dataset pertaining to the union of the distributions. Another property is that the reconstruction algorithm should be independent of the coordinate(s) x underlying the distribution. These general properties imply that S can be restricted to the form

$$S = \int dx\, m(x)\, \theta(f(x)/m(x)) \tag{6}$$

where m is a measure or "prior model" and θ is an arbitrary function.

A third property (Gull & Skilling, 1984b; Livesey & Skilling, 1985) is "system independence" (Shore & Johnson, 1980), meaning that marginal data on a two-dimensional rectangular distribution f(x,y) should yield the uncorrelated direct product of the data as the reconstruction f. Finally, if there are no data at all, f should default to the prior model m. These properties imply that θ can be restricted to the form

$$\theta(z) = z - 1 - z \log z \tag{7}$$

so that S must take the form (4) quoted above.

Note that this derivation is independent of all probabilistic and statistical arguments. It is possible to derive the entropy of a normalised distribution from combinatoric arguments (Jaynes, 1978, 1979) to illustrate the symbiosis between MaxEnt and probability theory, but it is not necessary to do this. Neither is it necessary to give entropy a quantitative combinatoric interpretation.

Because the DIMES entropy formula (4) can be applied to any positive distribution, regardless of normalisation and regardless of the type of data constraints, it can be applied in particular to the spectral analysis problem of recovering a positive spectrum from Fourier data (1). The result of maximising (4) over Fourier constraints (1) is a spectrum of exponential form

$$f(x) = m(x) \exp\left(\sum_t w_t \, e^{-2\pi i x t} \right) \tag{8}$$

where the Lagrange multipliers w are chosen to fit the constraint values A.

Any other method which produces a different reconstructed spectrum, whether or not it uses entropic formulae, must fail to satisfy at least one of the properties above. In particular, alternative methods such as MESA do not satisfy system independence, and hence induce extra correlation structure into the resulting spectrum, over and above what is required by the data.

Derivation of MESA

MaxEnt can, and should, be used to assign any positive distribution. In particular, it is the basis for assigning probability distributions, such as $p(\underline{u})$ in the time-series problem. The result of maximising its entropy

$$S(p) = - \int d^N u \; p(\underline{u}) \log p(\underline{u}) \; + \text{constant} \tag{9}$$

subject to the autocorrelation constraints (3) and to normalisation is that $p(\underline{u})$ should be assigned (Jaynes 1982, Skilling & Gull 1984b) as

$$p(u) = Z^{-1} \exp\left(- \sum_t w_t \sum_j u_j \, u_{j+t} \right) \tag{10}$$

where the Lagrange multipliers w are chosen to fit the constraint values and Z normalises p. This distribution is "the result" of the MESA technique. It is proper to use this probability distribution in the prediction problem - that of inferring future (or other un-observed) values of x from other values.

In order to develop the analysis further, it is conventional to assume that the time-series is N-periodic. If the autocorrelation data are derived from only part of this long time-series, as often happens in practice, then the N-periodic assumption may not be accurate, and Jaynes (1982) has argued that the consequential damage to the spectrum appears as line-splitting. However, the assumption is exact for the examples presented here. The Gaussian form (10) is diagonalised exactly by a Fourier transform:

$$p(u) = Z^{-1} \exp\left(- \sum_x W_x \, |U_x|^2 \right) \tag{11}$$

where

$$U_x = N^{-\frac{1}{2}} \sum_t u_t \, e^{-2\pi i x t/N} \quad , \quad W_x = \sum_t w_t \cos(2\pi x t/N) \tag{12}$$

and where only the "positive" frequencies x between 0 and

N/2 need be considered for a real time-series $\underline{u}$. Hence the power spectrum (2)

$$f_x = \int d^N u \; p(\underline{u}) \; |U_x|^2 = 1 \; / \; 2ZW_x \tag{13}$$

is a sufficient statistic for the MaxEnt probability distribution which can be written as

$$p(\underline{u}) = Z^{-1} \exp(- \sum_x |U_x|^2 / f_x) \tag{14}$$

Given this form, the autocorrelation components (3) become Fourier constraints

$$A = (2/N) \sum_x f_x \cos(2\pi xt/N) \tag{15}$$

on the power spectrum.

The entropy (9) of the probability distribution is

$$S(p) = H(f) + \text{constant} \quad , \quad H(f) = \sum_x \log f_x \tag{16}$$

Thus the MaxEnt probability distribution $p(\underline{u})$ can alternatively be found from (14) after first determining that power spectrum f which maximises the Burg entropy H(f) over the constraints, taken in their MaxEnt form (15). This MESA power spectrum takes the reciprocal form

$$f_x = (2 \sum_t w_t \cos(2\pi xt/N))^{-1} \tag{17}$$

as opposed to the exponential forms such as (8) which are produced by DIMES.

In summary, the MESA spectrum is obtained by maximising the entropy of the underlying probability distribution $p(\underline{u})$, and is a convenient way of displaying the selected $p(\underline{u})$. Through (14), it represents a MaxEnt probability distribution, but it is not a MaxEnt spectrum in its own right. Its derivation is restricted to ensemble average data for which the underlying probability distribution is (or is inferred by MaxEnt to be) separable in the spectrum channels x, and of Gaussian form.

General properties of DIMES and MESA spectra

A DIMES spectrum is obtained by maximising the entropy (4) directly over the constraints, whatever form they happen to take. The global maximum of S is obtained when f = m for each x (Fig. 1a) so that a DIMES spectrum is as "close" as possible to the pre-assigned model. The "strength" of the entropy maximisation is proportional to

$$\text{grad } S = \partial S / \partial f = \log m - \log f \tag{18}$$

which becomes logarithmically infinite both as f tends to zero and to infinity. Thus both positivity and finiteness are automatically assured.

Constraints are usually linear (at least in theoretical discussions), in which case the variational equations give

$$f(x) = m(x) \exp(\text{linear sum of constraint functions}) \quad (19)$$

of which (8) is a special case for Fourier data. Taking m(x) constant, spectrum maxima occur when this linear combination is maximum. Assuming the constraint functions to be differentiable, it follows that sharply resolved lines tend to have Gaussian profiles.

Particularising to Fourier data (1), the DIMES spectrum (8) is an entire function of complex x, with an essential singularity at infinity. Equivalently, the only singularities in the corresponding z-transform ($z=e^{2\pi ix}$) are essential singularities at the origin and at infinity, as far away from the unit circle as possible.

Figure 1.
(a) Direct entropy S(f) (b) Burg entropy H(f)

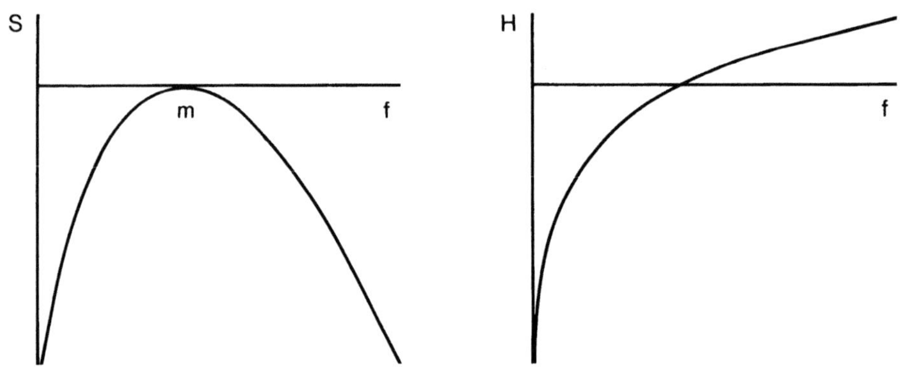

A MESA spectrum is obtained by maximising the Burg entropy (16) over the constraints. The Burg entropy is always increased by making f larger (Fig. 1b), so that a MESA spectrum is as "bright" as possible. This occurs because the MESA spectrum displays the variance of the underlying probability distribution p($\underline{u}$): MaxEnt applied to p ensures that p is spread out as widely as possible in $\underline{u}$, and this is accomplished by maximising its variance f(x). The "strength" of the entropy maximisation is

$$\text{grad } H = \partial H/\partial f = f^{-1} \tag{20}$$

which becomes infinite as f tends to zero, but not as f tends to infinity. Thus positivity is assured, but not finiteness. Indeed, grad H becomes small when f is large, so that there is no smoothing on bright regions of a MESA spectrum.

For linear constraint functions, the variational equations give

$$f(x) = 1 \ / \ (\text{linear sum of constraint functions}) \tag{21}$$

of which (17) is a special case. Spectrum maxima occur where this linear combination is minimum, so that sharply resolved lines tend to have Lorentzian profiles. In terms of the z-transform, the spectrum singularities are usually simple poles which can approach the unit circle to give sharp peaks.

Typical observational constraints are represented by integrals over the spectrum, which in turn tend to be dominated by the wings of Lorentzian lines, rather than the centres. This reflects the fact that the Burg entropy H is dominated more by the numerous background channels x than by the lines themselves, and can damage the visual appearance of MESA spectra.

DIMES and MESA reconstructions from small datasets

The ultimate test of any reconstruction technique is how it works in practice. However, it is important to remember that <u>any</u> continuous algorithm reconstructing a distribution must work perfectly in at least one case: the mapping from a spectrum through its Fourier or other data to its reconstruction by a particular algorithm will have at least one fixed point at which the algorithm works perfectly. For example, if the "true" spectrum were the exponential of an appropriately band-limited function like (8), then DIMES could recover it exactly, and if it were a reciprocal like (17), MESA could recover it. With this proviso in mind, consider the following examples, chosen for their simplicity and clarity.

1) <u>Single line, centre 0, width 1.</u>
The data are

$$\int_{-L}^{L} dx \ f(x) = 1 \ , \ \int_{-L}^{L} dx \ x \ f(x) = 0 \ , \ \int_{-L}^{L} dx \ x^2 f(x) = 1 \tag{22}$$

with L>>1. From these, DIMES recovers the unit Gaussian

$$f(x) = (2\pi)^{-\frac{1}{2}} \exp(-x^2/2) \tag{23}$$

to within corrections which are exponentially small in L. The MESA reconstruction is Lorentzian

$$f(x) = 1 \;/\; 2L(x^2+(\pi/2L)^2) \tag{24}$$

The width of this Lorentzian (defined as FWHM) is π/L, which is inversely proportional to the width 2L of the observing window. The MESA reconstruction can thus be made arbitrarily sharp by widening the observing window. This effect arises because the Burg entropy H is dominated by the wide background, which is made as large as possible, rather than by the line, which is correspondingly squeezed.

2) Single line, width 1, in asymmetric window.
The data are

$$\int_{-10}^{9990} dx\; f(x) = 1 \;, \quad \int_{-10}^{9990} dx\; x\, f(x) = 0 \;, \quad \int_{-10}^{9990} dx\; x^2 f(x) = 1 \tag{25}$$

The DIMES reconstruction is effectively the same Gaussian (23) as before, but the MESA Lorentzian is

$$f(x) = 0.0001000 \;/\; ((x+0.0006907)^2+0.0003142^2) \tag{26}$$

Not only is this narrow (FWHM = 0.0006284), but the line is displaced by more than its full width. This shows that a MESA spectrum can be sensitive to the symmetry as well as the width of the observing window.

3) Lorentzian doublet $f(x) = 1/((x+10)^2+1) + 1/((x-10)^2+1)$.
The data are

$$\int_{-100}^{100} dx\; f(x) = 6.2428, \quad \int_{-100}^{100} dx\; x\, f(x) = 0, \quad \int_{-100}^{100} dx\; x^2 f(x) = 1010.0 \tag{27}$$

From these, the DIMES reconstruction is a single Gaussian

$$f(x) = 0.1958 \exp(\,-\,0.003090\, x^2) \tag{28}$$

sufficiently wide (FWHM = 36.0) to span the positions $x = -10, +10$ of both of the original lines. The MESA reconstruction is a single Lorentzian

$$f(x) = 5.5142 \;/\; (\,x^2+ 2.7268^2) \tag{29}$$

Again, the MESA spectrum exhibits structure which is much narrower (FWHM = 5.4) than is required by the data. There is always a risk of such structure being misleading, as is the case here. Both the DIMES and the MESA reconstructions are wrong (because of inadequate data) in that they each have only one peak, but the resolution of the MESA spectrum is over-optimistic.

4) Disc in two dimensions.
The following measurements are taken of an object positioned in a circular field of view, with radius R>>1 :

$$\iint dxdy\ f(x,y) = 1 \quad , \quad \iint dxdy\ (x^2+y^2)\ f(x,y) = 2 \qquad (30)$$

The DIMES reconstruction is Gaussian

$$f(x,y) = (2\pi)^{-1} \exp(\ -\ (x^2+y^2)/2\ ) \qquad (31)$$

whereas the MESA reconstruction is

$$f(x,y) = (2/\pi R^2)/(x^2+y^2+a^2) \quad , \quad a = R\exp(-R^2/4) \qquad (32)$$

Here the width (linear FWHM = 2a) of the MESA reconstruction decreases exponentially with the size of the observation window.

5) Sphere in three dimensions.
The following measurements are taken of an object positioned in a spherical domain, with radius R>>1 :

$$\iiint dxdydz\ f(x,y,z) = 1 \ ,$$

$$\iiint dxdydz\ (x^2+y^2+z^2)\ f(x,y,z) = 3 \qquad (33)$$

The three-dimensional DIMES reconstruction is again Gaussian

$$f(x,y,z) = (2\pi)^{-3/2} \exp(\ -\ (x^2+y^2+z^2)/2\ ) \qquad (34)$$

However, the MESA reconstruction is yet sharper than in two dimensions. It is

$$f(x,y,z) = (1-9/R^2)\ \delta(x)\delta(y)\delta(z) + (9/4\pi R^3)\ r^{-2} \qquad (35)$$

with all but a small fraction $9/R^2$ of the intensity concentrated into a delta function at the origin.

This occurs because in three (or more) dimensions, the complex zeros of the denominator of a MESA reconstruction (21) can become real without giving a large contribution to the volume integrals which comprise typical data (Nityananda & Narayan, 1982). Delta functions give no contribution to the Burg entropy, so are not discriminated against, and can build up wherever the denominator of (21) vanishes: the phenomenon is akin to a Bose condensation. By contrast, delta functions give infinite contributions to the direct entropy, so that a DIMES reconstruction can never shatter in this way.

Infinities do not arise in one dimension, because the

data integrals would also be infinite. Two dimensions give an intermediate case in which the data integrals would become infinite, but only logarithmically, so that finite but exponentially sharp structure can then develop in a MESA reconstruction.

The Fougere simulation

Fougere (1986) proposed a test in which a spectrum having 63 sharp lines of known amplitudes a between 0.002 and 12.5, systematically separated in frequency x between 0 and 90 (first columns of Table 1) is to be reconstructed from its first 251 cosine Fourier coefficients

$$C_t = \sum_{j=1}^{63} a_j \cos(\pi x_j t/90) \quad , \quad t = 0,1,\ldots,250 \qquad (36)$$

There are almost twice as many data C as frequencies and amplitudes, so that the spectrum is well determined, and indeed the true spectrum of 63 lines is the only positive spectrum which fits the data. To avoid this singular solution, the t=0 component is artificially multiplied by 1.00004, as from a faint background of white noise.

The DIMES spectrum is of the form

$$f(x) = \exp\left(\sum_{t=0}^{250} w_t \cos(\pi x t/90) \right) \qquad (37)$$

Table 2 shows values of the Lagrange multipliers w for which the cosine components of the spectrum agree with the data to at least 10 decimal places: these were obtained by iterating the misfits towards zero with a suitably protected Newton-Raphson algorithm. The values are large, up to 12000, and the spectrum f has a correspondingly emormous dynamic range of some 75000 orders of magnitude. The top few orders of magnitude define the bulk of the intensity in the spectrum, which is confined to a relatively small fraction of the total width and is tolerably accurately represented (to about 1 part in 1000) by a set of Gaussian lines.

These lines were assigned the profile

$$g(x) = c \exp(- (x-\mu)^2/2\sigma^2) \qquad (38)$$

in which the parameters c, μ, σ were determined by least-squares fit to f(x) between successive minima, and the intensity A is defined as the integral of g(x) between the minima. Because the spectrum is defined by a polynomial of order 250 in $\cos(\pi x/90)$, there could be up to 125 maxima, each representing a line. In fact, there are 114.

63 of these (central columns of Table 1) correspond to the lines in the original simulation. The amplitude-weighted

rms proportional error in amplitude, largely due to forcing a Gaussian profile, is

$$\left(\sum_{j=1}^{63} a_j \, (A_j/a_j - 1)^2 \Big/ \sum_{j=1}^{63} a_j \right)^{1/2} = 0.0016 \qquad (39)$$

in frequency is

$$\left(\sum_{j=1}^{63} a_j \, (x_j - \mu_j)^2 \Big/ \sum_{j=1}^{63} a_j \right)^{1/2} = 0.0003 \qquad (40)$$

and the amplitude-weighted standard deviation width is

$$\sum_{j=1}^{63} a_j \, \sigma_j \Big/ \sum_{j=1}^{63} a_j = 0.0012 \qquad (41)$$

Plausibly, the errors and widths are less on the brighter lines, ranging from $\delta x = 0.00000005$, $\sigma = 0.0008$ on the brightest line to $\delta x = 0.0001$, $\sigma = 0.027$ on the dimmest.

With cosine data up to order 250, the formal resolution width in x is 90/250 = 0.36 . The DIMES standard deviation widths are around 300 times smaller than this, showing that the data are good enough to give super-resolution of this factor of 300. Line structure narrower than the DIMES widths could be present, but there is no evidence for it in the data. As it happens there is no asymmetry in the simulated lines, so there is no evidence for it in the data, so that the line centres are positioned considerably more accurately than the quoted line widths.

Finally, the remaining 51 satellite lines (first columns of Table 3) have clearly lower intensities.

The MESA spectrum is of the form

$$f(x) = 1 \Big/ \sum_{t=0}^{250} w_t \cos(\pi x t/90) \qquad (42)$$

where the multipliers w can be obtained to better than 10 significant figures by the standard Levinson recursion (Press et al. 1986). Again, the bulk of the intensity in the reconstruction is confined to a relatively small fraction of the total width, which is now represented by a set of Lorentzian lines. These lines are assigned the profile

$$g(x) = c \,/\, ((x-\mu)^2 + \sigma^2) \qquad (43)$$

in which the parameters c, μ, σ were determined by least-squares fit to f(x) between successive minima, and the intensity A is defined as the integral of g(x) between the minima. There are 97 lines out of the possible 125 maxima.

63 of these (last columns of Table 1) correspond to the lines in the original simulation. The amplitude-weighted

rms proportional error in amplitude is

$$\left(\sum_{j=1}^{63} a_j\,(A_j/a_j - 1)^2 / \sum_{j=1}^{63} a_j\right)^{\frac{1}{2}} = 0.00013 \qquad (44)$$

in frequency is

$$\left(\sum_{j=1}^{63} a_j\,(x_j-\mu_j)^2 \;/ \sum_{j=1}^{63} a_j\;\right)^{\frac{1}{2}} = 0.000057 \qquad (45)$$

and the amplitude-weighted half-width to half-amplitude is

$$\sum_{j=1}^{63} a_j\,\sigma_j \;/ \sum_{j=1}^{63} a_j = 0.0000055 \qquad (46)$$

(These widths are up to 100 times those quoted by Fougere). Again, the errors and widths are less on the bright lines than on the dim, and the remaining 34 satellite lines (last columns of Table 3) are of considerably lower intensity. Although the frequency errors on the 63 major lines are small, they are some ten times greater than the half-widths of the reconstructions, so that the lines are significantly misplaced.

Discussion of the Fougere simulation

In this example, the MESA reconstruction is closer to the original simulated spectrum than is the DIMES reconstruction, although the lines are reproduced more sharply than is warranted by the data. The answer to Fougere's question "Can any other method do better?" is, however, "yes".

The MUSIC algorithm (Schmidt 1986) clearly outclasses both DIMES and MESA for this dataset. MUSIC is a geometrically-inspired method which searches for sharp lines and lists them. It reproduces exactly 63 lines with frequencies correct to at least 14 significant figures (apparently limited only by arithmetic accuracy).

The Fougere simulation is, in fact, unusual not only because the data are very accurate, but also because there are four times as many data as significant lines in the spectrum. In the happy event that one knows that one is seeking a small number of simple lines, and has sufficient accurate data to over-determine the free parameters, then clearly one should use that information and seek the lines directly. The astronomers' CLEAN method (Hogbom 1974) would presumably also perform well: Tan (1986) has offered comments on this method.

MaxEnt is designed for the more difficult cases when one seeks an initially unknown distribution, and has inadequate data. It is not surprising that a technique like MESA which encourages the development of sharp structure should out-perform DIMES in a simulation limited to sparse, sharp structure. But is MESA reliable?

One simple test is to complicate the line profiles. A second simulation was run, in which each line was replaced by an equal-amplitude triplet at x-0.1, x, x+0.1 . The 251 data are too few to determine the amplitudes, frequencies, and widths of what are now 189 individual components.

DIMES finds all 63 major signals and separates 27 of them into double peaks (there is never sufficient resolution to find triplets). Thus the brightest signal, at frequencies 37.5586, 37.6586, 37.7586 is analysed as a doublet at 37.58±0.03, 37.74±0.03. Figure 2 shows an adjacent pair of unresolved but clearly non-Gaussian signals. All 7 satellite lines have intensities less than 0.0002, so are clearly distinguished from the major signals.

MESA also finds all 63 signals and splits 32 of them into double peaks. The brightest signal, at 37.5586, 37.6586, 37.7586 is analysed as a doublet at 37.577±0.003, 37.739±0.003. The sharp, narrow quality of MESA reconstructions has now become a more serious disadvantage, because the sharpness of the reconstruction is qualitatively misleading. Figure 3 plots the same adjacent pair of signals as before. These signals have similar strengths, frequencies, and neighbours, yet MESA reconstructs two very different shapes. This infelicity can not be due to incomplete sampling of the time-series, which Jaynes (1982) suggested was the cause of line-splitting, because the Fourier data (36) are exact.

There is only one satellite maximum, at frequency 66.0. Although its amplitude (0.10) is smaller than its immediate neighbours (8.8 and 7.7), it is fifty times brighter than the dimmest signals, so it could not be immediately rejected as spurious.

Figure 2. DIMES reconstructions of adjacent triplets

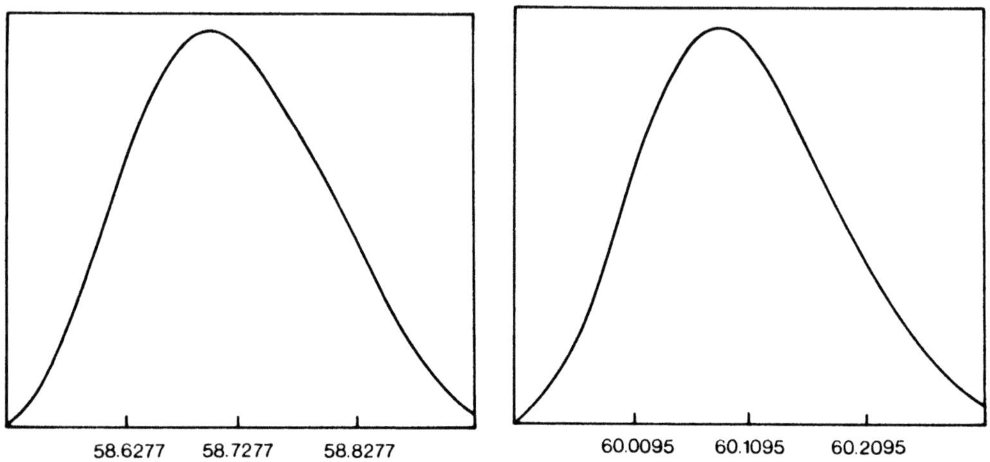

Figure 3. MESA reconstructions of adjacent triplets

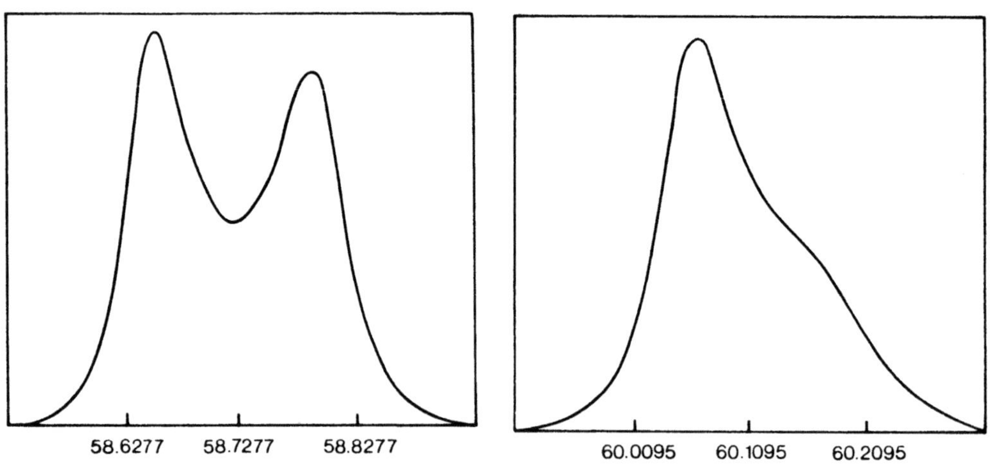

Table 1. The Fougere simulation

	Simulation		DIMES reconstruction			MESA reconstruction		
	Amplitude	Frequency	Ampl.	Freq.	Std.Dev	Ampl.	Freq.	Halfwidth
1	.12648	.4764	0.1266	0.47629739	.0037	0.1266	0.47634476	.000137
2	.19299	2.6503	0.1931	2.65020163	.0046	0.1931	2.65028590	.000113
3	.29195	4.8002	0.2920	4.80012403	.0049	0.2921	4.80020100	.000082
4	.43656	6.9262	0.4367	6.92615265	.0047	0.4367	6.92620677	.000059
5	.64329	9.0281	0.6434	9.02807647	.0042	0.6434	9.02810732	.000044
6	.93209	11.1060	0.9322	11.10599222	.0036	0.9322	11.10600484	.000034
7	1.3245	13.1599	1.3246	13.15990164	.0031	1.3246	13.15990092	.000025
8	1.8424	15.1897	1.8427	15.18970767	.0027	1.8425	15.18969867	.000017
9	2.5045	17.1954	2.5052	17.19541364	.0023	2.5046	17.19539925	.000011
10	3.3196	19.1769	3.3230	19.17689203	.0019	3.3197	19.17690036	.000008
11	4.2863	21.1343	4.2865	21.13429963	.0017	4.2864	21.13430029	.000007
12	5.3850	23.0674	5.3955	23.06740401	.0015	5.3851	23.06739972	.000005
13	6.5790	24.9764	6.5889	24.97639728	.0013	6.5791	24.97640006	.000004
14	7.8158	26.8611	7.8172	26.86110134	.0012	7.8159	26.86110006	.000003
15	9.0325	28.7215	9.0334	28.72149952	.0011	9.0326	28.72149992	.000003
16	10.155	30.5577	10.1553	30.55770011	.0010	10.1551	30.55770007	.000002
17	11.122	32.3694	11.1222	32.36940011	.0010	11.1221	32.36939995	.000002
18	11.870	34.1569	11.8701	34.15689995	.0009	11.8701	34.15690004	.000002
19	12.351	35.9199	12.3512	35.91990005	.0009	12.3511	35.91989998	.000002
20	12.518	37.6586	12.5184	37.65860007	.0008	12.5181	37.65860000	.000002
21	12.322	39.3728	12.3238	39.37279972	.0008	12.3221	39.37280000	.000002
22	11.720	41.0625	11.7202	41.06250027	.0009	11.7201	41.06250001	.000002
23	10.650	42.7277	10.6501	42.72770007	.0008	10.6501	42.72769997	.000002
24	9.0287	44.3685	9.0323	44.36849939	.0009	9.0288	44.36849996	.000002
25	6.7738	45.9847	6.7742	45.98469933	.0012	6.7739	45.98470009	.000003
26	3.7864	47.5763	3.7874	47.57630230	.0014	3.7865	47.57630036	.000005
27	1.9892	49.9176	1.9893	49.91759813	.0011	1.9893	49.91759874	.000008
28	5.3216	51.4477	5.3223	51.44769826	.0010	5.3217	51.44769945	.000003
29	7.8358	52.9531	7.8395	52.95309854	.0009	7.8359	52.95309965	.000002
30	9.6281	54.4338	9.6343	54.43379885	.0010	9.6282	54.43379988	.000002
31	10.817	55.8899	10.8219	55.88989928	.0010	10.8171	55.88990016	.000002
32	11.498	57.3212	11.4985	57.32119989	.0010	11.4981	57.32120048	.000002
33	11.757	58.7277	11.7588	58.72770075	.0011	11.7572	58.72770085	.000003
34	11.662	60.1095	11.6828	60.10950206	.0011	11.6622	60.10950117	.000003
35	11.266	61.4664	11.2944	61.46640468	.0012	11.2662	61.46640107	.000004
36	10.620	62.7985	10.6209	62.79849918	.0012	10.6202	62.79849990	.000005
37	9.7711	64.1058	9.8330	64.10579595	.0012	9.7713	64.10579898	.000004
38	8.7710	65.3882	8.7713	65.38819956	.0012	8.7711	65.38819961	.000003
39	7.6722	66.6457	7.7079	66.64570398	.0013	7.6723	66.64570086	.000004
40	6.5364	67.8782	6.5458	67.87819715	.0014	6.5366	67.87820031	.000006
41	5.4178	69.0858	5.4216	69.08579741	.0014	5.4179	69.08579878	.000005
42	4.3693	70.2684	4.3724	70.26840602	.0016	4.3694	70.26840114	.000006
43	3.4268	71.4261	3.4332	71.42609264	.0016	3.4269	71.42609930	.000010
44	2.6179	72.5586	2.6208	72.55860571	.0018	2.6180	72.55859966	.000008
45	1.9480	73.6662	1.9498	73.66618957	.0022	1.9481	73.66620167	.000017
46	1.4160	74.7487	1.4162	74.74870654	.0024	1.4161	74.74869855	.000014
47	1.0064	75.8061	1.0068	75.80608816	.0028	1.0065	75.80610171	.000031
48	.70104	76.8383	0.7012	76.83831579	.0035	0.7011	76.83830161	.000024
49	.47978	77.8455	0.4799	77.84549063	.0029	0.4799	77.84548729	.000052
50	.32331	78.8274	0.3235	78.82738693	.0056	0.3234	78.82742098	.000063
51	.21533	79.7842	0.2154	79.78422187	.0039	0.2154	79.78419302	.000063
52	.14209	80.7158	0.1422	80.71573720	.0044	0.1422	80.71574892	.000149
53	.093199	81.6222	0.0933	81.62197630	.0107	0.0933	81.62226627	.000255
54	.061051	82.5033	0.0612	82.50352304	.0080	0.0611	82.50339331	.000198
55	.040068	83.3591	0.0402	83.35919070	.0053	0.0402	83.35909505	.000215
56	.026455	84.1896	0.0266	84.18944545	.0054	0.0265	84.18939585	.000335
57	.017628	84.9949	0.0177	84.99424259	.0068	0.0177	84.99432611	.000654
58	.011891	85.7747	0.0120	85.77287144	.0099	0.0120	85.77349722	.001454
59	.0081315	86.5293	0.0082	86.52439897	.0164	0.0082	86.52723047	.003393
60	.0056404	87.2584	0.0056	87.24406124	.0314	0.0057	87.25584665	.007687
61	.0039674	87.9622	0.0038	87.91620770	.0788	0.0040	87.96170348	.015506
62	.0028267	88.6405	0.0027	88.71917967	.0839	0.0029	88.64815077	.024386
63	.0020357	89.2934	0.0019	89.33375156	.0268	0.0021	89.31215811	.019289

Table 2. DIMES Lagrange multipliers w_0 to w_{250} in row order

-12497.39250696992	1692.97672245856	11949.87190754904	-1234.44777361718
3712.24289517012	-3788.67776494679	-3713.83107302494	4795.31676135285
-1222.16105079234	-1921.90131312691	1563.13225653150	-985.48409691179
128.80472310321	1109.09145174933	-1856.95955468551	789.80517227866
2255.26250797851	-2567.05603912498	148.84865315894	2735.35736166389
-2721.98080856926	-844.94609465040	1951.12764770278	-1882.57671102029
427.96089576164	1492.89126285382	-1517.88567964258	2048.55631255353
844.11992239737	-3260.34486208850	1179.02048209850	16.00122895421
-2642.99240334474	2399.41380239206	111.99733868902	-339.17145101467
3451.04945843376	-2424.77271145785	-2170.89375742468	1756.89643883490
-2077.83851674924	1050.44487507791	2023.43669940568	-1951.92016309676
1514.56119491639	-244.32429004778	-2248.75281271838	1696.18779428191
-72.87121100201	81.15768980012	1322.52245072659	-1603.23539879206
-813.70717928800	-392.20760105290	212.71594610510	1509.33988228974
345.61444343750	863.58413706888	-739.43469587383	-1292.63601048712
-298.24159177504	-1429.54840320100	1282.82197253623	1387.26087983820
-575.41458639598	1360.65202995254	-744.34550517463	-1006.19260081130
832.53172502762	-1150.14514580327	-159.91548224010	322.22827916948
155.62706352636	635.92653820700	-48.06229768849	24.73608344305
-300.96355419907	967.47331069581	-725.73171584349	-1069.08608943677
1414.87897841668	-1050.55207671196	-122.13219430637	237.71121211767
-876.06691245246	944.07242001105	-480.81214028808	-239.64287537674
115.65595530707	136.00211615782	-18.82012548430	-995.25614982164
1417.78911726506	-928.57206767027	-546.45265965293	1989.50362523897
-2360.77906785266	1162.24984930618	1391.35991031357	-2967.36010281484
2523.97960041774	-57.54041186155	-2378.29954542402	2850.41506039540
-1728.52902209899	-844.93372106278	2771.40216633267	-2984.37152608938
725.35649553698	2521.69952494525	-3531.66494631032	1655.75409978357
2133.67285034009	-3668.33208489194	1341.69772548184	1640.44927742611
-3218.23396698984	1009.69711659860	1628.37979072883	-1850.39753491839
1087.43943470118	851.26255957251	-1325.19897197477	821.70869459439
-729.49263381824	-1285.39771331518	1866.11758665818	-933.21738491841
-860.32200844604	3376.91539323130	-1465.25223674648	-2145.31285221814
3007.39645638947	-1805.90283361549	-1883.84283161439	3393.41843686959
-1353.76548272206	-568.71802528243	2492.18612212338	-3178.61915297295
332.97798147272	2565.57100426308	-2633.68032936264	1509.99354345864
492.35397301049	-2408.02935362027	2154.02896604694	-1545.80488612585
-1017.17352608198	3152.64725330749	-1222.33856424949	321.39772324437
562.99073457800	-2109.31707062449	519.12565516770	-216.09641432494
69.42214500266	853.80428860526	-89.14727180621	316.09369237818
-598.81490381757	256.16485410084	-503.13930233420	-525.34170916737
1071.50758174249	-1186.90054203103	379.09482694393	641.37890663668
-1042.38975748098	910.51241608046	-270.74708778654	-566.17575760087
471.81705879426	-465.82294974472	-494.88375634012	253.07912135262
-422.04795984303	-341.22168576122	1182.44735986245	-98.56668968044
429.72082501078	1182.77366181615	-667.24865069768	-213.41519341532
-944.17491485627	-1446.53522134451	266.77417362482	601.08403463931
1198.59725953359	1818.48954286254	-38.94105972577	-1539.56685670958
-1784.13830805933	-1268.98988403902	1247.59002422075	2232.31859772303
1359.98646367719	-991.82616166322	-2832.85510176756	-831.90188404630
1551.84716182037	2666.51334767148	850.45921056435	-2818.11699878208
-2504.03780318962	770.10343112966	2623.91439582778	1616.18208003740
-2548.35017361418	-2551.38220016324	2290.39393921464	2763.94889166137
-716.97711761097	-3053.98785458420	-968.04703215407	2512.52951512651
1213.70937644639	-1834.79224665120	-946.10788081395	1801.91493881296
1403.61809779335	-1528.68681665296	-1614.28028917425	345.15729879750
1221.23553167508	82.30338686021	-1370.61131758095	846.65898951057
1643.49041682826	-1467.73623673798	-1339.85664844729	1117.25248099782
760.42156799991	-737.15147792876	-296.76108269748	722.01104861475
-269.83770871196	-1116.72781708794	153.70464604473	1728.20485144153
-244.18352400060	-1046.43265774706	1585.41168336593	-1574.73755868094
-1860.50908383208	2859.09903708727	-1077.85062851306	-1231.42490439611
3267.95724726926	-447.79994676018	-1997.19745804490	

Table 3. Satellite lines

DIMES reconstruction			MESA reconstruction		
Ampl.	Freq.	Std.Dev	Ampl.	Freq.	Halfwidth
.000093	1.0640	.0044	.000076	1.1530	.4639
.000096	1.7654	.0036	.000097	1.7721	.3228
.000083	3.1946	.0054	.000066	3.5352	.6128
.000096	3.8772	.0040	.000091	3.8613	.5051
.000066	5.3435	.0060			
.000096	5.9814	.0043	.000146	5.8442	.6164
.000048	7.5187	.0063			
.000093	8.0763	.0049	.000131	7.9370	.4273
.000039	9.7176	.0068			
.000087	10.1610	.0058	.000122	10.0285	.3139
.000045	11.9311	.0081			
.000071	12.2419	.0077	.000118	12.1151	.2647
.000081	14.1153	.0084	.000117	14.1841	.2595
.000027	14.3633	.0086			
.000102	16.1913	.0038	.000118	16.2193	.3063
.000000	16.6121	.0050			
.000101	18.1937	.0012	.000120	18.1923	.3744
.000012	19.8139	.0041			
.000093	20.1867	.0035	.000116	20.1400	.3089
.000087	22.0696	.0047	.000113	22.1075	.2725
.000015	22.3280	.0049			
.000097	24.0250	.0009	.000114	24.0356	.3257
.000024	25.7629	.0051			
.000076	25.9707	.0049	.000112	25.9091	.2994
.000095	27.7866	.0009	.000110	27.8021	.2798
.000093	29.6503	.0011	.000109	29.6378	.3100
.000092	31.4566	.0010	.000107	31.4685	.2724
.000091	33.2805	.0009	.000106	33.2639	.2962
.000089	35.0218	.0009	.000104	35.0438	.2720
.000089	36.8155	.0008	.000103	36.7893	.2787
.000088	38.5012	.0008	.000101	38.5197	.2730
.000084	40.2163	.0009	.000100	40.2219	.2688
.000086	41.9292	.0009	.000099	41.8982	.2580
.000086	43.5343	.0009	.000097	43.5443	.2605
.000080	45.1274	.0011	.000095	45.1803	.2746
.000078	46.8369	.0015	.000094	46.7989	.2587
.000093	48.4591	.0012	.000098	48.3976	.2421
.000091	49.1406	.0012	.000098	49.0961	.2317
.000088	50.6060	.0011	.000091	50.6455	.2414
.000080	52.1006	.0009	.000086	52.1533	.2830
.000070	53.5893	.0009	.000078	53.6511	.3563
.000058	55.0749	.0010	.000067	55.1647	.4700
.000045	56.5611	.0010	.000049	56.7771	.6486
.000033	58.0507	.0011			
.000025	59.5473	.0012			
.000019	61.0623	.0013			
.000010	64.5787	.0015			
.000010	66.1413	.0015			
.000003	69.7559	.0020			
.000014	71.9615	.0024			
.000630	88.2939	.2809			

Conclusions

When reconstructing an unknown spectrum or other positive distribution, theory indicates that one should use the direct DIMES method. This produces the "best" possible reconstruction, where "best" is defined in terms of practical properties which the reconstructions ought to possess. DIMES gives spectra which are necessarily finite and in which the structure is conservatively broad (while still fitting the data).

By contrast, MESA spectra can be misleadingly sharp, and can also be sensitive to the position of the boundary of the observing window. These effects are more severe in two and three dimensions. They occur because MESA maximises the Burg entropy, which is influenced more by the background than by the major spectral structure. MESA reconstructions always tend to infinity except where expressly prohibited by the type of data. In particular cases where the structure is sharp and over-determined by the data, MESA can out-perform DIMES. However, it is then likely to be itself out-performed by algorithms such as MUSIC.

As the theoretical derivation shows, a MESA spectrum is correctly used as a convenient display of the variance of an underlying MaxEnt probability distribution. It should not be used or presented as an optimal spectrum in its own right. At this meeting, whose purpose is to celebrate and acknowledge our debt to Edwin Jaynes, it seems appropriate to end with a quotation (Jaynes 1982) addressed to this very question: "We stress again; a method that is optimal in one class of problems can be dangerously misleading in another."

Acknowledgments

The MUSIC programs were written and run by N. Fenn, a Cambridge undergraduate student who also helped to develop the MESA calculations and the DIMES/MESA comparisons for the Fougere simulation.

References

Burg, J.P. (1967). Maximum entropy spectral analysis. Proc. 37th meeting Soc. Exploration Geophys., Oklahoma City.

Burg, J.P. (1972). The relationship between maximum entropy spectra and maximum likelihood spectra. Geophysics, 38, 375-376.

Fougere, P.F. (1977). A solution to the problem of spontaneous line splitting in maximum entropy power spectrum analysis. J. Geophys. Res., 82, 1051-1054.

Fougere, P.F. (1986). On the extreme accuracy of maximum entropy spectrum estimation from an error-free autocorrelation function. Presented at 1986 maximum entropy conference, Seattle, Washington

(this volume). Also (1987) IEEE-ASSP in press.
Fougere, P.F., Zawalick, E.J. & Radoski, H.R. (1976). Spontaneous line splitting in maximum entropy power spectrum analysis. Phys. Earth and Planet. Interiors,12, 201-207.
Gull, S.F. & Skilling, J. (1984a). Maximum entropy method in image processing. IEE Proc.,131(F), 646-659.
Gull, S.F. & Skilling, J. (1984b). The maximum entropy method. In Indirect imaging, ed. J.A. Roberts. Cambridge: Cambridge University Press.
Hogbom, J.A. (1974). Aperture synthesis with a non-regular distribution of interferometer baselines. Astron. Astrophys. Suppl.,15, 417-426.
Jaynes, E.T. (1978). Where do we stand on maximum entropy? Reprinted in E.T. Jaynes: Papers on Probability, Statistics and Statistical Physics, ed. R. Rosenkrantz, 1983 Dordrecht: Reidel.
Jaynes, E.T. (1979). Concentration of distributions at entropy maxima. Reprinted in E.T. Jaynes: Papers on Probability, Statistics and Statistical Physics, ed. R. Rosenkrantz, 1983 Dordrecht: Reidel.
Jaynes, E.T. (1982). On the rationale of maximum entropy methods. IEEE Proc.,70, 939-952.
Livesey, A.K. & Skilling, J. (1985). Maximum entropy theory Acta Cryst.,A41, 113-122.
Navaza, J. (1985). Maximum entropy estimate of the electron density function. Acta Cryst.,A41, 232-244.
Nityananda, R. & Narayan, R. (1982). Maximum entropy image reconstruction - a practical non-information-theoretic approach. J. Astrophys. Astron.,3, 419-450.
Press, W.H., Flannery, B.P., Teukolsky, S.A. & Vetterling, W.T. (1986). Numerical Recipes. Cambridge: Cambridge University Press.
Schmidt, R.O. (1986). Multiple emitter location and signal parameter estimation. IEEE Trans. Antennas and Propagation,34, 281-290.
Shore, J.E. & Johnson, R.W. (1980). Axiomatic derivation of the principle of maximum entropy and the principle of minimum cross-entropy. IEEE Trans.Info.Theory, IT-26, 26-39 and IT-29, 942-943.
Skilling, J. (1986). The Axioms of Maximum Entropy. Presented at 1986 Maximum Entropy conference, Seattle, Washington (this volume).
Skilling, J. & Gull, S.F. (1984). The entropy of an image. SIAM Amer. Math. Soc. proc. Appl. Math.,14, 167-189
Tan, S.M. (1986). An analysis of the properties of CLEAN and Smoothness Stabilised CLEAN - some warnings. Monthly Notices Royal astr. Soc.,220, 971-1001.
Ulrych, T.J. & Bishop, T.N. (1975). Maximum entropy spectral analysis and autoregressive decomposition. Rev. Geophys. Space Phys.,13, 183-200.

INFORMATION AND ENTROPY OF PATTERNS IN GENETIC SWITCHS

Thomas Dana Schneider
Department of Molecular, Cellular and Developmental Biology
University of Colorado, Boulder, Colorado 80309

Abstract. To turn genes on or off, cells use special molecules that recognize patterns in the genetic material. These patterns contain about as much information as one would predict from information theory. During evolution, mutations tend to destroy genetic patterns in the same way that thermal motion tends to destroy patterns in inanimate materials. I am suggesting here that the destructive tendency for entropy to increase in isolated systems may also apply to the genetic material.

Introduction

We will look at how the molecular machines in living things recognize patterns in the genetic material. We ask: how much information is contained in these patterns and what determines the amount?

You should be familiar with Shannon's uncertainty measure, (Shannon 1948; Pierce 1980) but you won't need to know much molecular biology. If you would like more background on molecular biology, I recommend the book by Watson *et al.* (1987). The original, more comprehensive paper is by Schneider *et al.* (1986).

Binding Sites and Recognizers

Living cells use a chemical code to store information about how to build the protein structural components of the cell and the enzymes that catalyze chemical reactions. The chemical, deoxyribonucleic acid (DNA), is a long polymer with 4 kinds of "bases" strung together by sugar and phosphate groups. For this paper the chemical structure of bases is not at all important, so they will just be called 'a', 'c', 'g' and 't'. In the cell, the DNA is usually made of two strands twisted around each other so that every 'a' on one strand is paired to a 't' on the other strand, and every 'g' on one strand is paired to a 'c' on the other. The strands have polarity and are read in opposite directions.

G. J. Erickson and C. R. Smith (eds.),
Maximum-Entropy and Bayesian Methods in Science and Engineering (Vol. 2), 147–154.

Long before the chemical nature of the genetic material was determined, geneticists discovered that there are distinct regions - genes - that have specific functions. In bacteria, genes are about 1000 base pairs long. (In humans and other higher organisms they can cover many thousands of base pairs.) Only by looking at the *pattern* of the bases can the cell machinery tell where the starting and stopping points are for reading each gene. These patterns are called "binding sites" because they are the place that special molecules, called "recognizers", can bind.

There are many known recognizers, and they do many things in cells. For example, a molecule called RNA polymerase binds to a pattern called the "promoter" near the starting point of genes. The RNA polymerase then moves along the DNA and copies it by polymerizing bases together to form a strand of RNA. Likewise, the ribosome binds to a "ribosome binding site" on the RNA, and then moves along the RNA while translating the RNA into protein. These two processes, called the "Central Dogma" of molecular biology, are performed in all cells.

A simple two-state molecular switch is formed when a protein, called a repressor, binds to the same region as a promoter, at a place called an operator. For example, the bacterium *Escherichia coli* has an enzyme for digesting the sugar lactose. Normally this protein is not synthesized because a repressor binds to the promoter. When there is lactose in the surrounding water, some enters the cell and binds to the repressor. This changes the shape of the repressor so that it no longer binds to the operator. The polymerase can then make the RNA for the digestive enzyme.

What determines the patterns of binding sites? Can we find laws to describe them?

The information content of binding sites

Figure 1 is an example of some binding site sequences. The numbers at the top give the coordinates of each base. Each horizontal line of letters represents one strand of the DNA, reading from left to right. If you look closely at all 14 sequences, you will see this pattern starting at position -7: ctgnnnnnnnnnncag (n means anything).

The protein that binds to these sequences is called LexA. When the DNA in an *E. coli* cell is damaged by chemicals or radiation, LexA, which normally sits at these sequences, is destroyed. This turns on a set of genes that repair

Figure 1: LexA binding sites

```
     -   -                   +   +
     1   1   -           +   1   1
     4   0   5   0   5   0   5
     .............................
 1   cttgatactgtatgagcatacagtataatt
 2   aattatactgtatgctcatacagtatcaag
 3   ccttttgctgtatatactcacagcataact
 4   agttatgctgtgagtatatacagcaaaagg
 5   agcataactgtatatacacccagggggcgg
 6   ccgccccctgggtgtatatacagttatgct
 7   tccaatactgtatattcattcaggtcaatt
 8   aattgacctgaatgaatatacagtattgga
 9   tgatgaactgtttttttatccagtataatt
10   aattatactggataaaaaaacagttcatca
11   ggatgtactgtacatccatacagtaactca
12   tgagttactgtatggatgtacagtacatcc
13   aatcatactgtgtatatatacagtattttg
14   caaaatactgtatatatacacagtatgatt
```

the damage. (The LexA system is another example of a molecular switch.) LexA is composed of two copies of a protein chain. The copies are identical and stick together strongly, like a yin-yang symbol. LexA is rotationally symmetric, and the pattern to which it binds is also symmetric.

To understand this, we write down the basic pattern we saw above, along with the other strand of DNA (recall that 'c' pairs with 'g' and 'a' with 't'):

```
ctgnnnnnnnnnncag   (reading left to right, the sequence)
gacnnnnnnnnnngtc   (reading right to left, the sequence's complement)
```

They are the same pattern because the two strands of DNA are always read in opposite directions (both by convention and by molecular machinery). The list of sequences in Figure 1 alternates between sequences and their complements, all written left to right. (For example, base 15 of sequence 1, a 't', is the complement of base -14 of sequence 2, an 'a'.) There are only 7 LexA binding sites represented in this example.

In position -5, there is always a 'g' in the pattern. So our *uncertainty* about what base to expect in another LexA binding site is low. Indeed, Shannon's uncertainty for this position is 0 bits. (A 'bit' is the information needed to choose between two equally likely states.) Position -4 gives us more uncertainty while positions outside the site have almost 2 bits of uncertainty because there are 4 possibilities. We can write this down as:

$$Hs(L) = -\sum_{B=a}^{t} f(B,L)\log_2 f(B,L) \quad (\textit{bits per base}), \tag{1}$$

where L is the position in the site, $f(B,L)$ is the frequency of base B at position L and the sum is taken over all bases (B is either 'a', 'c', 'g' or 't'). $Hs(L)$ is the uncertainty of what base to expect in another LexA binding site.

When LexA is searching the *E. coli* DNA for a binding site, it encounters all four bases at about the same frequency. That is, if we messed up the alignment of the sequences, we would have 2 bits of uncertainty about what base is at each position. As LexA finds sites (or, equivalently, as we bring the sequences into alignment) the uncertainty drops for each position. This difference is a measure of the information contained in the pattern (Tribus & McIrvine 1971):

$$Rsequence(L) = 2 - Hs(L). \quad (\textit{bits per base}). \tag{2}$$

(See Schneider *et al.* 1986, for an alternative formula.)

The curve for these differences, running across the site, is shown in Figure 2. On the bottom is a table of the number of bases at each position in the site. (The curve includes a correction for the small number of sample sequences. Shannon's formula is for probabilities, but we only have frequency data to work with, and this requires that we make a correction. See the appendix of Schneider *et al.* 1986 for more details.) The curve shows that the center of the sites do contain information of varying amounts at different positions. Outside the binding site the curve goes to zero.

By summing across this curve, we obtain a measure of the total information in the site. This is called Rsequence. For LexA, we get 21 bits per site.

How much information is needed to find binding sites?

If the genome (ie., the genetic material of an organism) has G positions at which a recognizer can bind and all the positions are equally available for inspection, then to pick one position out requires $\log_2 G$ bits of information.

If there are two positions to be picked out, it won't matter which one the recognizer binds to because once it has found one it may begin to do its job

Figure 2: Information curve of LexA binding sites

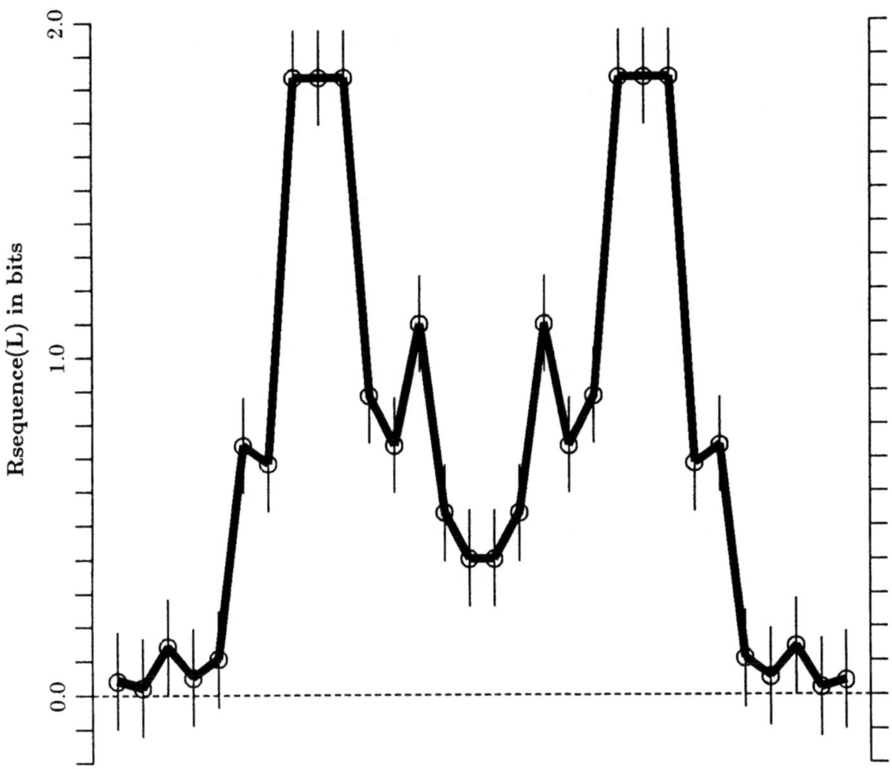

L	a	c	g	t	Rs(L)
-14	6	4	1	3	0.04
-13	5	3	5	1	0.02
-12	4	2	1	7	0.14
-11	3	2	2	7	0.05
-10	7	1	3	3	0.11
-9	3	1	0	10	0.74
-8	10	2	2	0	0.68
-7	0	14	0	0	1.83
-6	0	0	0	14	1.83
-5	0	0	14	0	1.83
-4	1	0	2	11	0.89
-3	10	0	3	1	0.74
-2	1	1	0	12	1.10
-1	7	0	6	1	0.54
0	3	1	1	9	0.40
1	9	1	1	3	0.40
2	1	6	0	7	0.54
3	12	0	1	1	1.10
4	1	3	0	10	0.74
5	11	2	0	1	0.89
6	0	14	0	0	1.83
7	14	0	0	0	1.83
8	0	0	14	0	1.83
9	0	2	2	10	0.68
10	10	0	1	3	0.74
11	3	3	1	7	0.11
12	7	2	2	3	0.05
13	7	1	2	4	0.14
14	1	5	3	5	0.02
15	3	1	4	6	0.04

there. In this case the recognizer does not need to know which site it is at, so one bit of uncertainty can remain after it has bound. In general, if there are γ sites, the uncertainty remaining after the sites are located is $\log_2\gamma$. (The full formula for H may be used if the sites vary in their binding capacity.)

The minimum information needed to find the set of sites in the genome is therefore:

$$Rfrequency = \log_2 G - \log_2 \gamma = -\log_2 \frac{\gamma}{G} = -\log_2 f \quad (bits\ per\ site) \tag{3}$$

where f is the frequency of sites in the genome.

Rsequence and Rfrequency are probably related

We can find out the size of the genome and the number of sites for certain recognizers, which means we can estimate Rfrequency. Because we can also determine the sequences of binding sites, we can estimate Rsequence. These measures are independent of each other, so we can ask how they are related. This is shown in Table 1.

Table 1: Two measures of binding site information (in bits per site).

Organism	Recognizer	Rsequence	Rfrequency	$\frac{Rsequence}{Rfrequency}$
E. coli	Ribosome	11.0	10.6	1.0
E. coli	LexA	21.1	18.4	1.1
E. coli	TrpR	23.4	20.3	1.1
E. coli	LacI	19.2	21.9	0.9
E. coli	ArgR	16.4	18.4	0.9
E. coli	AraC	19.3	19.3	1.0
λ	cI/Cro	17.1	19.3	0.9
λ	O (origin)	20.9	19.9	1.0
T7	RNA Polymerase	35.4	16.5	2.1
T7	Symmetry	16.4	17.8	0.9

The ratio of Rsequence to Rfrequency is close to 1 in the first 8 cases listed in the table. This suggests that the amount of information in the binding site patterns is generally the amount one would expect given the size of the genome and the number of sites.

Why then is there so much extra information in the T7 RNA polymerase binding sites? *Two* different recognizers could bind to the same place but not share information. This explains why the ratio is close to 2. Evidence for this hypothesis is a symmetrical pattern in these sites that has as much information as one would expect (last line of the table). This may be the binding site of another protein, possibly to form a molecular switch like LexA. This has not been proven experimentally.

Entropy Increase and the Evolution of Binding Sites

Shannon named his uncertainty measure H because it is almost the same formula as Boltzmann's formula for entropy. The only difference is the multiplicative constant that determines the units of measure. The Second Law of thermodynamics states that H will tend to increase to a maximum in a thermally isolated system. Because living things use solar energy and dump their waste heat into space, they are not part of an isolated system and one should not expect the entropy of their components to increase to maxima.

The formula for Rsequence used in this paper,

$$Rsequence = \sum_{L} \left(2 - Hs(L) \right) \quad (bits\ per\ site). \tag{4}$$

contains an uncertainty that depends on the sequences at binding sites. Mutations change the bases of the DNA, and if these are not deleterious they are passed on to the progeny. Thus mutation tends to increase the uncertainty of the pattern, $Hs(L)$. If the uncertainty increases too much, then Rsequence will become smaller than Rfrequency. Perhaps at this point the genetic control system will fail to work correctly because the binding sites don't have enough information to be found. These organisms would be less effective in genetic control than others with the correct amount of information, so during evolution selection would have eliminated them from the samples we can look at today.

The net effect is that the entropy of the binding sites increases to a maximum determined by the requirement for the system to function. The second law may indeed apply here because there are only 4 possible bases in DNA. Mutations are simply exchanges between these "states". Clearly it is not advantageous for a strong bias to exist in the direction of mutations. (Although the DNA of many organisms is not composed of equally probable bases, the reason is not known.) From this we might argue that each base position in a set of sites is reasonably well isolated from the cellular energy flux. To the degree that this is true, the nucleotide "states" will tend to spread out. A similar situation for thermally isolated quantum systems leads directly to the second law (Waldram 1985).

Darwinian variation, now known in part as mutation, has the same effect as thermal motion does in common materials such as two miscible liquids. Both tend to increase the entropy of the material. One difference is that liquids mix quickly compared to the slow accumulation of mutations over millions of years. But only the direction of time, not the amount of time, is relevant to the second law.

Thus the entropy of patterns in genetic material tends to increase in the same way that the entropy of isolated physical systems tends to increase in accordance with the second law of thermodynamics.

ACKNOWLEDGEMENT

This work was supported by NIH grant GM28755.

REFERENCES

Pierce, J. R. (1980). An Introduction to Information Theory: Symbols, Signals and Noise, Second Edition. Dover Publications Inc., New York.

Schneider, T. D., Stormo, G. D., Gold, L., & Ehrenfeucht, A. (1986). Information content of binding sites on nucleotide sequences. J. Mol. Biol., vol. 188, pp. 415-431.

Shannon, C. E., (1948). A Mathematical Theory of Communication. Bell System Tech. J., vol. 27, pp. 379-423, 623-656.

Tribus, M. & McIrvine, E. C. (September 1971). Energy and Information. Sci. Am., vol. 225, pp. 179-188.

Waldram, J. R. (1985). The theory of thermodynamics. Cambridge University Press, Cambridge.

Watson, J. D., Hopkins, N. H., Roberts, J. W., Steitz, J. A., & Weiner, A. M. (1987). Molecular Biology of the Gene, Fourth Edition. The Benjamin/Cummings Publishing Co., Inc., Menlo Park, California.

MAXIMUM ENTROPY AND THE PHASE PROBLEM IN PROTEIN CRYSTALLOGRAPHY.

R. K. Bryan
European Molecular Biology Laboratory, Meyerhofstrasse 1,
6900 Heidelberg, West Germany.

INTRODUCTION.

X-ray diffraction from a crystal enables the intensity of the Fourier transform of the scattering electron density to be measured. From these data we wish to deduce the atomic coordinates of the crystallised molecule. A crystal is an object which is translationally periodic, which allows a unit cell to be defined by three non-coplanar vectors $\mathbf{a}_1$, $\mathbf{a}_2$, $\mathbf{a}_3$, such that the density is the same after a translation by any of these vectors. Using the translation invariance, the Fourier transform $F(\mathbf{k}) = \int \rho(\mathbf{r}) \exp(2\pi i \mathbf{r}.\mathbf{k}) d^3r$ of the electron density ρ is then non-zero only at points $\mathbf{k}$ such that $h_i \equiv \mathbf{k}.\mathbf{a}_i$ are integers for each i. The vectors $\{\mathbf{a}'_i\}$ reciprocal to $\{\mathbf{a}_i\}$, so that $\mathbf{a}'_i.\mathbf{a}_j = \delta_{ij}$, form a basis for the space of $\mathbf{k}$, usually termed reciprocal space, with $\mathbf{k} = \sum_i h_i \mathbf{a}'_i$. The points given by integral h_i are called the reciprocal lattice. It is generally convenient to use fractional cell coordinates $\mathbf{x}$, so that $\mathbf{r} = \sum_i x_i \mathbf{a}_i$, and hence $\mathbf{r}.\mathbf{k} = \mathbf{x}.\mathbf{h}$.

For normal scattering, ρ is real, and $F_\mathbf{h}$ has complex conjugate (or *Friedel*) symmetry, $F_{-\mathbf{h}} = F^*_\mathbf{h}$. It is also possible for a crystal to have symmetries additional to translation, such as rotation axes, screw axes, mirror reflections, *etc.*, which in various combinations form the 230 possible space groups. Each symmetry leads to a corresponding symmetry in reciprocal space (Bienenstock & Ewald, 1962), and so gives no further independent information. Since only the diffraction intensity can be measured, and not the phase, it is not possible to calculate the electron density directly by an inverse transform. If the intensity of a continuous transform is known, is in principle possible, but rarely practicable, to deduce the phases by analytic means alone. However, crystallinity means that there is no redundancy within the data and such methods are totally inapplicable. On the other hand, it does enable the diffraction intensity at each reciprocal lattice point to be measured by a non-ideal instrument without corruption by the values at nearby points in reciprocal space. The phase problem is thus central to X-ray structural analysis.

It will be useful here to give some idea of the orders of magnitude involved. Bond lengths between atoms are about 1.5Å (1Å $= 10^{-10}$m), and proteins contain upwards of 200 atoms. Conventionally, CuK_α radiation, wavelength 1.54Å, is used, so that data to a spacing of 0.77Å could be collected, and thus enable individual atoms to be resolved. Usually, the resolution limit for proteins is determined by the degree of order within the crystal and by thermal vibration, so useful data can often be collected only to 2Å or worse. Thus even if the data could be accurately phased, we should have to apply some knowledge of molecular structure to the resulting electron density in order to deduce accurate atomic coordinates.

In the rest of the introduction, the principles of protein structure and the classical methods of phase determination will be outlined. The limitations of these methods enable us to identify areas where other methods, in particular maximum entropy, may prove useful in structure determination.

G. J. Erickson and C. R. Smith (eds.),
Maximum-Entropy and Bayesian Methods in Science and Engineering (Vol. 2), 155–169.

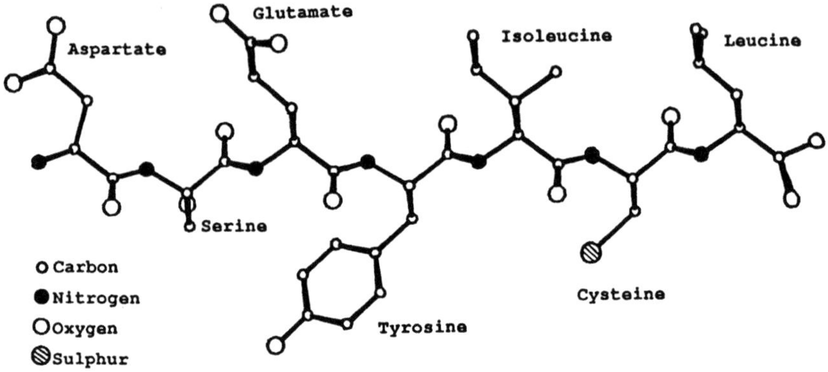

Figure 1. A β-strand in a 'ball and stick' representation, comprised of seven amino-acids, with various sidechains, which in this conformation protrude on alternate sides of the main chain. Hydrogen atoms omitted.

Protein structure.

Since about 200 distinct protein structures have been solved, much is known about their typical configurations (Blundell & Johnson, 1976). The basic building block is an amino acid, which has a main chain of two carbon atoms and one nitrogen, and a sidechain, which may contain up to 10 additional non-hydrogen atoms, branching off at the central (C_α) carbon. The four different bonds (2 main chain, sidechain and a hydrogen) to the C_α mean that it is chiral. All natural amino acids have the hand illustrated in figure 1. There are 20 different sidechains found in natural proteins, all consisting of atoms of hydrogen, carbon, nitrogen, oxygen or, in two cases, sulphur. A number of amino acids join together in a chain to give a protein. The sequence of amino acids, specified in the organism's gene which codes for that particular protein, is known as the primary structure. The various sidechains have different properties of electrical charge, hydrophobicity *etc*, and the interactions between them and with the aequeous environment determine how the chain folds to form the protein, although exactly how is unfortunately not yet understood sufficiently well for us to predict the three-dimensional structure of a protein from the primary structure.

The folding of an amino acid chain is hierarchically. At the first level, 'secondary structures' are formed, of which the α-helix (figure 2) and β-strand are the most common. The α-helix is stabilised by hydrogen bonds between successive turns. Such secondary structure elements are linked together by loops or turns, and pack together to form the 'tertiary' structure, which is stabilised by further hydrogen bonding between adjacent sidechains. Typically, a globular protein will consist mostly of a fairly rigid α, β, or mixed α-β, structure supporting an 'active site', perhaps of only a few amino acids, which interacts with other molecules. In larger proteins, several such domains, sometimes identical, may pack together to form the final product.

With this knowledge of the structure of proteins, it is not necessary to obtain an electron density map to atomic resolution in order to build a model. The main chain, particularly in helices, usually has lower thermal vibration than the sidèchains, and can be seen as a tube of density at about 4Å resolution. The degree of order in sidechains varies, with those protruding into solvent often very disordered and never visible at any resolution, although generally they can be seen at around 3Å resolution. The close atomic numbers of carbon, nitrogen and oxygen, means that they cannot be distinguished by their density alone, which complicates the identification of sidechains, and indeed often makes it difficult to see which way the backbone runs, whereas hydrogen atoms have insufficient density to be seen at all in this resolution range.

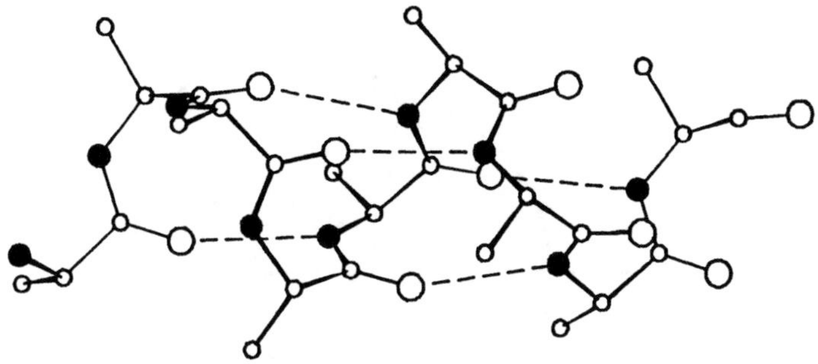

Figure 2. A length of α-helix, with sidechains omitted, illustrating the hydrogen bonds (dashed lines) between carboxyl groups and the NH group four amino acids along, which stabilise the structure.

Approaches to the phase problem.

Several methods are in use for deducing atomic structures from Fourier intensity data. For small molecules, less than 100 atoms, with data of sufficiently high resolution, 'direct' methods are the norm. It is assumed that the structure consists of equal, resolved atoms, and hence that the square of the electron density is proportional to the electron density itself. The Fourier transform of this expression gives relationships amongst the Fourier coefficients, and by assigning trial phases to a few strong reflections, the others may be estimated. Most implementations generate several different possible solutions, from which the user must select those agreeing with his ideas on the chemical structure.

The assumptions made for the derivation of direct methods are usually inapplicable for proteins. Additionally, the polymeric form means that the same structural motifs are repeated throughout the molecule, thus increasing the scope for ambiguous solutions. Except for those cases when there has been some initial information, for instance, the known structure of a similar protein, virtually all large molecules have been solved by the method of multiple isomorphous replacement (MIR). An isomorphous heavy atom derivative of a protein is one in which one or more atoms of large atomic number are added to each native molecule, without significantly changing the structure. This is often a considerable achievement in protein chemistry! If diffraction data are collected for the native and one or more derivatives, they can be used to calculate the phases in the following way. Let

$$F_{\mathbf{h}} = |F_{\mathbf{h}}| \exp(i\phi_{\mathbf{h}}) = \int_{\mathbf{x}\in \text{unitcell}} \rho(\mathbf{x}) \exp(2\pi i \mathbf{x}.\mathbf{h}) d^3x \tag{1}$$

be the Fourier coefficients of the electron density due to the protein alone, a superscript D denote those for the derivative, and $H_{\mathbf{h}}$ be the heavy atom Fourier coefficients, so ${F_{\mathbf{h}}}^D = {F_{\mathbf{h}}}^P + H_{\mathbf{h}}$. Taking the component in the direction of ${F_{\mathbf{h}}}^D$ gives

$$|{F_{\mathbf{h}}}^D| = |H_{\mathbf{h}}| \cos(\phi_{\mathbf{h}}^D - \phi_{\mathbf{h}}^H) + |F_{\mathbf{h}}| \cos(\phi_{\mathbf{h}} - \phi_{\mathbf{h}}^D). \tag{2}$$

If the heavy atom contribution is small, $\phi_{\mathbf{h}}^D \approx \phi_{\mathbf{h}}$, so

$$1 - \cos(\phi_{\mathbf{h}} - \phi_{\mathbf{h}}^D) \approx \frac{1}{2} \frac{|H_{\mathbf{h}}|^2 \sin^2(\phi_{\mathbf{h}}^D - \phi_{\mathbf{h}}^H)}{|{F_{\mathbf{h}}}^D|^2}, \tag{3}$$

and hence to lowest order

$$|{F_{\mathbf{h}}}^D| - |F_{\mathbf{h}}| \approx |H_{\mathbf{h}}| \cos(\phi_{\mathbf{h}}^D - \phi_{\mathbf{h}}^H). \tag{4}$$

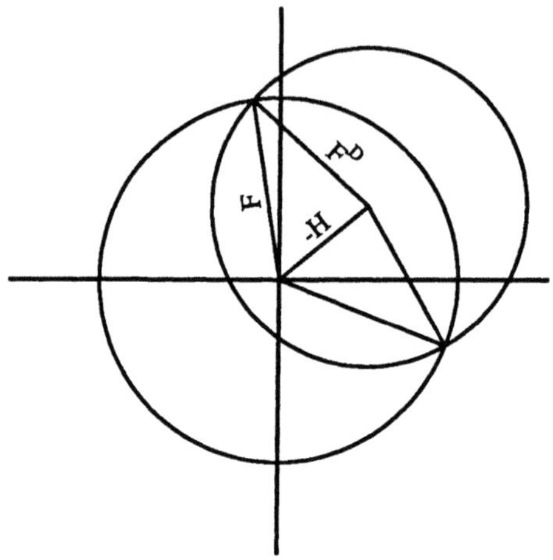

Figure 3. The Harker construction in the Argand diagram, illustrating the calculation of native phases, given native and derivative amplitudes and the heavy atom Fourier coefficients. A circle of radius $|F_\mathbf{h}|$ centred on the origin intersects one of radius $|F_\mathbf{h}{}^D|$ centred on $-H_\mathbf{h}$ at the two possible phase solutions.

The experimental data consist of measurements $|F_\mathbf{h}|^2$ and $|F_\mathbf{h}{}^D|^2$, which are conventionally used to compute the inverse Fourier transform of $||F_\mathbf{h}{}^D| - |F_\mathbf{h}||^2$, (the 'difference Patterson'), as an unfortunately rather poor approximation to the autocorrelation function of the heavy atom density. This map is corrupted by protein-heavy atom cross vectors, and is usually noisy since it is calculated from the differences of experimental quantities. Despite this, if the number of heavy atom sites is small, their approximate relative positions can often be deduced from this map. In some space groups, this will also fix their positions relative to crystal symmetry elements, otherwise, if more than one independent derivative is used, the relative origins must also be established, by, for instance, a difference Patterson between the various derivative data sets.

The transform of the located heavy atoms is then calculated, and used to phase the native amplitudes, illustrated by the Harker construction (Harker, 1956), figure 3, so that

$$\cos(\phi_\mathbf{h} - \phi_\mathbf{h}^H) = \left(|F_\mathbf{h}{}^D|^2 - |F_\mathbf{h}|^2 - |H_\mathbf{h}|^2\right) \Big/ \left(2|F_\mathbf{h}||H_\mathbf{h}|\right). \tag{5}$$

A single derivative clearly gives a 2-fold ambiguity in phase for each amplitude, which a second independent derivative is needed to resolve. The only exception to this is for those reflections, termed centrics, which must have a real phase due to space group symmetry, and hence can by phased by a single derivative. Not all reflections from a protein crystal can be centric, because amino acids are chiral, hence the protein cannot have a centre of symmetry, although the projection of the structure in certain directions (*e.g.,* down a 2-fold axis) may do. In practice, due to noisy data and uncertainties and errors in locating the heavy atoms, the phases are still not precisely defined, even with more than the theoretical minimum of two derivatives, and a probabalistic calculation is used to find the 'best' phase (Blow & Crick, 1959). It is assumed that the measured native amplitude is correct, that the probability of the phase being ϕ is proportional to

$$P(\phi) = \exp\left(-\frac{1}{2}w\left(|F_\mathbf{h}{}^D| - \big|\,|F_\mathbf{h}|\exp(i\phi) + H_\mathbf{h}\,\big|\right)^2\right), \tag{6}$$

and that the expectation value of $|F_{\mathbf{h}}|\exp(i\phi)$ over this distribution is an appropriate estimate of the Fourier coefficient. This expectation value, which may be written as $m|F_{\mathbf{h}}|\exp(i\phi_B)$, is known by crystallographers as the 'best' estimate, and m (always less than 1), as the 'figure of merit'. If the phasing of a Fourier coefficient is imprecise, m is small, so the measured value is effectively weighted down. Should there be data on more than one derivative, the net phase probability distribution is taken to be the product of those for each derivative. The electron density is then calculated by a straightforward Fourier transform of the phased amplitudes. Although the resultant map is best in a 'least squares' sense, one can criticise several aspects of this procedure:- there is truncation error, due to the (inevitable) finite resolution of the data, giving ripples on the map and areas of physically impossible negative density; the weighting scheme means that the amplitudes of the transform of the map are not the measured amplitudes, although at higher resolution the measured intensities tend to be smaller, and hence have a larger error and a lower weighting thus reducing the truncation ripple at the expense of resolution; and the native data often extends to higher resolution than the derivatives, so cannot be phased by this method.

Typically, a structure may be solved by obtaining isomorphous derivatives which give good phases to 2.5–3Å, and building an approximate atomic model into the resulting density, using idealised bond lengths and angles where appropriate. This model is then 'refined' against the native data alone, at the same time imposing geometrical constraints on the bonds. The key is still obtaining isomorphous derivatives to this resolution, which is not always easy, as the introduction of heavy atoms often disturbs the molecule, either at the sidechain level, so that isomorphism is only retained to 5–6Å, or such that the packing between molecules is affected, and the unit cell dimensions change. Delay in many protein structure determinations is frequently due to the difficulty of finding at least two derivatives which are sufficiently isomorphous to the native at 3Å resolution. Nevertheless, the majority of structural solutions have been achieved by these methods.

The most important variation on this method is the exploitation of anomalous scattering near the absorption edge of inner-shell electrons in heavy atoms. The resultant phase shifts cause an imaginary component of scattering to appear, and so the Friedel symmetry is lost. The differences in intensity between the $\mathbf{h}$ and $-\mathbf{h}$ reflections (usually termed a Bijvoet pair) can be measured, and exploited in a similar way to isomorphous differences, although yielding information on $\sin(\phi - \phi^H)$ rather than $\cos(\phi - \phi^H)$. Thus, in principle, if the anomalous differences of a single isomorphous derivative are measured, there is sufficient information to calculate the phases. However, since the anomalous scattering component is typically around 10% of the normal part, the anomalous differences are comparable with the noise level on the intensity measurements, and the precision of the phase information is much less than that of isomorphous differences. On the other hand, the anomalous differences are free from any additional problems of non-isomorphism. One of the finest exploitations of this effect has been the solution of the structure of Crambin (Hendrickson & Teeter, 1981), a small protein of 46 amino acids, but containing 6 cysteines, each with one sulphur atom in its sidechain. The anomalous scattering from the sulphur atoms, about 1.4% of the total, was sufficient for their positions to be found, and hence, using also the normal part of the sulphur scattering, the protein phases to be deduced. The use of tuneable synchrotron radiation is likely to enable more structures to be solved in this manner, as data can be collected near to and far from the absorption edge of a particular atomic species, and thus allow one effectively to turn on and off anomalous scatterers whilst preserving perfect isomorphism. Another approach has been to combine isomorphous replacement and direct methods (Hauptman, 1982), which, like all direct methods, requires the assumption of atomicity, and will possibly lose power at lower resolution.

In fortuitous cases, a protein may crystallise in such a way that some of the symmetries are purely local, and not crystal space-group symmetries. This must happen if the structure has an internal symmetry which is not a possible space group symmetry, most notably in the case of icosahedral viruses, whose coat proteins have 60-fold symmetry (Harrison *et al.*, 1978, Hogle *et al.*, 1985, Rossmann *et al.*, 1985). The additional symmetry means that there is considerable redundancy in the intensity data, which may be exploited if the local symmetry elements can be determined and positioned in the crystal unit cell. Most successful applications of this method have, however, needed some isomorphous derivatives to position the molecule and to initiate the phasing, although recent results have been achieved with only one low resolution derivative.

The usual method of phase determination is seen to rely on the collection of sufficient isomorphous derivative data, without in any way using the extensive knowledge of protein structure that we have. The effort required is thus several times that required to make crystals and collect diffraction data for the native protein alone. Any computational method which can aid the determination of phases when complete derivative data is lacking would therefore be of great advantage. For this reason, much interest has been shown in recent years in applying the maximum entropy method to this problem.

MAXIMUM ENTROPY THEORY.

Several authors have described the theory of the maximum entropy method as applied to the reconstruction of maps from incomplete and noise data, with conceptually clean accounts being given by Gull & Daniell (1978), Gull & Skilling (1984), and in a more specifically crystallographic context, by Wilkins *et al.* (1983), Bricogne (1984), and Livesey & Skilling (1985). The latter two papers also discuss the relationship of the maximum entropy formulation that will be used here to some others which have been suggested for use in crystallography. In essence, the maximum entropy method involves selecting the map having the greatest configurational entropy, and hence least configurational information, from the 'feasible set' of maps which fit the experimental data to within the noise limits. This technique has been proved to be the unique extremum principle which does not introduce correlations in the map which are not required by the data (Shore & Johnson, 1980). Selecting the map with least configurational information confers many advantages. For example (Gull & Daniell, 1978), there must be evidence in the data for any structure seen in the reconstruction. Noise is automatically suppressed, as are artifacts, such as sidelobes due to incomplete coverage of reciprocal space by the data.

First, the 'feasible set' of maps is defined as those which are consistent with the observed intensity data. If the differences between the intensities calculated from a given trial map ρ and the observed native intensities can be attributed solely to noise on the data, then we claim that our trial map agrees with the data, and is thus a 'feasible' map. If isomorphous replacement or anomalous difference data are also available, then, assuming that the positions of the heavy atoms have already been established, the calculated derivative intensities can be similarly compared with the observations. It is not necessary to go via the intermediate step of explicitly calculating the phase probability distribution for each Fourier coefficient. The usual comparison measure, already used for many types of data, is the logarithm of the likelihood, giving the χ^2 test (Abels, 1974, Gull & Daniell, 1978). Assuming uncorrelated noise of known variance, the appropriate form for crystallographic data is

$$\chi^2(\rho; I, I^{D_i}, F^P) = \sum_{\mathbf{h}} \Big\{ \quad w_{\mathbf{h}}\big(|F_{\mathbf{h}}|^2 - I_{\mathbf{h}}\big)^2 \tag{7a}$$

$$+ \sum_i w_{\mathbf{h}}^{D_i}\big(|F_{\mathbf{h}} + H_{\mathbf{h}}^i|^2 - I_{\mathbf{h}}^{D_i}\big)^2 \tag{7b}$$

$$+ w_{\mathbf{h}}^P |F_{\mathbf{h}} - F_{\mathbf{h}}^P|^2 \quad \Big\}, \tag{7c}$$

where $I_{\mathbf{h}}$ are the observed native intensities, $I_{\mathbf{h}}^{D_i}$ the observed intensities for the ith derivative, weighted by $w_{\mathbf{h}}$ and $w_{\mathbf{h}}^{D_i}$ (usually inverse variances) respectively, the $F_{\mathbf{h}}$ are the Fourier coefficients calculated from a trial map ρ, $H^i_{\mathbf{h}}$ the transform of heavy atom contribution to the ith derivative, and the $F^p_{\mathbf{h}}$ phased data, included to take account of reflections which can be phased reliably by conventional isomorphous replacement, such as centrics when a single isomorphous derivative is used.

The form of constraint function we use can be extended to include data from further derivatives, simply by adding extra terms like (7b). Anomalous differences can be incorporated most easily by treating the $I_{\mathbf{h}}$ and $I_{-\mathbf{h}}$ measurements separately, each being compared with the appropriate protein Fourier coefficient plus heavy atom contribution. Conventionally, crystallographers work with the average and difference of the measurements of a Bijvoet pair, and if the weights of the pair are equal, it is possible to rearrange the equations to give a comparison on the average intensity plus a further contribution on the anomalous difference itself, so that $|F + H|^2 + |H''|^2$ is compared with I^{av}, and $4\Im((F + H)H''^*)$ with ΔI^{anom}. Here H is the Fourier coefficient calculated from the sum of the normal and the real part of the anomalous scattering factors of the heavy atoms, and H'' the corresponding quantity for the imaginary component of the anomalous scattering factors.

As in any form of statistical testing, the χ^2 test enables one to determine whether a trial map ρ is an acceptable fit to the data at a given confidence level. If the total number M of intensity measurements is large, the distribution of the χ^2 statistic can be approximated by a $\mathcal{N}(M, 2M)$ Gaussian, and so a trial map ρ which gives a χ^2 value significantly greater than M can be rejected as incompatible with the data. The feasible set therefore consists of those maps that give $\chi^2 \leq M$. The maximum entropy solution is then the member of this set having the greatest configurational entropy S, where

$$S(\rho) = -\sum_j p_j \log p_j/m_j, \qquad p_j = \rho_j/\sum \rho. \tag{8}$$

$\mathbf{m}$ is the normalised prior map, which is the estimate of the solution before the data are considered, and $\mathbf{p} = \mathbf{m}$ has the global unconstrained entropy maximum. If $\mathbf{m}$ lies within the feasible set, it will be the maximum entropy solution, which should only happen if the data introduce no new information. Usually, $\mathbf{m}$ is taken as the completely unbiased flat map, which will only lie within the feasible set if the data are so noisy that they convey no information whatsoever. Otherwise, since the entropy function is convex, any feasible maximum entropy map will lie on the surface of the feasible set.

If constraint (7c) alone is used, the map and data are linearly related, the feasible set of maps defined by the χ^2 test is an ellipsoidal cylinder in the space of all maps, and there is a unique entropy maximum. The topology of the feasible set if intensity constraints are also used is, however, more complicated. Native intensity data constrain the corresponding Fourier coefficient to lie within an annulus in the complex plane. If native and derivative intensity constraints are both used, there may be disconnected regions of high likelihood. Figure 4a shows an example of this. If, however, there is little phase information for a reflection in the SIR data, the χ^2 distribution will remain more nearly circularly symmetric (figure 4b). Thus, depending on the data, a given Fourier coefficient may be constrained to lie within an annulus, in one or two simply connected regions of the complex plane, or be completely unconstrained. The feasible set, if data are provided for M reflections, is the M-dimensional product of such regions, extending to infinity in those directions for which there are no data. The entropy function automatically restricts maps to the positive orthant $\rho_j \geq 0$, and so eliminates many of the disconnected regions, but may also introduce extra topological complications if the orthant edges intersect parts of the feasible set. We may thus expect to find several local entropy maxima when intensity constraints are used, corresponding to the possible phase ambiguities of the problem.

A complete solution to the problem would require that all the possible local entropy maxima be examined in order to find the global maximum. The numerical algorithm used to find

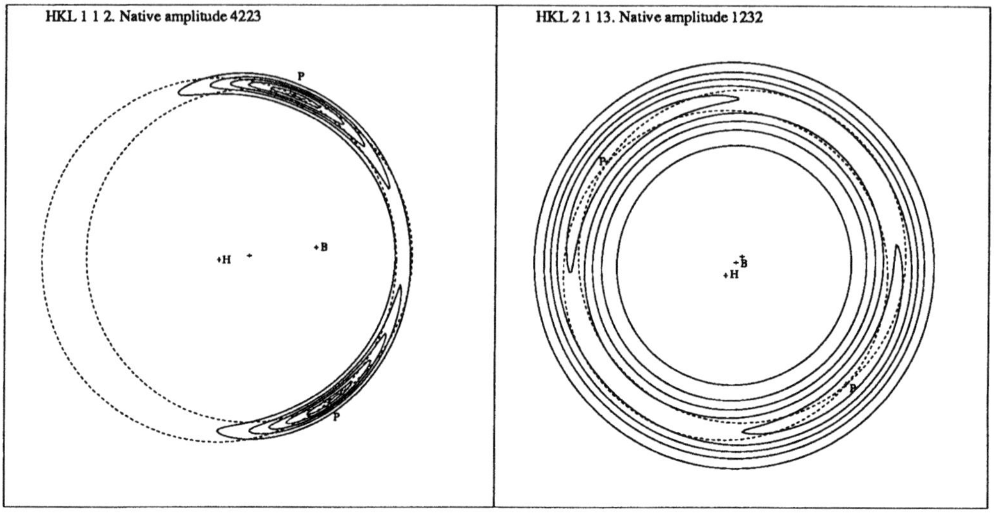

Figure 4. Contour maps in the complex plane of contributions to the likelihood, $\exp(-\frac{1}{2}\chi^2)$, for single reflections. Contour levels are 0.1, 0.3, 0.5, 0.7, 0.9. H denotes the heavy atom, B the 'best', and P the 'most probable' Fourier coefficients. The measured native and derivative amplitudes (dotted circles) intersect at the points P. a) Bimodal phase distribution, with each possible phase fairly precisely defined. b) Almost circularly symmetric phase distribution.

the solutions shown in later sections can reliably locate a local entropy maximum, and has been described in detail in Skilling & Bryan (1984) for the case of convex constraints, with the extension to non-convex constraints in Bryan & Skilling (1986). Despite criticisms of this algorithm (Bricogne, 1984), there is no difficulty in employing it to find local maxima of the non-convex constrained problem, although which of the possible maxima (assuming that there is more than one) is reached will necessarily depend on the starting map.

APPLICATION TO PROTEIN CRYSTALLOGRAPHY.

Several authors (*e.g.* Collins, 1982, Bricogne, 1984, Wei, 1985, Wilkins & Stuart, 1986, Navaza, 1986) have presented work showing the effect of maximum entropy when used with Fourier coefficient data, although with minor variations in the exact formulation of the problem. The resolution of the data used varied from atomic to 4Å, but improvements in map quality were uniformly reported. These calculations are analogous to those of Gull & Daniell (1978), in which the data were Fourier coefficients obtained by radio interferometer observations. The phase problem itself has also been tackled by maximum entropy. Gull & Daniell (1978) showed that if suitable starting phases are provided, a good maximum entropy reconstruction of a 2-dimensional field of point sources can be obtained, and Bryan & Skilling (1986) presented an algorithm that could solve a similar problem without assuming the starting phases, but also showed that data from a diffuse object could give a maximum entropy solution very different from the original object. Livesey (1984) obtained successful reconstructions of a small molecule with atomic resolution data. On the other hand, Navaza (1986) reports that the density calculated using intensity data only, at unspecified resolution, was uninterpretable, and suspects that an alternative entropy maximum was found. Moreover, Bryan (1984) and Bryan & Banner (1986) found an example which shows that at 3Å resolution, a protein map which is an entropy maximum when subject to Fourier coefficient constraints, is not one when constrained by intensities only, and that when the entropy is maximised using this as a starting map, the phases shift by an average of 40°, resulting in an uninterpretable map. Following the logic of Jaynes (1968, 1982), there is no inconsistency here, but we have simply assumed uniformity in the prior distribution, thus ignoring the very strong underlying constraint of atomicity, and then, perhaps not surprisingly, obtain a non-atomic map.

These results show that the phase problem is approachable by maximum entropy if the data demand point-like or atomic features, but perhaps not otherwise. Therefore, the problem of interest in protein crystallography, that of using intensity data at 2.5-3Å to produce interpretable maps, seems to be outside the current scope of the method. However, this is not to say that maximum entropy is of no use in this field. The relative entropy expression provides a way of combining prior information with new data. As already described, we have a very good understanding of molecular structure, and one aspect of incorporating this into a prior is discussed in the last section. Another application is to problem of inadequate isomorphous replacement data. If MIR phases are only obtained to low resolution, we may consider using higher resolution native intensity data as a further constraint, the 'phase extension' problem. However, the problem that will be considered in the rest of this section will that of single isomorphous replacement (SIR).

Single isomorphous replacement data.

Whilst in some circumstances a conventional figure of merit weighted SIR Fourier map (Blow & Rossmann, 1961) may be interpretable, this is not always the case. As outlined in a previous section, each Fourier coefficient is estimated to be the average of the two possible phase solutions. In particular, if the heavy atoms form a centrosymmetric array, the heavy atom Fourier coefficients will all be real when the point of symmetry is taken as the phase origin, hence the SIR protein Fourier coefficients will also be real, and the density will consist of the superposition of the true density and its inverse. By using maximum entropy, we can attempt to select between the possible phases, so that our map will at least agree with the data. The number of acceptable phase solutions will be considerably reduced by positivity alone, although it will probably not resolve the ambiguity in phase of weaker reflections, where an incorrect phase choice would not cause this criterion to be violated. Maximum entropy has already been used to solve the structure of a filamentous virus from SIR data at 4Å resolution (Bryan, 1983, Bryan *et al.*, 1983). The virus had helical symmetry, so its Fourier transform is sampled in one dimension only, and is continuous in the other two. The phases at adjacent data points in reciprocal space are thus related, and the possible phase ambiguities therefore very much reduced compared with a similar-sized crystallographic problem.

A calculation using simulated crystallographic data is described here. A representative structure was obtained by extracting a fragment of 20 amino acids (160 non-hydrogen atoms) forming an α-β motif from Triose Phosphate Isomerase (Banner *et al.*, 1970). This was placed it in the asymmetric unit of a $P2_12_12_1$ unit cell, $a = b = 24$Å, $c = 64$Å, and Fourier coefficients to 3Å resolution were calculated. Part of the density synthesised from these Fourier coefficients is shown in figure 5. A derivative data set was simulated by adding the Fourier coefficients of a single zinc atom per asymmetric unit to the native Fourier coefficients, giving a heavy atom contribution of about the same size as that expected from a real heavy atom in a medium sized protein, *e.g.* one mercury atom in a protein of molecular weight 16000, using the formula of Crick & Magdoff (1956). The data used were the intensities of these two sets of Fourier coefficients. A conventional SIR synthesis from these data is shown in figure 6. The average figure of merit for all reflections, including centrics, was 68%, so not surprisingly the density is generally weaker than the original, and the average amplitude weighted phase change for acentrics was 43° from the original phases.

Applying maximum entropy to these 'best' SIR Fourier coefficients would be inappropriate; they do not necessarily correspond to a positive structure. Instead, the native and derivative intensity data sets were both used as constraints, in χ^2 terms (7a) and (7b). The single derivative is

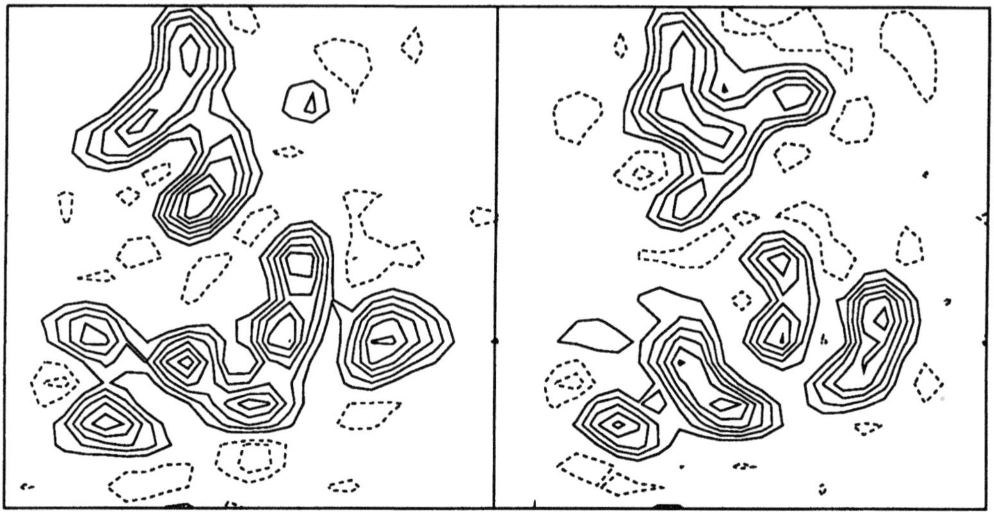

Figure 5. Density calculated from model protein fragment. Contour map of two successive sections at 1Å intervals in z. This, and all subsequent contour maps, have the same contour intervals. The zero contour is suppressed, and negative contours dashed. An α-helix lies with its axis almost horizontal in the lower half, and another part of the molecule is intersected obliquely in the upper half.

sufficient to phase the centric reflections, which were therefore incorporated in the phased term (7c), except that those with a low figure of merit, below 0.75, were ignored, as their sign is then unreliable. The heavy atom Fourier coefficients used were the same as in the SIR synthesis. Since this problem exhibits the full complexity of optimisation with non-convex constraints, it takes much more CPU time to solve than a similar sized convex problem. The resulting density, figure 7, shows all the correct features and no artifacts. Indeed, due to the constraint of positivity, the structure is sharper than the original 3Å Fourier synthesis, because of a reduction in series termination effects. The average amplitude weighted phase change from the original calculated Fourier coefficients was only $8\frac{1}{2}^\circ$, so it is clear that the phasing is considerably improved over the conventional 'best' map, and that most of the strong reflections must be correctly phased.

However, there is no assurance that this solution is unique, since without a multisolution algorithm, the possible presence of other local maxima cannot be investigated exhaustively. Nevertheless, running the algorithm from a variety of starting maps (Bryan & Banner, 1986), has not discovered any solutions significantly different from figure 7, although the phases of the weak reflections vary somewhat. In the same work, the solution was also found to be stable with respect to noise on the data, as would be expected for a maximum entropy solution.

The constraint functions used here can clearly be used with any number and quality of derivative data sets. If some reflections are judged to be reliably phased by conventional MIR phasing, perhaps low resolution ones if there are several derivatives good to low resolution, the MIR phase can be used in a Fourier coefficient constraint (7c). It is then a waste of computer time to put in all the intensity data as constraints (7a) and (7b). Less reliably phased reflections can still be constrained by the intensities of the native and whatever derivatives provide data. There are still several aspects worth investigating, for instance, whether success is also possible if the heavy atoms form a centrosymmetric array, what effects errors in the heavy atoms positions have, and also whether χ^2 is the most appropriate test, since the residuals $|F_{\mathbf{h}}|^2 - I$ are found to be nearly all negative.

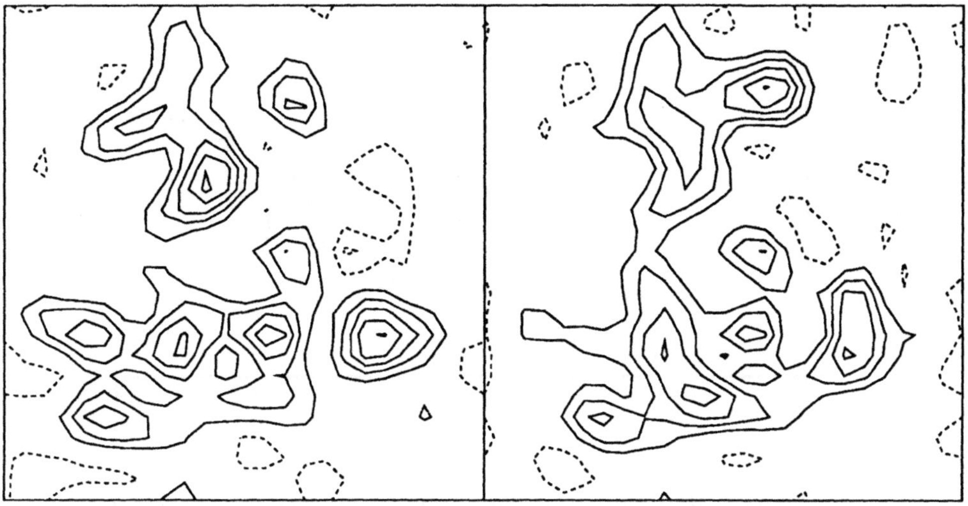

Figure 6. Density from SIR data by classical 'best' method.

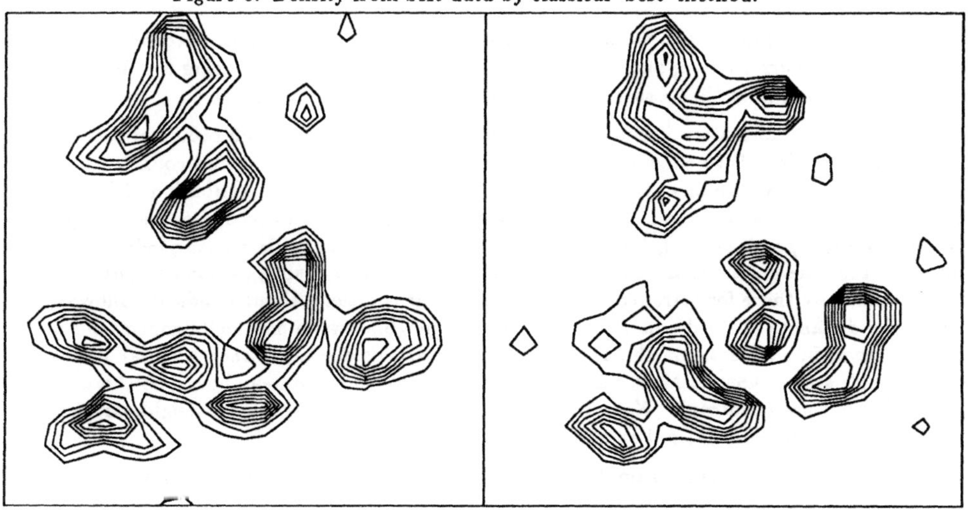

Figure 7. Density from SIR data by maximum entropy.

INCORPORATING CORRELATIONS VIA A SECOND ORDER PRIOR.

We now turn to the question of incorporating prior information, in terms of the molecular structures discussed previously, into the problem. Translation invariance means that in the first place we must use a flat initial model $\mathbf{m}$. Perhaps if we have partial or low resolution phase information, some of the structure, such as the backbone of an α-helix, can be picked out visually. An atomic model of this part of the structure can be built, and used as $\mathbf{m}$ in a further maximum entropy calculation. However, this will not always be possible, particularly in the pure phase problem. The information we wish to use is in terms of atomic bond lengths and angles, which means that it is in the form of correlations of atomic positions. A mechanism for dealing with such information was proposed by Skilling (1986). The idea is to work in the space of N samples from the map, and to apply the entropy to the N-sample joint distribution. It was

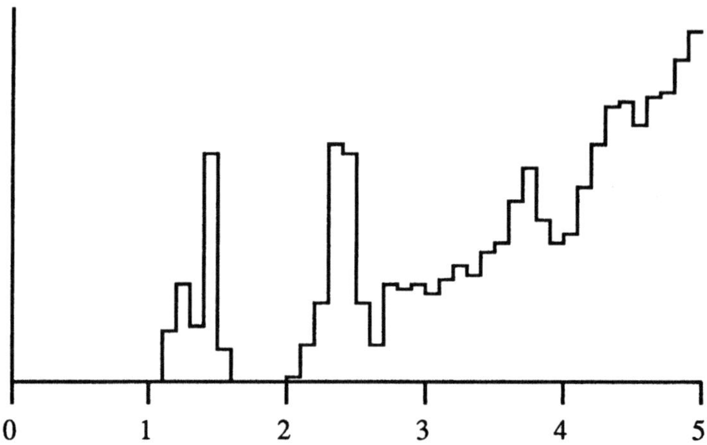

Figure 8. Histogram of pairwise interatomic distances for a typical globular protein, in 0.1Å bins, calculated by D. W. Banner.

shown that this is equivalent to working in the 1-sample space, but with an effective prior which depended on the current map, and was illustrated with a simple example using the position-independent information that a star (point source) was present in the image.

Clearly, the prior information in molecular structures extends to very high orders of correlation, and will be very difficult to encode. We shall therefore start most simply, and consider 2-point correlations. Do they give any useful information. Figure 8 shows the distribution of interatomic distances for a protein. It is seen that the nearest, at about $1\frac{1}{2}$Å, and second nearest neighbour distances are fairly well defined, but beyond that the curve becomes close to a quadratic, indicating that the positions are uncorrelated. This is quite reasonable, as bond angles are all approximately those for tetrahedral coordination, but rotations about bonds are allowed, so that the distances of third and subsequent nearest neighbours are not well defined, and in addition atoms in adjacent side chains start to appear at this distance. Such a radial average is not the whole story, though. Most atom have three or fewer bonded neighbours, so the angular distribution is very anisotropic. Nevertheless, we shall first try using nearest-neighbour distance information in a one-dimensional example.

Following Skilling (1986), the 2-sample entropy is defined on the 2-sample distribution p_{ij} as

$$S^{(2)} = -\sum_{ij} p_{ij} \log\left(p_{ij}/m_{ij}\right). \tag{9}$$

Since we want m_{ij} to represent the correlation of positions, it must be a function of $|i-j|$ only, and if successive samples from the map are independent, $p_{ij} = p_i p_j$. Hence

$$\begin{aligned} S^{(2)} &= -\sum_{ij} p_i p_j \left(\log p_i p_j / m_{i-j}\right) \\ &= -2\sum_i p_i \left(\log p_i - \tfrac{1}{2}(p * \log m)_i\right), \end{aligned} \tag{10}$$

and by comparison with (8), the expression

$$m_{\text{eff}} = \exp\left(\tfrac{1}{2} p * \log m\right) \tag{11}$$

can be identified as the 'effective prior' in the 1-sample space.

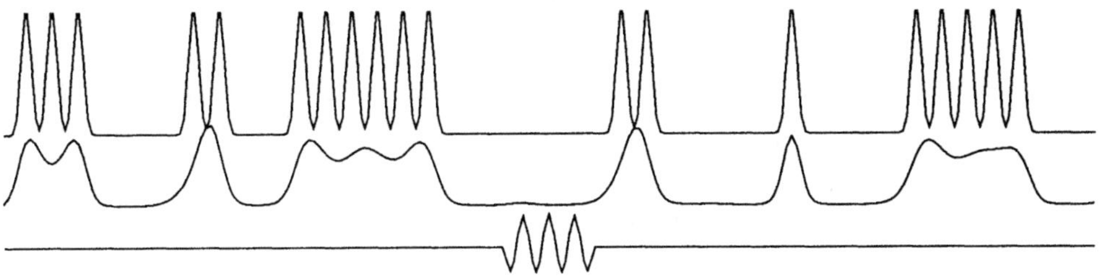

Figure 9. a) Top curve: simulated distribution of atoms. b) Middle curve: reconstruction of the density by maximum entropy using only the 18 lowest order Fourier coefficients calculated from the top curve. c) Bottom curve: Autocorrelation function of a regular array of atoms similar to (a), but set to a constant beyond $1\frac{1}{2}$ atoms from the centre. This function was used as the second-order prior for the reconstructions in figure 10.

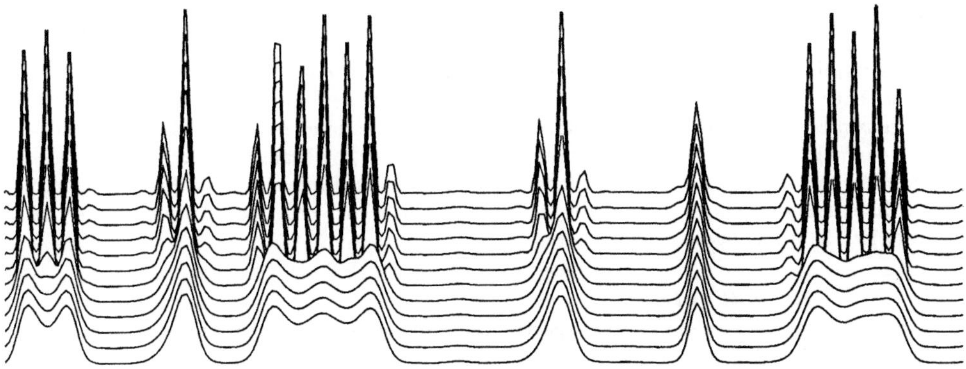

Figure 10. Maximum entropy densities as a function of the strength s of the second-order prior, with s starting from zero (bottom curve) and increasing in equal intervals, again using the 18 lowest order Fourier coefficients as data.

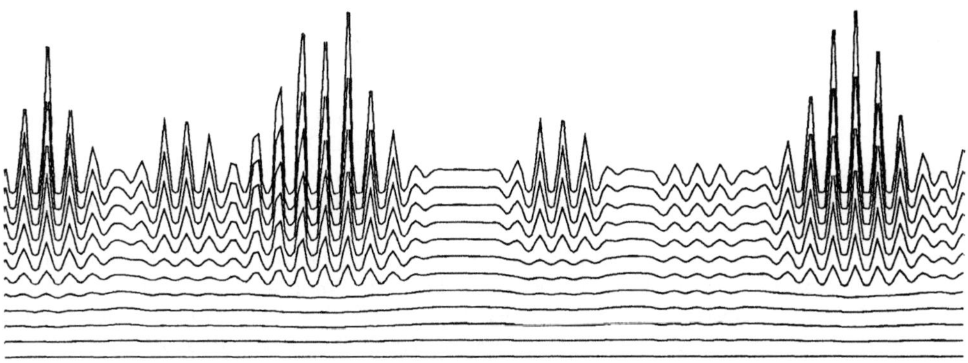

Figure 11. The effective priors $\exp(sp * \log m)$, corresponding to the solutions in figure 10.

To test this method, a model molecule was constructed in one dimension on a 256 pixel array as groups of Gaussian atoms, with uniform spacing of 6 between atoms within a group, but no positional correlation between groups (figure 9a). Fourier coefficients to about the 40th order are required to resolve such atoms classically. Using maximum entropy on the Fourier coefficients to the 18th order and a flat prior gave figure 9b, where the atoms are obviously not resolved, but the positions of the groups are revealed. The second order prior was taken as the correlation function of a regular array of such atoms, but set constant at distances from the centre greater than that of the second minimum (figure 9c). One further modification was made to expression (11), by making the 'strength' of the prior variable, so $m_{\text{eff}} = \exp(sp * \log m)$. $s = 0$ gives a flat prior, just as if no second-order correlations were included. The programming was fairly crude, in that the effective prior was recalculated at the beginning of each iteration, but then left fixed, so that the existing 1-sample entropy maximisation program could be used with minimal changes. The solution was investigated as s was increased from zero. Figure 10 shows the result, and figure 11 the corresponding effective priors. It can be seen that there is little effect until s reaches a critical value, when the solution undergoes a 'phase change', and turns into a set of correctly-spaced spikes within the envelope defined by the data. Where many atoms are in contact, they strongly reinforce each other through the prior. The singles and doubles are much less affected, except where they are close to a larger group, when the relative positioning may even cause unfavourable interactions. Beyond the edges of groups, there are further ripples in the prior, which cause a smaller peak in the map, but only to the amount allowed by the data.

Brief investigation has shown that as s is increased and then decreased again the effect is reversible, but the sampling of s has not yet been sufficiently dense to determine whether hysteresis is present, nor whether the solution for a given s is unique. Indeed, the second derivative of S is no longer negative definite, so 'maximising' S is fraught with danger. Nevertheless, this simple example has shown that introducing correlations via the prior can produce interesting effects, and no doubt incorporation of higher-order correlations is likely to be a very rich field for investigation.

REFERENCES.

Abels, J. G. (1974). Maximum entropy spectral analysis. *Astr. Astrophys. Suppl.*, **15**, 383–393.

Banner, D. W., Bloomer, A. C., Petsko, G. A., Phillips, D. C. & Wilson, I. A. (1976). Atomic coordinates for Triose Phosphate Isomerase from chicken muscle. *Biochemical and Biophysical Research Communications*, **72**, 146–155.

Bienenstock, A. & Ewald, P. P. (1962). Symmetry of Fourier Space. *Acta Cryst.*, **15**, 1253–1261.

Blow, D. M., & Crick, F. H. C. (1959). The treatment of errors in the isomorphous replacement method. *Acta Cryst.*, **12**, 794–802.

Blow, D. M., & Rossmann, M. G. (1961). The single isomorphous replacement method. *Acta Cryst.*, **14**, 1195–1202. Correction. *Acta Cryst.*, **15**, 1060.

Blundell, T.L., & Johnson L.N. (1976). Protein Crystallography. New York: Academic Press.

Bricogne, G. (1984). Maximum Entropy and the Foundations of Direct Methods. *Acta Cryst.*, **A40**, 410–445.

Bryan, R. K. (1983). Maximum entropy in structural molecular biology – the fibre diffraction phase problem. Paper presented at the Third Workshop on Maximum Entropy and Bayesian Methods in Applied Statistics, University of Wyoming, August 1-4.

Bryan, R. K. (1984). Application of the maximum entropy method in the fibre and crystallographic phase problems. Paper presented at the EMBO workshop on maximum entropy methods in the X-ray phase problem, Orsay, France.

Bryan, R. K. & Banner, D. W. (1986). The maximum entropy method with native and single isomorphous derivative data. Submitted to *Acta Cryst.*

Bryan, R. K., Bansal, M., Folkhard, W., Nave, C. & Marvin, D. A. (1983). Maximum entropy calculation of the electron density at 4Å resolution of Pf1 filamentous bacteriophage. *Proc. Natl. Acad. Sci. USA*, **80**, 4728–4731.

Bryan, R. K., & Skilling, J. (1986). Maximum entropy image reconstruction from phaseless Fourier data. *Optica Acta*, **33**, 287–299.

Collins, D. M. (1982). Electron density images from imperfect data by iterative entropy maximisation. *Nature*, **254**, 49–51.

Crick, F. H. C. & Magdoff, B. S. (1956). The theory of the method of isomorphous replacement for protein crystals. *Acta Cryst.*, **9**, 901–908.

Gull, S. F. & Daniell, G. J. (1978). Image reconstruction from incomplete and noisy data. *Nature*, **272**, 686–690.

Gull, S. F. & Skilling, J. (1984). The maximum entropy method. *In* Indirect Imaging, ed. J. A. Roberts, pp. 267–279. Cambridge: Cambridge University Press.

Harker, D. (1956). The determination of the phases of the structure factors of non-centrosymmetric crystals by the method of double isomorphous replacement. *Acta Cryst.*, **9**, 1–9.

Harrison, S. C., Olson, A. J., Schutt, C. E., Winkler, F. K. & Bricogne, G. (1978). Tomato bushy stunt virus at 2.9Å resolution. *Nature*, **276**, 368–373.

Hauptman, H. (1982). On integrating the techniques of direct methods and isomorphous replacement I. The theoretical basis. *Acta Cryst.*, **A38**, 289–294.

Hendrickson, W. A. & Teeter, M. M. (1981). Structure of the hydrophobic protein crambin determined directly from the anomalous scattering of sulphur. *Nature*, **290**, 107–113.

Hogle, J. M., Chow, M. & Filman, D. J. (1985). Three-dimensional structure of poliovirus at 2.9Å resolution. *Science*, **229**, 1358–1365.

Jaynes, E. T. (1968). Prior probabilities. *IEEE Trans.*, **SCC-4**, 227–241.

Jaynes, E. T. (1982). On the rational of maximum-entropy methods. *Proc. IEEE*, **70**, 939–952.

Livesey, A. K. (1984). Paper presented at the EMBO workshop on maximum entropy methods in the X-ray phase problem, Orsay, France.

Livesey, A. K. & Skilling, J. (1985). Maximum entropy theory. *Acta Cryst.*, **A41**, 113–122.

Navaza, J. (1986). The use of non-local constraints in maximum-entropy electron density reconstruction. *Acta Cryst.*, **A42**, 212–223.

Rossmann, M. G., *et al.* (1985). Structure of a human common cold virus and functional relationship to other picornaviruses. *Nature*, **317**, 145–153.

Shore J. E. & Johnson R. W. (1980). Axiomatic derivation of maximum entropy and the principle of minimum cross-entropy. *IEEE Trans.*, **IT-26**, 26–37. Comments and corrections. *IEEE Trans.* **IT-29**, 942–943.

Skilling, J. (1986). Theory of Maximum Entropy Image Reconstruction. *In* Maximum Entropy and Bayesian Methods in Applied Statistics, Proceedings of the Fourth Maximum Entropy Workshop, University of Calgary, 1984, ed. James H. Justice, pp. 156–178. Cambridge: Cambridge University Press.

Skilling, J. & Bryan, R. K. (1984). Maximum entropy image reconstruction: general algorithm. *Mon. Not. R. astr. Soc.*, **211**, 111–124.

Wei, W. (1985). Application of the maximum entropy method to electron density determination. *J. Appl. Cryst.*, **18**, 442–445.

Wilkins, S. W. & Stuart, D. (1986). Statistical geometry. IV. Maximum-Entropy-Based Extension of Multiple Isomorphously Phased X-ray Data to 4Å Resolution for α-Lactalbumin. *Acta Cryst.*, **A42**, 197–202.

Wilkins, S. W., Varghese, J. N., & Lehmann, M. S. (1983). Statistical geometry. I. A self-consistent approach to the crystallographic inversion problem based on information theory. *Acta Cryst.*, **A39**, 49–60.

CONTRAST TRANSFER FUNCTION CORRECTION IN ELECTRON MICROSCOPY.

R. K. Bryan
European Molecular Biology Laboratory, Meyerhofstrasse 1,
6900 Heidelberg, West Germany.

1 *INTRODUCTION.*

Whilst the only practicable technique for determining the structure of biological molecules to atomic resolution is X-ray crystallography, many specimens of considerable interest cannot be formed into three-dimensional crystals. Much useful data can nevertheless be obtained from smaller amounts of material, for example, single viruses, or two-dimensional crystals of proteins, by electron microscopy, because electrons interact much more strongly with atoms than do X-rays. However, because of the stronger interaction, and the smaller amount of material used, the specimen is destroyed by the electron dose necessary to form an image. Traditionally, the specimen has been preserved by negative staining - embedding in a salt of a heavy atom, such as uranyl acetate - and thus the image only records the shape of the specimen, and the resolution achieved is limited by (*inter alia*) the grain size of the stain. The staining also protects the specimen from dehydration in the vacuum of the microscope. In order to increase the resolution, and also to image the internal structure of the specimen, the specimen has been embedded in other media, such as glucose (Unwin & Henderson, 1975), and more recently in amorphous ice (Lepault *et al.*, 1983). This necessitates using a lower electron dose, decreasing the signal to noise ratio, and the object also scatters primarily as a 'phase object'. In section 2 we discuss the image formation theory for such objects, and show that the recorded image is, to a good approximation, the convolution of the projected atomic potential of the specimen with a space invariant point-spread function. To obtain a full 3-dimensional reconstruction of an object therefore requires a complete set of projections to be obtained. This is usually done by means of a tilting specimen holder, but if the object has internal symmetry, there will a proportionate reduction in the number of views required, and, for example, a helical virus may be reconstructed from a single view.

Electron microscopes may also be used to give diffraction patterns directly. However, only the intensity of the diffraction pattern can be recorded, and quantitatively good data can only be obtained for crystalline samples. The problem of deconvolution from a transfer function is replaced by the considerably harder phase problem. However, a combination of data from the two modes has been used, taking phases from computed transforms of recorded images and amplitudes from electron diffraction patterns (Unwin & Henderson, 1975).

The Fourier transform of the point-spread function (the contrast transfer function or CTF) is an oscillating function of reciprocal space radius, and it is essential to correct for its effects if the image is to be interpreted correctly. If the specimen is 2-D periodic, few of the reciprocal lattice points fall near zeros of the CTF, and correction is comparatively simple (Lepault & Pitt, 1984). For non-periodic specimens, such as the helical particles considered here, the complete specimen transform cannot be deduced from the recorded image due to the CTF zeros, and so no simple 'filter' algorithm can be used.

We show here how the deconvolution can be achieved using the maximum entropy method, and prior knowledge of a maximum radius for the particle is imposed. The deconvolution is combined with the Radon problem, so that the averaged radial density distribution of the particle

G. J. Erickson and C. R. Smith (eds.),
Maximum-Entropy and Bayesian Methods in Science and Engineering (Vol. 2), 171–179.

is calculated, and the parameters defining the CTF's are also refined as they are not known precisely.

2 *IMAGE FORMATION THEORY.*

We shall now develop briefly the theory of image formation in the 'weak phase-object' approximation, mostly following the treatment of Saxton (1978).

The magnetic field of the electron lenses has axial symmetry about the microscope axis, which is taken as the z-axis of our coordinate system. We consider the propagation of the electron wave along the z-axis as a function of a 2-dimensional vector $\mathbf{x}$ perpendicular to the z-axis. Then, if $\psi_z(\mathbf{x})$ is the wave function at z, with the subscript 0 referring to the starting plane z_0, a propagation equation in the paraxial approximation, with rotation about the axis removed, may be derived from the Schrödinger equation as

$$\psi_z(\mathbf{x}) = \tfrac{1}{i\lambda h} \exp\left(\tfrac{ikh'}{2h}|\mathbf{x}|^2\right) \int \psi_0(\mathbf{x}_0) \exp\left(\tfrac{ikg}{2h}|\mathbf{x}_0|^2\right) \exp\left(-\tfrac{ik}{h}\mathbf{x}.\mathbf{x}_0\right) d^2x_0, \tag{1}$$

where k is the electron momentum, $\lambda = \frac{2\pi}{k}$ the wavelength, and $h(z)$, $g(z)$ are the two independent classical solutions for transverse displacement of the trajectories of electrons in the magnetic field of the microscope lenses, with

$$\begin{aligned} h(z_0) &= g'(z_0) = 0, \\ h'(z_0) &= g(z_0) = 1. \end{aligned} \tag{2}$$

There are two sets of values of z of particular interest. At any z such that $g(z) = 0$, say at $z = z_d$, the quadratic phase shift within the integral disappears, giving

$$\psi_d(\mathbf{x}) = \frac{1}{i\lambda h(z_d)} \exp\left(\tfrac{ikh'(z_d)}{2h(z_d)}|\mathbf{x}|^2\right) \tilde{\psi}_0(k\mathbf{x}/h(z_d)),$$

where a $\tilde{}$ denotes a Fourier transform, showing that the diffraction pattern of the wave function at z_0 is formed. Furthermore, if $h(z) = 0$, the integral (1) may be evaluated by the method of stationary phase, but more insight is gained by considering the propagation in two stages, first from the specimen plane to the diffraction plane, and then by a second application of (1), from the diffraction plane to the final plane. In the second propagation, h and g are replaced by solutions of the electron trajectory which obey conditions similar to (2), but at z_d instead of z_0, say h_d and g_d, related to h and g by

$$\begin{aligned} h_d &= g/g'(z_d), \\ g_d &= h/h(z_d) - gh'(z_d)/h(z_d)g'(z_d), \end{aligned} \tag{3}$$

so that the net phase change applied to $\tilde{\psi}_0$ is

$$k|\mathbf{x}|^2\left(\frac{g_d}{h_d} + \frac{h'(z_d)}{h(z_d)}\right) = k|\mathbf{x}|^2\frac{hg'(z_d)}{h(z_d)g}.$$

This expression is clearly zero whenever $h(z) = 0$, at $z = z_i$, say, so that

$$\psi_i(\mathbf{x}) = \frac{-1}{\lambda^2 h_d(z_i)h(z_d)} \exp\left(\tfrac{ikh_d'(z_i)}{2h_d(z_i)}|\mathbf{x}|^2\right) \int \tilde{\psi}_0(k\mathbf{x}_d/h(z_d)) \exp\left(-\tfrac{ik}{h_d(z_i)}\mathbf{x}.\mathbf{x}_d\right) d^2x_d, \tag{4}$$

is just an inverse Fourier transform. After some manipulation of the h's and g's this gives

$$\psi_i(\mathbf{x}) = \frac{1}{g(z_i)} \exp\left(\tfrac{ikg'(z_i)}{2g(z_i)}|\mathbf{x}|^2\right) \psi_0(\mathbf{x}/g(z_i)), \tag{5}$$

an image of ψ_0, magnified by $g(z_i)$. The expression (4) also allows us to consider the effects of slight defocus, which reintroduces the quadratic phase factor due to a missmatch in the conditions (3), of spherical aberration of the lens, giving a quartic phase shift, and of thermal spread of electron energies causing a reduction of response at high resolution. Effects due to lack of spatial coherence have been shown to be negligible under typical imaging conditions (Frank, 1973, Henderson & Glaeser, 1985). The final expression for the image–plane wave function can be written conveniently as

$$\psi_i \propto \mathcal{F}^{-1} T e^{-i\gamma} \mathcal{F} \psi_0, \tag{6}$$

up to a constant, where

$$\gamma = \tfrac{1}{2}\pi\lambda^3 C_s \mathbf{t}^4 - \pi\lambda D \mathbf{t}^2, \tag{7}$$

$\mathbf{t}$ is a suitably scaled reciprocal space vector, and (Frank, 1976, Wade & Frank, 1977)

$$T = \exp - \tfrac{\pi^2}{16 \log 2} C_c^2 (\tfrac{\Delta E}{E})^2 \lambda^2 \mathbf{t}^4. \tag{8}$$

C_s and C_c are the coefficients of spherical and chromatic aberration respectively, D the defocus, and ΔE the energy spread about the mean electron energy E. Most of the parameters are essentially fixed for a given microscope at a given magnification, but the defocus is at the control of the experimenter.

We now consider the effect of a thin specimen in an incident plane-wave electron beam. The typical operating conditions of a microscope are such that the specimen thickness is large compared with the electron wavelength ($\approx$ 0.004 nm), and small compared with the defocus (of the order of 1000 nm), so the variation in defocus through the specimen may be ignored. For elastic scattering of electrons, the phase of the incident wave is changed by an amount $\eta(\mathbf{x})$, proportional to the projected atomic potential of the specimen,

$$\eta(\mathbf{x}) \propto \int \phi(\mathbf{x}, z) dz, \tag{9}$$

giving $\exp i\eta(\mathbf{x})$ for the wave function just after the specimen. In the "weak phase" approximation, η is assumed to be small, so

$$\psi_0(\mathbf{x}) = \exp i\eta(\mathbf{x}) \approx 1 + i\eta(\mathbf{x}). \tag{10}$$

This approximation is accurate for the light atoms (hydrogen, carbon, nitrogen, oxygen) which make up the bulk of biological material, up to a total specimen thickness of about 50 nm. Using (6), the image wave function is thus

$$\psi_i \propto \mathcal{F}^{-1} T e^{-i\gamma} \mathcal{F} e^{i\eta} \tag{11}$$

$$\approx \mathcal{F}^{-1} T e^{-i\gamma} (\delta + i\mathcal{F}\eta) \tag{12}$$

$$= 1 + \mathcal{F}^{-1} T(i\cos\gamma + \sin\gamma) \mathcal{F}\eta, \tag{13}$$

where δ is the Kronecker delta. The detecting medium, usually photographic film, records the intensity of the incident beam,

$$|\psi_i|^2 \approx 1 + 2\mathcal{F}^{-1}(T \sin\gamma \mathcal{F}\eta), \tag{14}$$

ignoring terms of second order in η, thus showing that the deviation of the recorded image from uniformity is linearly related to the projection of the object potential η, via a convolution with the transform of the contrast transfer function $T \sin\gamma$.

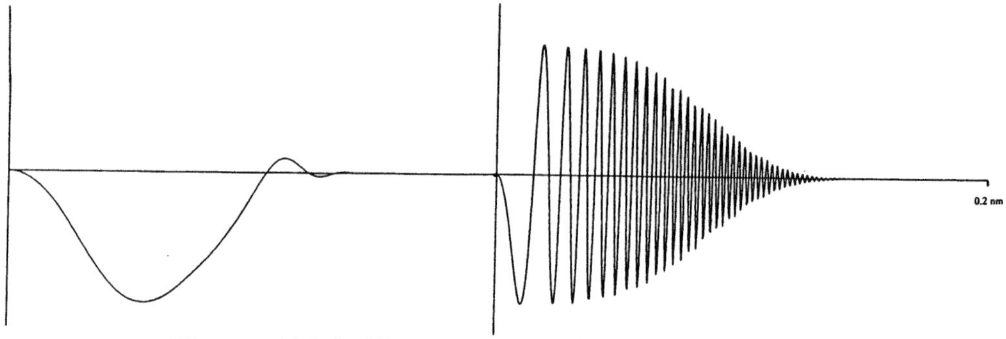

Figure 1. (a) Left. Phase contrast transfer function at Scherzer focus. Parameters $C_s = 1.3 \times 10^6$ nm, $C_c = 2 \times 10^6$ nm, $\lambda = 4 \times 10^{-3}$ nm, $D = 72$ nm, $\frac{\Delta E}{E} = 2.5 \times 10^{-5}$ (b) Right. Phase contrast transfer function at large defocus. Parameters as in (a), except $D = 1.8 \times 10^3$ nm.

Other mechanisms, as well as weak elastic scattering, are usually significant in image formation. If any scattering process causes an imaginary η in (13), an image will again be formed, but with $\sin\gamma$ replaced by $\cos\gamma$ in the CTF. This is usually termed an amplitude contrast image, and can be attributed to two effects in particular. If the phase shifts due to the specimen are large, because of a thick specimen or because the specimen contains heavy atoms, higher order terms in the expansion (10) are significant, and give an imaginary component. Furthermore, some electrons will be scattered though an angle sufficiently large that they do not pass through the microscope aperture and are absorbed, equivalent to an imaginary scattering potential in the object. For stained objects, where the scattering is from the heavy atoms, amplitude contrast is the dominant imaging mode, and thus at low resolution, where $\cos\gamma \approx 1$, the images are interpretable without correction for CTF effects. Representing the amplitude contrast of the specimen by $a(\mathbf{x})$ then gives a total image of

$$|\psi_i|^2 \approx 1 + 2\,\mathcal{F}^{-1}\,T(\sin\gamma\,\tilde{\eta} - \cos\gamma\,\tilde{a}).$$

A positive amplitude contrast corresponds to removal of electrons from the beam. A positive defocus (underfocus) gives a negative $\sin\gamma$, so that at low resolution any amplitude contrast will reinforce the phase contrast image. This effect has been exploited (Lepault & Leonard, 1985) to reconstruct images when the two contrast modes are effective in different resolution ranges.

Inelastic scattering is of comparable magnitude to elastic scattering, but is more strongly peaked in the forward direction. Since the position of the focal plane for inelastically scattered electrons also depends on the energy loss and on the chromatic aberration of the lens, the net effect is to increase the background density, and reduce image contrast. This is also a further mechanism for removal of electrons from the unscattered beam, and so will also enhance the amplitude contrast image.

The form, $\sin\gamma = \sin(\frac{1}{2}\pi\lambda^3 C_s \mathrm{t}^4 - \pi\lambda D\mathrm{t}^2)$, of the phase–contrast CTF is in general rapidly oscillating, and correction for its effects is clearly essential. At exact focus, if the lens were aberration–free, $\sin\gamma = 0$, and no image would be formed. At low resolution, $\sin\gamma \approx -\pi\lambda D\mathrm{t}^2$, which corresponds to the Laplacian operator in real space, giving an 'edge enhancement' effect. The optimum focus for use without correction is the Scherzer focus (Scherzer, 1949), with $D = \sqrt{C_s\lambda}$, which gives the greatest resolution range without a sign reversal in the CTF (figure 1a). However, the response at resolutions below 2 nm, a range of great importance for positioning domains in macromolecules, is then negligible. Higher values of defocus give a greater response at lower resolution, at the expense of an earlier onset of ringing (figure 1b). We shall consider the deconvolution problem in this region of defocus values.

3 *MAXIMUM ENTROPY.*

The application of the maximum entropy method to deconvolution and similar inverse problems has been described sufficiently in the past (Frieden, 1972, Abels, 1974, Gull & Daniell, 1978, Bryan & Skilling, 1980) that only a brief outline will be given here. The object is described at suitable sampling by a normalised set of positive numbers ρ, $\sum \rho = 1$. In the absence of noise, such an object would result in data $\mathbf{P} = \mathcal{T}(\rho)$ being observed, where $\mathcal{T}$ defines the response of the observing equipment, in our case, projection and convolution with the transfer function. If the actual experimental data $\mathbf{D}$ is subject to noise, and $\mathbf{P}$ and $\mathbf{D}$ agree to within the noise, then ρ could indeed represent the observed object. Agreement between $\mathbf{P}$ and $\mathbf{D}$ is usually measured with a χ^2 test,

$$\chi^2 = \sum (P - D)^2/\sigma^2,$$

where σ is the noise standard deviation. If the value of χ^2 is sufficiently small that the discrepancy between $\mathbf{P}$ and $\mathbf{D}$ can be accounted for by the noise, then we may conclude that ρ could indeed be the object actually observed. The χ^2 test thus defines the 'feasible set' of all reconstructions agreeing with the data. From this set, the one with the greatest configurational entropy, $S = -\sum \rho \log \rho/m$, is chosen as the preferred reconstruction, where $\mathbf{m}$ is an initial model of ρ before taking the data into account, and usually chosen to be uniform. This has been shown to be the only consistent way to select a single reconstruction by a variational method (Shore & Johnson, 1980, Johnson & Shore, 1983, Gull & Skilling, 1984). If $\mathcal{T}$ is linear, the set of feasible reconstructions is an ellipsoid (or ellipsoidal cylinder), and hence convex. Since S is a convex function, it will have a unique maximum in this set. The problem therefore becomes one of constrained non-linear optimisation, and in the work described here the algorithm of Bryan (1980) and Skilling & Bryan (1984) was used.

We have so far assumed that the response function is known exactly. If not, but its functional form, defined in terms of a few parameters, is known, then we can attempt to optimise the fit to the data with respect to these parameters as well. We make an initial estimate of their values, and attempt to solve the problem by maximum entropy. The data may be fitted successfully if the estimate is sufficiently good, or perhaps not, due to a large number of points being forced down to zero. In either case, the values of the parameters are then adjusted so as to minimise the value of χ^2, keeping the current reconstruction fixed. A new maximum entropy solution can then be calculated using the new parameters, and the process iterated.

This technique of parameter refinement has been used in other maximum entropy applications, such as calibration phases in radio astronomy (Scott, 1981) and NMR spectroscopy (Sibisi, 1983), and refinement of heavy atom positions (Bryan *et al.*, 1983) in isomorphous replacement calculations using fibre diffraction data.

4 *MAXIMUM ENTROPY AND THE CTF PROBLEM.*

As outlined in section 2, the image is related to the projected atomic potential via a convolution with the CTF. The full problem of determining a three-dimensional structure therefore combines the deconvolution problem and the Radon problem. Here we shall consider a simplified form of the problem, that of calculating the averaged radial density distribution of a helical particle from a single view. We must therefore assume that the particle is undistorted. The frozen-hydrated technique has the further advantage over conventional staining that the particle is less likely to be distorted by flattening against the supporting grid.

Let $f(r)$ represent the averaged radial potential. Then the Fourier transform at radius R is given by the zeroth order Hankel transform

$$F(R) = \int_0^\infty f(r) J_0(2\pi Rr) r \, dr, \tag{15}$$

which, by the projection theorem, is the Fourier transform of the transaxial projection of the structure. This function, multiplied by the CTF and inverse Fourier transformed, will give the simulated data for the trial structure $f(r)$. We shall also include a multiplication by $\exp(i\delta R)$ to allow for a possible miss-centering of the data by an offset δ. Thus the simulated data $P(x)$ is given by

$$P(x) = \int_{-\infty}^{\infty} F(R) e^{i\delta R} e^{2\pi i x R} dR, \tag{16}$$

where we define $F(-R) = F(R)$, $R \geq 0$.

For practical calculation, a descretised form of these relations is required. Part of our prior knowledge in this problem is an upper bound to the particle diameter, which conceptually can be applied to the entropy expression by taking $m(r) = const.$, $r \leq r_{\max}$, $m(r) = 0$, $r > r_{\max}$, but of course in practice is implemented by using a grid extending only to $r_{\max}$. For simplicity, equal grid spacing was used, and the Hankel transform performed by matrix multiplication using stored coefficients. Equal grid spacing also enables the Fourier transform to be performed by a Fast Fourier Transform, but requires the entropy expression to be weighted by the size of the pixels, proportional to radius. Thus, in discrete form

$$P_l = \mathcal{F}_{lk} T_k \sin\gamma_k \, e^{i\delta k} \, \mathcal{H}_{kj} f_j, \tag{17}$$

where $\mathcal{F}$ and $\mathcal{H}$ represent the Fourier and Hankel transforms as matrices, and $T_k \sin\gamma_k$ is the CTF. This image is superimposed on a uniform background α, so the final χ^2 test with the data D_l for this problem is

$$\chi^2 = \sum_{l \in \text{data set}} w_l (P_l + \alpha - D_l)^2. \tag{18}$$

α must also be estimated from the data. The value of the defocus may be estimated from the positions of the zeros of the CTF in the Fourier transform of an image (Thon 1968). The noise standard deviation can be estimated from parts of the image adjacent to the object. We assume that it is uniform over the image, since the object itself causes only small deviations from uniformity. The parameters α, δ and defocus D were refined as described in the previous section, and the other microscope parameters set to the appropriate values for the operating conditions of the microscope (see figure captions).

We first show the results of a calculation using simulated data. Figure 2a shows the model radial distribution of ϕ, based on the idea of a hollow rod virus or phage tail, and figure 2b shows the radial distribution after applying the CTF, which would be the result if the Radon problem were solved *without* correction for the CTF effects. This is not really interpretable as the original object.

Figure 2c shows the resulting image data, with the microscope parameters as described in the caption. Application of a Wiener filter to this data (by a Fourier transform, filtering with the CTF, and an inverse Hankel transform) gives a disastrous restoration (figure 3a), since data has been lost over a large proportion of Fourier space.

The maximum entropy result (figure 3b), with a maximum particle radius considerably larger than the real radius (which would be a reasonable guess from the data, given a knowledge of typical CTF effects) is much better. In the analysis of actual experimental data, one could then identify the maximum particle radius more accurately, and repeat the calculation with a lower maximum radius, which would conceivably improve the restoration of detail. Provided an initial defocus value is chosen such that the first three or four zeros of the CTF have not crossed over, the parameters α, δ and D have been found to converged reliably to close to their correct values.

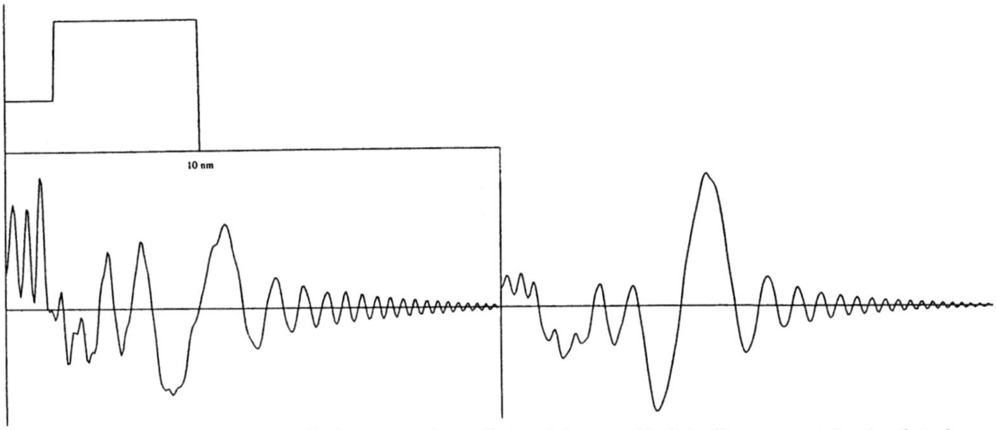

Figure 2. Radial distributions derived from cylindrically symmetric simulated object. (a) Above. Initial radial distribution. (b) Below left. Radial distribution after convolution with CTF. (c) Below right. Projected data from (b).

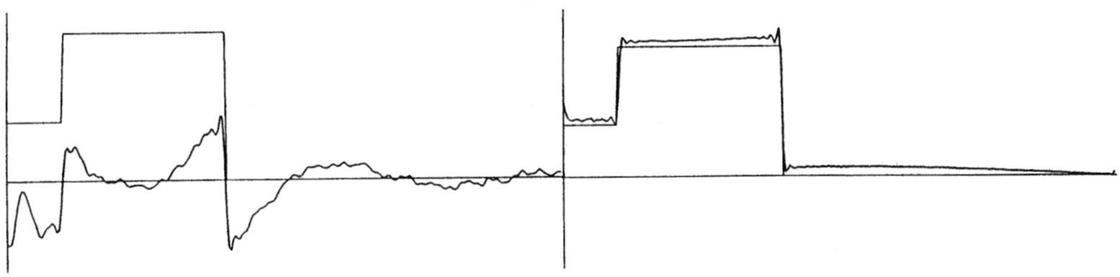

Figure 3. Reconstructions of object from data of figure 2. (a) Left. By Wiener filter. (b) Right. By maximum entropy, using a maximum radius of 15 nm.

Figure 4a shows the recorded distribution across a Tobacco Mosaic Virus (TMV) rod, obtained by projecting a two-dimensional image in the direction of the virus axis. The structure of TMV is known accurately from X–ray analysis (Stubbs, Warren & Holmes, 1977) and we impose a 15 nm maximum radius, well above the known 9 nm radius. The maximum entropy reconstruction, figure 4b, shows that the amount of information in the data is comparatively small, and only the particle edge is shown with fairly low resolution.

The lack of detail in this reconstruction may be attributed to several causes. The combination of the Radon problem with a deconvolution gives an extremely ill-conditioned problem, with a correspondingly large set of feasible reconstructions, and hence greater scope for a reconstruction with little detail. Secondly, a good background estimate is essential, otherwise it is impossible to estimate the low resolution components of the reconstruction. In practice, it is difficult to obtain a particle isolated sufficiently that a large surrounding area can be used to calculate the background. The simulated data derived from our reconstruction indeed shows significant variation outside the interval for which data was provided. Finally, although we have treated the problem as one of restoring a positive image, there is no reason that the function ϕ should not take either sign, or indeed that the image also have amplitude contrast components. In fact, the assumption of positivity may not be so bad, since the density of TMV derived from X-ray analysis relative to a aqueous solvent is known to be positive (Franklin & Holmes, 1958).

We can envisage extensions to the method described here which could take account of these possibilities. Data from several images taken at different values of defocus (which will therefore have different CTF's), can be used, with the χ^2 test now the sum over the χ^2 values for the

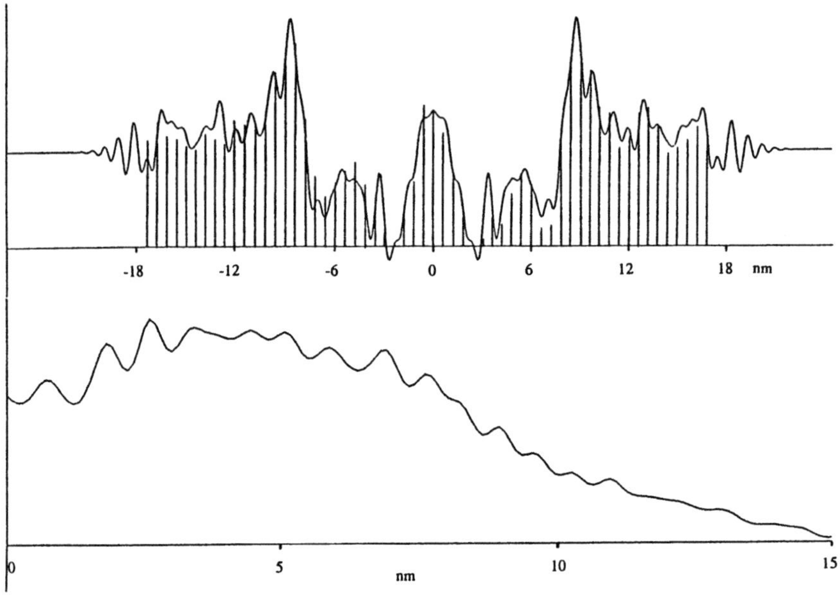

Figure 4. a) Above. Measured density of TMV virus (vertical bars), with the transform (continuous line) of reconstructed radial distribution, which is shown in (b), below.

individual images, thus providing data over a greater range of Fourier space. Such a focus series could also enable estimates of both the amplitude and the phase contrast contributions to the image to be made, analogous to the 'colour' images in astronomy (Gull & Skilling, 1984). Initial trial calculations show, though, that because of the much larger response for amplitude contrast than phase contrast at low resolution, any amplitude contrast present may completely dominate the problem. Perhaps a better way to proceed would be to work with the number density of atoms of a given type in the object, and then calculate the resultant image taking all relevant scattering effects into account, for comparison with the recorded data.

ACKNOWLEDGMENTS.

I thank Jean Lepault for the use of his TMV data in this calculation, and also several other colleagues at EMBL for many interesting discussions of this problem.

REFERENCES.

Abels, J. G. (1974). Maximum entropy spectral analysis. *Astr. Astrophys. Suppl.*, **15**, 383–393.

Bryan, R. K. (1980). Maximum Entropy Image Processing. PhD Thesis, University of Cambridge.

Bryan, R. K., Bansal, M., Folkhard, W., Nave, C. & Marvin, D. A. (1983). Maximum entropy calculation of the electron density at 4Å resolution of Pf1 filamentous bacteriophage. *Proc. Natl. Acad. Sci. USA*, **80**, 4728–4731.

Bryan, R. K. & Skilling, J. (1980). Deconvolution by maximum entropy as illustrated by application to the jet of M87. *Mon. Not. R. astr. Soc.*, **191**, 69–79.

Frank, J. (1973). The envelope of electron microscope transfer functions for partially coherent illumination. *Optik*, **38**, 519–536.

Frank, J. (1976). Determination of source size and energy spread from electron micrographs using the method of Young's fringes. *Optik*, **44**, 379–391.

Franklin, R. E. & Holmes, K. C. (1958). Tobacco Mosaic Virus: Application of the method of isomorphous replacement to the determination of the helical parameters and radial density distribution. *Acta Cryst.*, **11**, 213–220.

Frieden, B. R. (1972). Restoring with maximum likelihood and maximum entropy. *J. Opt. Soc. Am.*, **62**, 511–518.

Gull, S. F. & Daniell, G. J. (1978). Image reconstruction from incomplete and noisy data. *Nature*, **272**, 686–690.

Gull, S. F. & Skilling, J. (1984). The maximum entropy method. *In* Indirect Imaging, ed. J. A. Roberts, pp. 267–279. Cambridge: Cambridge University Press.

Henderson, R. & Glaeser, R. M. (1985). Quantitative analysis of image contrast in electron micrographs of beam-sensitive crystals. *Ultramicroscopy*, **16**, 139–150.

Lepault, J., Booy, F. P. & Dubochet, J. (1983). Electron microscopy of frozen biological suspensions. *J. Microsc.*, **129**, 89–102.

Lepault, J. & Leonard, K. (1985). Three-dimensional structure of unstained, frozen-hydrated extended tails of bacteriophage T4. *J. Mol. Biol.*, **182**, 431–441.

Lepault, J. & Pitt, T. (1984). Projected structure of unstained, frozen-hydrated T-layer. *The EMBO Journal*, **3**, 101–105.

Saxton, W. O. (1978). Computer Techniques for Image Processing in Electron Microscopy. Academic Press: NY.

Scherzer, O. (1949). The theoretical resolution limit of the electron microscope. *Journal of Applied Physics*, **20**, 20–29.

Scott, P. F. (1981). A 31 GHz map of W3(OH) with a resolution of 0.3 arcsec. *Mon. Not. R. astr. Soc.*, **194**, 25P–29P.

Shore J. E. & Johnson R. W. (1980). Axiomatic derivation of maximum entropy and the principle of minimum cross-entropy. *IEEE Trans.*, **IT–26**, 26–37. Comments and corrections. *IEEE Trans.* **IT–29**, 942–943.

Sibisi, S. (1983). Two-dimensional reconstructions from one-dimensional data by maximum entropy. *Nature*, **301**, 134–136.

Skilling, J. & Bryan, R. K. (1984). Maximum entropy image reconstruction: general algorithm. *Mon. Not. R. astr. Soc.*, **211**, 111–124.

Stubbs, G., Warren, S. & Holmes, K. C. (1977). Structure of RNA and RNA binding site in Tobacco Mosaic Virus from 4-Å map calculated from X-ray fibre diagrams. *Nature*, **267**, 216–221.

Thon, F. (1968). Zur Deutung der Bildstructuren in hochaufgelösten elektronenmikroskopischen Aufnahmen dünner amorpher Objeckte. PhD Thesis, Eberhard-Karls-Universität, Tübingen.

Unwin, P. N. T. & Henderson, R. (1975). Molecular structure determination by electron microscopy of unstained crystalline specimens. *J. Mol. Biol.*, **94**, 425–440.

Wade, R. H., & Frank, J. (1977). Electron microscope transfer functions for partially coherent axial illumination and chromatic defocus spread. *Optik*, **49**, 81–92.

CLIMATICALLY INDUCED CYCLIC VARIATIONS IN UNITED STATES CROP PRODUCTION: IMPLICATIONS IN ECONOMIC AND SOCIAL SCIENCE

Robert Guinn Currie
Laboratory for Planetary Atmospheres Research
State University of New York
Stony Brook, NY 11794

> 'The "correctness" of mathematics must be judged by its applicability to the physical world. Mathematics is an empirical science much as Newtonian mechanics. It is correct only to the extent that it works and when it does not, it must be modified. It is not a priori knowledge even though it was so regarded for two thousand years'.
>
> Morris Kline (1980)

Abstract: Analysis of United States corn yield data discloses evidence for signals near 19 years and 10-11 years in the Corn Belt states, and a large number of other states, which are significant in economic terms. Yield minima in the 19-year wave are in phase with dates or epochs of maxima in 18.6-year luni-solar tidal forcing, and also in phase with the 19-year wave in maximum drought measured from tree-ring data to the west of the Great Plains (Currie, 1984c). Yield minima in the 10 to 11 year wave are also in phase with maxima in solar cycle drought as measured from tree-rings (Currie, 1984c). Both waves closely aligned near the 1936.1 and 1954.7 tidal epochs, thereby enhancing yield shortfalls. These signals are also found in most of the major crops grown as well as in livestock and poultry production. We propose that the systematic modulation of agricultural output is the principal cause of the "Kuznets' long swings" in the American economy (so named to honor Nobel laureate Simon Kuznets) with period about 20 years on average (Soper, 1978). Time series on population and the building industry are reexamined and they yield a bandlimited signal between 16 to 21 years in period, in agreement with earlier studies by the economists Burns and Mitchell (1946), Kuznets (1961), and Abramovitz (1964). Precipitation data in the northeastern U. S. yield the 19-year luni-solar term which is out-of-phase with the western U. S. for our century. This explains why evidence for the Kuznets effect in aggregate economic data began to deteriorate after the turn of the century.

G. J. Erickson and C. R. Smith (eds.),
Maximum-Entropy and Bayesian Methods in Science and Engineering (Vol. 2), 181–241.

1 INTRODUCTION

The first recognition that there existed a recurrent clustering of severe drought years on a 20-year time scale in the western United States occurred during those of the 1950s. Borchert (1971) documented the serious impact on the American economy of the prolonged widespread droughts of the 1910s, 1930s, 1950s, and warned that another clustering of drought years was imminent, a warning echoed by a few other agronomists in the Midwest. This drought began in 1970 in northern Mexico and southwest Texas, advanced northward expanding in space with growing severity, and abruptly ended in 1977 shortly after Governors of western states had met in emergency session (Thompson, 1973; Rosenberg, 1978). Those of the 1950s ended in a similar fashion. President Eisenhower convened a conference to decide what the government should do and the drought thereupon ended so promptly that proceedings were not published (Borchert, 1971).

By 1980 it was widely accepted in meteorology that the phenomenon was due, somehow, to the 20 to 22-year magnetic Hale cycle of the sun where the polarity of the field switches by 180° every 10 to 11 years. Indeed, the entry of my colleague Sultan Hameed into this field began after he viewed a public television program on the subject. I had never examined the evidence of Mitchell et al. (1979) because if the Hale cycle polarity switch of the solar magnetic field can modulate solar luminosity then Maxwell's laws of electrodynamics are invalid and that is not plausible. Currie (1981d), from experimental evidence in air temperature, air pressure, and height of sea level records, and from study of the Mitchell et al. (1979) paper, concluded that the 20-year drought phenomenon is induced by a highly resonant 18.6-year standing wave in the atmosphere. This wave is forced by the 18.6-year constituent luni-solar tide in Newton's tidal potential which is the 12th largest of all the tides induced on earth by the moon and the sun (Godin, 1972).

There now exists evidence worldwide for the 18.6 year term (and 10-11 year signal) in instrumental air temperature and pressure records (Currie, 1979; 1981b,d; 1982, 1987a). In terms of drought proxy data, as for example tree-rings, there is experimental evidence in northeastern China for 500 years (Hameed et al., 1983; Currie and Fairbridge, 1985), in the Patagonian Andes for four centuries (Currie, 1983), in regions of Africa drained by the River Nile for 1400 years (Hameed, 1984; Currie, 1987b), in South Africa since the 19th century (Tyson, 1980, 1981), in India since the late 19th century (Campbell, 1983; Currie, 1984a), and in the United States and Canada since A.D. 1700 (Currie, 1984b, c; Hameed and Currie, 1986).

Currie (1976, 1980, 1981a, c) has also given evidence for one or both of the terms in other geophysical data such as height of sea level and length of the day . In this paper we will present evidence for the signals in precipitation data from the northeastern United States since 1840. Experimental evidence shows that in terms of barometer data the amplitude of the highly resonant 19 year atmospheric standing wave is above Newton's equilibrium value by more than an order of magnitude

(Currie, 1982, 1987a); in terms of 300 mbar wind anomalies over the Himalayas amplification is four orders of magnitude (Campbell, 1983).

In the late 1970s and during the 1980s NASA spacecraft have detected variations in solar luminosity of order 0.1% on the time scale of the solar cycle (Eddy, 1983), in agreement with Currie's (1979) estimate based on study of ground based air temperature records which show a modulation of about 0.2^{o}C. There is therefore now a recognized forcing function for this term. Thus, between 10 and 20 years the atmosphere is a resonating physical system at two discrete periods, one periodic and the other cyclic.

In the writer's published work (Currie, 1984c; Currie and Fairbridge, 1985) there are numerous references to recent evidence obtained by others for the luni-solar and/or solar cycle terms in geophysical data. The data encompass tree-rings, air temperature, air pressure, height of sea level, sea surface temperature, fish catches, and the length of the day. In addition, after a historical debate among seismologists, the leading opponent has now concluded that the luni-solar solid earth tide does trigger large earthquakes (e.g., in southern California). Historically there was considerable interest in the possibility of this tide's influence on climate in the 19th century, but by about 1925 work had apparently completely ceased, only to be revived in the 1950's by scientists in the Soviet Union led by I. V. Maximov (Lisitzin, 1974).

This paper surveys in some detail evidence for both signals in U.S. corn production and closely related commodities such as hog and chicken production. We then discuss evidence for what is known in economic science as the Kuznets' 20-year long swings in the American economy, named in honor of Nobel laureate Simon Kuznets (Soper, 1978). Our proposal is that the long swings in economic variables are principally caused by modulation of agricultural output by climate cycles which are induced by an orbital characteristic of the moon and sun, and the sunspot cycle. A degradation in cyclic behavior occurs as the 18.613 year gravitational force is transmitted by Newton's tidal potential to complex systems in climate where cyclical regularity is maintained, and eventually to sectors of the economy discussed by Kuznets where complex interactions and shocks (e.g., wars or gold discoveries or a collapse of the banking system as in 1932 following the financial crash of 1929) produce larger departures from the cyclical pattern. The relative phasing between the luni-solar and solar cycle terms would also contribute to departures.

2 DATA BASES

Fig. 1 shows the yield in bushels per acre for corn since 1866. Until about 1940 the yield was flat and then began increasing rapidly due to the use of chemical fertilizers and advancing technology. The lower panel displays the fluctuations of interest after the trend and bias have been removed by an N=6 high pass filter (see Fig. 2).

To familiarize ourselves with the magnitudes involved consider 1984. In that year 80.4 million acres of corn were planted of which 79.6

Fig. 1. Upper panel is yield in bushels per acre for corn since 1866. Lower panel shows the fluctuations after a high pass N=6 filter (see Fig. 2) was applied.

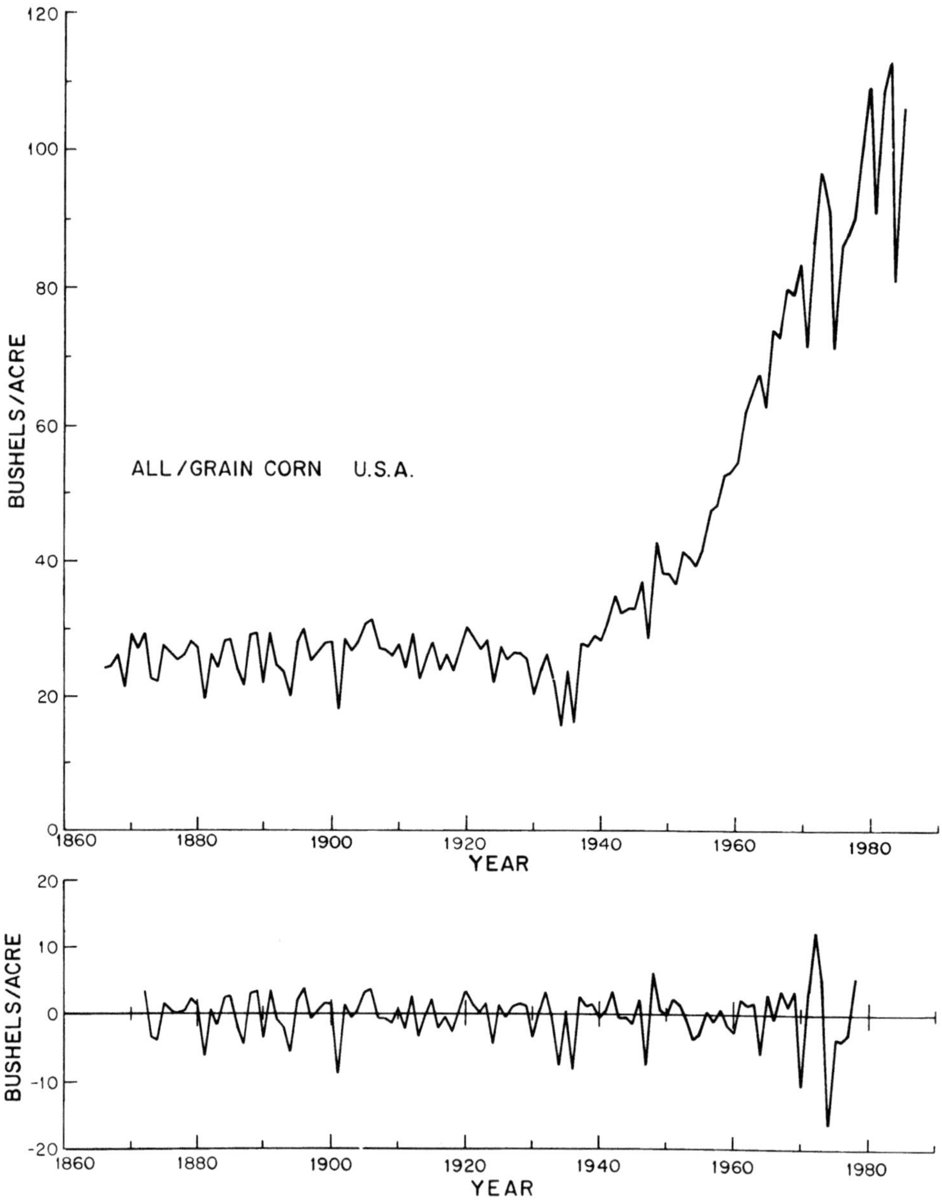

million acres were harvested and 0.8 million acres abandoned. Of the area harvested 71.8 million acres, that is, 90 percent, produced corn for grain with an average yield of 106.6 bushels per acre. The market price of corn averaged $2.69 per bushel making the value of the grain crop to farmers equal to $20.6 billion. Of the remaining acres most were harvested for silage and the rest, comprising about 0.3 million acres, was used for foraging, grazing, and hogging. Corn is the largest cash crop grown and seven states accounted for 68 percent of the output in 1984. Corn also has a shallow root structure and thus is least resistant to drought of all the grains.

Historical data on corn production exists for all 48 contiguous states and are given in a series of statistical bulletins issued yearly and published by the U. S. Department of Agriculture. We reference only two of these (1954, 1984) and have also utilized compilations of data in bulletins 101, 108, 185, 290, 384, 498, 582, and 646. Estimated yield for all corn, that is, the sum of corn for grain and for silage plus hogged and siloed, as well as the total acreage harvested begin in 1866; the time series for all corn was discontinued in 1961 due to difficulty in converting silage yield in tons per acre to bushels per acre. Separate estimates for corn for grain and corn for silage date from 1919, while data on total acres planted begin in 1926. We have analyzed four types of data bases: corn for silage 1919-83, corn for grain 1919-84, percentage of total planted acres abandoned 1926-83, and a 119 year long hybrid series comprising all corn yield from 1866 to 1943 and corn for grain yield from 1944 to 1984. Prior to maximum entropy spectrum analysis a high pass 2N+1 (N=6) weight filter was applied to each record so that the four data bases varied in length from 46 to 107 years when the spectra were constructed.

A variety of time series on U. S. livestock and poultry production are available (U. S. Dept. Commerce, 1970). The longest records date from 1867 whereas the shortest start in 1909. For hog production there are four series, number of hogs on farms in January, live hogs on farms in pounds, number of hogs slaughtered, and dressed pork in pounds.

We have also reexamined data bases on population studied by Kuznets (1961), and on building industry variables studied by Abramovitz (1964). The population data are various series for U. S. census and mid-census dates from 1870 through 1950 while the building sector data are sampled annually. The longest series begin in 1853 while the shortest start in 1915.

Finally, using data only from 1932 and much simpler analysis procedures, Marshall (1972) discovered that luni-solar precipitation in the eastern U. S. is out-of-phase with the western U. S. (he incorrectly ascribed the phenomenon to the 22-year Hale magnetic cycle of the sun). In order to confirm Marshall's work we have analyzed 1,104 monthly precipitation records for the northeastern U. S. The longest record dates from 1814 while the shortest begins in 1915. Results of this analysis has been published (Currie, 1987c) and is included in this paper to illustrate why evidence for the Kuznets 20-year swings in aggregated American economic data should have begun to deteriorate in the 20th century.

3 PROCEDURES OF ANALYSIS

In order to eliminate trends and ultra low frequency components we applied a high pass 2N+1 weight filter to each time series. The frequency response of the N=6 high pass filter over our range of interest is shown in Fig. 2, where we see that at a period of 19 years a unit amplitude would be reduced to 0.4.

Fig. 2. Amplitude response as a function of frequency for a 2N+1 (N=6) weight high pass filter, and a N=30 weight bandpass filter centered at 19 years. The design of the filters is a combination of two side lobe suppression techniques (see Currie, 1967).

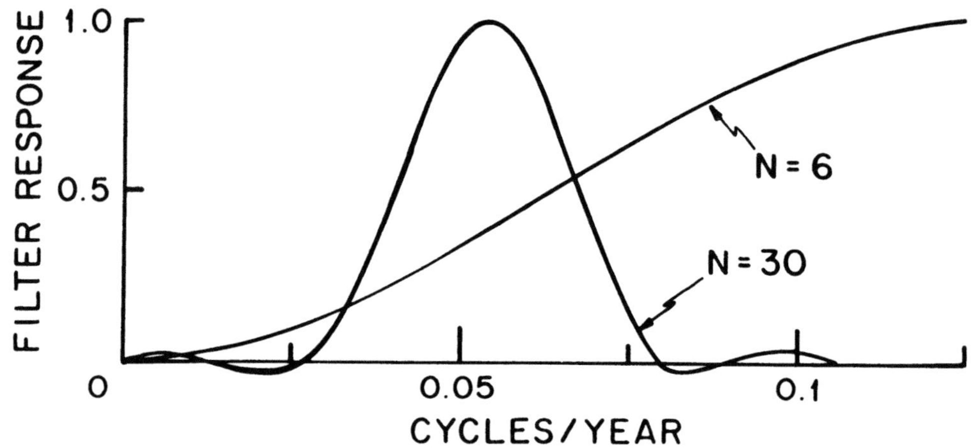

Spectrum analysis is now done using the standard maximum entropy method (MEM) of modern spectrum analysis (Childers, 1978; Kay, 1987; Marple, 1987). The power spectrum gives the power at frequency f as:

$$P(f) = \frac{2P_{L-1}}{\left|\sum_{j=0}^{L-1} a_j \exp(-2\pi i f j)\right|^2} \tag{1}$$

where the sequence $a_0,\ldots,a_{L-1}$ in Eqn. 1 is termed a prediction error filter of order L, and P_{L-1} is the prediction error power. In Jaynes' (1982) derivation of MEM the coefficients are shown to be Lagrange multipliers and we shall so term them henceforth.

For j=0, a zero order spectrum, Eqn 1 yields $P(f)=2P_0$ where P_0 is the zero lag variance of our data. The spectrum is a constant, appropriate to white noise. Higher values of j give distribution of power among different frequencies in the data.

The denominator in Eqn 1 is the power transfer function for frequency f. Once the Lagrange multipliers a_j are obtained we can calculate any number of spectral estimates desired. We computed fifty estimates from f=0 to f=0.125 cycles per year and reduced the power spectrum to an amplitude spectrum (we multiply each estimate by 1/200 and take the square root). We then divided by the transfer function of the high pass filter (see Fig. 2) to restore lower frequencies because, recall, a unit amplitude at 19 years had been reduced to 0.4 by the high pass filter.

It is important to understand that fifty spectral estimates from f=0 to f=0.125 cpy (period 8 years) could be computed even if the number of annual value data points were as low as 20. In order to obtain the resolution given in this paper with the unsmoothed classical fast Fourier transform (FFT) one would need 400 years of data. And even that would not suffice for Brillinger (1981) who insists an unsmoothed FFT is unstable and must be smoothed with low pass filters or "windows". Such insistence takes one back to the state of the art in the 1950's (Blackman and Tukey, 1959). Depending on the length of the filter it would require a wait of 500 to 1000 years to obtain evidence that Brillinger (1981) and other orthodox statisticians would accept.

The order L of the Lagrange multipliers is determined empirically because in our experience the criteria available to determine it do not work with real world data (see Marple, 1987). From experience with meteorological data we know that if all the time series are of equal length, then a pilot study can establish the optimal L for a few series and be applied to all to yield reasonable results. This indicates the noise characteristics of such instrumental data are nearly the same for each record. However, it is known experimentally that when the noise level for a synthetic sine wave is progressively increased then the order of the Lagrangian must also be progressively increased to detect the wave (Chen and Stegen, 1974). For the long 107 year corn records we found optimal L varied from 29 to 44, that is, we had to study each record individually. This indicates the noise characteristics vary substantially from state to state; this is not surprising because the corn data are not numerated by instruments but estimated by aggregating reports from farmers.

Orthodox statisticians, who work in time series, complain that this is "arbitrary" and thus not valid yet Brillinger (1981), and all orthodox statisticians who work in spectrum analysis, advocate applying low pass filters of different lengths to unsmoothed FFT and then choosing the optimal one. Judgment in this case is not considered "arbitrary". Later in the paper we conclude that the entry of orthodox statisticians into time series analysis in the 1920's and 1930's was one of the greatest disasters ever to befall the sciences, and that the very foundations of Brillinger's (1981) book and all the rest are physically invalid.

Colleagues and reviewers have asked why we do not plot the entire 200 spectral estimates out to the Nyquist frequency f=0.5 cpy. The reason has to do with the problem of "over sampling" which is never discussed in textbooks on signal processing because it is entirely a matter of judgment. In principle nothing prevents us from using daily or even hourly data values (if available) and computing a spectrum from $f = 0$ to $f = 0.5$ cycles per day or $f=0.5$ cycles per hour. One difficulty is that this would yield far too many estimates at higher frequencies than we want. But more fundamentally, as acoustic engineers who model human speech find, it is not possible to model a spectrum over too wide a frequency range with a single set of Lagrange multipliers (or prediction error coefficients as they are known in engineering).

In practice we find that with annual values and short records it is not possible to model the spectrum of meteorological data from $f = 0$ to $f = 0.5$ cpy with a single set of Lagrange multipliers, that is to say, the data are "over sampled". In order to obtain line structure between $0 = f \leq 0.125$ cpy the order L of the Lagrangians must be high, and this results in numerous peaks in the spectrum for $f > 0.125$ cpy which are spurious on physical grounds. When modeling the spectrum from $0.125 < f \leq 0.5$ cpy we decrease the order and compute about 60 spectral estimates. The writer has discussed the problem of "over-sampling" with A. Papoulis, author of numerous textbooks on Fourier analysis, and we agree it is one of the most common mistakes in judgment made. Strictly speaking, we should low pass filter the yearly sampled data and decimate by 4 before spectrum analysis, but that would eventually yield a waveform sampled at every 4 years which we do not want.

In Brillinger's (1981) text on pure mathematics in spectrum analysis he takes the long air temperature records from Europe sampled monthly, computes an FFT, applies low pass filters of various lengths, and declares that only an annual term exists. This is an example of gross oversampling combined with invalid procedures. Currie (1987a) processed the same records and found the 18.6-year and solar cycle terms. Unhappily, even modern textbooks on modern methods contain serious errors in procedure. Marple (1987) analyzed a U. S. air temperature record supplied by the writer and tabulated in an appendix. He sampled every six months (mistake 1), tried to model the spectrum from $f = 0$ to $f = 0.5$ cycles per six months (mistake 2), and declares there is no evidence for the signals!

Spectra are necessary to establish the likelihood of the bandlimited terms but their waveforms are more informative. In order to extract the waves, two $2N+1$ ($N=30$) weight bandpass filters centered at 10.5 and 19 years were designed (see Fig. 2 for the 19 year filter). These cannot immediately be applied to the data because that would entail loss of 60 data points, and most of the series are less than 60 years in length. However, the Lagrange multipliers used to construct the spectra embody the characteristics of the available data, that is, two narrow bandlimited signals plus noise. Therefore, they can be used to generate data outside the range of observation and this constructed data will also have two narrow bandlimited terms plus noise (Ulrych and Clayton, 1976). We thus generated thirty points off each end of every record whose spectrum showed the signal near 19 years, and convolved each record with the bandpass filter centered at 19 years; in this manner no data are loss. Each resultant wave was also divided by 0.4

because, recall from discussion of Fig. 2, that was the unit amplitude response at 19 years of the N=6 high pass filter originally applied before analysis began.

Modern methods of spectrum analysis (Childers, 1978; Kay, 1987; Marple, 1987) have been widely accepted by engineers because in terms of performance they are far superior to the classical methods. Acceptance in the physical and social sciences has been very slow because modern methods, such as MEM, stand in conflict with ideas advocated by orthodox statisticians in time series analysis. In a field such as economics, which is completely under their sway, modern techniques are unknown and have, to our knowledge, never been used within the profession. Later we discuss how this situation arose in economics and physical science during the 1920's and 1930's.

4 EXPERIMENTAL RESULTS

4.1 Hybrid long series 1866-1984

Fig. 3 (see p. 190) displays the maximum entropy spectra for aggregate U.S. data and for the three Corn Belt states of Illinois, Iowa, and Missouri. Each spectrum shown from $0 = f \leq 0.125$ cycles per year was constructed from 50 spectral estimates and shows two narrow bandlimited signals near 10 and 20 years. The rapid rise of amplitude at low frequencies is an artifact due to the amplitude response of the high pass filter originally applied which rapidly approaches zero as frequency goes to zero. The signal near 20 years is present in spectra from 24 states while the smaller 10 to 11 year term is detected in data from 30 states. Fig. 4 schematically shows the detection number within the four data bases for each of the two terms. In Kansas both signals were evidenced in all four data bases so the sum of the two numbers shown for Kansas is 8; detection was also especially good in Nebraska, Iowa, Missouri, Illinois, and Indiana, a region comprising most of the Corn Belt.

Fig. 5 shows the 19 year waveforms for the two states of Illinois and Missouri and aggregate U.S. corn. Vertical bars mark the dates of maximum short falls in corn yield. Paired with each bar is a downward pointing arrow and date for the U.S. aggregate which marks an epoch of maximum in 18.6 year tidal forcing of the atmosphere (see Table 1); as is visually evident the paired bars and arrows are generally closely coincident. Fig. 6 shows the average time march for all 24 records where the mean discrepancy between vertical bars and dates of epoch for five epochs is 0.9 ± 1.0 years as shown in Table 1. The mean discrepancy between celestial mechanics and modulation of U.S. corn production is thus about 5%.

Table 1 also displays from 1843.0, for comparative purposes, a mean discrepancy of 1.4 ± 1.0 years between dates of maximum tide and dates of maximum drought D for a tree-ring chronology in Texas. This chronology begins in A.D. 1700 and for the whole record the mean discrepancy is 0.8 ± 1.0. Table 1 also shows dates or epochs of maximum flood F as found from tree-rings in southwestern Canada with a mean discrepancy of

Fig. 3. Maximum entropy spectra for U. S. corn yield, and yields from three Corn Belt States.

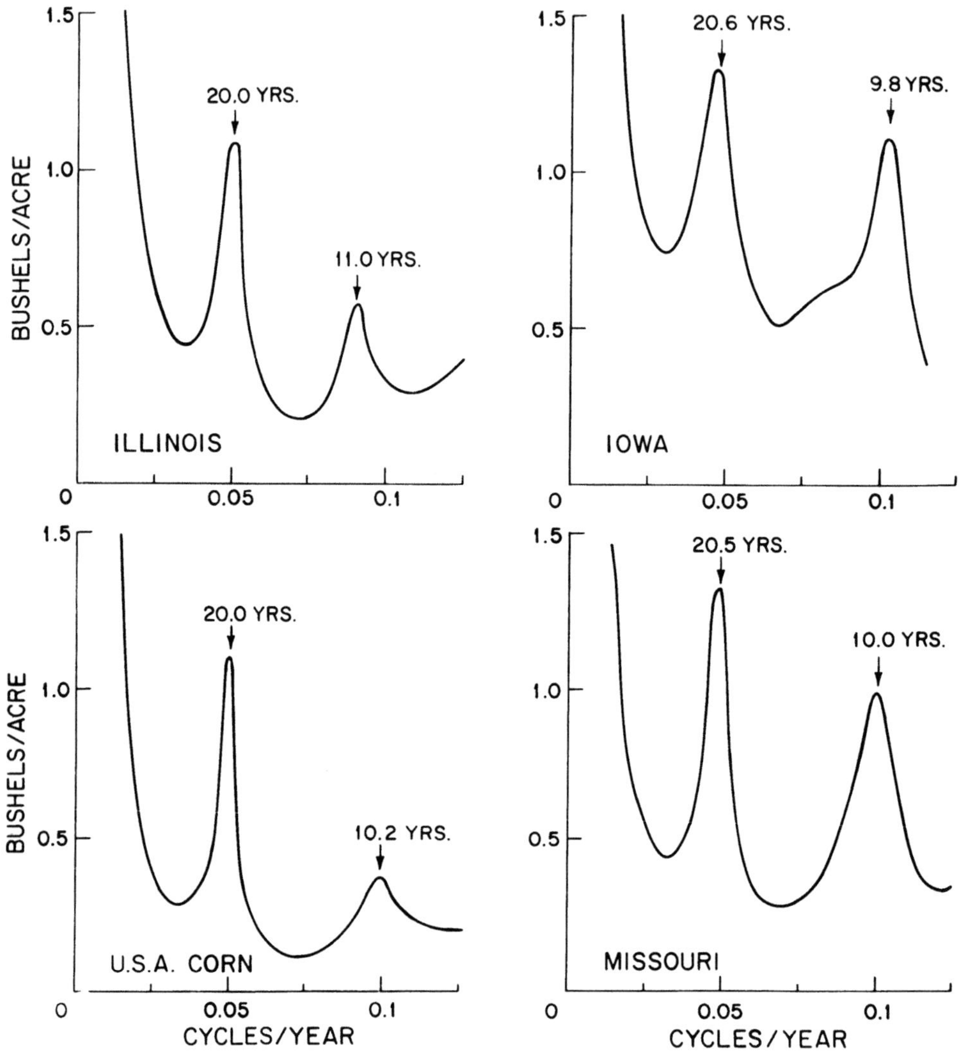

-0.6$\pm$1.5 years. These begin in A.D. 1670 and for the entire record the mean discrepancy is 0.8$\pm$1.1 years. Thus, for the past three centuries tidally induced luni-solar drought maxima in southwestern Canada have been out-of-phase with those in the western United States.

Fig. 4. The two detection numbers per state for four corn data time series, three of which are independent. Number on left is that for signal near 19 years.

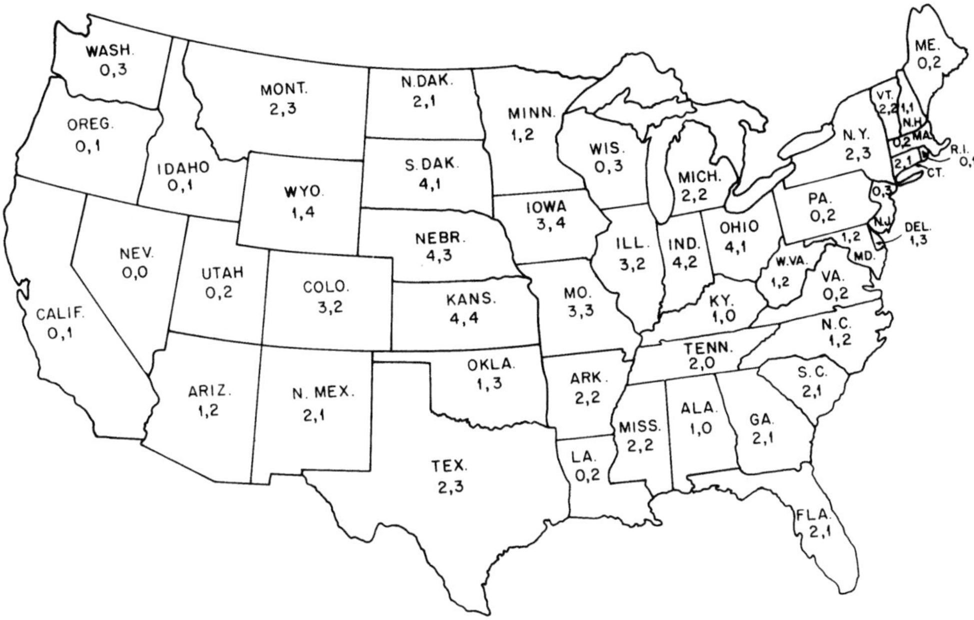

Such nodes in terms of geography are expected for a standing wave in the atmosphere and have been found in the Patagonian Andes since A.D. 1600 in tree-rings (Currie, 1983), in India since 1890 in an areal flood index (Currie, 1984a), in northern China in a drought index for Beijing since A.D. 1500 (Currie and Fairbridge, 1985), and in Africa in the flood levels of the Nile River since A.D. 650 (Currie, 1987a). Since early in the 20th century drought maxima in all of these regions have been out-of-phase with those in the western U. S., and are therefore in-phase with southwestern Canada. Thus, in any region drought maxima can be either in-phase or out-of-phase with dates of maxima in tidal force.

Examination of the individual waveforms making up Fig. 6 for the 19th century indicated that noise levels vary from about 0.25 to 0.5 bushels per acre. Integrating the product of bushels per acre and acres harvested over five year intervals centered at the vertical bars gives a mean value of 6.5×10^8 bushels for the shortfall in U.S. corn production

Fig. 5. The 19-year waves for aggregate U. S. corn and two Corn Belt states.

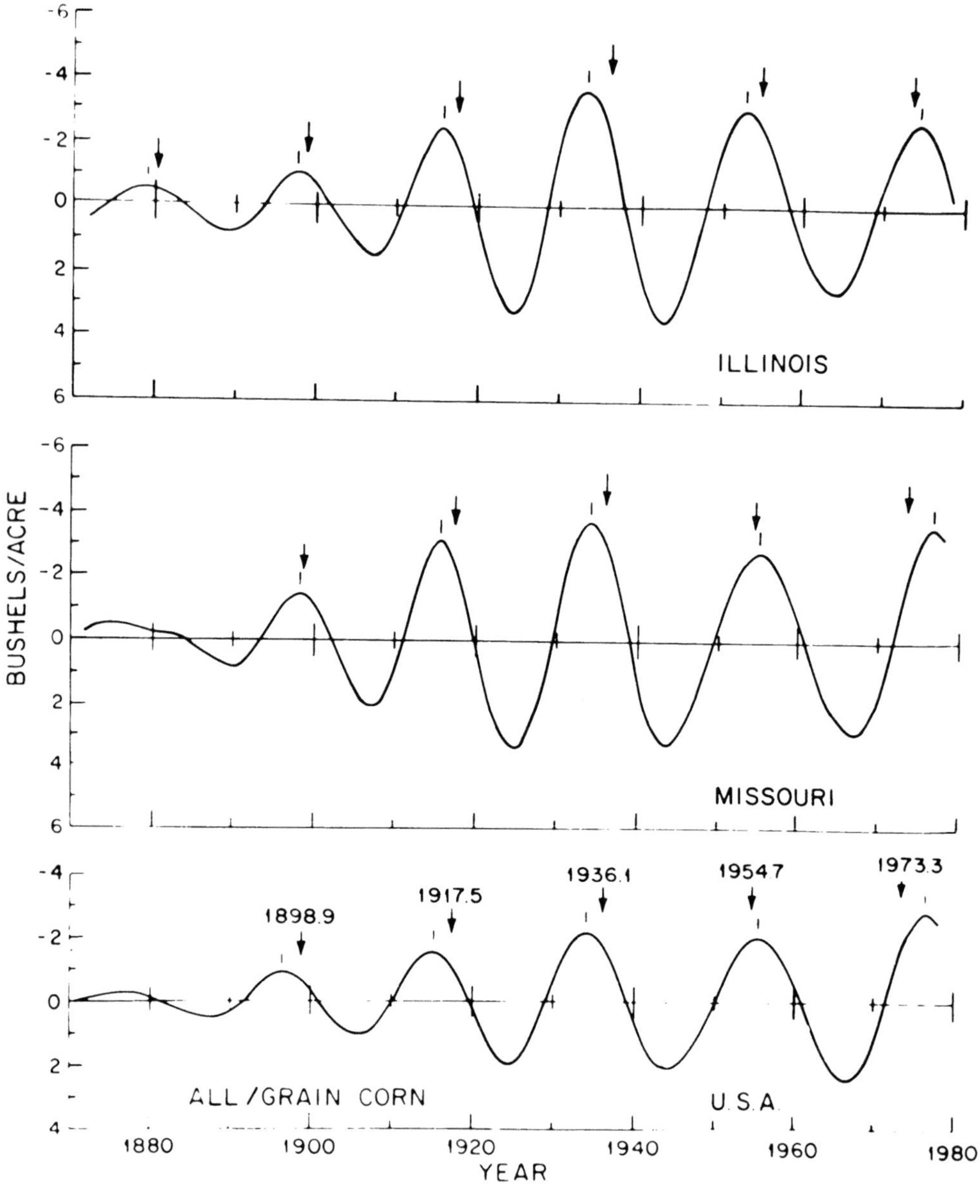

Fig. 6. Arithmetically averaged 19-year wave for 24 states where spectra showed evidence for the signal. Arrows and associated dates are epochs of maxima in the luni-solar tide (see Table 1).

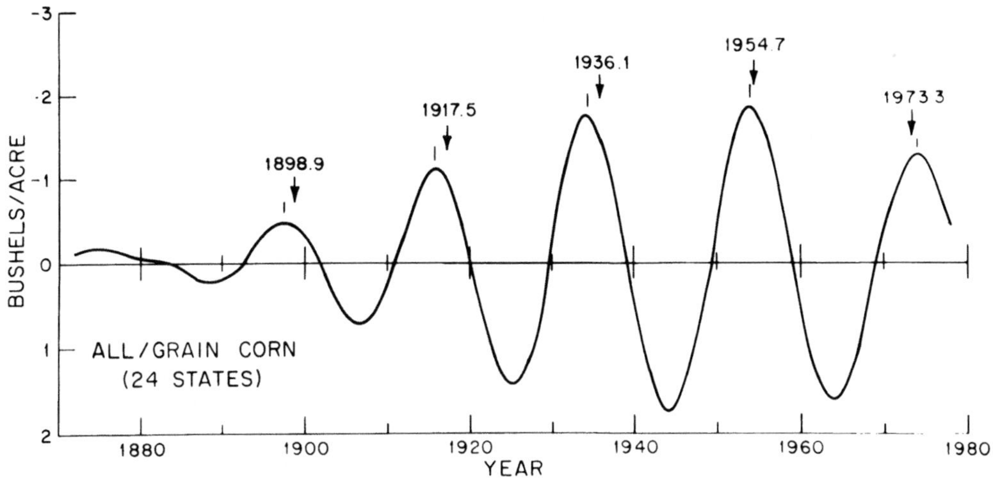

TABLE 1. Summary of Luni-Solar 18.6-Year Epochs for Data in North America

Nodal Tide(a)	U.S.A. All Corn(b)	Texas Tree-Rings(c)	S. Western Canada(d)	N.E. U.S.A. Precipitation(e)
1843.0		D 1842.7 0.3	F 1842.3 0.7	D 1843.0 --
61.6		D 59.8 1.8	F 58.8 2.8	D 62.8 -1.2
80.3		D 78.8 1.5	F 80.5 -0.2	D 1880.2 0.1
98.9	M 1897.8 1.1	D 96.9 3.0	F 99.9 -1.0	SWITCH
1917.5	M 1915.9 1.6	D 1915.6 1.9	F 1916.0 1.5	F 1919.6 -2.1
36.1	M 34.3 1.8	D 35.9 0.2		F 37.3 -1.2
54.7	M 54.1 0.6	D 53.9 0.8		F 56.0 -1.3
73.3	M 74.0 -0.7			74.2 -0.9
	0.9±1.0	1.4±1.0	-0.6±1.5	-1.1±0.7

(a)Dates of tidal maxima for the luni-solar 18.6-yr tide.

(b)Dates or epochs of minima M in all corn yield based on data from 24 states, see Fig. 6.

(c)Dates of maximum in drought D for one tree-ring chronology in Oak Park, Texas, (Hameed and Currie, 1986). Currie (1984c) has obtained similar results from 50 tree-ring chronologies to the west of the Great Plans.

(d)Dates or epochs of maxima in flood F for 8 tree-ring chronologies in southwestern Canada (Hameed and Currie, 1986).

(e)Dates of maximum in drought D or flood F from 739 monthly precipitation records in the northeastern United States (see Figs. 28-29).

in an upswing of the wave, if the noise level is taken to be 0.5 bushels per acre. The loss is $7.6\text{x}10^8$ bushels if the noise is taken as 0.25 bushels per acre. Comparable values are found for windfalls in corn production for epochs represented by downswings of the wave.

Results for the 10 to 11 year solar cycle wave in all/grain corn yield are shown in Fig. 7 for the states of Nebraska, Iowa, and Maine. Estimates for Maine were discontinued after World War II because corn

Fig. 7. The 10-11 year waves for two Corn Belt states and Maine. Arrows represent dates of solar cycle drought maxima from tree-rings (see Table 2).

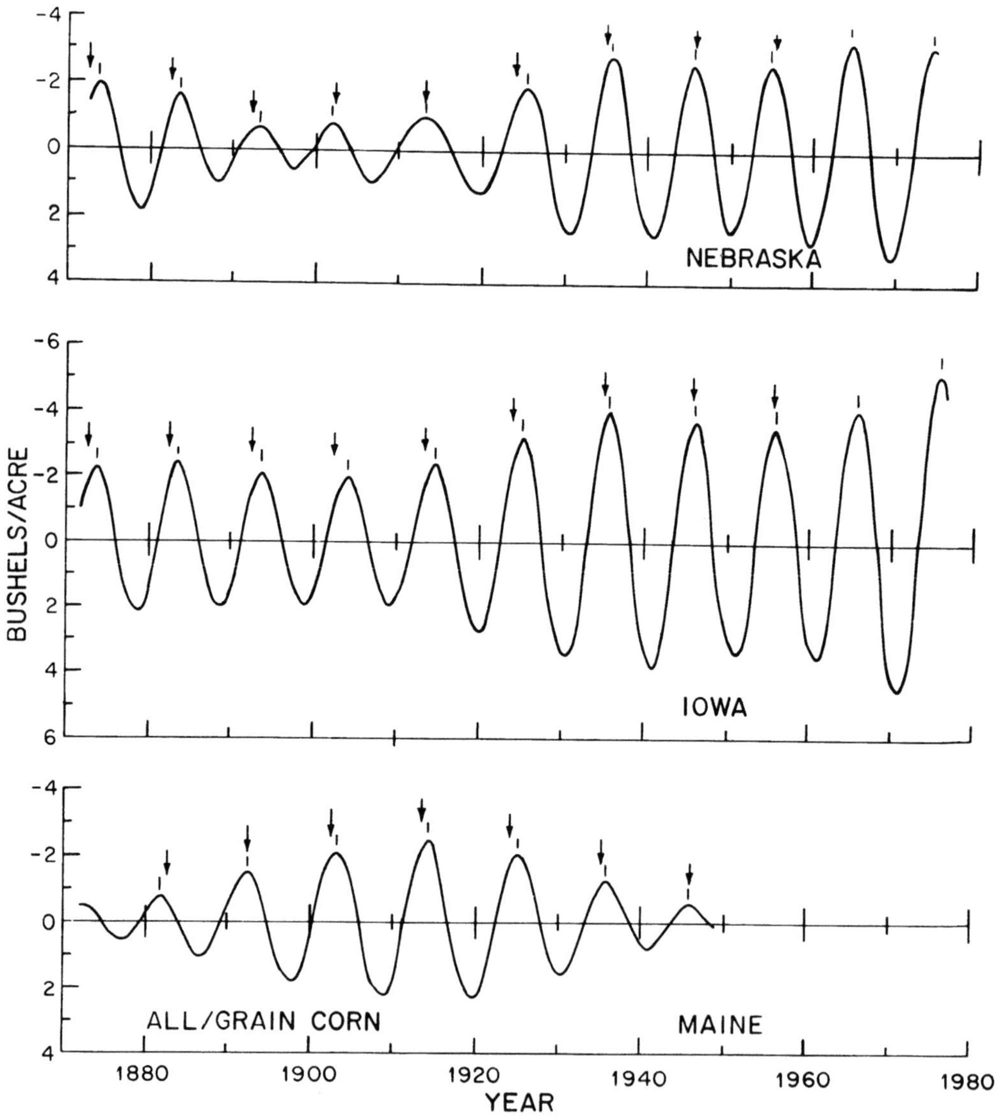

production was no longer a significant factor in Maine's economy. In this instance downward pointing arrows are mean epochs of maximum solar cycle drought measured from tree-rings to the west of the Great Plains (see Table 2); as is evident the paired bars and arrows are nearly coincident. Fig. 8 shows the solar cycle time march for 30 states where dates have been added to the arrows; for this signal the mean discrepancy between solar drought maxima in tree-rings and corn is 0.2±0.6 years as given in Table 2. The two asterisks above the dates 1935.1 and 1955.6 mark times when the 10 to 11 year and 19-year waves were closely in phase and the effect of drought on corn yield was enhanced. Currie (1984c) has discussed in detail the solar cycle drought wave in U.S. tree-rings for the past four centuries.

For the waveform in aggregate U.S. series (not shown) integrating the product of bushels per acres and acres harvested over three year intervals centered at the vertical bars in Fig. 8 gave mean values of 1.3×10^8 and 2.2×10^8 bushels for noise levels of 0.5 and 0.25 bushels per acre, respectively. As was seen in Fig. 3 the amplitude of the solar cycle term in aggregate U.S. data is smaller than in Corn Belt states. Therefore, the impact of this cyclic modulation of corn production in the Corn Belt region is likely to be more pronounced than

TABLE 2. Summary of Solar Cycle Epochs for Data Sets Shown.

U.S.A. Tree-Rings(a)	U.S.A. All-Corn(b)		U.S.A. Hog Production(c)	
D 1882.7	M 1882.1	0.6	M 1883.6	-1.5
D 92.5	M 92.1	0.4	M 92.5	-0.4
D 1902.5	M 1902.7	-0.2	M 1902.7	0.0
D 13.3	M 13.7	-0.4	M 16.1	-2.4
D 24.0	M 24.7	-0.7	M 26.7	-2.0
D 35.1	M 34.9	0.2	M 37.0	-2.1
D 45.9	M 44.8	1.1	M 47.1	-2.3
D 55.6	M 54.8	0.8	M 56.6	-1.8
	M 65.0		M 66.0	-1.0
	M 75.5		M 76.4	-0.9
		0.2±0.6		-1.5±0.8

(a)Dates or epochs of maximum solar cycle drought (D) in tree-ring chronologies west of the Great Plains. The standard deviations on the dates averaged ±0.7 years. See Currie (1984c).

(b)Dates or epochs of minimum (M) in all corn yield from 30 states. The standard deviations on the dates averaged ±0.4 years. See Fig. 8.

(c)Dates or epochs of minimum (M) in hog production from four series. The standard deviations on the dates average ±0.3 years. See Fig. 15.

Fig. 8. Arithmetically averaged 10-11 year wave for 30 states where spectra showed evidence for the signal. Arrows and associated dates are epochs of solar cycle drought maxima from tree-rings (see Table 2).

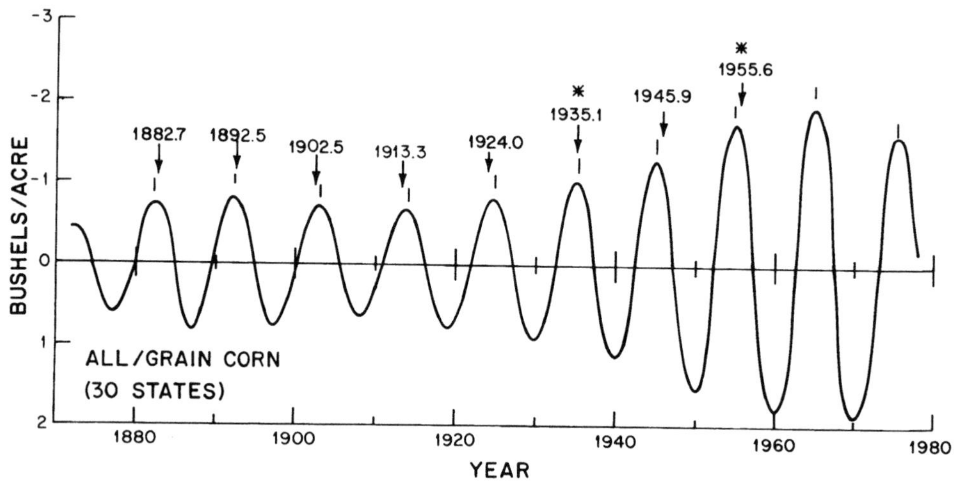

the numbers given. At the luni-solar epoch of 1936.1 the solar wave was closely in phase as shown by the asterisks in Fig. 8 and the combined shortfall for both terms amounted to 11.1×10^8 bushels, for a noise level of 0.25, resulting in an economic loss to farmers of approximately 0.7 billion contemporary dollars. To put this 0.7 billion dollar loss into perspective the total receipts of the Federal Government in 1936 were only 5 billion dollars. The two waves were also nearly in phase during the 1954.7 lunar epoch and the similarly estimated combined loss was 8.3×10^8 bushels or 1.6 billion contemporary dollars. Comparable values are found for epochs of corn yield maxima and represent windfalls for the economy. It should be noted that food from the field to the supermarket is a highly leveraged product in a modern economy. Corn in the field selling for $2.50 a bushel will cost 40 to 50 dollars a bushel when bought as a breakfast food in a supermarket.

In the upper panel of Fig. 9 is displayed the arithmetically averaged lunar and solar cycle waves each of which are replotted in the two bottom panels. Due to phasing the maximum combined shortfalls in corn production centered near 1915, 1935, 1954, and 1974. It is also virtually certain that, given only the composite wave and no prior information on how it was obtained, the human eye could not decompose it into the two constituents.

Let us digress for a moment. By 1970 almost all geographical regions were grain deficit and imported a portion of their food from North America, Argentina, and Australia. This year foreshadowed the lunar epoch 1973.3 and passed with food reserves as days of world consumption standing at three months. In 1974 fifty million acres of idled American cropland, the only major reserves left on the planet, were put into production but could not stem the tide. World reserves had fallen to 1 month, 130 nations met in Rome to discuss the world food crisis, and in three out of four consecutive years the U.S. limited exports due to soaring domestic food prices and consequent political pressure. It was not until the 1980s that grain again became a glut on world markets. For extensive discussion see Currie (1984b, Section 7) and Currie (1984c, Section 9).

Fig. 9. Bottom two panels replotted from Figs. 6 and 8. Top panel is the arithmetic average. When closely in-phase as at 1936.1 and 1954.7 the adverse effect on corn yield is enhanced.

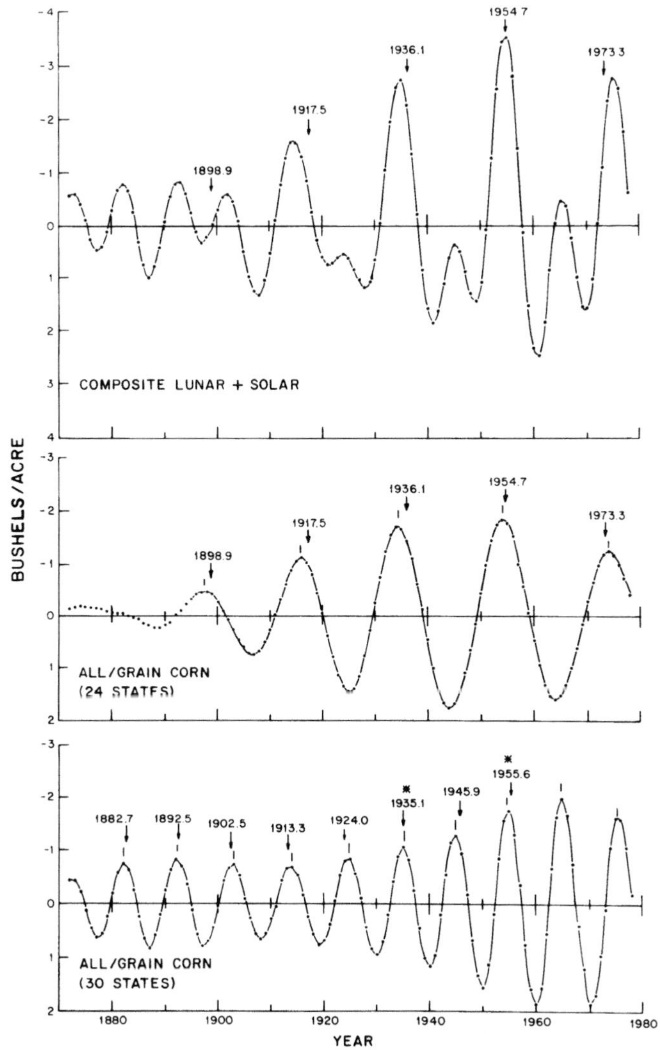

4.2 Short time series

A certain percentage of land which is planted is not harvested each year. These time series are only 46 years in length when spectra are constructed but Fig. 10 shows that both signals are evident in aggregate U.S. percentage of acres abandoned. The two panels on the right show the two waves where it is evident they were closely in phase at lunar epoch 1936.1; they were also approximately in phase at epoch 1954.7. The two sets of downward pointing arrows above the solar wave are dates of maximum shortfall in all corn yield seen earlier and maxima in drought as experienced by trees to the west of the Great Plains. Fig. 11 displays the 19-year wave for U.S. aggregate yield of silage in tons per acre, as well as for the states of South Dakota and Nebraska.

The 19-year term in percentage of acres abandoned was evident in only 11 states and their waveforms did not average in a coherent fashion. The bottom panel of Fig. 12 displays the wave for Nebraska where it is exceptionally large. The upper two panels display the luni-solar wave in corn for grain from 15 states and corn for silage from 14 states. In both instances it is evident the waves are closely in phase. The consistency in phase between these two independent data sets and tree-rings to the west of the Great Plains strongly indicate that a common

Fig. 10. Left panel is spectrum of percentage of planted acres abandoned, while right panels are the two waves (see text).

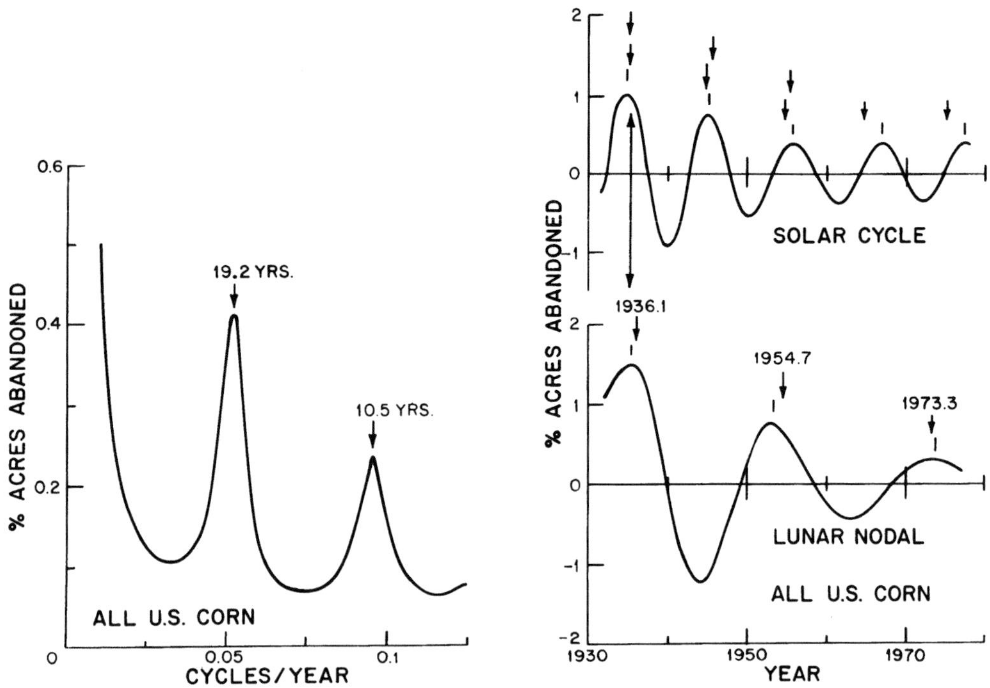

Fig. 11. The 19-year luni-solar wave for two states and U. S. aggregate data for silage yield in tons per acre.

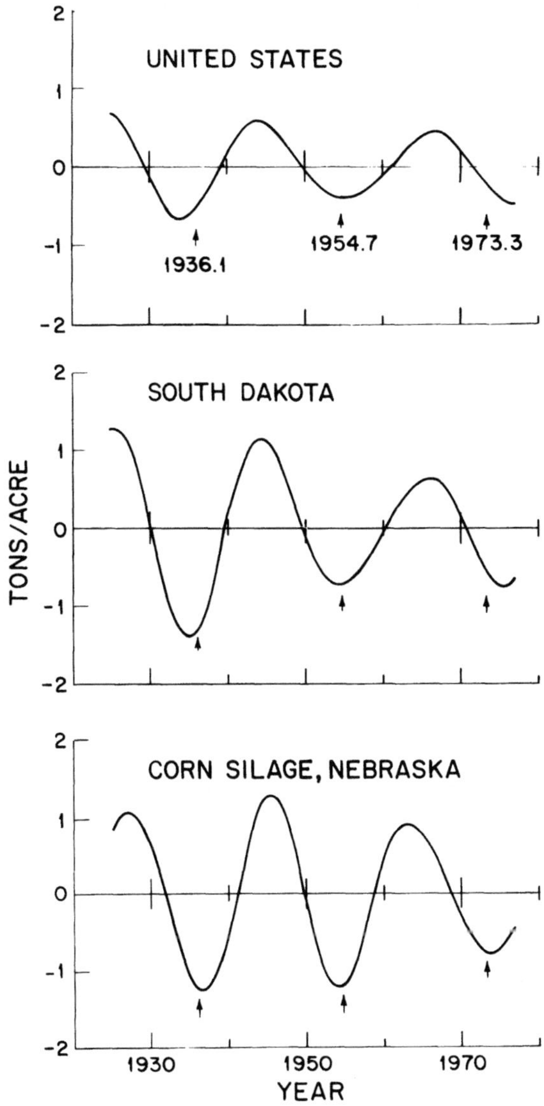

Fig. 12. The two top panels display arithmetically averaged luni-solar waves for corn for grain (15 states) and corn for silage (14 states). The bottom panel is wave for percentage of planted acres abandoned in Nebraska.

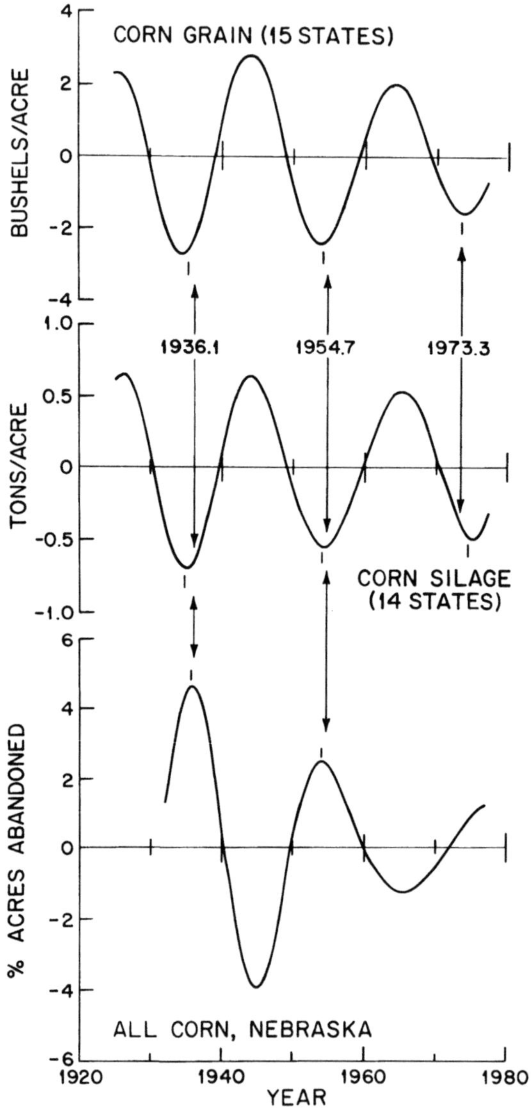

causative agent, namely a highly resonant standing wave in the atmosphere modulating precipitation, affects these time series with a period of about 19 years.

Fig. 13 displays three panels for solar cycle modulation of corn production. The top panel is corn for grain from 18 states, the middle is corn for silage from 16 states, and the bottom is percentage of

Fig. 13. The two top panels display arithmetically averaged 10-11 year waves for corn for grain (18 states) and corn for silage (16 states). The bottom panel is the average wave for percentage of acres abandoned (17 states).

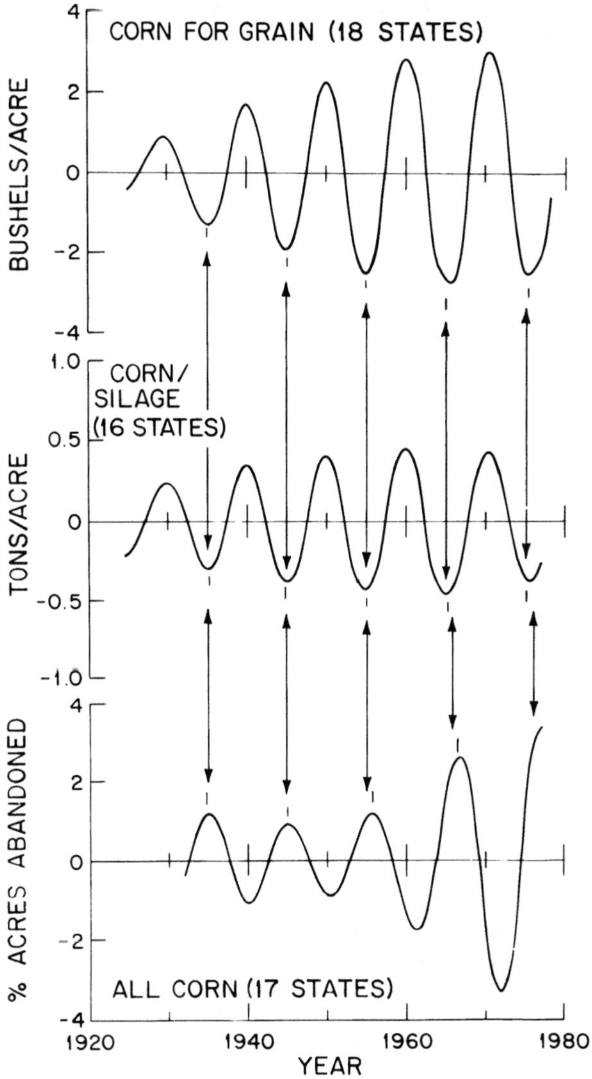

abandoned acres for 17 states. Evidence for the solar wave in abandoned acres is more clear cut than for the longer period luni-solar term. The two sets of arrows linking the three panels are dates of solar cycle drought maxima measured from tree-rings (see Table 2). All four waves are closely in phase providing strong evidence that a common causative agent, in this case a thermally driven tide in the atmosphere induced by a solar cycle variation in solar luminosity of order 0.1% (Eddy, 1983), affect these time series with a period of 10 to 11 years.

We see that each of the two signals under discussion can contribute to the abandonment of 2 to 3 percent of the planted acreage at its maximum. To estimate the impact of the luni-solar wave we again integrated over five year windows at epochs of 1936.1, 1954.7, and 1973.3 and found that a mean value of 2.5 million acres were abandoned due to upswings of this wave. A similar integration over three year intervals gave an average of 1 million acres abandoned due to the upswing of the solar wave. All of the capital and labor expended in planting and caring for these abandoned acres were lost to the economy. These losses are in addition to those caused by synchronous decreases in corn yield discussed earlier.

4.3 Livestock and poultry production

It is well known that a close association exists between corn production and production of hogs and chickens. Moreover, since numbers of hogs and chickens can be rapidly increased or decreased we might expect production would track, fairly closely, the modulation of corn output. Market forces internal to the economy could be a factor so we should not expect to find phasing as exact as exists between corn and growth of trees.

The left panels of Fig. 14 display spectra for number of hogs on farms, number of hogs slaughtered, and dressed pork in pounds. Clearly evident are two signals near 10 and 18 years. For lunar epochs 1917.5 through 1954.7 the panels on the right for the 18-year wave show that hog production closely tracked corn production, that is, when corn production reached a minimum so did production of hogs. The wave is small and seriously distorted at epoch 1973.3.

The single left panel in Fig. 15 displays the spectrum for number of live hogs on farms as measured in pounds. It shows the solar cycle term but no evidence for the luni-solar signal. The four panels on the right are the solar cycle wavetrains for hog production, three of whose spectra were shown in the previous figure. Table 2 tabulates the average dates or epochs of solar cycle minima (M) in hog production seen in Fig. 15 and compares them with solar minima (M) in corn production (see Fig. 8). The mean discrepancy in minima between corn and hogs is -1.5 ± 0.8 years as shown. Thus, minima in hog production lags that in corn by a mean of 1.5 years which is physically realistic. Also shown in Table 2 is the comparison of solar cycle drought maxima as measured by tree-rings to the west of the Great Plains and minima in corn production. The mean discrepancy is 0.2 ± 0.6 years.

Fig. 14. Left panels are spectra for hog production, while right panels are waves for the long period luni-solar term.

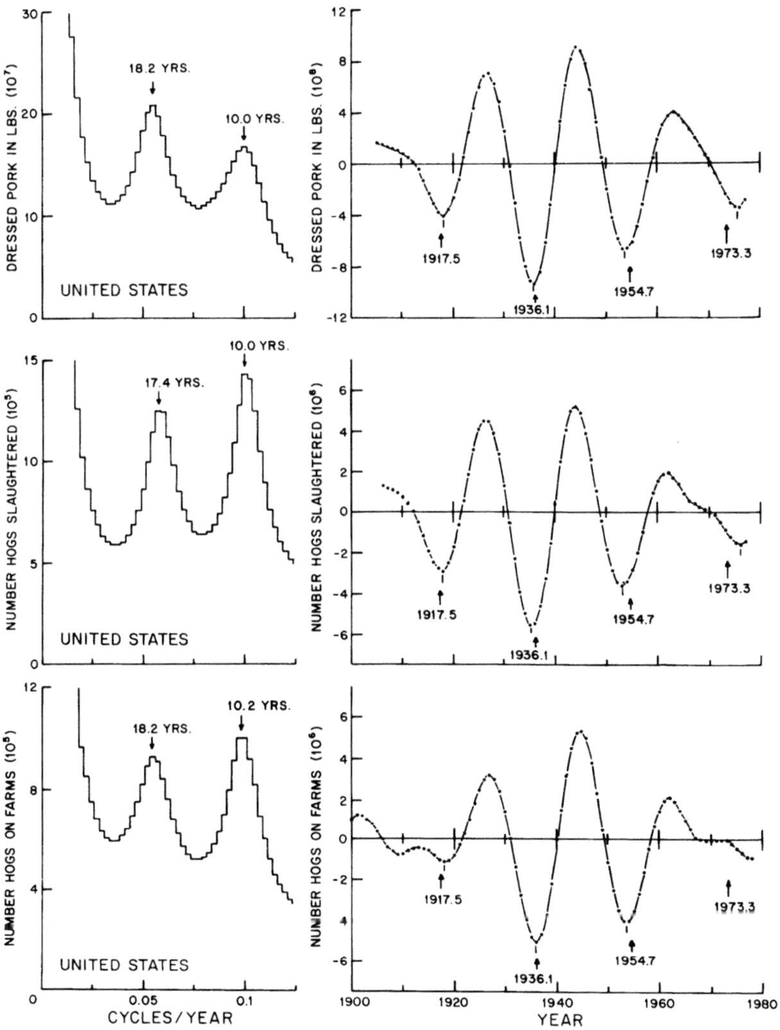

Fig. 15. Two top panels and bottom panel are 10-11 year waves seen in spectra in previous Fig. 14. Remaining panel is spectra and corresponding wave.

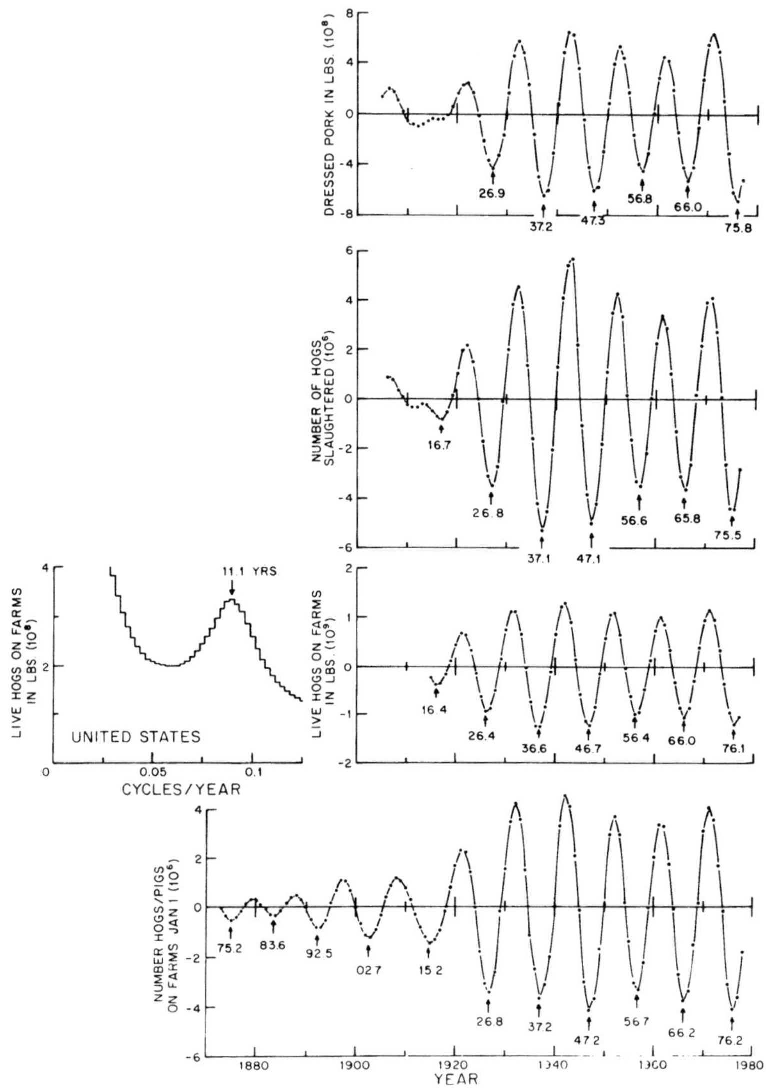

The left panels of Fig. 16 displays the spectra for number of chickens on farms, number of chickens produced, and number of eggs produced. All three spectra display a term near 18 years, but no evidence for the 10-11 year signal. The corresponding wavetrains are shown in the right panels of Fig. 16 and are very similar to those for hogs seen in Fig. 14. Minima in chicken production are highly correlated with minima in hog production through 1960. We mentioned earlier the world food crisis and soaring food prices of the 1970s which, according to Volcker

Fig. 16. Left panels are spectra for chicken and egg production, while right panels are waves for the luni-solar term.

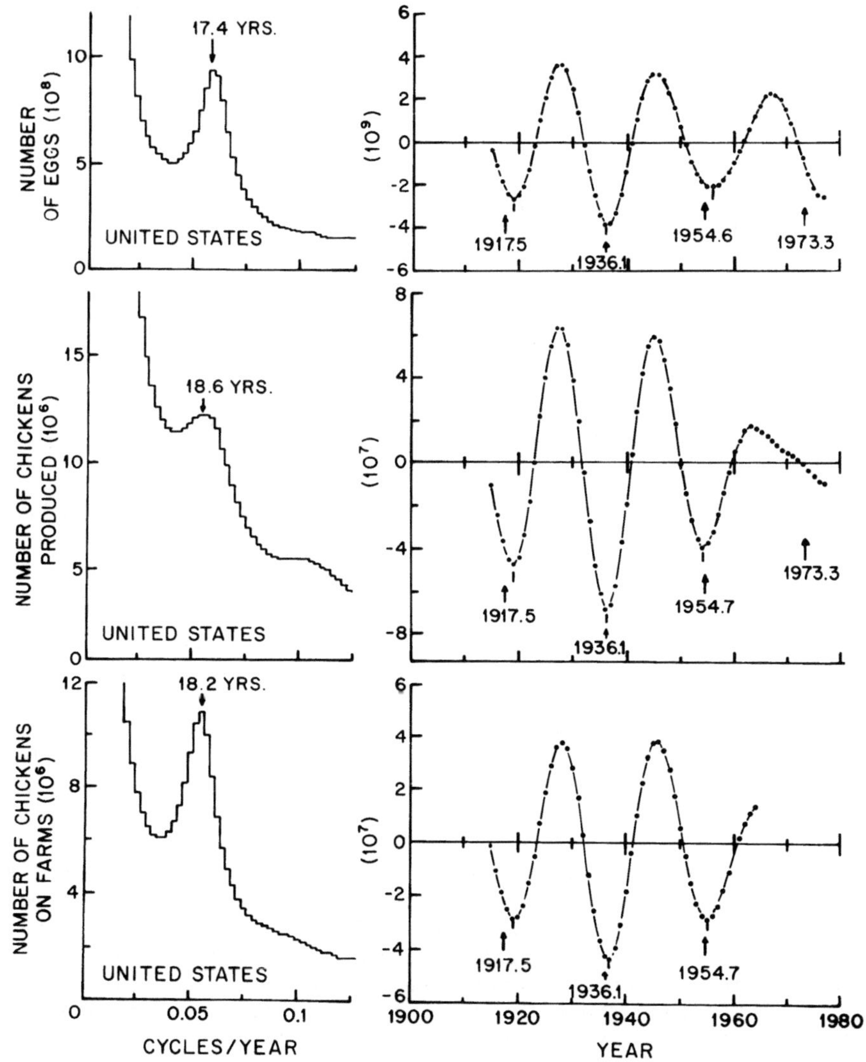

(1978), along with soaring oil prices contributed to the severe recession and economic shocks that befell the nation during that decade. In this instance the historic pattern of hog, chicken, and egg production was significantly distorted.

We would not expect other livestock production to be tied to that of corn, but rather to production of grass, fodder, and hay. There were six time series on cattle, cow, calve, and heifer production all of which had a term between 10 and 11 years. However, only one spectrum, shown in Fig. 17 (left panel), showed a term near 18 years. The waveform (right panel) displays perplexing behavior. For epochs 1880.3 and 1898.9 minima in cattle/calves lead maxima in drought which is not realistic. The remaining four epochs are correlated more closely with maxima in cattle/calves which is not realistic either.

Fig. 17. Spectrum and wave of long period term in a production series for cattle and calves.

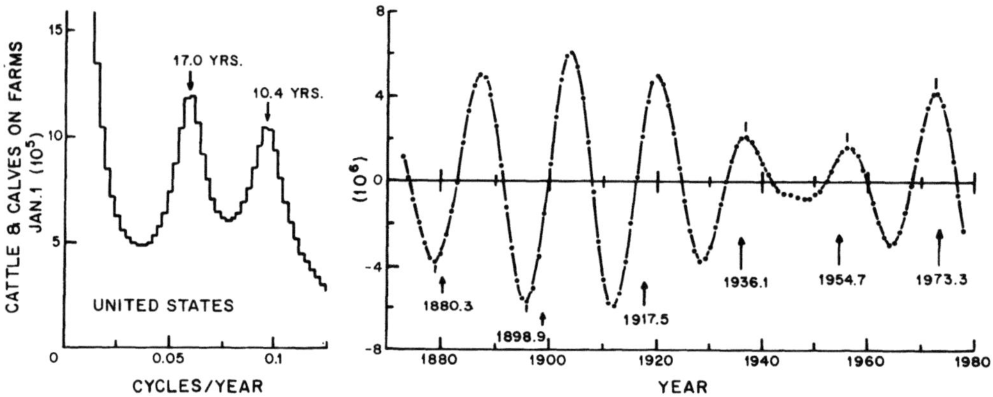

Fig. 18 (left panels) show the spectra for the remaining five series and all display a term from 10 to 11 years. The wavetrains and dates of minima are shown in the right panels. After 1915 the phasing among the six waves are closely in-phase and calculations show minima in production lag maxima in solar drought by 4.6$\pm$1.0 years. We think this lag realistic because a cattleman thinks very seriously before drastically reducing his herd (because a cow produces only one calve each year so it takes several years to rebuild the herd).

We segregated the dates in terms of minima in cattle slaughtered versus minima in live cattle on farms and find the former lags with a mean of 1.4 years. Again this is consistent with the sound business practice of a cattleman. The standard deviations on the dates of solar minima and maxima for slaughtered cattle and dressed beef are smaller by a factor of three than those for live cattle on farms which is also what one expects on ordinary thought since there are very few slaughter houses compared to number of farms and records are accurate and easily collected.

Fig. 18. Top four panels and bottom panel are spectra and waves for the solar cycle 10-11 year signal. Remaining panel is wave for spectra seen in previous Fig. 17.

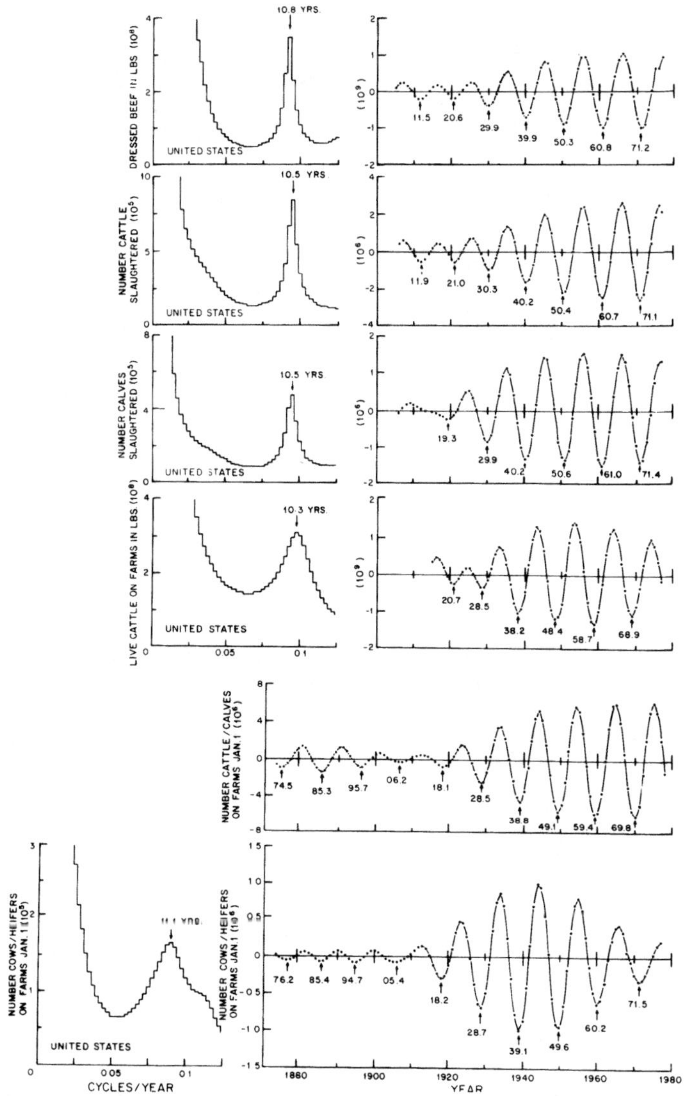

For sheep and lambs there were four series. Three showed a term near 18 years and three a term near 10. As for cattle, the phasing for sheep/lambs for the luni-solar term did not seem realistic. However, for the 10 year signal results in terms of phasing were consistent as shown by the three solar waves in Fig. 19. Moreover, the dates are closely coincident with those for cattle. The mean discrepancy between minima in cattle and minima in sheep production is -0.5 ± 0.8 years.

Fig. 19. Three solar cycle waves from spectra (not shown) for sheep and lambs.

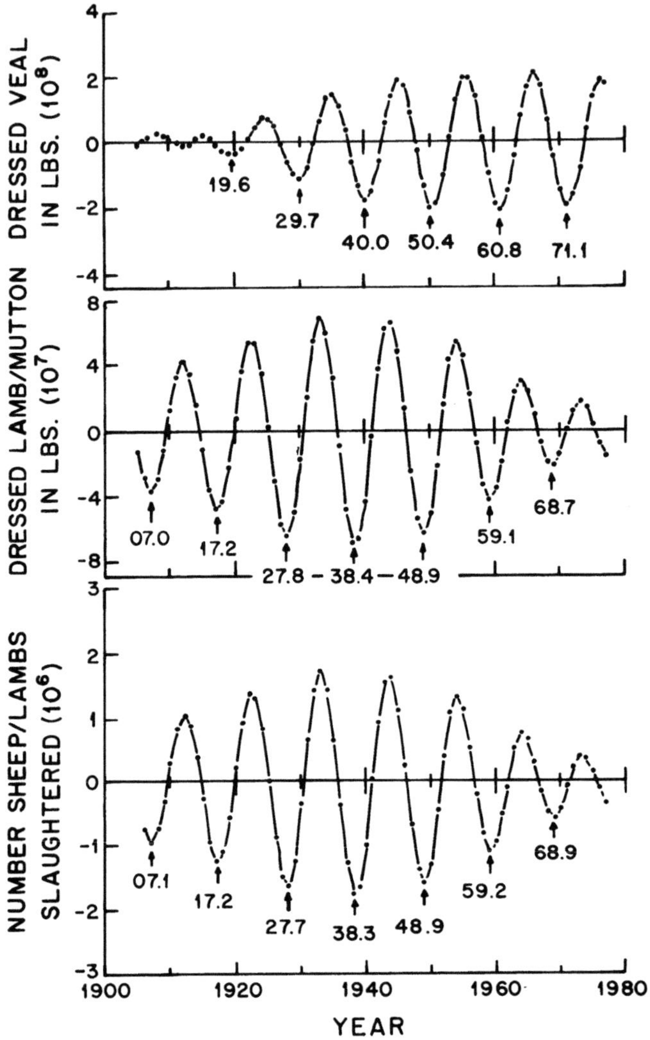

We also find the two signals under discussion in most of the major crops grown in the United States. For reasons unknown the terms are clear cut in yield of spring wheat but difficult to establish in yield of winter wheat. However, the evidence in percentage of acres of abandoned winter wheat is good. Fig. 20 shows the spectra for Colorado and Texas where we note that the scale for Texas is twice as large as for Colorado.

The next major clustering of drought years in the western United States is likely to center near the next lunar epoch of 1991.9. The luni-solar and solar cycle waves will not be closely in phase so, all other

Fig. 20. Maximum entropy spectra for percentage of planted winter wheat acres abandoned in Colorado and Texas.

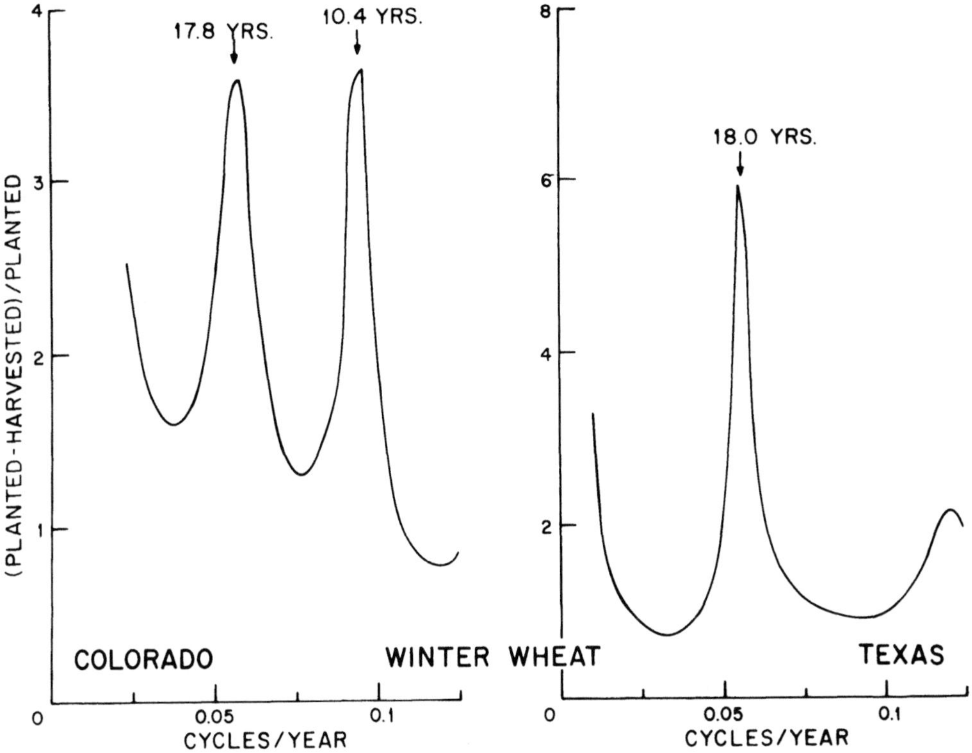

things being equal, these droughts should not be as serious as those of the 1930s and 1950s.

5 IMPLICATIONS IN ECONOMICS

Climatically induced cyclic variations in United States corn, livestock, and poultry production are significant in economic terms, especially for the five Corn Belt states that produce about half the total output. We have noted that these variations are also found in most major crops grown (R. G. Currie, unpublished data, 1986).

As pointed out earlier the luni-solar droughts of the past have seriously affected the U.S. economy due to their impact on agriculture and water resources as documented by Borchert (1971). In the analysis presented we have shown that the years of drought correspond to decreases in corn yield and increases in acreage abandoned, all three responding in a systematic fashion to the atmospheric standing tidal wave of period 18.6 years and the atmospheric solar cycle thermal wave of period 10 to 11 years. This leads us to propose that the 19-year wave in crop output, induced ultimately by an orbital characteristic of the moon and the sun, is a significant factor in causing the "Kuznets' long swings" in the American economy, characterized by Kuznets (1961) as follows: "These swings, approximately twenty years long, are quite distinct from business cycles and must be considered in any discussion of long-term changes in the economy". And further, these swings are "...long alternations in the rate of growth, which we designate 'long swings'-not cycles".

It should be noted that farming is only one component of a modern agricultural economy. Modern agriculture consists of the input supply sector (machinery, fertilizer, power, pesticides, etc.), the marketing sector (truck, barge, and railcar transport, processing, packaging, storing, retail distribution in supermarkets, fast food outlets, etc.), and public agricultural service (Wilcox et al., 1974). It has been estimated that even today about one quarter of the labor force is involved in the agricultural economy (Perelman, 1977). Obviously, the dependence of the economy on agricultural output was much greater in the 19th century when industrial production was much smaller. According to our proposal, climatically induced cyclic variations in crop output propagate into major sectors of the national economy and largely cause the Kuznets' long swings.

Kuznets (1961) has pointed out that the long swing affects diverse sectors of the economy. He states: "These long swings, about twenty years in duration on the average, are clearly observable in additions to population, immigration, gross nonfarm residential construction, gross durable capital expenditures by railroads, net changes in claims against foreign countries (capital imports and exports), and indexes of stock market prices for some groups of securities. We find them also in other components of national product, capital formation, and financing".

Fig. 21 schematically illustrates the long swings in an economic variable. Superimposed on the trend are systematic zig-zag changes in slope. Study of these fluctuations has been almost completely eclipsed since World War II by neo-Keynesian line of analysis because Keynes' theory was not a business cycle theory; concern has focused almost entirely on the trend and how to keep it rising as steeply as possible. We see that as long as the trend keeps rising the minima labeled A, B, and C are not much lower than the preceding peaks of prosperity labeled D, E, and F. The economy is always pretty good and at the peaks D, E, and F it is much better. But, if the trend levels off we see that this is no longer true and the economy is either very good at point G or very bad at point H.

We have applied modern signal processing techniques to all of the data bases examined by Kuznets (1961). The data sets presented in this paper are population time series for U.S. census and mid-census dates from 1870 through 1950 so there are a total of 16 data points. Two points were lost due to the high pass filter applied leaving 14 points to construct the spectrum. Fig. 22 displays maximum entropy spectra of total U.S. population for different orders of the Lagrange multipliers. For L=6 no structure is apparent, while for L=7 through 9 a

Fig. 21. Trend with a zig-zag wave of 20% amplitude superimposed for an economic variable.

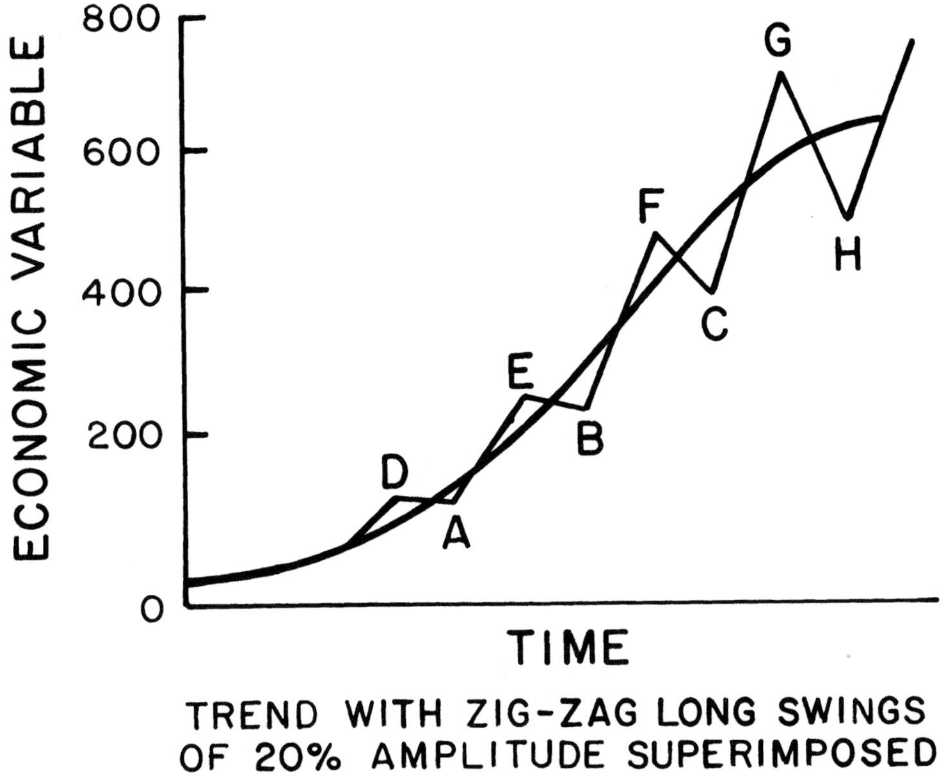

strong bandlimited signal with period 17 to 20 years is clearly evident. The spectrum becomes unstable when L is increased to 10. Fig. 23 shows spectra for six subsets of total population data and clearly there is a bandlimited signal whose period is from 15 to 18 years.

We have analyzed all the yearly sampled data published by Kuznets (1961) which begin in 1869 or 1871, and most of which end in 1953. These series encompass the value of finished commodities and construction materials, balance of exports over imports, GNP and its major components, flow of various type goods to consumers, capital formation and consumption, gross and net construction by type as well as depreciation, changes in inventories and claims against foreign countries, population of native and foreign born (male and female) and nonwhite, and labor force (male and female). We found that during the World War II and post-War era the distortion in the Kuznets wave is so serious that in terms of a spectrum we could not establish a bandlimited signal in the period range 16 to 22 years. However, restricting analysis from 1870 to 1939 a bandlimited term between 16 and 22 years was found in 78 of the 96 time series, and a bandlimited signal between 9.5 and 10.5 years was found in 61 series. The spectra and phasing of the waveforms, the long swings of Simon Kuznets, will be presented elsewhere.

Fig. 22. Maximum entropy spectra for total U. S. population for Lagrange multipliers of order 6 through 10. Data from Kuznets (1961.

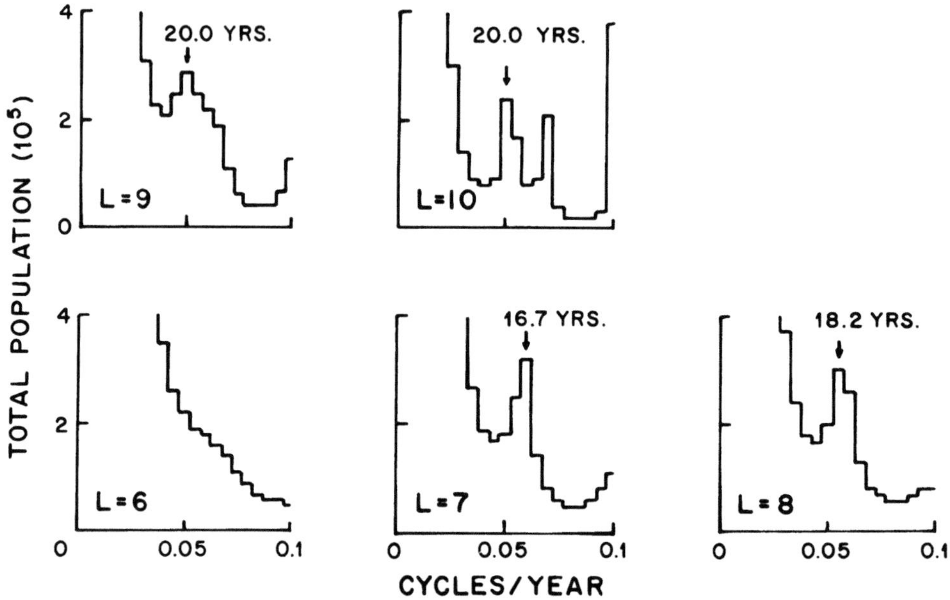

Fig. 23. Maximum entropy spectra for subsets of total U.S. population. Data from Kuznets (1961).

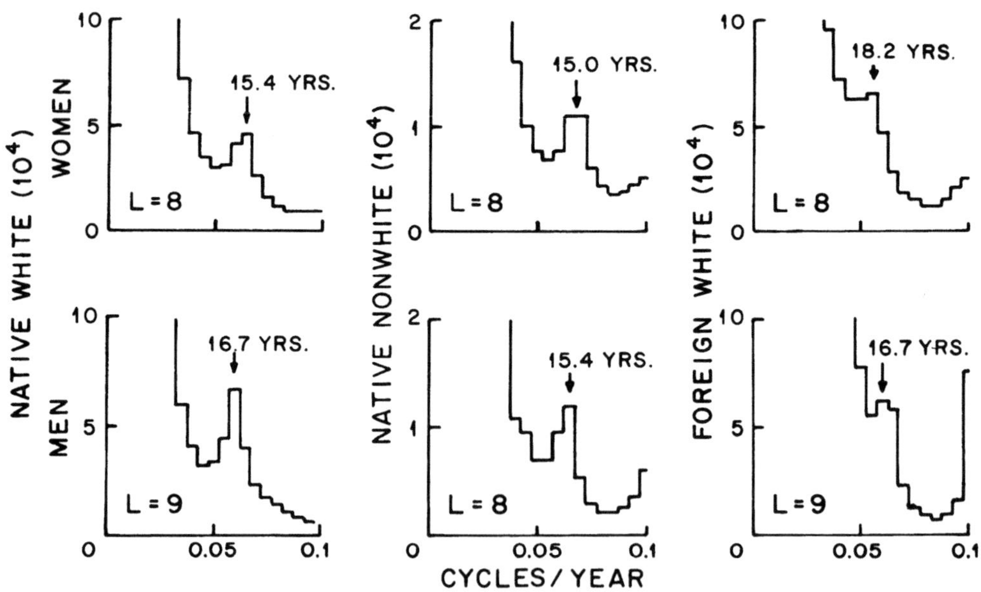

As an example of a parallel phenomenon we discuss the so called "Building Cycle" studied by Burns and Mitchell (1946) who state: "Our studies indicate that building construction is characterized by long cycles of remarkably regular duration. They run usually from about fifteen to twenty years; they are clear-cut in outline, attain enormous amplitudes, and are paralleled by long cycles in other real estate processes". We have calculated the average cycle duration for the 25 U.S. time series given by Burns and Mitchell (1946, Table 161) and find it to be 18.9 ± 3.2 years.

Fig. 24 shows an index of building construction from 1830 through the 1930s. We have superimposed upward pointing arrows and dates which mark maxima in the luni-solar tide; minima in the building wave are highly correlated with lunar epochs and thus with maximum shortfalls in

Fig. 24. The long swings of Kuznets in U. S. building construction from 1830 through the 1930's. Data from Warren and Pearson (1937).

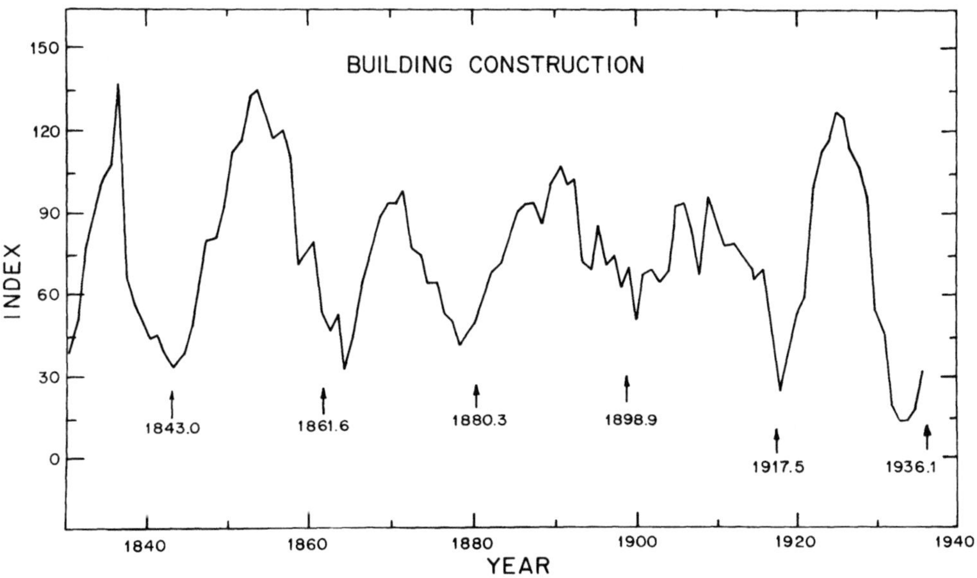

crop production; and vice-versa with respect to mid-epochs. Burns and Mitchell (1946) measured the turning points and the mean discrepancy between their dates on building and our dates on corn is 0.2 ± 1.6 years (see Currie and Hameed, 1986). Building accounts for roughly 25% of total investment and like hog production has a fast response time to changing economic conditions. Fig. 25 displays an index of real estate activity since 1795 and the same wave with the same phasing is evident.

Abramovitz (1964) examined numerous sectors of the building industry and found that the pattern we have seen held in all cases through the 1930s. During the World War and postwar era the wave was seriously

Fig. 25. The long swings of Kuznets in U. S. real estate activity. Data after Roy Wenzlick in the Real Estate Analyst. See Dewey and Dakin (1947, p. 116).

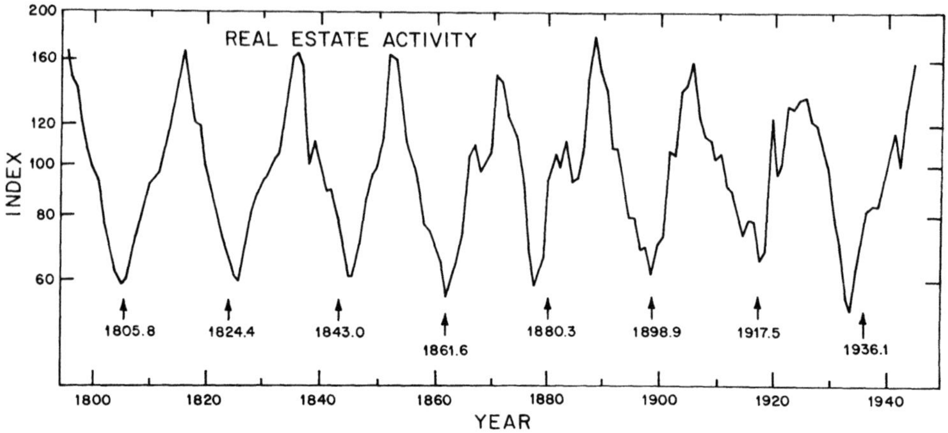

Fig. 26. Autocorrelation functions for six of the U. S. building series examined by Abramovitz (1964). Series number is same as given by Abramovitz.

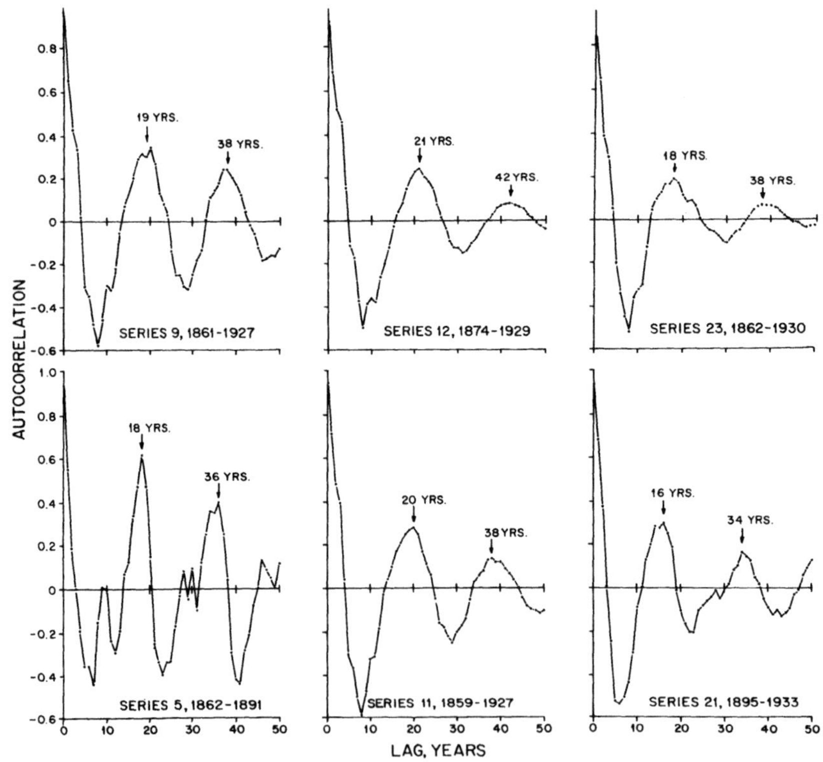

distorted and Abramovitz (1968) later came to believe that the Kuznets phenomenon was restricted to the period between 1840 and 1914. We think we have the explanation as to why aggregate economic data for the 20th century would begin to cast doubt on the Kuznets' effect, but first let us discuss some building data through to the end of the 1930s.

Fig. 26 (see p. 214) shows autocorrelation functions for some of the building series examined by Abramovitz (1964). The functions peak at between 16 to 21 years and at between 34 to 42 years. Note that for series 21 there were only 49 data points so clearly these functions were not computed in the conventional manner; instead, the Lagrange multipliers employed in constructing the frequency domain spectra were used to construct the time domain correlation functions as well.

In Fig. 27 the left panels show unsmoothed fast Fourier transforms for three building series and an average of the transforms for fourteen series. In order to obtain the resolution shown, available data were padded at each end with enough zeros to get a 400 point record. The signal in the available data is so enormous that even after unrealistically padding with zeros it is evident. The right panel displays the fast Fourier transforms after the data were extrapolated to 400 points by the Lagrange multipliers. Since these embody the statistical characteristics of the available data they should do a better job than padding with zeros and we see that they do.

Soper (1978), in a 1970 Ph.D. thesis later published, has presented the most extensive discussion of the long swings in the economy of the United States and other nations. Studies were so successful that by the 1940's Davis (1941), in a textbook on economic time series, extensively discussed the "10" and "20" year waves in economic data. Other books containing evidence in terms of graphs include Warren and Pearson (1937), Long (1940), Dewey and Dakin (1947), Hoffmann (1955), Matthews (1959), Easterlin (1960), Abramovitz (1964), Lewis (1965), and Dewey (1970).

The economic downturn in the 1950s to be expected on the basis of our proposal was mild, yet a retardation in economic growth did occur of such persistence that Abramovitz (1959) testified before Congress on the phenomenon of Kuznets. However, the downturn to be expected in the 1970s was severe. Building investment, residential and non-residential, reached a deep trough in 1975 (van Duijn, 1983). Soaring food prices combined with shocks in oil prices led to a 6% plunge in GNP, the most severe recession since the 1953-1954 downturn. The resulting pervasive economic distress and schisms within the economics profession led Paul Volcker (1978), present Chairman of the Federal Reserve Board, to announce the rediscovery of the business cycle, that is to say, the long swings of Kuznets.

We have been unable to find books on economics which discuss the contribution of the agricultural economy to business fluctuations. There were differences from situation to situation, but Kindleberger (1973,

Fig. 27. Left panels are fast Fourier transforms (FFT) for three of the Abramovitz (1964) building series, padded with zeros to yield 400 points. Right panels are FFT after the series were extrapolated to 400 points by the Lagrange multipliers (see text).

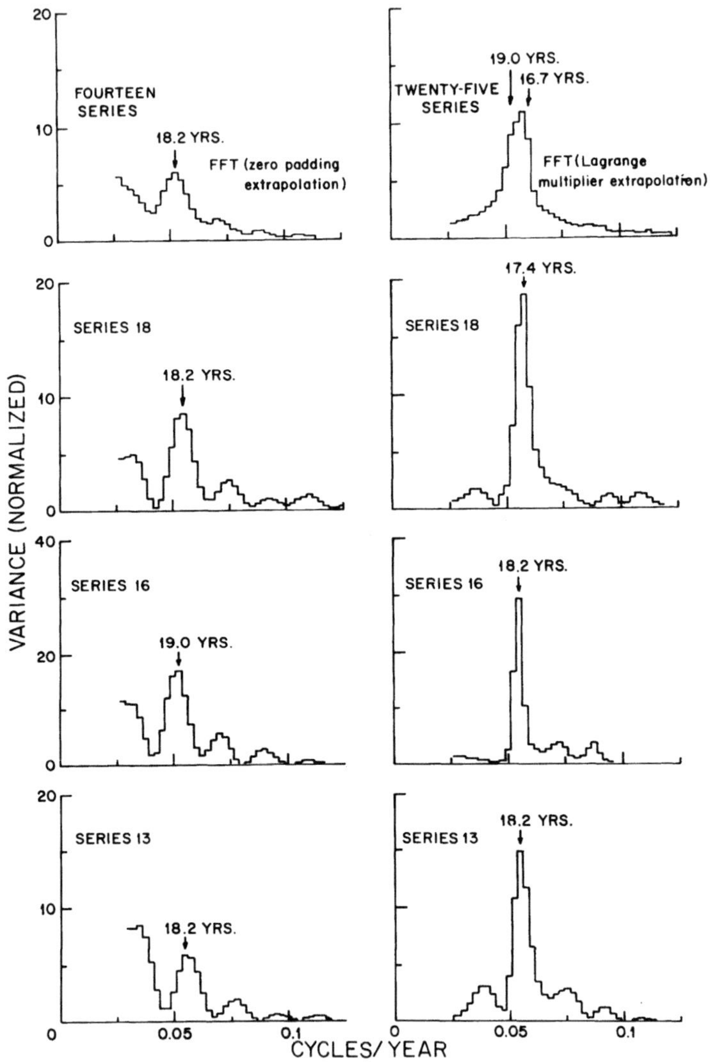

p. 85) states that generally until 1857 or perhaps 1866, the harvest was the measure of business conditions. A bumper crop lowered the price of bread, and hence industrial wages, and simultaneously provided an outlet for industrial produce by enlarging farm income. Crop failure, on the other hand, led to depression. Subsequently, he says, there was a tendency for economists to forget agriculture altogether when seeking explanations for business fluctuations; and to relate them to financial conditions or the state of industrial inventories, expenditure for plant and equipment, or population movements. Most economists seem to equate agricultural workers with number of farmers on farms and this leads to estimates a factor of five or six times lower than that given by Perelman (1977) as discussed earlier.

6 ANALYSIS OF CLIMATE DATA

Numerous precipitation records in the northeastern U.S. exist of sufficient length to investigate the signals under discussion. In Fig. 28 the left panel is the average spectrum of yearly total precipitation data from 74 rain gauge stations while the right panel is the average 19 year wave for these data. For some reason the solar cycle signal in rainfall is weak in this region (see Currie, 1987c). From 1843 to 1880 shortfalls in precipitation were approximately in phase with lunar epochs, that is, lunar induced drought was in phase over the entire United States. A polarity switch of 180° then occurred and by epoch 1917.5 windfalls in precipitation were approximately in

Fig. 28. Maximum entropy spectra and luni-solar wave for yearly total precipitation data over northeastern U. S. Seventy four station records were employed. See Currie (1987c) for details.

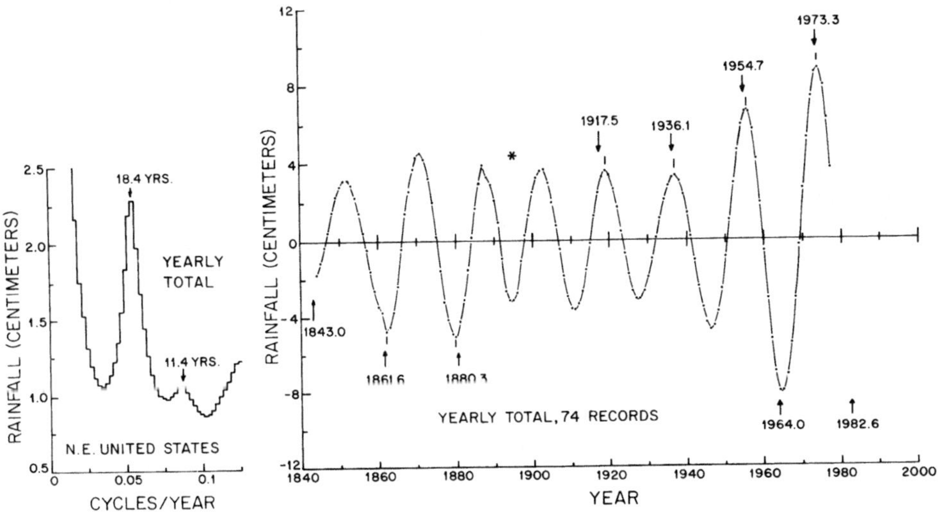

phase with epochs. Bistable 180° phase switches with respect to time in drought/flood data have also been found in South America (Currie, 1983), India (Currie, 1984a), China (Currie and Fairbridge, 1985), and in Africa (Currie, 1987c). Once a polarity is established, the time it remains in the same state varies from 100 to at least 300 years (Currie, 1984c). Bistability in physical systems has been extensively studied by physicists and engineers and can occur when the parameters in the equations of motion are periodically forced (Stoker, 1950).

Under our proposal once the bistable switch in eastern U.S. precipitation data occurred, then aggregate economic time series would begin to present a confused picture with regard to the long swings and this is what Abramovitz (1968) observed. Fig. 28 thus has important implications in future research by economists on the long swings, both in the United States and in other nations.

Prior to the 1940s the amplitude of the wave is fairly constant and an upper bound on the effect is obtained by integrating over 5 year windows. We find an average upperbound value of 17.8±3.5 centimeters. Since this area is humid with a mean annual rainfall of 100 to 130 centimeters the effect might be moderate. However, for some reason as yet unknown, the amplitude after the 1940s rapidly increased. At the 1964 mid-epoch the integrated shortfall was 36 centimeters which amounts to a wall of water 14 inches high over this portion of the country. This region did experience a severe water crisis near mid-epoch 1964 (Barksdale et al., 1966; Palmer, 1965-67). According to Namais (1966, 1967) the drought lasted nearly half a decade from 1961 to September 1966.

Given our present state of knowledge, but placing ourselves back in the 1970's, we would on the basis of Fig. 28 (truncated during the 1970s) confidently anticipate a serious water crisis near mid-epoch 1983. This crisis occurred. For example, by 1985 the cumulative deficit in rainfall had reached such a magnitude that New York City was forced to draw some of its drinking water supplies directly from the Hudson River.

Although not visually evident in Fig. 28 the dates of lunar rainfall minima in the 19th century and rainfall maxima in the 20th century systematically lag lunar epochs by 1.1±0.7 years (see Table 1). Thus at current mid-epoch 1982.6 the window of drought potential should center at about 1983.7. The water crisis in the northeast peaked in 1985 whereas in the southeast it peaked in 1986, drought and high temperatures so extreme that the situation received extensive national media attention.

We thought that perhaps the amplitude might be unevenly distributed over the year but Fig. 29 shows this not to be the case. The amplitude is uniformly distributed over April, May, and June and for each of the other 9 months as well. In retrospect we realized the amplitudes should be uniform because the forcing function is long in relation to 1 year. The mean period for monthly rainfall spectra is 18.324±0.133 years so the discrepancy from celestial mechanics is 155 parts in 10,000 or 1.5%. It is interesting to note that Newton established the lunar theory with a discrepancy between observation and his theory of 400 parts in 10,000 or 4%.

Currie (1979, 1981b, d) has exhaustively studied air temperature records in the United Sates from the world records, and also those for the whole world (Currie, 1987a). We were thus surprised to learn that there are available an order of magnitude more records for the U. S.

Fig. 29. Left panels are spectra and right panels are luni-solar precipitation waves for months of April, May, and June in northeastern United States. See Currie (1987c) for details.

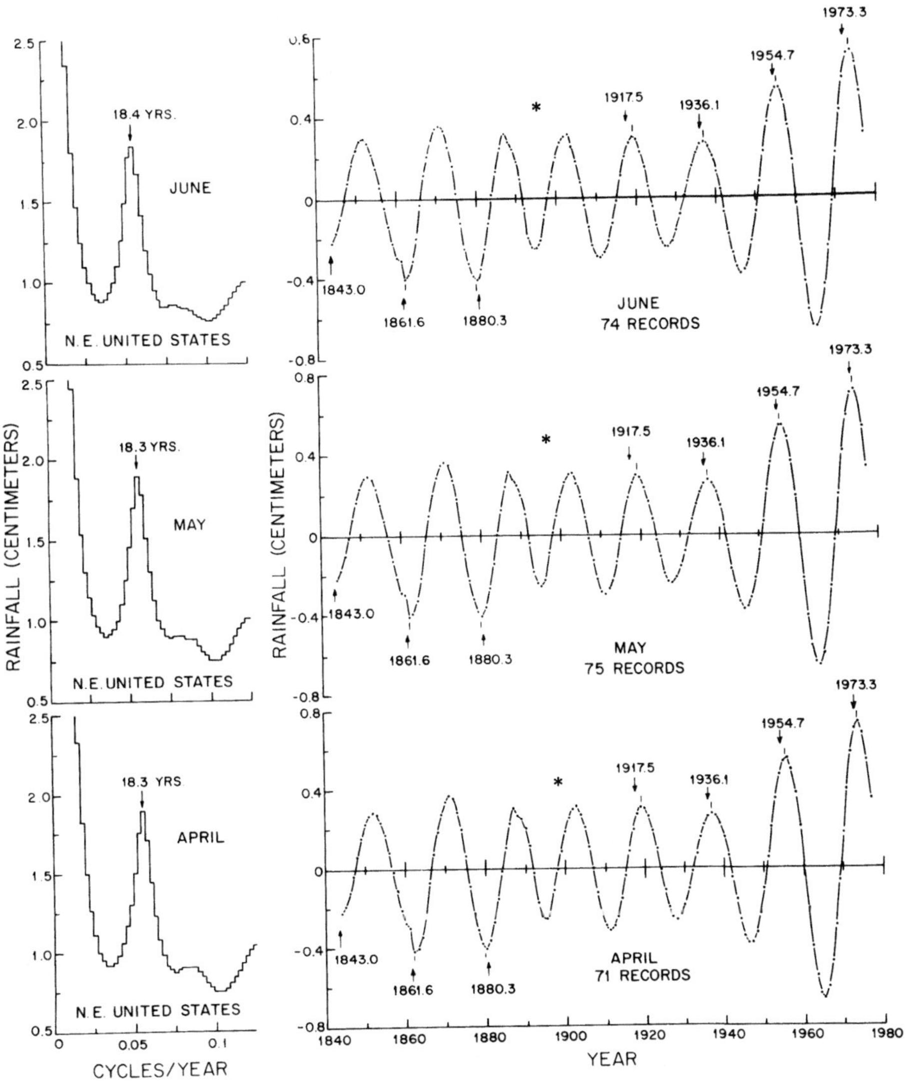

than we had thought, although they are considered to be of lesser quality.

Using monthly data we detected the luni-solar and solar cycle signals in 994 air temperature records over the northeastern U. S. The left panels of Fig. 30 show arithmetically averaged spectrums for the months of October, November, and December. Those for other months are very

Fig. 30. Left panels are spectra and right panels are luni-solar air temperature waves for three months in the northeastern U. S.

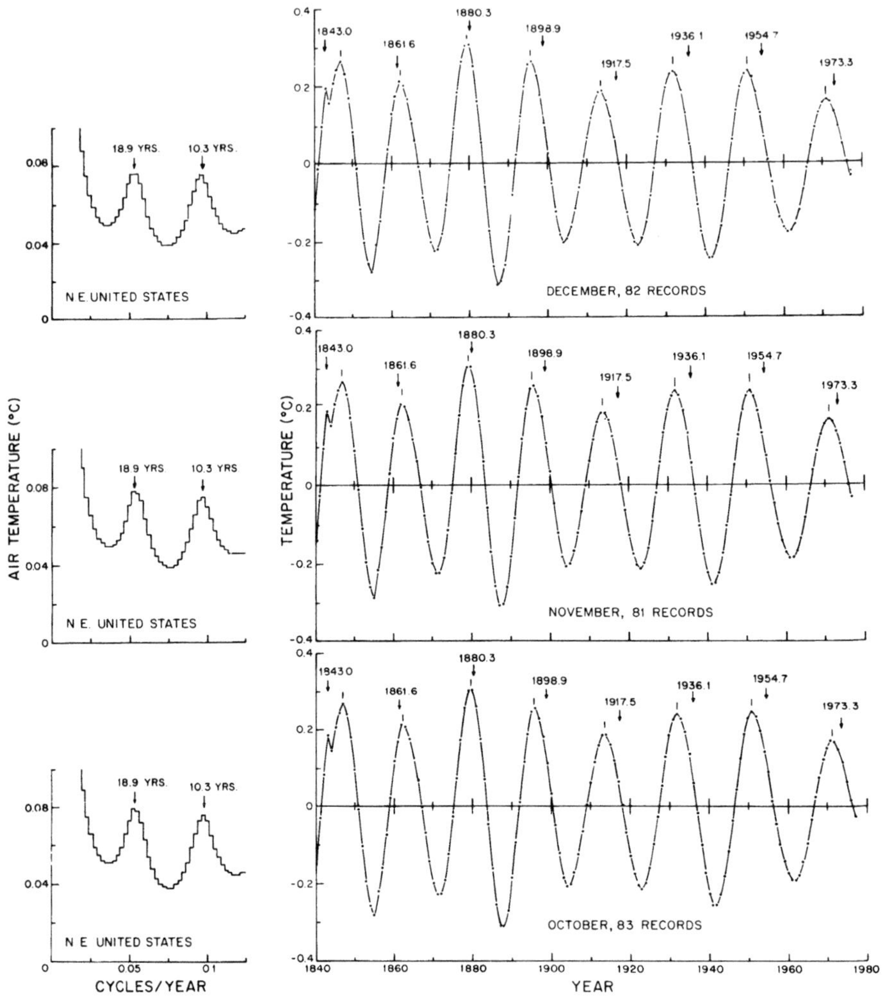

similar. The right panels display the averaged wavetrains for the luni-solar term. Results are similar to those presented by Currie (1981d), that is, maxima in air temperature came closely into phase with the tidal epoch of 1880.3 and thereafter systematically lagged epochs by an average of 3.4 years. Such systematic phase shifts have also been found elsewhere in air temperature (Currie, 1987a), but are not observed to occur in data directly related to precipitation. Integrated amplitudes over five year intervals centered at wave maxima are upwards of 1°C.

7 IMPLICATIONS IN SOCIOLOGY

Soper (1978) proposes that the relevance of the long swings extend beyond economic history into the realms of sociology and political history. And that the pressures of a prolonged downturn in economic conditions might well go far in explaining profound intellectual pessimism, political unrest, and generalized social discontent over a sustained length of time. Conversely, a long upswing would be conducive to an era of relaxed sociopolitical tensions, buoyant expectations, and an enterprising or "progressive" spirit within a nation. We have obtained experimental evidence which supports Soper's (1978) thesis.

During almost all of human existence people have lived by hunting or by subsistence agriculture. This continues to be true for the majority in Africa, South America, and Asia. Any economy based largely or entirely on subsistence farming is inherently poor and highly susceptible to economic distress caused by shortfalls in food production due to lack of rain. The unhappy events in recent years in Africa with regard to prolonged luni-solar drought and consequent famine at mid-epoch 1983 is a good example and widely known. Luni-solar drought also led to serious famine in Brazil at the most recent mid-epoch but received little notice in the U.S. because Brazil did not ask for food assistance.

Although much human conflict has been rooted in opposing religious ideologies, it is probable that most civil unrest and conflict between nations has been the result of economic adversity. We have analyzed two indices for the past 26 centuries, one for civil war battles and one for international battles. These were constructed by R. H. Wheeler (see Dewey, 1970) giving value one to a mild engagement, value two to a moderately severe encounter, and value three to a very exceptionally heavy engagement. The index for each year is the sum of the values so determined.

We computed 26 spectra for each series in two hundred year overlapping swaths and found for the total 52 spectra evidence for a bandlimited term between 16 and 21 years in 45 instances; 38 spectra displayed a term between 10 and 11 years. Fig. 31 displays spectra of combined international and civil war battles from the third through the sixth centuries after the birth of Christ. The earlier two centuries witnessed the decline of the Roman Empire, while the following two centuries witnessed its fall when the last Emperor was deposed in A.D. 472; we see that the term near 20 years is enormous, larger than anytime during the past twenty six hundred years.

Fig. 32 displays the 'long swing' waveform from A.D. 200 to 600. The recurrent wave is especially strong during the sixth century, the century following the collapse of the Roman Empire and its central authority. It appears that, contrary to general belief, mankind is a part of nature and subject to natural law; and that more than one tide exists in the affairs of men and woman.

The moon has long played an important role in the lives of people. Each Jewish month begins with the appearance of a new moon. In the

Fig. 31. Maximum entropy spectra for international and civil war battles for two 200 year intervals.

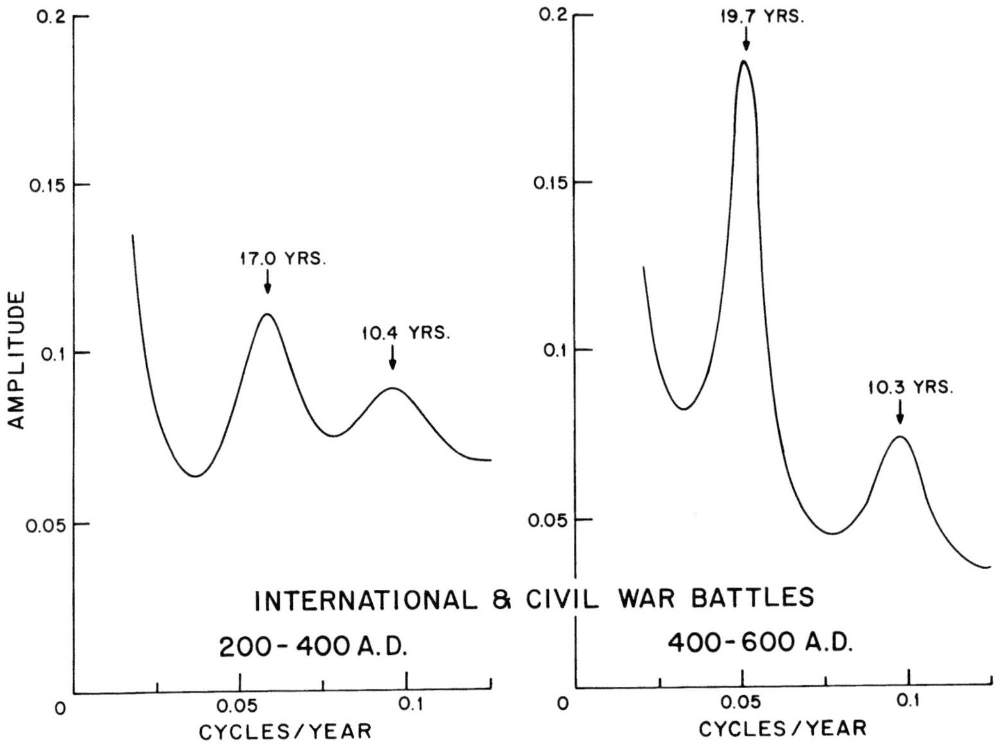

Fig. 32. Recurrent wave for the long period term seen in the previous Fig. 31.

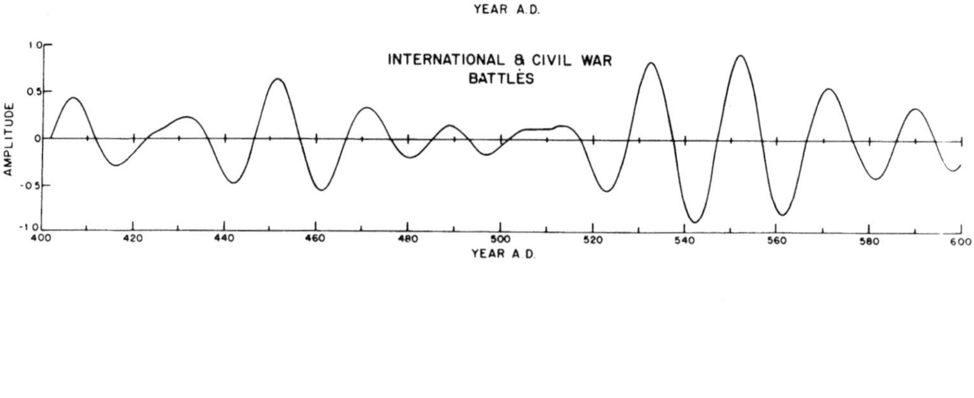

English Book of Common Prayer: "Easter-Day is always the first Sunday after the Full Moon...". This led, in part, to the first ecumenical (worldwide) council of the Christian Church held at Nicaea in Asia Minor in A.D. 325. One of the world-unifying questions the council addressed was the date of Easter. Islam continues to live by the cycles of the moon for the Prophet said: "Do not fast until you see the new moon, and do not break the fast until you see it" (Boorstin, 1983).

Everywhere we find relics of mythic, mystic, romantic meanings--in "moonstruck" and "lunatic", in "moonshine", and in the moonlight settings of lovers' meeting. When Romeo begins to swear by the moon his love for Juliet she interrupts and asks that he not swear by the inconstant moon. Today we know different. Two billion years ago this celestial object was much closer to the earth and the Kuznets' phenomenon much stronger, and two billion years from today the effect will be much smaller as the moon continues to recede from earth. But in the short run of human existence its orbit has been very constant. And it appears that it has been responsible in large measure for a great deal of human joy when the rains came and a great deal of suffering when they didn't.

Emlyn Williams (1968), the distinguished British actor and author, wrote a book entitled "Beyond Belief", an account of a particularly brutal and gruesome series of murders in the 1960s. Near the end, Williams in poetic fashion attributes human emotion to the moon who has witnessed these grisly events and writes: "And if after a moment the clouds move on, the moon's face will tell nothing. Too old for shock, or even sorrow. If it were not, tonight the puzzled world might stare skyward, and point in awe".

8 WHAT WENT WRONG?

Part of what went wrong in economics was accurately described by the Nobel Prize winner Wassily Leontief (1982): "Not having been subjected from the outset to the harsh discipline of systematic fact-finding, traditionally imposed on and accepted by their colleagues in the natural and historical sciences, economists developed a nearly irresistible predilection for deductive reasoning. As a matter of fact, many entered the field after specializing in pure or applied mathematics. Page after page of professional economic journals are filled with mathematical formulas leading the reader from sets of more or less plausible but entirely arbitrary assumptions to precisely stated but irrelevant theoretical conclusions". Instead of mathematics being a servant to this science it became the master, and the same thing has happened to all other inexact non laboratory sciences.

The other part is that the very axioms and postulates on which the mathematics is based are not physically valid. For example, the modern method of spectrum analysis employed in this paper is unknown to economists because it stands in opposition to the orthodox statistical views on time series analysis espoused by John Wilder Tukey and his numerous disciples such as Fishman (1969) and Brillinger (1981). This is ironic for, in his book on probability theory, Keynes (1921) strongly attacked the orthodox view of this subject. Jaynes (1979) presents at length a non-technical history of how Bayes' theorem and Laplace's outlook on the subject were rejected in the 19th century when physicists abandoned the field to pure mathematicians, and we shall later discuss this. It led to the greatest disaster ever to befall modern science.

Adam Smith (1776) and his successors erected an imposing conceptual edifice based on the notion of the national economy as a self-regulating system of a great many different but interrelated, and therefore interdependent, activities known as Classical Economics. Later, two mathematically trained engineers (Leon Walras and Vilfredo Pareto) translated Smith's common sense edifice into the concise language of algebra and calculus, and called it the General Equilibrium Theory. This theory posits an equilibrating interplay in demand, supply and price, and postulates that when supply equals demand one has the price. Here again there is irony, for Keynes writing to a colleague said: ".... I shall hope to convince you some day that Walras's theory and all the others along those lines are little better than nonsense" (Eichner, 1983, p. 133).

W. S. Jevons became in 1862 the first economist to insist on the rhythmical character of business fluctuations and sought to explain business cycles not by any economic cause, but by an 11-year modulation of crop production caused by the sunspot solar cycle. It is pleasing to note that Keynes (1936), in what many claim to be the most influential book of the 20th century, defended Jevon's hypothesis. Thus began empirical studies of economic time series which were so successful that, in a 1940's book on economic time series analysis, Davis (1941) discusses at length the 10 and 20 year cycles discussed in this paper. Such people as Mitchell (1927), Kuznets (1930a), and Burns (1934) posited no mathematical theory but relied on measurement; this brought them into conflict with pure economic theorists.

Classical economic theory assigned a small place to the problem of business crises because they were associated with causes external to the theory such as wars, technical changes, crop failures, and speculative manias. The persistence of booms and busts led some critics, most notably Karl Marx, to insist that the simple explanations of classical economics were misleading. But until the regularity of the disturbances was established, until it was made certain that the fluctuations are ever present and quite pervasive, that they are not mere exceptions, these fluctuations presented no question, and their explanation was no challenge to theoretical economics (Kuznets, 1930b).

Broadly speaking there are two views on what causes recurring waves in economic activity, and thus two methodologies arose; classic papers are reprinted in Haberler (1944), Hansen and Clemence (1953), and Gordon and Kline (1965). The first is our proposal, namely that climate cycles induce cyclic modulation of crop output and the economic consequences ripple through the economy with a multiplier effect. This possibility is not discussed in the post 1930s literature, possibly because Burns (1934) could not find the long swings in American commodity series. But, in any event, work along this line was always minute. A 1944 classified bibliography of books and papers lists only 24 vis-a-vis 690 devoted to theory (Haberler, 1944, p. 443). Kuznets (1930b) opines that to explain business cycles in this way is to confess the failure of economic science to explain their appearance, but this is surely too harsh a judgment. A science that produced such figures as Adam Smith, Karl Marx, John Maynard Keynes, and Simon Kuznets cannot be rated a failure.

The other view is that the recurrent waves are self generated within the economy by some means. This was the belief not only of Western economists, but also of Karl Marx who considered them endemic to a capitalistic market economy. And it is true that the unrestrained laissez faire capitalism of the 19th century was an excellent system for allowing the Kuznets' long swings to have maximum impact.

Western economists sought to explain the waves by means of lags between different economic variables. For example, let us suppose (Davis, 1941, p. 340) that I(t) is the total volume of orders for capital goods per unit of time t, and that L(t) is the corresponding volume of deliveries of orders for capital goods. It is clear the relationship between these two quantities is expressed by the equation:

$$L(t) = I(t - t_o)$$

where t_o is the lag between order and delivery. This quantity of lag is extremely fundamental in the theory of economic oscillations according to Davis. The problem is, it is always assumed to be constant for a particular industry, but different investigators get wildly different lags.

In 1927 the orthodox statisticians G. U. Yule and E. Slutsky (see Davis, 1941) showed mathematically that a linear combination of uncorrelated random variables can generate damped wavetrains which possess an oscillatory aspect. Accordingly, business cycles could be explained by cumulative random shocks hitting the economy. This random shock theory is the basis of econometric modeling which has been, and still is, Establishment economics in the United States since about 1947.

Irony enters yet again, for Keynes (1936) was very critical of such explanations and wrote: "Too large a proportion of recent mathematical economics are merely concoctions as imprecise as the initial assumption they rest on, which allow the author to lose sight of the complexities and interdependencies of the real world in a maze of pretentious and unhelpful symbols". A good example of what Keynes objected to is Slutsky's sinusoidal limit theorem (see Davis, 1941, p. 57). If from a random time series we form a new series by n iterated summations by 2, followed by the forming of the mth differences, then if m/n is kept constant, the difference series will tend to a sine curve of period $T = 2\pi/(\mathrm{Cos}^{-1} r_1)$, where $r_1 = (1-m/n)/(1+m/n)$ as n tends toward infinity.

This theorem is nothing more than numerology and one could, without violating logic, use such theorems to explain Newton's lunar theory, the celestial mechanics of Laplace, or the emission spectra of atoms.

An example of econometric modeling to "explain" the building wave discussed earlier in this paper (see Figs. 24-25) is Derksen (1940). He starts with a restatement of Tinbergen's (1939) stock-adjustment equation:

$$df/dt = -af(t-t_o)$$

where f(t) is the total available supply of housing at time t, expressed as deviations from a normal level; df/dt is the net incremental increase in total supply; and the product $at_o = c$ is the solution value of the equation. For $c = e^{-1}$ the solution is an exponential function, for $c > \pi/2$ one has explosive oscillations, and for $c = \pi/2$ one has a sine wave with period $T = 4t_o$. For $e^{-1} < c < \pi/2$ the solution is a damped oscillation with $T > 4t_o$. Econometric modelers always choose the damped oscillation as the solution, and then assume that a stream of cumulative shocks (never identified) introduce energy into the system to maintain the swings which are not observed to damp out.

Using multiple regression analysis he used the equation:

$$b = 16.3r - 8.5c_{-1} + 5.8(i_{+1}-i) + 0.44p_{-2} - 545$$

to "explain" the wave in U.S. residential building over the period 1914-1938. Such data existed of course from 1830 as we have seen earlier (see Figs. 24-25) but that is ignored.

In the above equation b is the number of dwelling units in thousands on which construction started annually, r is an index of rents, c_{-1} is an index of building costs lagged one year, i is the average nonfarm family income, and p_{-2} is the annual increase in the number of nonfarm families lagged two years. Derksen (1940) claimed the explanatory factors fell into two groups. First, the "incentive to build" consisting of rents and building costs, and second, the "acceleration principle" consisting of incomes and number of families. All the regression coefficients were claimed to be significant and the coefficient of multiple correlation was 0.96.

By some additional computational equations he found c = 0.72 and concluded "The period of the cycle is about 12 years. The oscillations are very damped and the cyclical fluctuations disappear practically after completion of one cycle". But the historical data on U.S. building, one series extending back to 1795 as we have seen (see Figs. 24-25), clearly show that every statement of Derksen's is false. The period is near 19 years and the wave does not damp out.

Derksen's work followed the method advocated in Tinbergen's (1939) book which Keynes (1939) harshly denounced as follows: "In practice Prof.

Tinbergen seems to be entirely indifferent whether or not his basic factors are independent of one another", and later concludes, "In plain terms, it is evident that if what is really the same factor is appearing in several places under various disguises, a free choice of regression coefficients can lead to strange results. It becomes like those puzzles for children where you write down your age, multiply, add this and that, subtract something else, and eventually end up with the number of the Beast in Revelation". Keynes concludes his lengthy polemic with a fear that Tinbergen's "reaction will be to engage another ten computors and drown his sorrows in arithmetic. It is a strange reflection that this book looks likely, as far as 1939 is concerned, to be the principal activity and raison d'etre of the League of Nations". Clearly, Keynes would be most unhappy that the econometric modeling advocated by Tinbergen (1939) became mainstream American economics after World War II, and that those who practiced it called themselves "neo-Keynesians".

Mitchell (1927, p. 265), a distinguished American economist and founder of the National Bureau for Economic Research, sounded the alarm much earlier by writing "Once started upon this career of transforming time series into new shapes for comparison, statisticians have before them a limitless field for the exercise of ingenuity. They are beginning to think of the original data, coming to them in a shape determined largely by administrative convenience, as concealing uniformities which it is theirs to uncover. With more emphasis upon statistical technique than upon rational hypothesis, they are experimenting with all sorts of data, recast in all sorts of ways. Starting with two series having little resemblance in their original shape, they can often transmute one series into 'something new and strange,' which agrees closely with the other series. In work of this type, they rely upon the coefficient of correlation to test the degree of relationship between the successive transformations."

The Kuznets' long swings have been attacked on various grounds and all are invalid (see Soper, 1978 for a survey). In one case (Adelman, 1965), conventional Blackman and Tukey spectra were computed for short economic time series and 17 spectral estimates obtained from $0 \leq f = 0.5$ cpy. Only four estimates were in the frequency range shown in the figures of this paper so it is self evident the signal could not be resolved. The author nevertheless asserts that the spectrum shows no evidence for the Kuznets wave (which is true) and solely on this basis questions it's existence. The assertion is not relevant.

In a second case, Fishman (1969) and Sargent (1979) compute the frequency transfer response of filtering operations carried out with an imaginary time series sampled yearly and a peak occurred near 20 years. They then charged that Kuznets had performed a "Slutsky effect" on the data he analyzed; that is, the filtering procedure itself generated the cycle. But we examined the data and found it had been originally sampled at 5 year intervals (census and mid-census data on population) and the spectra of the data are shown in Figs. 22-23. Since the 1930's the

"Slutsky effect" has been elevated to the status of a "law of economics" in popular books (Silk, 1978), yet economists know that the effect need never occur unless you want it to occur (Granger and Hatanaka, 1964).

In a third case (Bird et al., 1965), the investigators start with a sequence of random numbers, apply filtering operations, obtain a fabricated series with fluctuations on a time scale of 20 years, and claim this explains the long swings. No one seems to have pointed out that the data Kuznets examined were not fabricated in such a manner, and again such exercises date back to the mathematical theorems deduced by Slutsky in the 1930's (see Davis, 1941). Kuznets (1961) stresses that the long swings can be seen by eye in the raw data which is true (see Figs. 24-25).

Mainstream American economics since World War II is termed the "neoclassical synthesis", the wedding of Keynes ideas to 19th century pure economic theory, a theory Keynes himself rejected. This synthesis fared very poorly in coping with the economic shocks of the 1970s caused by the Kuznets' long swing, and economics schismed into several competing schools of thought. The polemics are harsh (see Balogh, 1982; Eichner, 1983), and the heretics place most blame on the enormous influence Paul Samuelson's (1947) textbook, and its successive editions issued as recently as 1985, have had on American economists.

It is apparent that after the War a great deal went wrong in economics and from Keynes own words quoted earlier we know his revolution was aborted, just as the heretics claim. But in a broader sense Jaynes (1979) presents in a non-technical manner the history of what went wrong in all the sciences when the physics community abandoned probability theory in the 19th century to pure mathematicians, and there came into existence "orthodox mathematical statisticans" who rejected Bayes theorem and Laplace's outlook on this subject.

Until about 1925 time series analysis was still under the influence of Newtonian determinism and many believed that time series arising in economics (and meteorology and geophysics) might contain deterministic components (Kendall, 1973). But because trade "cycles" of the 19th century were not perfect sine waves (of fixed amplitude and period) orthodox statisticians rejected the possibility and eventually obtained a contrary consensus. This was a gross error in judgment, as is self-evident to any scientist, because although the sunspot cycle as measured by number of spots, for example, does not have fixed amplitude and period, it is a bandlimited signal and thus has considerable predictive value (Currie, 1980, 1981a).

It was in 1927 that Udny Yule (Kendall, 1973, p. 4) changed the outlook of economic science (as well as meteorology and geophysics) in an investigation of sunspot numbers. His "fresh" approach included a "classic" illustration: if we have a clock pendulum its motion is harmonic. But if we pelt it randomly with peas the motion is disturbed; it will still swing, but with irregular amplitudes and intervals. The pelting peas provide a series of random shocks which are incorporated into the future motion of the pendulum. This concept of Yule (1927) led to the theory of stochastic processes and stochastic time series, which became the cornerstone of mainstream economics (and meteorology and geophysics) after the strong attack on spectrum analysis by Kendall (1945).

The result, half a century later, is shelves of books on "stochastic processes" and "stochastic time series analysis". The books start with a series of postulates which a mere scientist is not allowed to question; and the world view expressed by these postulates completely dominates economics and meteorology, and are dominant in all the other sciences. In order to escape these postulates one has to abandon science and become an electronic or acoustic signal processing engineer (Childers, 1978). I shall discuss three of the sacred scriptures of pure mathematics in time series analysis as posited by Brillinger (1981).

I am not allowed, as a mere scientist, to consider just the time series observed and get on with the calculations. I must believe that the observed series is only one example of an infinite set of time series (the ensemble) which I might have observed but did not. Such a reasoning format is completely foreign to the scientist, the detective, the attorney, the medical doctor, and the housewife. As Jaynes (1985) notes, if you go to a doctor and tell him your symptoms, he does not start thinking about the class of all symptoms you might have but don't have. He thinks about the class of all disorders that might cause the symptoms you do have. The first one he will test for is the one which, in that class, appears to be a priori most likely from your medical history.

Second, I am forced to believe that time-series are "stationary". Broadly, this means there is no systematic change in the mean (no trends), there is no systematic change in the variance, and all periodic variations have been removed. To put it baldly, orthodox statisticians are not interested in signals but rather in noise, in systems driven by "stochastic" processes. The writer has analyzed time series arising in the fields of geomagnetism, seismology, oceanography, astronomy, weapons detection, meteorology, and economics, and has never encountered the posited "stationary" series. Yet, outside of acoustic and electrical engineering, the mainstream view in time series analysis is that if the postulate does not describe observed data I must, perforce, accept it anyway.

And finally, I am forced to believe that time series satisfy "ergodic theorems" that arise, and are an active area of inquiry, in pure mathematics. That is, the time average over the one observed infinite record is equivalent to an ensemble average of an infinite number of records not observed. Such theorems are completely without relevance to the real world because our observed time series are of finite, not infinite, length. And they do not apply unless all bandlimited signals have been removed from the time series. Again we have the fact that orthodox statisticians are not interested in signals but rather noise.

The proposal of this paper will eventually be accepted in the science of economics. Aspects of classical theory which are valid in the short run, such as the equilibrating interplay of countervailing changes in demand, supply and price, will be retained. The distinction drawn between production and trade vis-a-vis distribution should be dropped because economics is also about "who gets what". To this should be added the concept of circular causation in the economy with cumulative effects (Myrdal, 1957), driven by modulation of crop output. Tools such as "input-output" analysis pioneered by W. Leontief are available to learn how the long swing (and the 10-11 year wave) ripples through the economy; one can then apply counter measures to smooth them.

But until then, Keynes' (1936) most famous words hold true: "The ideas of economists and political philosophers, both when they are right and when they are wrong, are more powerful than is commonly understood. Indeed the world is ruled by little else. Practical men, who believe themselves to be quite exempt from any intellectual influences, are usually the slaves of some defunct economist. Madmen in authority, who hear voices in the air, are distilling their frenzy from some academic scribbler of a few years back". Writing today Keynes would add that the defunct economists, in turn, have become the slaves of defunct orthodox mathematical statisticians, the very same people Keynes (1921) denounced.

It was self evident to Keynes (1936, p. 300), Karl Marx, and the great classical economists that if economics was to mean anything at all it would have to be political economy, and that is what it should become when the proposal of this paper is accepted. Economics was not created by Adam Smith to delight the aesthetic sense of pure mathematicians who refuse to adopt the reasoning format of scientists. For science, unlike mathematics, is played on a logical field that is open at both the bottom and the top. We cannot start at ground level with axioms and a set of propositions which cannot be questioned. We start in the middle and are thus, of necessity, creatures of the bog (Jaynes, 1986). And in the quest, mathematics is supposed to be the servant and not the master it has become in economics and other inexact non laboratory sciences.

Finally, if Keynes and Marx came back they would have no difficulty in accepting our proposal. Marx gave credence to a 10 year wave in economic activity and Keynes (1936) defended the hypothesis that such a wave existed. Keynes almost single-handedly built the modern theory of national income and justified a policy of government intervention in economic affairs. Their disagreement would be over the degree of intervention needed to effectively apply counter-cyclical measures to smooth out the long swings of Simon Kuznets.

9 CONCLUDING REMARKS

It is clear that in every department of economics in the United States the writer would not only be regarded as a heretic, but most probably worse, an outsider. But economics is a healthy science. When the long swing of Kuznets struck in the 1970's this science schismed, as any healthy science would have done, into numerous schools of thought. And in the process the Kuznets' phenomenon was rediscovered by Volcker (1978), who was then Chairman of the Federal Reserve Bank of New York.

In climatology and meteorology Pittock (1978) has become the spokesman for denying evidence for cycles in climate. He adopted the principle that no effects exist until they are proven to exist beyond a reasonable doubt since this is the traditional hypothesis which safeguards the purity of established scientific theory. This principle is echoed again when Pittock (1983) speaks of `scientific truth'. But on what basis has this `truth' been established?

Campbell (1983) and Currie (1983, 1984a, b, c) state that the characteristics of long period highly resonant atmospheric tides do not exist in scientific literature. There is a present-day consensus, with the support of a vast literature, concluding that beyond about a week weather is inherently unpredictable. But since none of this deals with characteristics of the aforementioned tides and current experimental data it completely fails to address the subject of prediction. In reply to a request in 1983 that Pittock provide scientific papers establishing the `truth' he sent the writer a long list of references with discussion. These papers, part of the vast literature, treat the equation ma=0, whereas the relevant equation is ma=F, Newton's second law.

In virtually all papers the `scientific truth' that Pittock claims he is defending is implicit. A rare exception is Shutts (1983), who gives no references, but simply states that atmospheric tides have a negligible impact on the global transfer of energy and momentum within the troposphere and can be safely neglected. Such statements, whether explicit or implicitly implied, are based on nothing more than an assumption which has been transformed into a `scientific truth' and defended as such. There is no evidence, experimental or theoretical, to support the current paradigm or weltanschauung. It is a unique situation in the history of science, and especially deplorable since the problem of climate and its variations is becoming one of the leading problems in world science, because the variations affect a great variety of human activities of an economic and social character.

In addition, the assumption was written into international law in the 1930's. In that decade the forerunner of the present day World Meteorological Organization, then attached to the League of Nations, obtained a legal agreement from all national weather agencies that "normal" weather is to be defined in terms of a 30 year moving average. Such a procedure applied to yearly data precludes the detection, and thus the existence, of the luni-solar and solar cycle signals. As recently as 1985 the best known climatologist in Europe took all the long air temperature records from that continent, applied the legally defined 30-year moving average, and then discussed the remaining fluctuations with respect to the CO_2 greenhouse effect.

The first forecast along the lines given in this paper was made in the Bible. Joseph was brought before the Pharaoh who listened to the dreams of seven healthy and seven sick cows and of seven strong and seven withered ears of corn. Joseph had forecast a fourteen year agricultural cycle. Pharaoh was so pleased that he said unto Joseph:

> "Foreasmuch as God has shown you all this, there is none so discreet and wise as you. You shall be over my house and according to your word shall all my people be ruled....". And Pharaoh took off his signet ring from his hand and put it upon Joseph's hand, and arrayed him in vestures of fine linen, and put a gold chain around his neck. And he made him to ride in the second chariot which he had (Genesis 41:30).

Talk about the declining status of today's economists after the economic debacle of the 1970's! Souvenir jackets from Camp David are scarcely a match for the Pharaoh's signet ring. Our evidence for modulation of crop production differs from that of Joseph in two respects. First, the primary forcing function has a period of 18.613 years. And second, that systematic modulation in agricultural output will occur is not a forecast, but a prediction due to the work of a man who lived three centuries ago, Sir Isaac Newton.

Since this paper, in a real sense, weds economic and social science to physical science it is proper to discuss what the Master of the first thought of the Master of the second. Keynes (1972) wrote two essays on Isaac Newton. The second was completed two weeks before his death 21 April, 1946. He had just returned from the Savannah Conference in the United States where the World Bank and International Monetary Fund were established.

When Newton left Cambridge for London he stored his unpublished writings in a trunk, and upwards of 1,000,000 words survive of which Keynes had acquired and studied much. Most people think of Newton as the first and greatest rationalist, one who taught us to think in the lines of cold and untinctured reason. Keynes saw him differently: "He was the last of the magicians, the last of the Babylonians and Sumerians, the last great mind which looked out on the visible and intellectual world with the same eyes as those who began to build our intellectual inheritance rather less than 10,000 years ago".

Keynes believed that Newton was a most extreme example of a profoundly neurotic man and that his deepest instincts were "occult, esoteric, semantic--with profound shrinking from the world". For it turned out among those 1,000,000 words that perhaps the most absorbing of Newton's occupations were devoted to topics wholly magical and wholly devoid of scientific value. Studies and experiments in alchemy--transmutation, the philosopher's stone, the elixir of life. Studies of apocalyptic writings--the measurements of Solomon's Temple, the Book of Daniel, the Book of Revelations. Along with hundreds of pages of Church History and the like, designed to discover the truth of tradition.

If it is true, as the author F. Scott Fitzgerald (1956) believed, that the test of a first-rate intelligence is the ability to hold two opposed ideas in the mind at the same time, and still retain the ability to function, then it is true as Keynes stated: "His peculiar gift was the power of holding continuously in his mind a purely mental problem until he had seen straight through it. I fancy his pre-eminence is due to his muscles of intuition being the strongest and most enduring with which a man has ever been gifted". Newton held two opposed ideas in mind for a quarter century working on both simultaneously in dynamo fashion, and Keynes tells us how: "There was extreme method in his madness. All his unpublished works on esoteric and theological matters are marked by careful learning, accurate method and extreme sobriety of statement. They are just as sane as Principia, if their whole matter and purpose were not magical".

The collected writings of Lord John Maynard Keynes encompass twenty seven volumes and, as we noted, his second essay on Isaac Newton was written two weeks before his death. Except for official notes, it was the last thing Lord Keynes wrote.

Notes Added in Press

On p. 224 and p. 230 we stated that Keynes (1936) defended Jevons' hypothesis that the 10 to 11-year sunspot cycle induced modulation of crop output. This inference was based on a strange book, which I cannot now locate, where the author assaulted every aspect of Keynes (1936), devoting a whole chapter to ridiculing his remarks on the "harvest cycle" of W. S. Jevons. Keynes (1936, p. 330) actually adopted the position that the question of physical causes for a regular cycle of good and bad harvests is one with which he is not concerned.

Keynes (1936, p. 329) is, however, emphatic that if cyclic modulation of crop output exists then his General Theory explains the resulting recurrent wave in economic activity, for he says that Jevons' hypothesis:

> ...appears as an extremely plausible approach to the problem. For even to-day fluctuation in the stocks of agricultural products as between one year and another is one of the largest individual items amongst the causes of changes in the rate of current investment; whilst at the time when Jevons wrote--and more particularly over the period to which most of his statistics applied--this factor must have far outweighed all others.
>
> Jevons' theory, that the trade cycle was primarily due to the fluctuations in the bounty of the harvest, can be re-stated as follows. When an exceptionally large harvest is gathered in, an important addition is usually made to the quantity carried over into later years. The proceeds of this addition are added to the current incomes of

> the farmers and are treated by them as income; whereas the increased carry-over involves no drain on the income-expenditure of other sections of the community but is financed out of savings. That is to say, the addition to the carry-over is an addition to current investment. This conclusion is not invalidated even if prices fall sharply. Similarly when there is a poor harvest, the carry-over is drawn upon for current consumption, so that a corresponding part of the income-expenditure of the consumers creates no current income for the farmers. That is to say, what is taken from the carry-over involves a corresponding reduction in current investment. Thus, if investment in other directions is taken to be constant, the difference in aggregate investment between a year in which there is a substantial addition to the carry-over and a year in which there is substantial subtraction from it may be large; and in a community where agriculture is the predominant industry it will be overwhelmingly large compared with any other usual cause of investment fluctuations. Thus it is natural that we should find the upward turning-point to be marked by bountiful harvests and the downward turning-point by deficient harvests.

Keynes (p. 331) points out that agricultural causes of investment fluctuation are much less important in the modern world for two reasons. First, agricultural output is now a much smaller proportion of total economic output. And second, that international markets for agricultural products tend to average out the effects of good and bad seasons in the harvests of individual countries. He then states:

> But in old days, when a country was mainly dependent on its own harvest, it is difficult to see any possible cause of fluctuations in investment, except war, which was in any way comparable in magnitude with changes in the carry-over of agricultural products.

It is rather evident that Lord Keynes, the pre-eminent economist of the twentieth century, would be a supporter of our proposal regarding the cause of the long swings.

In Section 8 (p.228-229) we pointed out that Yule's (1927) concept of a low order autoregressive scheme to model sunspot numbers changed the history of economics and meteorology, a change dictated by his eyeball judgment of two figures. He concludes: "...it seems desirable to break away from the periodogram method. The problem is, in fact, no longer one merely of determining the period, but also...". Also of what? The also problem of determining the "disturbances" or random shocks as they are universally known today in economics. The joke on today's economists is that Yule's "disturbances" were not at all random but highly systematic with a half-period of "some 40 to 42 years in duration", a bandlimited signal in sunspot numbers known today as the "Gleissberg cycle"!

Yule's eyeball judgment was seconded by Kendall (1945, p. 100) who set up the usual strawdata (the Beveridge wheat-price index reproduced in countless books since) and says: "To my eye this diagram clearly indicates an autoregressive scheme...". He later (p. 119) announces: "Where, therefore, there is any serious possibility that a series is of autoregressive type, the periodogram may not only be worthless, but extremely dangerous in suggesting periods of no reality. The effect of this work has been to lead me to the conclusion that for economic series, and probably for meteorological and geophysical series as well, periodogram analysis is simply not worth the trouble of the arithmetic which it involves". Spectrum analysis was thus simply written off and dismissed by the eyeball judgment of two orthodox statisticians from a community who, for over a century, have regarded their methods as "objective" and scorned those of everyone else, including the thought of Lord Keynes (1921). The ancient Greeks had a word for it: hubris--overwhelming pride or self-confidence; arrogance.

The final glance at this endless landscape of postulates, axioms, assumptions, mathematical theorems, and mathematical modeling in time series analysis is the text by Kendall (1973), purchased by the writer in a London bookstore April, 1976. Kendall (1973, p. 13) tabulates annual immigration data into the United States from 1820 through 1962, on p. 14 the raw data are graphed, and the long swings of Simon Kuznets (Figs. 22-23) are clearly evident to the eye; and on p. 104 he presents the power spectrum of the detrended data which contains only seven spectral estimates for $0 < f \leq 0.125$ cpy. The spectrum displays a broad peak near 20 years and an inflection point in the curve as one approaches 10 years. Kendall (1973, p. 18) calls such a systematic effect an "oscillation" to avoid: "describing it as a cycle unless it can be shown to be genuinely cyclic in the pattern of recurrence, and in particular that its peaks and troughs occur at equal intervals of time. Very few economic series are cyclic in this sense". As a matter of fact, no time series arising in the real world, including those in astronomy, are cyclical in Kendall's sense. His dictum could only arise in pure mathematics which is where it arose.

Modern methods of signal processing invented by engineers to solve real world problems were applied to the immigration data tabulated by Kendall (1973) and yielded two bandlimited signals (Currie, unpublished data, 1987); one enormous signal is near 20 years (the long swings of Simon Kuznets), and another much smaller one is near 10 years (the "harvest cycle" of W. S. Jevons).

In the methodological issues raised by Keynes' economics (Lawson and Pesaran, 1985), it is pointed out (p. 134) that rather like military historians re-fighting Waterloo, econometricians dissect the savage attack of Keynes (1939) on Tinbergen (see p.226-227 of this volume) for its lessons; and that just as Napoleon loses the battle every time it is re-fought so does Keynes. Well, the battle begins again with the advent of this paper and this time around Keynes is not going to lose.

Keynes used econometrics of course--informal and sensible calculations for the short run done by hand on envelopes. What Keynes objected to most vehemently (and correctly as we now see) was the premise of structural stability of the autoregressive equations over long periods of time. Keynes argued (correctly as we now see) that the material was non-homogeneous through time and that the coefficients were not constant (Lawson and Pesaran, 1985, p. 147).

References

Abramovitz, M. (1959). "Statement: to U. S. Congress, Joint Economic Committee. Employment, Growth, and Price Levels, Hearing, 86th Congress, 1st Session, Part II, Washington, D. C.

Abramovitz, M. (1964). Evidences of Long Swings in Aggregate Construction since the Civil War. New York: Nat. Bureau Econ. Res.

Abramovitz, M. (1968). The passing of the Kuznets cycle. Economica, 35, 349-367.

Adelman, I. (1965). Long cycles--fact or artifact?. Amer. Econ. Rev., 55, 444-463.

Balogh, T. (1982). The Irrelevance of Conventional Economics. New York: Liveright.

Barksdale, H. C., O'Bryan, D., and Scheider, W. J. (1966). Effect of Drought on Water Resources in the Northeast. Washington D. C.: U. S. Geological Survey, March.

Blackman, R. B., and Tukey, J. W. (1959). The Measurement of Power Spectra. New York: Dover.

Bird, R. C., Desai, M. J., Enzler, J. J., and Taubman, P. J. (1965). 'Kuznets cycles' in growth rates: the meaning. Internat. Econ. Rev., 6, 229-239.

Boorstin, D. J. (1983). The Discoverers. New York: Random House.

Borchert, J. R. (1971). The dust bowl of the 1970s. Annals Assoc. Amer. Geogr., 61, 1-22.

Brillinger, D. R. (1981). Time Series: Data Analysis and Theory. San Francisco: Holden Day.

Burns, A. F. (1934). Production Trends in the United States Since 1870. New York: Nat. Bureau Econ. Res.

Burns, A. F., and Mitchell, W. C. (1946). Measuring Business Cycles. New York: Nat. Bureau Econ. Res.

Campbell, W. H. (1983). Possible Tidal Modulation of the Indian Monsoon Onset. Ph. D. thesis, Univ. Wisconsin, Madison.

Chen, W. Y., and Stegen, G. R. (1974). Experiments with maximum entropy power spectra of sinusoids, J. Geophys. Res., 79, 3019-3022.

Childers, D. G. (Ed.) (1978). Modern Spectrum Analysis. New York: IEEE Press.

Currie, R. G. (1967). Magnetic shielding properties of the earth's mantle. J. Geophys. Res., 79, 2623-2633.

Currie, R. G. (1976). The spectrum of sea level from 4 to 40 years. J. Roy. Astron. Soc., 84, 753-761.

Currie, R. G. (1979). Distribution of solar cycle signal in surface air temperature over North America. J. Geophys. Res., 84, 753-761.

Currie, R. G. (1980). Detection of the 11-year sunspot cycle signal in earth rotation. Geophys. J. Roy. Astron. Soc., 61, 131-139.

Currie, R. G. (1981a). Solar cycle signal in earth rotation: Non-stationary behavior. Science, 211, 386-389.

Currie, R. G. (1981b). Solar cycle signal in air temperature in North America: Amplitude, gradient, phase and distribution. J. Atmos. Sci., 38, 808-818.

Currie, R. G. (1981c). Amplitude and phase of the 11-year term in sea level: Europe. Geophys. J. Roy. Astron. Soc., 67, 547-556.

Currie, R. G. (1981d). Evidence for 18.6-year signal in temperature and drought conditions in North America since A.D. 1800. J. Geophys. Res., 86, 11055-11064.

Currie, R. G. (1982). Evidence for 18.6-year term in air pressure in Japan and geophysical implications. Geophys. J. Roy. Astrono. Soc., 69, 321-327.

Currie, R. G. (1983). Detection of 18.6-year nodal induced drought in the Patagonian Andes. Geophys. Res. Lett., 10, 1089-1092.

Currie, R. G. (1984a). On bistable phasing of 18.6-year induced flood in India. Geophys. Res. Lett., 11, 50-53.

Currie, R. G. (1984b). Evidence for 18.6-year lunar nodal drought in western North America during the past millennium. J. Geophys. Res., 89, 1295-1308.

Currie, R. G. (1984c). Periodic 18.6-year and cyclic 11-year induced drought and flood in western North America. J. Geophys. Res., 89, 7215-7230.

Currie, R. G. (1987a). Examples of 18.6- and 11-year terms in world weather records: the implications. In Climate: History, Periodicity and Predictability, Eds. M. Rampino, W. Newman, J. Sanders, and L. K. Konigsson. New York: Van Nostrand Reinhold, in press.

Currie, R. G. (1987b). On bistable phasing of 18.6-year induced drought and flood in Africa since A.D. 650. J. Climatol. (in press).

Currie, R. G. (1987c). Periodic 18.6-yr signal in northeastern United States precipitation data. J. Climatol. (submitted).

Currie, R. G., and Fairbridge, R. W. (1985). Periodic 18.6-year and cyclic 11-year induced drought and flood in northeastern China and some global implications. Quat. Sci. Revs., 4, 109-134.

Currie, R. G., and Hameed, S. (1986). Climatically induced cyclic variations in United States corn production and possible implications in economics. In Proceedings of the Canadian Hydrology Symposium No. 16-1986, pp. 661-674.

Davis, H. T. (1941). The Analysis of Economic Time Series. Bloomington, Ill.: Principia Press.

Derksen, J. B. D. (1940). Long cycles in residential building: An explanation. Econometrica, 8, 97-116.

Dewey, E. R. (1970). Cycles: Selected Writings of Edward R. Dewey. Pittsburgh: Foundation for the Study of Cycles.

Dewey, E. R., and Dakin, E. F. (1947). Cycles: The Science of Prediction. New York: Holt.

van Duijn, J. J. (1983). The Long Wave in Economic Life. London: George Allen and Unwin.

Easterlin, R. A. (1960). Long Swings in the Growth of Population and Labor Force. New York: Nat. Bureau Econ. Res.

Eddy, J. A. (1983). The solar constant, climate and some tests of the storage hypothesis. In NASA Workshop on Solar Irradiance Variations on Active Region Time Scales, Eds. G. A. Chapman, H. S. Hudson and B. J. LaBonte. Pasadena, California.

Eichner, A. S. (Ed.) (1983). Why Economics is not yet a Science. Armonk, New York: Sharpe.

Fishman, G. S. (1969). Spectral Methods in Econometrics. Cambridge, MA.: Harvard University Press.

Fitzgerald, F. S. K. (1956). The Crack-Up, with other uncollected pieces. E. Wilson (Ed.). New York: New Directions.

Godin, G. (1972). Analysis of Tides. Toronto: Univ. Toronto Press.

Gordon, R. A. and Klein, L. R. (1965). Readings in Business Cycles. Homewood: Irwin.

Granger, C. W. J., and Hatanaka, M. (1964). Spectral Analysis of Economic Times Series. Princeton: Princeton Univ. Press.

Haberler, G. (Ed.) (1944). Readings in Business Cycle Theory. Philadelphia: Blakiston.

Hameed, S. (1984). Fourier analysis of Nile flood level. Geophys. Res. Lett., 11, 843-845.

Hameed, S., Yeh, W. M., Li, M. T., Cess, R. D., and Wang, W. C. (1983). An analysis of periodicities in the 1470 to 1974 Beijing precipitation record. Geophys. Res. Lett., 10, 436-439.

Hameed, S., and Currie, R. G. (1986). Cyclic variations in Canadian and United States drought. In Proceedings of the Canadian Hydrology Symposium No. 16-1986, pp. 113-122.

Hansen, A. H., and Clemence, R. V. (Eds.) (1953). Readings in Business Cycles and National Income. New York: Norton.

Hoffmann, W. C. (1955). British Industry, 1700-1950. New York: Kelly and Millman.

Jaynes, E. T. (1979). Where do we stand on maximum entropy?, R. D. Levine and M. Tribus (Eds.) The Maximum Entropy Formalism. Cambridge, MA.: M.I.T. Press. Reprinted in Jaynes, E. T. (1983). Papers on Probability, Statistics, and Statistical Physics. Hingham, MA.: Reidel.

Jaynes, E. T. (1982). On the rationale of maximum entropy methods. Proc. IEEE, 70, 939-952.

Jaynes, E. T. (1985). Where do we go from here?. C. R. Smith and W. T. Grandy, Jr. (Eds.). In Maximum-Entropy and Bayesian Methods in Inverse Problems. Dordrecht: Reidel.

Jaynes, E. T. (1986). Bayesian methods: General background and introductory tutorial. J. H. Justice (Ed.). In Maximum Entropy and Bayesian Methods in Applied Statistics. London: Cambridge Univ. Press.

Kay, S. M. (1987). Modern Spectral Estimation. Englewood Cliffs: Prentice-Hall.

Kendall, M. G. (1945). On the analysis of oscillatory time-series. J. R. Statist. Soc., A, 108, 93-141.
Kendall, M. G. (1973). Time-Series. London: Griffin.
Keynes, J. M. (1921). A Treatise on Probability. London: Macmillan.
Keynes, J. M. (1936). The General Theory of Employment, Interest, and Money. London: Macmillan.
Keynes, J. M. (1939). Professor Tinbergen's method. Econ. J., 49, 558-568. Reprinted in A. H. Hansen and R. V. Clemence (1953). Readings in Business Cycles and National Income. New York: Norton.
Keynes, J. M. (1972). The Collected Writings of John Maynard Keynes, Vol. X, Essays in Biography. London: Macmillan.
Kindleberger, C. P. (1973). The World in Depression 1929-1939. Berkeley: Univ. Calif. Press.
Kline, M. (1980). Mathematics: The Loss of Certainty. New York: Oxford Univ. Press.
Kuznets, S. (1930a). Secular Movements in Production and Prices. Reprinted by Kelley, New York, 1967.
Kuznets, S. (1930b). Equilibrium economics and business-cycle theory. Quart. J. Econ., 44, 381-415. Reprinted in Kuznets, S. (1953). Economic Change, New York: Norton.
Kuznets, S. (1961). Capital in the American Economy: Its Formation and Financing. Princeton: Princeton Univ. Press.
Lawson, T. and Pesaran, H., Eds. (1985). Keynes' Economics: Methodological Issues. London: Croom Helm.
Leontief, W. (1982). Academic economics. Science, 217, 104-105. Reprinted in Eichner, A. S. (Ed.) (1983). Why Economics is not yet a Science. Armonk, New York: Sharpe.
Lewis, J. P. (1965). Building Cycles and Britains' Growth. London: Macmillan.
Lisitzin, E. (1974). Sea-Level Changes. New York: Elsevier.
Long, C. D. Jr. (1940). Building Cycles and the Theory of Investment. Princeton: Univ. Princeton Press.
Marple, S. L. Jr. (1987). Digital Spectral Analysis with Applications. Englewood Cliffs: Prentice-Hall.
Marshall, J. R. (1972). Precipitation Patterns in the United States and Sunspots. Ph. D. thesis, Univ. Kansas, Lawrence.
Matthews, R. C. O. (1959). The Business Cycle. Chicago: Univ. Chicago Press.
Mitchell, J. M. Jr., Stockton, C. V., and Meko, D. M. (1979). Evidence of a 22-year rhythm of drought in the western United States related to the Hale solar cycle since the 17th century. B. M. McCormac and T. A. Seliga (Eds.). In Solar-Terrestrial Influences on Weather and Climate. Hingham, MA: Reidel.
Mitchell, W. C. (1927). Business Cycles: The Problem and its Setting. New York: Nat. Bureau Econ. Res.
Myrdal, G. (1957). Rich Lands and Poor: The Road to World Prosperity. New York: Harper.
Namias, J. (1966). Nature and possible causes of the northeastern United States drought during 1962-65. Mon. Weather. Rev., 94, 543-554.

Namias, J. (1967). Further studies of drought over northeastern United States. Mon. Weather Rev., 95, 497-508.

Palmer, W. C. (1965-1967). "The northeast drought situation"--a running commentary and set of charts on the state of the drought in terms of an index of severity, appearing frequently in the Weekly Weather and Crop Bulletin, National Summary. Washington, D. C.: Environmental Data Service, ESSA.

Perelman, M. (1977). Farming for Profit in a Hungry World. New York: Landmark Studies.

Pittock, A. B. (1978). A critical look at long-term sun-weather relationships. Revs. Geophys. Space Phys., 16, 400-420.

Pittock, A. B. (1983). Solar variability, weather, and climate: an update. Quart. J. Roy. Meteor. Soc., 109, 23-55.

Rosenberg, N. J. (Ed.) (1978). North American Droughts. Boulder, Col.: Westview Press.

Samuelson, P. (1947). Foundations of Economic Analysis. Cambridge: Harvard Univ. Press.

Sargent, T. J. (1979). Macroeconomic Theory. New York: Academic Press.

Shutts, G. J. (1983). Parameterizaton of traveling weather systems in a simple model of large-scale atmospheric flow. B. Saltzman (Ed.), In Advances in Geophysics, Vol. 25. New York: Academic.

Silk, L. (1978). Economics in Plain English: All You Need to Know About Economics. New York: Simon and Schuster.

Smith, A. (1776). An Inquiry into the Nature and Causes of the Wealth of Nations. E. Cannan (Ed.), Chicago: Univ. Chicago Press. Reprint of 1904 edition published by Methuen, London.

Soper, J. C. (1978). The Long Swing in Historical Perspective: An Interpretive Study. New York: Arno Press.

Stoker, J. J. (1950). Nonlinear Vibrations in Mechanical and Electrical Systems. New York: Interscience.

Thompson, L. M. (1973). Cyclical weather patterns in the middle latitudes. J. Soil Water Conser., 28, 87-89.

Tinbergen, J. (1939). A Method and its Application to Investment Activity. Vol. 1, Geneva: League of Nations.

Tyson, P. D. (1980). Temporal and spatial variation of rainfall anomalies in Africa south of latitude 22° during the period of meterological record. Climatic Change, 2, 363-371.

Tyson, P. D. (1981). Atmospheric circulation variation and the occurrence of extended wet and dry spells over Southern Africa. J. Climatol., 1, 115-130.

Ulrych, T. J., and Clayton, R. W. (1976). Times series modeling and maximum entropy. Phys. Earth Planet. Inter., 12, 188-200.

U. S. Dept. Agriculture (1954). Corn: Acreage, Yield and Production by States, 1866-1943. Washington, D. C.

U. S. Dept. Agriculture (1984). Field Crops by States, Revised Estimates 1978-1982. Washington D. C.: Statistical Bulletin 708.

U. S. Dept. Commerce (1976). Historical Statistics of the United States: Colonial Times to 1970, Part 1, Washington, D. C.

Volcker, P. A. (1978). The Rediscovery of the Business Cycle. New York: Free Press

Warren, G. F., and Pearson, F. A. (1937). World Prices and the Building Trades: Index Numbers of Prices of 40 Basic Commodities for 14 Countries in Currency and in Gold, and Material on the Building Industry. New York: Wiley and Sons.

Wilcox, W. W., Cochrane, W. W., and Herdt, R. W. (1974). Economics of American Agriculture. Englewood Cliffs: Prentice-Hall.

Williams, E. (1968). Beyond Belief: A Chronicle of Murder and its Detection. New York: Random House.

Yule, G. U. (1927). On a method of investigating periodicities in distrubed series, with special reference to Wolfer's sunspot numbers. Phil. Trans., A, 226, 267-294. Reprinted in 1971 by A. Stewart and M. G. Kendall (Eds.). Statistical Papers of G. Udney Yule. London: Griffin and Co.

A Maximum Entropy Method for Expert System Construction

ALAN LIPPMAN
Division of Applied Mathematics
Brown University
Providence, R.I. U.S.A. 02912

Abstract We consider the maximum entropy method of expert system construction. We show that the construction of the expert system is equivalent to the minimization of a convex function in as many dimensions as there were pieces of knowledge supplied the system. We show that in the case where the knowledge presented the system is self-contradictory, the minimization of this function creates an expert system for a set of constraints that is consistent and 'close' to the original inconsistent constraints. Monte Carlo methods for minimizing the function are discussed, and illustrated by computer experiment. One of the examples given suggests an approach to the problem of invariant optical character recognition.

Introduction An expert system is designed to answer questions. We consider probabilistic expert systems — if the system is given an event, it should be able to calculate its probability. Such an expert system is actually a distribution on the set of all events we wish to consider. Typically the knowledge the system is based on will be insufficient to answer all questions. In many cases we wish to consider, the sheer size of the state space precludes such knowledge. A medical expert system could be asked for the probability of a disease given some combination of symptoms, yet the set of all possible combinations of symptoms is huge, and the knowledge base can not be expected to contain all the different probabilities. We desire our system to answer questions even in such cases, and to do so in a reasonable manner, much like a human expert would. For this purpose we consider 'the principle of maximum entropy'. Of all the distributions which satisfy the knowledge supplied the system, we will pick the one with maximum entropy to be our expert system. The entropy H of a distribution p is defined as

$$H(p) = -\sum_{\omega\in\Omega} p(\omega)\log p(\omega)$$

where ω is an event, Ω is the set of all events, and $p(\omega)$ is the probability of the event ω. Entropy has an information-theoretic meaning; the distribution with maximum entropy can be viewed as the one containing the least knowledge. By maximizing the entropy over all distributions that agree with the knowledge base, we are picking as our expert system the distribution that makes the fewest unnecessary 'assumptions'. For more information regarding the justification of the principle of maximum entropy, we refer the reader to [1].

1. Knowledge The construction of a probabilistic expert system begins with knowledge. We classify as knowledge anything that answers probabilistic questions; we think of a probabilistic question as a function of probabilities and we consider an answer to be the value we say the function will take on. (This brings up a more general way to view knowledge; we could view an answer as specifying that the function, that defines the question, has a value in a certain range. We will not be using this type of answer.) Using these ideas, we see that the knowledge supplied the system can be broken up into distinct 'pieces' of knowledge, each of which corresponds to a distinct probabilistic question and its answer. Each piece of knowledge is a constraint that must be satisfied by the expert system; if the answer to the question we ask the system is included in the knowledge base, then the expert system's answer is constrained to duplicate it. We can write a constraint in its most general form as

$$B(p) = c$$

G. J. Erickson and C. R. Smith (eds.),
Maximum-Entropy and Bayesian Methods in Science and Engineering (Vol. 2), 243–263.

We consider a restriction of the above to the case where B is a linear function of p, our constraint can thus be written as

$$\sum_{\omega\in\Omega} b(\omega)p(\omega) = c$$

The above constraint is the same as specifying that the expected value of b is c. Since $c = c\sum_{\omega\in\Omega} p(\omega)$, we consider the function a, where $a(\omega) = b(\omega) - cp(\omega)$, and we re-write the above constraint as

$$\sum_{\omega\in\Omega} a(\omega)p(\omega) = 0$$

It may seem that this form is very restricted, but it is sufficient for several important types of constraints ([3],[4]). It is capable of representing any piece of knowledge that can be put in terms of the expected value of a function; it can thus represent knowledge about marginal, joint and conditional probabilities. To illustrate this consider the following example:

$$p(\omega \in S_1 | \omega \in S_2) = .5$$

Using Bayes's rule, we can re-write the above as

$$\frac{p(\omega \in S_1 \cap S_2)}{p(\omega \in S_2)} = .5$$

This can be written as

$$p(\omega \in S_1 \cap S_2) - .5p(\omega \in S_2) = 0$$

which is the same as

$$\sum_{\omega\in\Omega} \left(\mathcal{X}_{S_1\cap S_2}(\omega) - .5\mathcal{X}_{S_2}(\omega)\right) p(\omega) = 0\,,$$

where $\mathcal{X}_S$ denotes the indicator function on the set of events in S; when $\omega \in S$ we will have $\mathcal{X}_S(\omega) = 1$, when $\omega \notin S$ we will have $\mathcal{X}_S(\omega) = 0$.

1. Lagrange Multipliers Recall our goal. We wish to find a distribution that satisfies a set of constraints, and has higher entropy than any other such distribution. Using the form for knowledge that we introduced in the previous section, we can state the problem as follows:

$$\max \left(-\sum_{\omega\in\Omega} p(\omega)\log p(\omega)\right)$$

over all p satisfying (1)

$$\sum_{\omega\in\Omega} a_i(\omega)p(\omega) = 0 \qquad i = 1,\ldots m$$

$$\sum_{\omega\in\Omega} p(\omega) = 1$$

$$p(\omega) \geq 0 \qquad \omega \in \Omega$$

With suitable care, we can use Lagrange multipliers to reduce the above, constrained, problem to an unconstrained problem. In order to apply the theory of Lagrange multipliers we must add some assumptions. For the complete details, we refer the reader to [2]; here we will note that the required assumptions are that

$$\begin{array}{ll} \{a_i(\cdot)\} & \text{are linearly independent vectors} \\ \exists p(\omega) & \text{such that} \quad \sum_{\omega\in\Omega} a_i(\omega)p(\omega) = 0 \,\forall\, i \quad \text{and} \quad p(\omega) > 0 \,\forall\, \omega \in \Omega \end{array} \tag{2}$$

The Lagrangian is

$$\sum_{\omega\in\Omega} p(\omega)\log p(\omega) \quad + \quad \sum_{i=1}^{m} \lambda_i \Big(\sum_{\omega\in\Omega} a_i(\omega)p(\omega) \Big) \quad + \quad \delta \Big(\sum_{\omega\in\Omega} p(\omega) - 1 \Big)$$

We know from Lagrange multiplier theory that there exist $\bar{\lambda} = \{\bar{\lambda}_1, \ldots, \bar{\lambda}_m\}$ and $\bar{\delta}$ such that the derivative of the Lagrangian (with respect to λ_i, δ and $p(\omega)$) at $\bar{\lambda}, \bar{\delta}$ is zero, and that such $\bar{\lambda}$, $\bar{\delta}$ define local extrema. Performing some algebraic manipulations, we arrive at the following equations

$$\begin{array}{rcll} p(\omega) & = & \exp\Big(-\sum_{i=1}^{m} \bar{\lambda}_i a_i(\omega)\Big) \Big/ \sum_{\hat{\omega}\in\Omega} \exp\Big(-\sum_{i=1}^{m} \bar{\lambda}_i a_i(\hat{\omega})\Big) & \forall\, \omega \in \Omega \\ \frac{\partial}{\partial \lambda_k} \sum_{\omega\in\Omega} \exp\Big(-\sum_{i=1}^{m} \lambda_i a_i(\omega)\Big)\Big|_{\lambda=\bar{\lambda}} & = & 0 \qquad k = 1, \ldots, m & \end{array} \tag{3}$$

The function $\sum_{\omega\in\Omega} \exp(-\sum_{i=1}^{m} \lambda_i a_i(\omega))$ will be denoted $Z(\lambda)$, where $\lambda = \{\lambda_1, \ldots, \lambda_m\}$. The function Z is sometimes called the partition function.

Notice that Z is a convex function. Hence there is at most one $\bar{\lambda}$, corresponding to the global minimum of Z, at which Z has an extremal point (*i.e.*, $\frac{\partial}{\partial \lambda_i} Z(\lambda)|_{\lambda=\bar{\lambda}} = 0 \forall i$). Under the assumptions (2) we know that such a $\bar{\lambda}$ must exist, hence the maximum entropy distribution exists and is unique. A interesting property of Z is that when the assumptions (2) do not hold, Z has no extremal point (see [2] for the details). Hence, if we try to minimize Z and succeed, we have found the maximum entropy distribution (since the maximum entropy distribution is defined, through (3), by the $\bar{\lambda}$ at which the minimum occurs). Our computational goal (section 4) will thus be to minimize Z. We note that there have been many ideas and methods for the computation of the maximum entropy distribution, some involving Lagrange multipliers, others not; some examples are [1],[5]-[7]. The method we use is based on work by Geman[3] and Geman[4].

3. Contradictions Let us consider the case where the assumptions (2) do not hold. We will still assume that the a_i are linearly independent. a_i will usually be a simple function of ω (for example a_i is often a combination of indicator functions), in such cases independence is relatively easy to verify. If the constraints are dependent, some can be removed so as to provide independence. Hence, the restriction that the a_i be independent is often easy to satisfy.

More hazardous is the assumption that

$$\exists p(\omega) \quad \text{such that} \quad \sum_{\omega\in\Omega} a_i(\omega)p(\omega) = 0 \,\forall\, i \quad \text{and} \quad p(\omega) > 0 \,\forall\, \omega \in \Omega$$

This assumption can fail in two fundamentally different ways. The first occurs when there exists distributions p that satisfy the constraints (so $\sum_{\omega\in\Omega} a_i(\omega)p(\omega) = 0$), but all such distributions assign probability zero to some events. The other way this assumption can fail is when there exists no p that satisfies the constraints. This is the case when the constraints are self-contradictory. In both the above situations we can show that trying to minimize the partition function (*i.e.*, driving the gradient of Z close to zero) and using the form for the probabilities found by the use of Lagrange multipliers (3), does something useful.

The original constraints we supplied the system with were

$$\sum_{\omega\in\Omega} a_i(\omega)p(\omega) = 0 \qquad i = 1,\ldots,m \tag{4}$$

Consider the system of constraints

$$\sum_{\omega\in\Omega} a_i(\omega)p(\omega) = \epsilon_i \qquad i = 1,\ldots,m \tag{5}$$

When $\|\epsilon\| = (\epsilon_1^2 + \ldots \epsilon_m^2)^{\frac{1}{2}}$ is small enough, we would expect the two systems of constraints to be interchangeable. Now, let p_λ be defined as follows

$$p_\lambda(\omega) = \exp\Big(-\sum_{i=1}^{m} \lambda_i a_i(\omega)\Big) \Big/ Z(\lambda)$$

where Z is defined with respect to the constraints (4). If Z has an extrema at $\bar{\lambda}$ then $p_{\bar{\lambda}}$ is the maximum entropy distribution for the constraints (4). For any λ, we can show (see [2]) that p_λ is the maximum entropy distribution for the system of constraints

$$\sum_{\omega\in\Omega} a_i(\omega)p(\omega) = \sum_{\omega\in\Omega} a_i(\omega)\exp\Big(-\sum_{i=1}^{m} \lambda_i a_i(\omega)\Big) \Big/ Z(\lambda) = -\nabla_i Z(\lambda) \Big/ Z(\lambda)$$

where ∇_i is the ith component of the gradient. So, if, for a given λ, $\|\nabla Z(\lambda)/Z(\lambda)\|$ is small (corresponding to $\|\epsilon\|$ being small in (5)), then p_λ is the maximum entropy distribution for a system of constraints that is close to the original system of constraints.

Hence, our desire is to find a λ such that $\|\nabla Z(\lambda)/Z(\lambda)\|$ is small. In light of this, let us examine the cases where the assumptions (2) do not hold. When a system of constraints has as its only solutions distributions p that assign probability zero to some events, we can show (see [2]) that $Z(\lambda)$ is bounded below by 1. Hence, all we need to do is make the gradient of Z arbitrarily small, and we will have found a λ that defines a maximum entropy distribution which satisfies constraints arbitrarily close to those originally supplied. When the constraints are contradictory, we can show (see [2]) that when ∇Z goes to zero, Z will also. But, we can also show (see [2]) that using a continuous gradient descent method (define $\lambda(t)$ by the O.D.E. $d/dt\lambda_i(t) = -\nabla_i Z(\lambda(t))/\|\nabla Z(\lambda(t))\|$, with the initial condition $\lambda_i(0) = 0 \forall i$) to minimize Z yields a path $\lambda(t)$ such that $\|\nabla Z(\lambda(t))/Z(\lambda(t))\|$ decreases as t increases. In this sense, we get a maximum entropy distribution for a consistent set of constraints that approximates the inconsistent set.

4. Minimizing the Partition Function In this section we consider the computational side of finding a maximum entropy distribution. Recall that finding the maximum entropy distribution is equivalent to minimizing a convex function, the partition function $Z(\lambda)$, as we showed in section 2. Recall also that $Z(\lambda)$ and $\nabla Z(\lambda)$ are defined by sums over all elements in Ω. When Ω has a small number of elements, computation is simple. The gradient of Z

can be calculated exactly, and Z minimized by gradient descent. However, even in a small letter recognition problem, for example, we may have our letters described by ten features, each feature being able to take on thirty different values. This yields a state space with 30^{10} elements, and a sum of 30^{10} terms (each of which involves the exponential of a sum of m terms), as would be necessary to explicitly compute ∇Z, is beyond the practical limits of computation. This difficulty can be overcome by estimating the direction of ∇Z, instead of calculating it exactly. Before we delve too deeply (for more details see [2]), let us first outline the general idea. The crucial observation is that we can find a distribution p_λ, such that $\nabla Z(\lambda)/Z(\lambda)$ is just an expected value (with respect to the distribution p_λ) of some simple function. Notice that $\nabla Z(\lambda)/Z(\lambda)$ supplies us with both the direction of the gradient (so we can minimize Z by gradient-descent type methods) and also tells us how close we are to satisfying our constraints (see section 3). Since by using Monte Carlo type methods we can simulate such a distribution p_λ, and since the sample mean from a simulation is close to the real expected value, we can actually approximate $\nabla Z(\lambda)/Z(\lambda)$ without doing a size of Ω number of calculations. The idea of using sampling to find the direction of the gradient of Z, was first proposed by Geman [3].

Consider a distribution p_λ on the space Ω where the probability of the event ω, $p_\lambda(\omega)$ is defined, as before, as

$$p_\lambda(\omega) = \frac{\exp(-\sum_{i=1}^{m} a_i(\omega)\lambda_i)}{\sum_{\omega\in\Omega} \exp(-\sum_{i=1}^{m} a_i(\omega)\lambda_i)}$$

The expected value of the function $f(\omega)$ with respect to the distribution p_λ is

$$E_\lambda(f) = \sum_{\omega\in\Omega} f(\omega)p_\lambda(\omega) = \frac{\sum_{\omega\in\Omega} f(\omega)\exp(-\sum_{i=1}^{m} a_i(\omega)\lambda_i)}{\sum_{\omega\in\Omega} \exp(-\sum_{i=1}^{m} a_i(\omega)\lambda_i)}$$

and for the function $-a_i(\omega)$ we have

$$E_\lambda(-a_i) = \frac{-\sum_{\omega\in\Omega} a_i(\omega)\exp(-\sum_{i=1}^{m} a_i(\omega)\lambda_i)}{\sum_{\omega\in\Omega} \exp(-\sum_{i=1}^{m} a_i(\omega)\lambda_i)} = \frac{\nabla_i Z(\lambda)}{Z(\lambda)}$$

Since all a gradient descent method needs is the direction of the gradient, we can use the above. Also note that a measure of how close we are, at a certain λ, to satisfying the original constraints is just $\|E_\lambda(a_i)\| = \|\nabla Z(\lambda)/Z(\lambda)\|$ (see section 3).

Now that we have $\nabla Z/Z$ in terms of an expected value we come to the problem of simulation. The goal is to find an ergodic sequence ω^i with marginal distribution p_λ. In this way the sample expected value of $f(\omega)$ using S samples is

$$\frac{1}{S}\sum_{j=1}^{S} f(\omega^j) \tag{6}$$

and for S large this should be close to the true value of $E_\lambda(f)$.

The method we use to find an ergodic sequence requires that Ω have some sort of neighborhood structure, we will thus revise our view of the state space Ω. For the sake of clarity we will consider each event in Ω as the state of a 1-dimensional lattice with N_1 elements. An element ω in Ω will thus be of the form $\omega = \{\omega_1, \ldots, \omega_{N_1}\}$. Furthermore, each component of ω, ω_k will be restricted to N_2^k different values (Ω will thus have $\prod_{k=1}^{N_1} N_2^k$ elements). We note that with a lattice structure Ω can get very large, without much effort.

Our ergodic sequence will start at a random point ω^1 in Ω. We will then pick ω^{l+1} given ω^l as follows; we will fix most of the components of ω^{l+1} to be the same as the value for ω^l. The values we don't fix will allow ω^{l+1} to be any element in some subset of Ω, call it T, containing τ elements, $\{t_1, \ldots, t_\tau\}$. We will then randomly pick an element from T, according to the probabilities $p_\lambda(t_i)$, to be ω^{l+1}. This is done as follows: we first calculate $Z(\lambda) \cdot p_\lambda(t_i)$ for every t_i in T, then we randomly (according to a uniform distribution) pick a number between 0 and $Z(\lambda) \cdot \sum_{i=1}^{\tau} p_\lambda(t_i)$. This randomly chosen number will be between $Z(\lambda) \cdot p_\lambda(t_j)$ and $Z(\lambda) \cdot p_\lambda(t_{j+1})$ for some t_j in T, and we will let $\omega^{l+1} = t_j$. Note that $Z(\lambda) \cdot p_\lambda(t_i)$ equals $\exp \sum_{i=1}^{m}(-a_i(t_i)\lambda_i)$, which just requires an order of m operations to calculate. Since we usually have m less than several hundred, we are in good computational shape. Of course, one has to be carefull when picking T at each step l (*i.e.*, decide which components of ω^l to fix) in order to avoid creating numerical artifacts. This is not too difficult; one approach is to fix components in a random order and with equal likelihood. This method of finding an ergodic sequence is known as Stochastic Relaxation [3] and is closely related to the Metropolis Algorithm [8].

Now let us say a word about the minimization of Z, given that we have estimates for the gradient. We note that finding the gradient is still a computationally difficult task, and hence we desire to use a method that requires the direction of the gradient at as few a number of points as possible. A discrete analog of the continuous gradient descent scheme, suggested in section 3 for handling contradictions, would prove too costly; we will therefore assume that the constraints are not self-contradictory, so any method that drives ∇Z to zero will be acceptable.

We implement a modification of the standard gradient descent method. Typically, gradient descent refers to constant small steps in the direction opposite to the gradient. Instead, to minimize the number of times we need to compute the gradient, we employ a slight modification. We will still move in the direction opposite to the gradient, but the size of the step we take will not be constant. When we begin we will pick a value for our step-size δ (positive). We will always start at $\lambda^0 = 0$, since this corresponds to the uniform distribution on Ω, a logical starting point. At the point λ^i we will find a λ^{i+1} such that $\lambda^{i+1} = \lambda^i - \hat{\delta}\nabla Z(\lambda^i)/Z(\lambda^i)$ where $\hat{\delta}$ is picked as follows. Since $Z(\lambda^i - \delta\nabla Z(\lambda^i)/Z(\lambda^i))$ is a convex function of δ its derivative (with respect to δ) can only be zero for at most one δ, which we shall call $\bar{\delta}$. If $\bar{\delta}$ does not exist, then $Z(\lambda^i - \delta\nabla Z(\lambda^i)/Z(\lambda^i))$ is a decreasing function of δ, and since $Z(\lambda)$ is bounded below by zero, we would have $\nabla Z(\lambda^i - \delta\nabla Z(\lambda^i)/Z(\lambda^i))$ going to zero; so for δ large enough $\lambda^i - \delta\nabla Z(\lambda^i)/Z(\lambda^i)$ would define an adequate solution (section 3). When $\bar{\delta}$ does exist, we see that the derivative of $Z(\lambda^i - \delta\nabla Z(\lambda^i)/Z(\lambda^i))$ (with respect to δ) is negative for all δ greater than zero and less than $\bar{\delta}$. Hence, picking $\hat{\delta}$ between 0 and $\bar{\delta}$ would yield $Z(\lambda^{i+1})$ less than $Z(\lambda^i)$. However, the closer $\hat{\delta}$ is to $\bar{\delta}$ the smaller $Z(\lambda^{i+1})$ will be. We will pick $\hat{\delta}$ between $\bar{\delta}$ and $\bar{\delta}/\epsilon$ (ϵ around 2) by doing a binary search: if the dot product $\nabla Z(\lambda^i) \cdot \nabla Z(\lambda^i - \hat{\delta}\nabla Z(\lambda^i)/Z(\lambda^i))$ (remember that Z is positive, so the sign of this term is computable even without normalizing) is negative we try $\hat{\delta} = \hat{\delta}/\epsilon$, if positive we try $\hat{\delta} = \hat{\delta}\epsilon$. When $\nabla Z(\lambda^i) \cdot \nabla Z(\lambda^i - \hat{\delta}\nabla Z(\lambda^i)/Z(\lambda^i))$ switches sign from the last $\hat{\delta}$ to the current $\hat{\delta}$, we will have completed our search in the direction $\nabla Z(\lambda^i)$; we will define λ^{i+1} using the $\hat{\delta}$ (choosing from either the current or the last) for which the sign was positive. In this manner we will be sure that $Z(\lambda^{i+1}) < Z(\lambda^i)$. Making ϵ smaller (close to, but above, one) yields higher accuracy, but since our gradients are not exact, and since we would need to find many more gradients, a computationally expensive task, it is not worth it. We save the value of $\hat{\delta}$ that we used last, for the next step, since it is usually of the correct magnitude.

A useful feature of the above method is that it provides a means to test our sampling method. As we increase δ, the dot product $\nabla Z(\lambda^i) \cdot \nabla Z(\lambda^i - \delta\nabla Z(\lambda^i)/Z(\lambda^i))$ should be a decreasing function. Likewise as we decrease δ it should be an increasing function (keeping δ positive). If this is true for the sampled values of the dot product, we can have more confidence in the sampling method. If it is not, we know that our estimate of the gradient is wrong, in which case we can take appropriate action. We can increase the number of samples we are averaging over (S in (6)), we can start at multiple begining points $(\omega^1, \bar{\omega}^1, \ldots)$ and then average over the different trials $(\{\omega^1, \ldots, \omega^l, \bar{\omega}^1, \ldots, \bar{\omega}^l, \ldots\})$, we can discard the first n elements of the series to get rid of the effects of the random starting point, etc... There are many things that can, at the expense of increased computation, be done to improve the accuracy of the sampling method.

The next section is composed of two examples. The first is a test of our simulation methods. We construct a distribution and extract statistics. We then find the maximum entropy distribution. We then calculate $\nabla Z/Z$ exactly, and see that the simulation was successful (since the values for $\nabla Z/Z$ are quite small). The state space for this example is of size 2^{24}, so the exact calculation of ∇Z was quite lengthy.

The second example we consider is the problem of letter recognition. Sample letters were presented and features extracted from them. Statistics of the features conditioned on the letter served as our knowledge. The maximum entropy distribution was found and used to identify the sample letters. Considering the primitiveness of the features the results are encouraging.

5. Results

Example 1: A test of our method

In this section we conduct a test of our simulation methods. We will consider a distribution on a large state space and use the statistics generated by the distribution to form constraints. We use sampling methods to conduct the gradient descent (section 4), and find a point $\bar{\lambda}$ that will serve as our guess for the extremal point of the partition function. We then compute $\nabla Z(\bar{\lambda})/Z(\bar{\lambda})$ exactly. This will serve to tell us how the statistics generated by the distribution generated by $\bar{\lambda}$ differ from the statistics of the original distribution. We will present (on the following pages) the statistics of the original distribution, the estimated value of $\nabla Z(\bar{\lambda})/Z(\bar{\lambda})$ and the true value of $\nabla Z(\bar{\lambda})/Z(\bar{\lambda})$.

We will have as our state space, Ω, the set of all strings of length 24 composed of 1's and -1's, so Ω has 2^{24} elements. We picked this state space so that the exact calculation of ∇Z and Z is possible, although quite lengthy. The distribution $\hat{p}$ we use to generate statistics is

$$\hat{p}(\omega) = \frac{\exp(-\sum_{i=1}^{24}\sum_{k=i}^{24}\omega_i W(i,k)\omega_k - \sum_{i=1}^{24} T(i)\omega_i)}{\sum_{\omega\in\Omega}\exp(-\sum_{i=1}^{24}\sum_{k=i}^{24}\omega_i W(i,k)\omega_k - \sum_{i=1}^{24} T(i)\omega_i)}$$

Where $W(i,j)$ was picked randomly to be either $+c$ or $-c$, and $T(i)$ was picked randomly to be either $+d$ or $-d$. The values of d and c were chosen so that the distribution $\hat{p}$ is neither too flat nor too sharp. We used $c = 1/5$ and $d = 1/2$. Our constraints are the expected values of ω_i and $\omega_i\omega_j$ with respect to $\hat{p}$. Our constraints are thus

$$E(\omega_i) - \sum_{\omega\in\Omega}\omega_i p(\omega) = 0 \quad \text{for all } i$$

$$E(\omega_i\omega_j) - \sum_{\omega\in\Omega}\omega_i\omega_j p(\omega) = 0 \quad \text{for all } i,j \text{ with } i \geq j$$

where $E(\omega_i)$ and $E(\omega_i\omega_j)$, the expected values with respect to $\hat{p}$, were computed exactly.

We can write the partition function, $Z(\lambda)$ (where $\lambda = \{\lambda_1, \ldots \lambda_{300}\}$) as

$$Z(\lambda) = \sum_{\omega \in \Omega} \exp\Big(\sum_{i=1}^{24}\sum_{k=i}^{24} \lambda_{(i-1)\cdot(25-(i/2))+k-i+1}(\omega_i\omega_k - E(\omega_i\omega_k)) + \sum_{i=1}^{24} \lambda_{276+i}(\omega_i - E(\omega_i))\Big)$$

On the computational side of things we used 20 starting points in the sampling. Each sample involved 80 steps, $\{\omega^1, \ldots, \omega^{80}\}$, the last 50 being kept to form the expected value. Each step was composed of randomly dividing the 24 components of the string into six groups (four in each). We then chose a group and, holding the other groups fixed, picked a value for it according to the distribution p_λ (see section 4). We repeated this procedure until each of the six groups had been allowed to vary once.

We note that the computational time taken to conduct all the steps of the gradient descent (involving the estimation of $\nabla Z/Z$ several hundred times) was less that that needed to do one exact computation of $\nabla Z/Z$.

Recalling that $\bar{\lambda}$ was our estimate, we have (see section 3 and 4)

$$E_{\bar{\lambda}}(\omega_i\omega_k) = E(\omega_i\omega_k) - \nabla_{(i-1)\cdot(25-(i/2))+k-i+1} Z(\lambda)/Z(\lambda)$$
$$E_{\bar{\lambda}}(\omega_i) = E(\omega_i) - \nabla_{276+i} Z(\lambda)/Z(\lambda)$$

$E_{\bar{\lambda}}$ being the expected value under the distribution generated by $\bar{\lambda}$. The percent error in $E_{\bar{\lambda}}$ is

$$\frac{\text{the true value of } \nabla_i Z/Z}{\text{the value of the associated statistic in the original system}}$$

One measure of the "fit" of the maximum entropy distribution generated by $\bar{\lambda}$ is the median value of the percent error, which was, for the $\bar{\lambda}$ we found, .07. So, compared to the original statistics, the errors in the statistics for the maximum entropy distribution generated by $\bar{\lambda}$ were typically small.

The results on the following pages contain more detailed information about the behavior of the maximum entropy distribution generated by $\bar{\lambda}$. They are the statistics of the original distribution, the estimated value of $\nabla Z(\bar{\lambda})/Z(\bar{\lambda})$ and the true value of $\nabla Z(\bar{\lambda})/Z(\bar{\lambda})$. The results are presented in ten row, thirty column tables. (On the first row we will have $\{\nabla_1 Z/Z, \ldots, \nabla_{10} Z/Z\}$, on the second $\{\nabla_{11} Z/Z, \ldots, \nabla_{20} Z/Z\}$, etc.) They are presented in such a way that the statistics in the first table have the same position in their table as the gradient associated with that statistic has in its own table. The statistics concerning the $E(\omega_i\omega_j)$ are thus on the top of the table.

0.0806	0.4255	-0.5389	0.4194	-0.0658	0.1038	-0.2769	-0.1695	-0.0471	-0.4299
-0.2675	0.1587	-0.4326	-0.1210	-0.0087	-0.5435	-0.2733	-0.1378	0.0675	-0.2034
-0.1561	0.1139	-0.2435	0.1101	-0.2267	0.0143	-0.1123	-0.1405	-0.0727	0.0058
0.2384	-0.1810	0.1295	0.1596	-0.0861	-0.2118	-0.1079	-0.2927	-0.1603	-0.0798
0.1153	-0.2382	0.0820	0.0935	0.0209	-0.3710	0.1078	-0.0310	0.0578	-0.2163
-0.2566	-0.0483	-0.3900	0.0760	-0.0324	-0.1739	-0.1297	0.1580	-0.5501	-0.3170
-0.0611	0.1214	-0.3010	0.1056	0.2089	-0.2999	-0.4132	-0.0192	-0.2100	0.2303
-0.0626	0.0139	0.3400	0.2605	-0.3386	0.5548	0.3427	-0.1447	0.4944	0.2766
0.1199	0.1702	0.4366	0.1741	-0.0355	0.1152	-0.1655	0.0974	-0.0307	0.1474
-0.0506	-0.1471	-0.3767	0.1875	-0.4381	-0.2993	-0.1297	-0.2945	-0.0289	-0.1637
0.0578	-0.1158	-0.2955	0.1512	-0.1849	0.3361	0.2909	-0.0898	0.1631	0.0670
0.0747	0.2564	-0.0816	-0.1574	-0.0171	0.2316	0.2813	0.0894	-0.2684	-0.1118
0.1283	-0.3382	0.2184	0.1454	0.0056	0.0848	0.1245	-0.1346	0.2802	-0.1257
-0.1996	0.1433	0.0238	-0.0056	0.0272	-0.3076	-0.1960	-0.1132	-0.2609	0.1727
0.1411	0.0855	0.1670	-0.1307	-0.0344	0.0115	-0.0492	-0.1137	0.3927	0.4310
-0.1274	-0.1325	0.1253	0.0540	-0.3286	0.0894	-0.1906	0.2345	-0.1451	0.1649
-0.1359	-0.1806	0.1306	0.2210	-0.0150	0.0971	-0.2461	-0.0658	-0.1536	-0.0483
-0.0332	0.0306	0.1210	0.1267	-0.0879	-0.0752	-0.2189	-0.0902	0.1518	-0.1577
0.1225	-0.1729	-0.0798	0.0621	0.1784	0.1587	0.0587	0.2528	0.0247	0.0028
0.4491	0.1652	0.0265	-0.1744	0.0841	0.0775	-0.2494	0.1086	-0.0153	0.3228
0.0290	0.1046	0.0138	-0.0861	0.2085	0.1336	-0.0243	0.2994	0.1015	-0.0400
-0.3274	-0.3316	0.1087	-0.0558	0.0223	0.1356	-0.3317	-0.1941	0.0738	-0.1794
0.0815	0.3093	-0.1086	0.3438	0.0098	0.1848	0.1888	0.3935	0.1908	0.0734
0.1954	0.1346	0.1088	-0.0214	-0.0820	0.0429	0.2873	0.1646	-0.1003	-0.0602
-0.0993	-0.1701	-0.0309	-0.2673	-0.2077	0.1756	-0.2132	-0.1913	0.3897	0.1054
-0.1502	0.3540	-0.0165	-0.2966	0.3280	-0.1534	0.0292	0.0823	0.0419	-0.2966
0.0808	-0.1662	0.1384	0.2146	0.1215	-0.0259	0.1721	-0.1384	0.3438	-0.0841
0.0553	0.0776	0.1831	-0.1699	-0.1746	-0.0239	-0.6172	-0.3503	-0.5228	0.6238
-0.4085	0.1994	0.0478	0.3981	0.1448	-0.0911	0.4876	0.1390	-0.2291	0.4696
0.2816	0.0204	0.7344	0.3949	-0.0182	-0.0226	0.3911	0.0047	-0.3163	0.3157

Table of $E(\omega_i \omega_j)$, $E(\omega_i)$

0.0181	0.0044	0.0103	-0.0576	0.0205	0.0128	0.0354	-0.0573	-0.0112	-0.0207
0.0526	-0.0433	0.0646	0.0483	-0.0115	0.0238	0.0049	-0.0582	0.0352	0.0075
0.0459	-0.1033	-0.0199	0.0230	-0.0314	-0.0270	-0.0332	0.0172	0.0277	0.0107
0.0089	-0.0006	0.0140	0.0496	-0.0075	0.0112	0.0152	-0.0301	-0.0083	-0.0053
-0.0019	-0.0347	0.0754	-0.0593	-0.0005	-0.0170	-0.0432	-0.0054	-0.0363	-0.0052
-0.0027	0.0156	0.0104	0.0496	-0.0028	0.0264	0.0278	0.0160	-0.0186	-0.0580
-0.0189	0.0001	-0.0323	0.0506	-0.0181	-0.0392	0.0497	-0.0578	-0.0270	-0.0151
0.0076	-0.0114	0.0145	-0.0614	-0.0072	-0.0147	0.0072	-0.0130	-0.0220	-0.0437
0.0466	0.0200	0.0033	-0.0422	0.0862	-0.0520	0.0346	0.0025	0.0689	0.0136
0.0134	0.0287	0.0575	-0.0071	0.0232	0.0174	-0.0222	0.0441	0.0726	-0.0476
0.0137	0.0614	0.0371	-0.0227	0.0176	0.0547	-0.0176	0.0389	-0.0030	0.0127
0.0183	0.0545	-0.0850	-0.0454	0.0182	0.0142	0.0230	0.0347	-0.0603	-0.0647
-0.0331	0.0112	0.0113	0.0776	-0.0295	0.0342	0.0191	-0.0370	0.0079	0.0057
-0.0248	0.0096	0.0523	0.0355	-0.0403	-0.0448	-0.0213	-0.0560	-0.0749	0.0524
0.0834	-0.0371	0.0368	-0.0412	0.0673	-0.0717	-0.0453	0.0335	-0.0058	-0.0162
0.0240	-0.0730	-0.0221	-0.0160	0.0146	0.0747	0.0260	0.0130	-0.0213	-0.0192
-0.0263	-0.0442	-0.0489	0.0198	0.0603	-0.0337	0.0176	-0.0155	-0.0457	0.0396
0.0524	-0.0090	-0.0063	0.0133	-0.0652	-0.0188	-0.0057	-0.0195	0.0259	0.0063
0.0252	-0.0466	-0.0233	0.0099	-0.0020	-0.0233	0.0006	-0.0170	-0.0228	-0.0263
0.0516	0.0211	-0.0190	-0.0321	-0.0085	-0.0110	0.0063	0.0075	0.0340	-0.0543
-0.0406	-0.0124	-0.0615	-0.0315	0.0615	0.0441	-0.0648	-0.0273	0.0446	-0.1162
0.0178	-0.0483	-0.0144	0.0403	0.1209	-0.0200	-0.0304	0.0106	0.0494	-0.0770
0.0250	0.0564	0.0125	-0.0317	-0.0714	0.0483	0.0162	0.0110	-0.0118	0.0310
-0.0513	-0.0110	-0.0483	-0.1200	0.0366	0.0231	0.0208	-0.0155	0.0306	0.0119
0.0016	0.0281	-0.0513	0.0288	-0.0188	-0.0227	-0.0221	0.0133	-0.0107	0.0304
-0.0363	-0.0077	-0.0792	0.0160	0.0388	0.0365	-0.0204	-0.0463	-0.0299	0.0240
0.0323	0.0200	-0.0552	0.0275	0.0142	-0.0724	0.0058	-0.0108	-0.0077	-0.0415
-0.0188	0.0411	0.0186	0.0623	-0.0883	-0.0424	0.0195	-0.0346	-0.0599	-0.0191
0.0643	0.0026	0.0519	0.0355	0.0610	-0.0524	0.0213	-0.0826	0.0023	-0.0123
-0.0108	0.0230	0.0034	-0.0136	-0.0389	0.0168	0.0367	-0.1384	0.0327	0.0507

Table of estimated $\nabla Z(\bar{\lambda})/Z(\bar{\lambda})$

-0.0233	0.0004	0.0116	-0.0173	0.0168	0.0138	-0.0258	-0.0143	-0.0304	-0.0050
0.0088	0.0029	0.0204	0.0477	0.0017	0.0095	-0.0259	0.0077	0.0056	0.0386
0.0042	-0.0317	-0.0117	-0.0152	-0.0069	-0.0288	-0.0116	-0.0147	0.0091	0.0114
0.0049	-0.0140	0.0130	0.0087	0.0282	-0.0004	-0.0071	-0.0014	-0.0171	0.0035
0.0081	-0.0035	0.0215	-0.0188	0.0214	-0.0027	-0.0297	0.0244	0.0249	0.0135
-0.0135	-0.0196	-0.0130	0.0241	-0.0123	0.0262	0.0044	0.0198	0.0049	-0.0208
0.0026	-0.0019	0.0015	0.0301	-0.0010	-0.0112	0.0188	-0.0195	-0.0145	0.0258
-0.0129	0.0259	0.0057	-0.0064	0.0000	-0.0185	-0.0125	-0.0276	-0.0139	0.0189
0.0029	0.0281	-0.0028	-0.0001	0.0279	-0.0257	-0.0197	0.0059	0.0068	0.0244
-0.0388	0.0301	0.0052	0.0022	0.0205	0.0093	0.0014	0.0183	0.0120	-0.0009
0.0000	0.0401	0.0010	0.0006	-0.0094	0.0058	-0.0060	0.0031	0.0023	-0.0132
0.0213	0.0045	-0.0235	-0.0115	0.0045	0.0065	-0.0008	0.0152	-0.0188	-0.0323
0.0065	-0.0001	-0.0046	0.0057	-0.0079	0.0034	0.0077	-0.0069	0.0025	0.0225
-0.0036	0.0243	-0.0186	-0.0180	0.0083	-0.0144	-0.0173	-0.0136	-0.0128	0.0019
0.0324	-0.0094	-0.0030	-0.0055	-0.0238	-0.0071	-0.0105	0.0030	0.0184	0.0057
-0.0095	-0.0159	0.0169	0.0132	-0.0046	-0.0230	-0.0011	0.0057	-0.0165	0.0074
-0.0256	-0.0285	-0.0130	0.0238	0.0146	-0.0065	0.0043	-0.0159	-0.0240	0.0129
-0.0079	0.0203	0.0315	0.0142	0.0012	0.0306	0.0129	-0.0136	-0.0041	0.0124
-0.0097	-0.0131	0.0050	0.0218	0.0098	0.0016	0.0128	0.0009	-0.0316	-0.0288
0.0036	0.0049	-0.0353	-0.0014	-0.0258	-0.0010	-0.0069	-0.0120	0.0307	-0.0157
-0.0292	0.0070	-0.0269	-0.0089	-0.0022	0.0189	-0.0253	-0.0034	0.0195	-0.0142
-0.0053	-0.0039	0.0180	0.0210	0.0176	0.0094	-0.0033	0.0144	0.0152	-0.0099
0.0133	-0.0107	-0.0375	-0.0175	-0.0173	-0.0007	0.0311	0.0001	0.0005	0.0259
0.0120	0.0049	-0.0465	-0.0200	-0.0194	-0.0037	-0.0021	0.0017	-0.0100	-0.0199
-0.0240	-0.0005	-0.0344	0.0008	-0.0130	0.0004	-0.0170	-0.0019	0.0167	-0.0035
-0.0038	-0.0087	-0.0272	0.0035	0.0050	-0.0045	0.0021	-0.0245	0.0099	-0.0006
-0.0143	0.0111	0.0005	0.0045	0.0213	-0.0234	0.0275	-0.0046	0.0039	-0.0108
0.0041	0.0229	0.0031	0.0017	-0.0148	0.0223	-0.0016	-0.0240	-0.0092	-0.0002
0.0066	0.0200	0.0274	0.0393	0.0226	-0.0146	0.0205	0.0019	-0.0198	-0.0030
-0.0054	0.0204	0.0006	0.0308	-0.0525	0.0347	0.0002	-0.0478	-0.0175	0.0187

Table of true $\nabla Z(\bar{\lambda})/Z(\bar{\lambda})$

Example 2: Letter recognition

Let us consider the problem of invariant letter recognition. We will be presented with a picture of a letter of unknown size, orientation and font, and we wish to find out which letter it is. For the sake of simplicity we will use simple images with just two grey levels (black and white), and we will just consider the capital letters $A, \ldots, G$.

There are many ways to approach this problem, the one we will consider is based on feature extraction. We will deal with the invariance of the problem by extracting features (scalars) that are independent of the orientation or size of the letter. Our expert system will be a distribution on the space of features and labels, where the latter identify the letter. This distribution will be used to find the probabilities of the labels conditioned on the observed features (*e.g.*, $P($ 'the letter is an A' $\mid$ $\text{feature}_1 = 5,\ \text{feature}_2 = 6) = .3)$

Choosing the features is a crucial task, and should be given as much consideration as the construction of the expert system that uses them. Features can be roughly separated into two groups, local and global. Global features deal with the whole picture and are what we used in the results presented in the following pages. Local features deal with the local behavior of the picture elements. Hence, local features are ideal for occluded pictures. Local features seem more powerful and, it is our belief, will be essential for a true solution to the invariant character recognition problem.

The features we used in our example were non-standard. They were picked because they seemed reasonable and not too difficult to compute. They mostly deal with holes and indentations. A hole being a white (non-letter) region completely surrounded by the letter (typically A has a hole, C does not), and an indentation being a white region that is connected, is in the convex hull (the convex hull of the set S is the smallest convex set containing S) of the letter, and yet not a hole. Some thought will show that this is exactly what we mean by an indentation (typically O has no indentations, T has two). Below are listed twelve of the features we use.

1	The size of the largest hole	/ The size of the convex hull of the letter
2	The size of the second largest hole	/ The size of the convex hull of the letter
3	The size of the third largest hole	/ The size of the convex hull of the letter
4	The size of the largest indentation	/ The size of the convex hull of the letter
5	The size of the second largest indentation	/ The size of the convex hull of the letter
6	The size of the third largest indentation	/ The size of the convex hull of the letter
7	The ratio of longest to shortest axis of the largest hole	
8	The ratio of longest to shortest axis of the largest indentation	
9	The ratio of longest to shortest axis of the second largest indentation	
10	The ratio of longest to shortest axis of the third largest indentation	
11	The total area of the indentations in the largest hole	/ The size of the letter
12	The total area of the indentations in the largest indentation	/ The size of the letter

We also have several other features that deal with the points that span the convex hull. We construct these features as follows. Let our original set of points be the smallest set that spans the convex hull of the letter. At every step remove one point from our set of points, picked to maximize the area spanned by the remaining points. Continue doing this until no points are left. Our final features shall be

13	The number of points that span the convex hull of the letter	
14	The area spanned by six remaining points	/ The area of the convex hull
15	The area spanned by five remaining points	/ The area of the convex hull

16 The area spanned by four remaining points / The area of the convex hull
17 The area spanned by three remaining points / The area of the convex hull

These features are useful since they tell us how curved the letter is. For example the letter 'E' is square so the area left after removing all but four points should be large, but when only three points remain the number should be much smaller.

The knowledge we used to form our expert system deals with the expected values of the features, the features squared, and the products of selected features, all conditioned on the letter (eg. $E(\text{feature}_1|$ the letter is an $A) = C_1$, $E((\text{feature}_1)^2|$ the letter is an $A) = C_2$, $E(\text{feature}_1 \cdot \text{feature}_2|$ the letter is an $A) = C_3$). We also give our system the very important piece of knowledge that the letters are of equal probability (each of probability 1/7). It would be nice to use all the products of features as constraints, but with seven letters and 17 features we would have several thousand constraints and this is computationally difficult. In the experiment that yielded the results on the following pages we used the conditional probabilities of only fifteen different products. The total number of constraints was thus 350.

On the computational side we use several techniques to improve the behavior of the gradient descent. The first technique deals with the constraints themselves. In general a constraint is of the form $E(G_i) = C_i$, where G_i is some function and C_i is a constant. Typically we find C_i by taking a test sample, and using the sample mean of G_i. However this does have problems. Since our gradient descent is not exact we typically end up with $\nabla_i Z/Z$ small for all i and typically of the same order of magnitude, but not exactly zero. Thus if we have as a constraint $E(100 \cdot G_j) = 100 \cdot C_j$ for some j, then $E(G_j) = C_j$ will come much closer to being true in the resulting maximum entropy distribution than if we had $E(G_j) = C_j$ as the constraint. Also, there is a problem caused by wanting our expert system to recognize things not in the sample that formed our constraints. In the following results we trained our system with 5 samples of each letter. Now, what will happen if we try to recognize a letter that was not in the population we used to train the system? We would like it to be recognized, especially if it is similar to the original population. This does not always happen. One particular case of this problem is caused by boundry effects. If a feature has range 0 to 1, and all the sample letter C's had value 0 for this feature, then the only way to satisfy the constraint $E(\text{feature} \mid \text{letter is a C}) = \text{sample mean} = 0$, is to have $P(\text{feature} = 0 \mid \text{letter is a C}) = 1$. If we present a C which has value .001 for this feature, it will not be recognized. While this problem could be cured by having a large sample (and should be), it and the previous problem can both be dealt with by scaling and slightly modifying the constraint functions. For the full details we refer the reader to [2].

Now let us consider the sampling method itself. In this problem the constraints have a rather odd form, almost all of them are conditioned on the letter. This can lead to difficulties in the sampling method. When the letter is an A, for example, the features tend to have certain values, as the constraints specify. At every step in the sampling we go through the feature vector, holding most of the features fixed and then picking those that are not fixed according to a distribution. However when the label is 'A', the features tend to stay within a certain range. When it comes time to fix the features and vary the label, the distribution that we use to pick a label, being generated by features that correspond to an A, will emphasize the label 'A'. This is to be expected, since we can think of the label 'A' as corresponding to some region in the state space, and forming a sort of 'well' in the energy landscape (a region of very likely events, corresponding to 'A's, surrounded by a region of low probability that corresponds to feature values not associated with any letter). Once such a 'well' is entered it can be difficult to get out of. So, if the label 'A' is turned on it tends to stay on, and our sample will quite possibly over-emphasize one particular letter at the expense of the rest.

Since all the constraints involving the features are conditional, we can use the following change in the sampling method to cure the problem mentioned above. We will first fix label at 'A'. We will then conduct gradient descent on Z until the values of $\nabla_i Z/Z$ are small for all the i's corresponding to constraints conditioned on the label being 'A'. Then we fix the label 'B' and continue. After we have gone through all the letters (in our case A,. . .,G) we start sampling normally (letting the label vary). Since the values of $\nabla_i Z/Z$ are small for all i's corresponding to conditional constraints, we need only conduct gradient descent until the constraint that all letters be equally likely is (close to being) satisfied.

Now let us present the results. The letters we wished to identify are on the following pages. They are the same letters that were used to find the sample means in the constraints. The probabilities of the labels conditioned on the observed features, given by the maximum entropy expert system, is provided underneath the letters. Only the top three probabilities are listed for each letter, in the interest of saving space. The energies are also listed, where the energy is $\sum_{i=1}^{m} - a_i(\omega)\bar{\lambda}_i$ (where ω is the element of Ω corresponding to the feature vector plus the hypothesized label, and $\bar{\lambda}$ is the result of our minimization of Z). The energies are given to provide some comparison between different letters ("this E looks more like an E than that E").

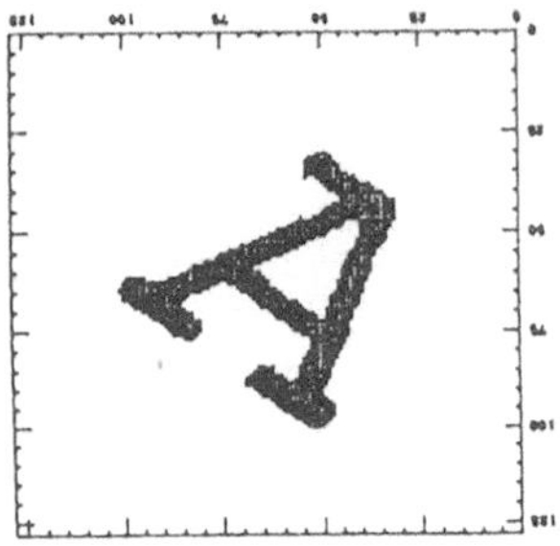

1

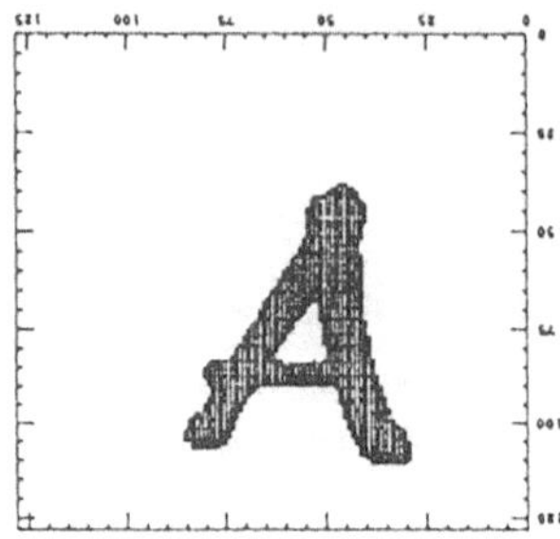

2

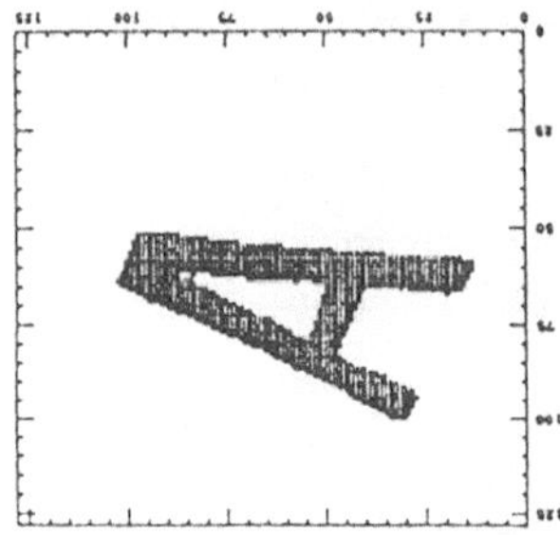

3

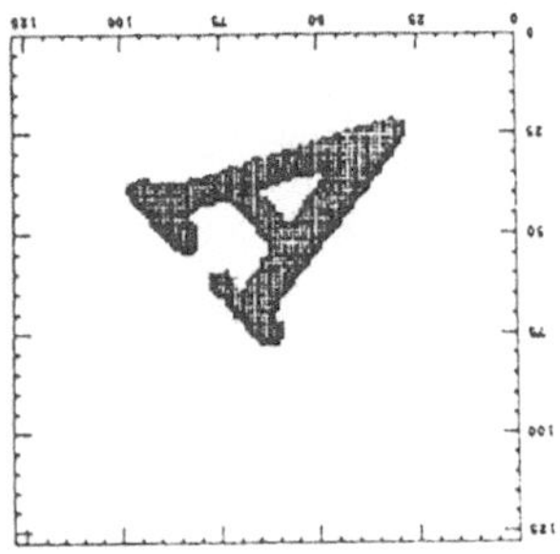

4

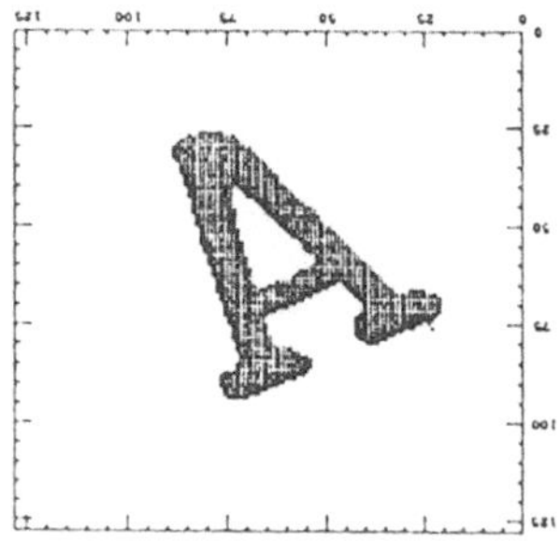

5

Picture	Hypothesis	Probability of Hypothesis	Energy of Hypothesis
A1	it is an A	0.6719273	3.8487754
A1	it is an E	0.0002583	11.7124262
A1	it is an F	0.3277985	4.5665264
A2	it is an A	0.9809950	2.1446524
A2	it is an E	0.0012927	8.7764645
A2	it is an F	0.0177039	6.1594334
A3	it is an A	1.0000000	0.2035229
A3	it is an B	0.0000000	18.8109589
A3	it is an E	0.0000000	23.3654041
A4	it is an A	0.9920666	1.6209198
A4	it is an E	0.0000050	13.8176394
A4	it is an F	0.0079282	6.4502902
A5	it is an A	0.9974482	5.4865880
A5	it is an E	0.0001584	14.2343798
A5	it is an F	0.0023898	11.5205803

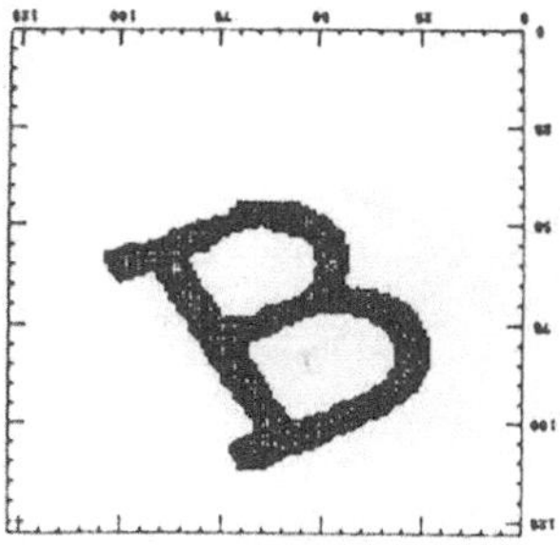

1

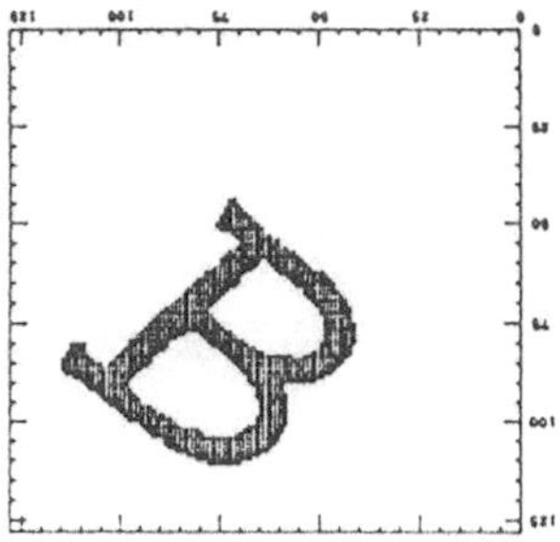

2

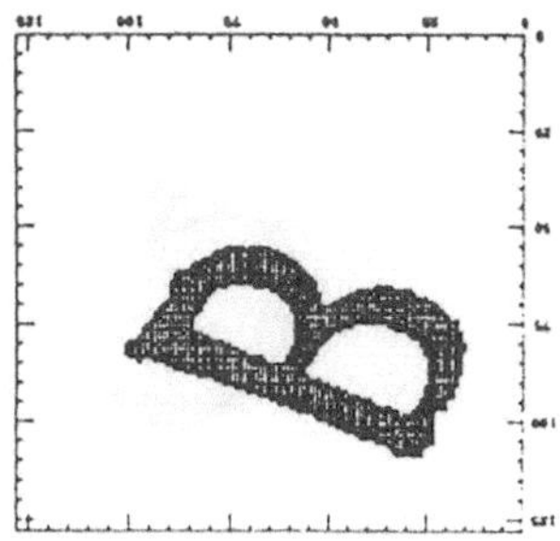

3

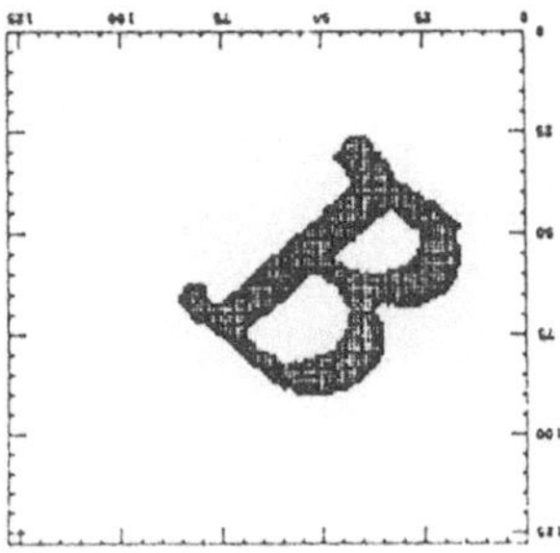

4

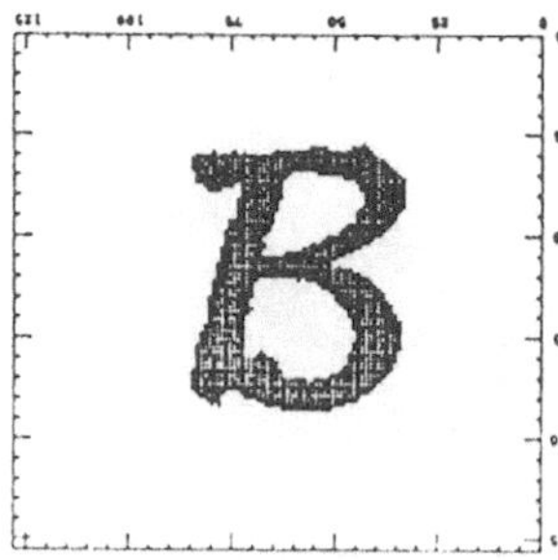

5

Picture	Hypothesis	Probability of Hypothesis	Energy of Hypothesis
B1	it is an B	1.0000000	5.7593293
B1	it is an C	0.0000000	26.1249256
B1	it is an D	0.0000000	26.0998402
B2	it is an B	1.0000000	6.2038503
B2	it is an C	0.0000000	26.3224182
B2	it is an E	0.0000000	26.6758900
B3	it is an A	0.0000000	29.8165340
B3	it is an B	1.0000000	9.1290588
B3	it is an E	0.0000000	30.7371597
B4	it is an B	0.9999986	5.9010286
B4	it is an E	0.0000006	20.2056236
B4	it is an F	0.0000003	20.8549023
B5	it is an B	0.9999995	5.5978689
B5	it is an E	0.0000000	23.7445774
B5	it is an F	0.0000004	20.2729683

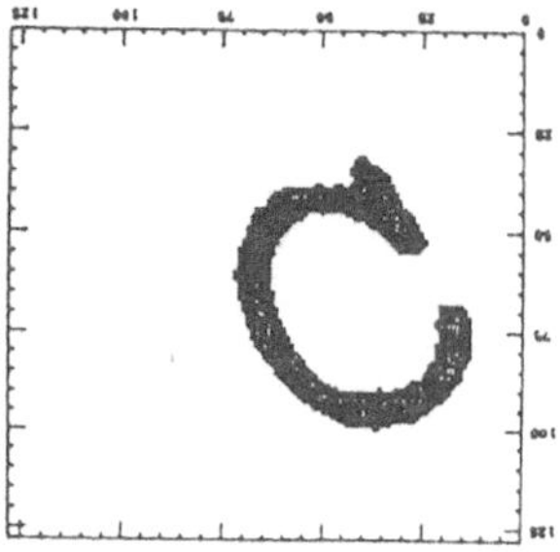

1

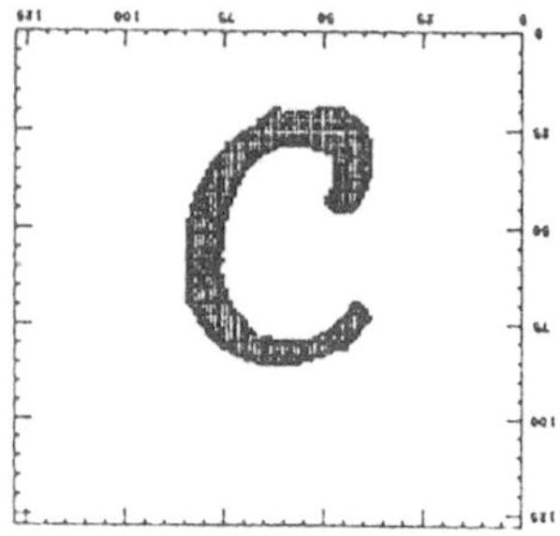

2

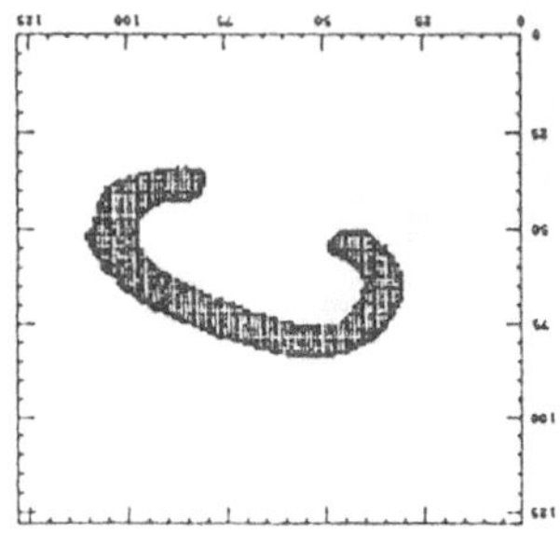

3

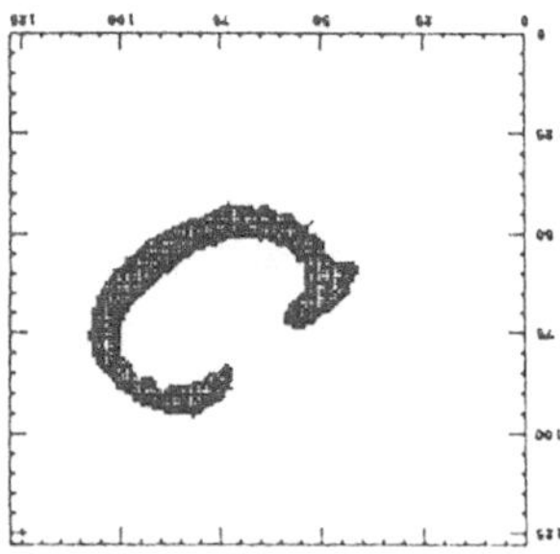

4

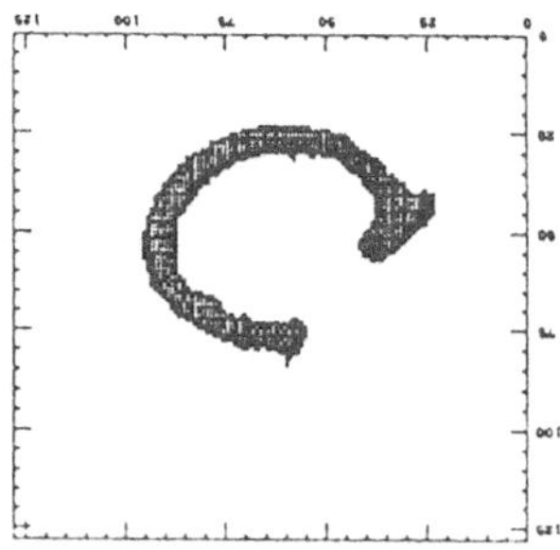

5

Picture	Hypothesis	Probability of Hypothesis	Energy of Hypothesis
C1	it is an C	0.6604182	-1.9814487
C1	it is an E	0.0184070	1.5986940
C1	it is an G	0.3191914	-1.2543660
C2	it is an C	0.6313014	-1.3120894
C2	it is an E	0.1306149	0.2634406
C2	it is an G	0.2368994	-0.3319414
C3	it is an C	0.5955555	-1.6064481
C3	it is an E	0.0437944	1.0035404
C3	it is an G	0.3588811	-1.0999445
C4	it is an C	0.6606517	-0.5129181
C4	it is an E	0.0676684	1.7656897
C4	it is an G	0.2695387	0.3835969
C5	it is an C	0.6386604	-1.4192295
C5	it is an E	0.0491251	1.1457740
C5	it is an G	0.3097548	-0.6956378

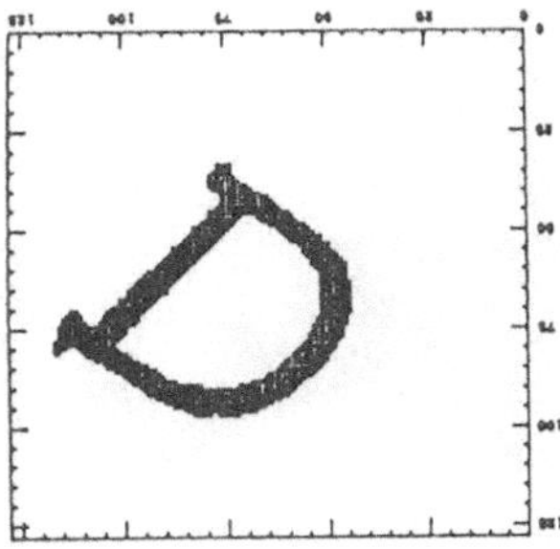

1

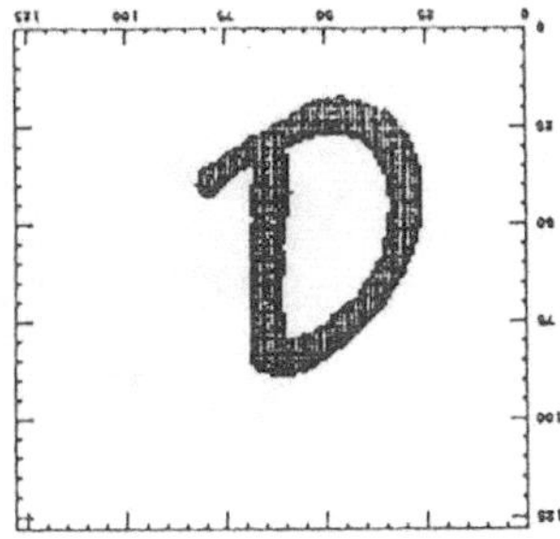

2

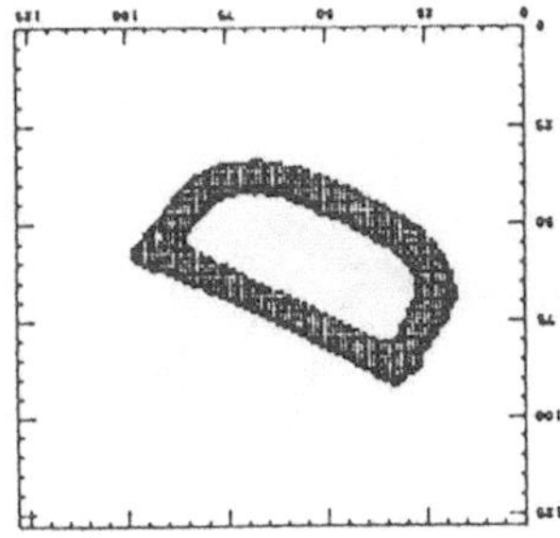

3

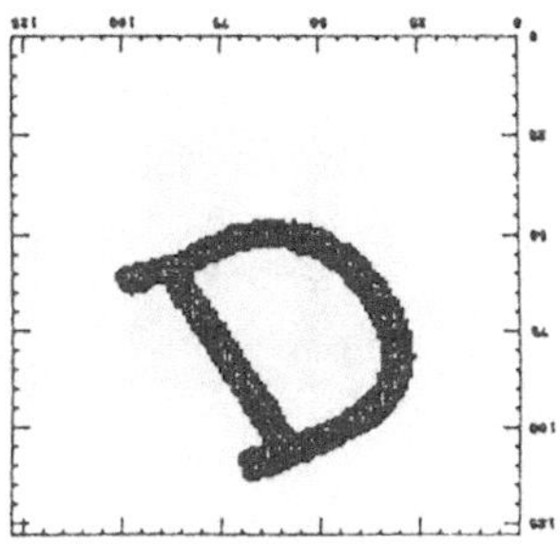

4

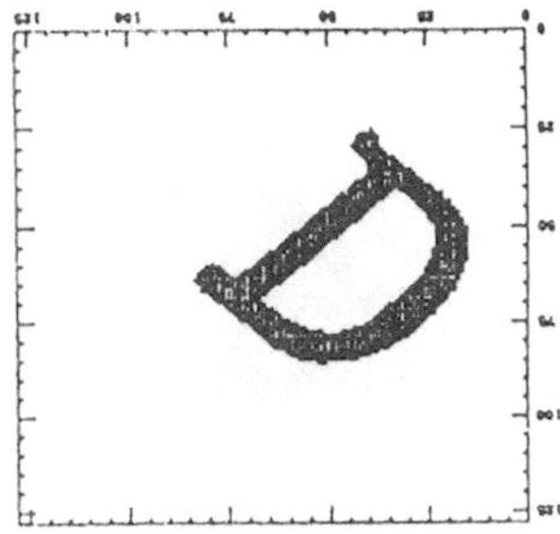

5

Picture	Hypothesis	Probability of Hypothesis	Energy of Hypothesis
D1	it is an A	0.0000051	19.0304546
D1	it is an B	0.0011396	13.6119480
D1	it is an D	0.9988480	6.8360224
D2	it is an B	0.0003877	12.2251072
D2	it is an D	0.9994236	4.3704495
D2	it is an G	0.0000976	13.6045828
D3	it is an B	0.0000000	89.0601807
D3	it is an D	1.0000000	-12.7670460
D3	it is an F	0.0000000	82.0110168
D4	it is an B	0.0005200	12.7396383
D4	it is an D	0.9994300	5.1784315
D4	it is an G	0.0000154	16.2575455
D5	it is an A	0.0000039	17.1760368
D5	it is an B	0.0003292	12.7477312
D5	it is an D	0.9996585	4.7292614

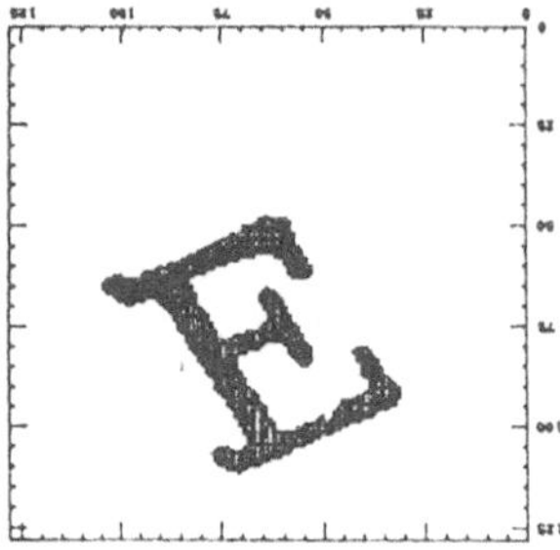

1

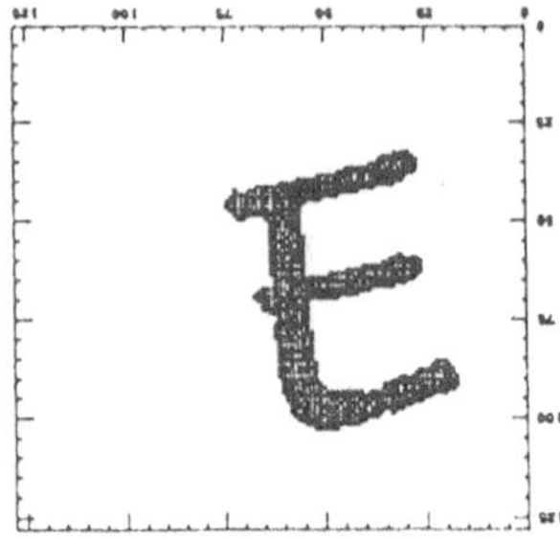

2

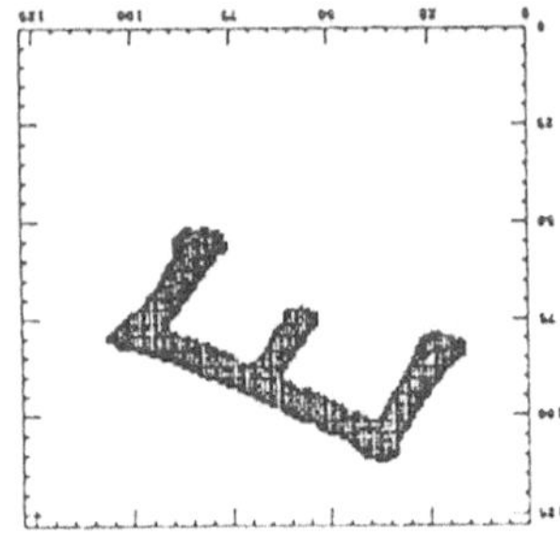

3

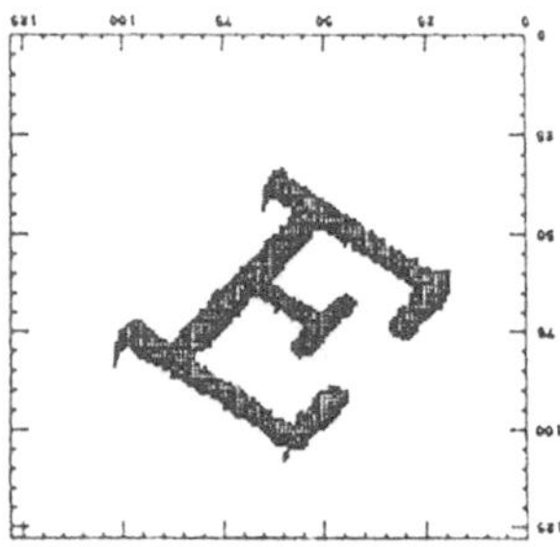

4

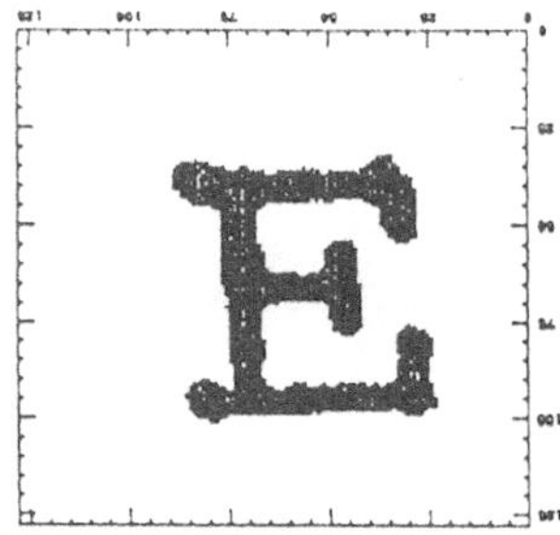

5

Picture	Hypothesis	Probability of Hypothesis	Energy of Hypothesis
E1	it is an A	0.0000728	7.3868618
E1	it is an E	0.9531993	-2.0928898
E1	it is an F	0.0467241	0.9226741
E2	it is an C	0.0000507	7.1914301
E2	it is an E	0.9832692	-2.6817935
E2	it is an F	0.0166214	1.3983997
E3	it is an A	0.0003051	8.8423738
E3	it is an C	0.0000295	11.1771812
E3	it is an E	0.9996634	0.7478167
E4	it is an A	0.0003749	6.4038081
E4	it is an E	0.9340211	-1.4169102
E4	it is an F	0.0655886	1.2391865
E5	it is an A	0.0000976	6.7060304
E5	it is an E	0.9618846	-2.4899280
E5	it is an F	0.0380000	0.7413796

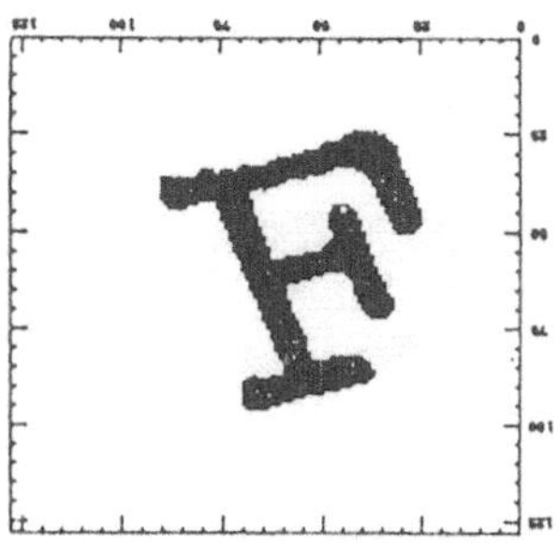

1

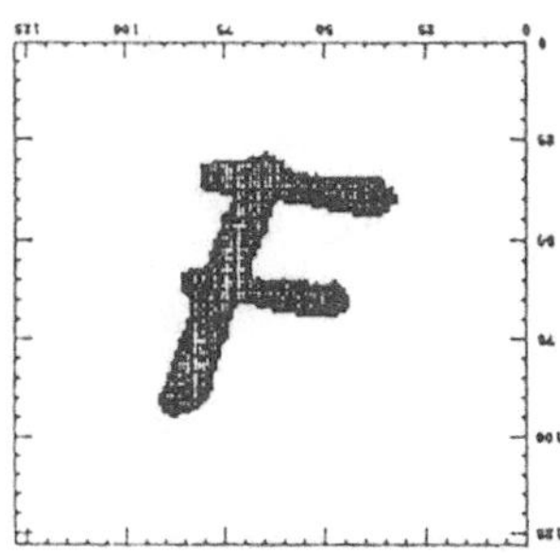

2

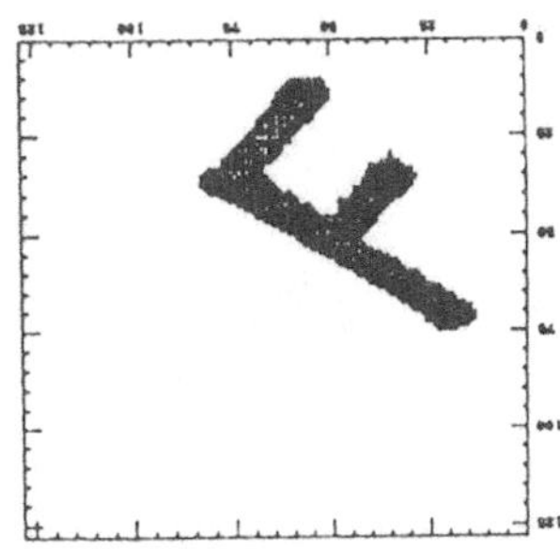

3

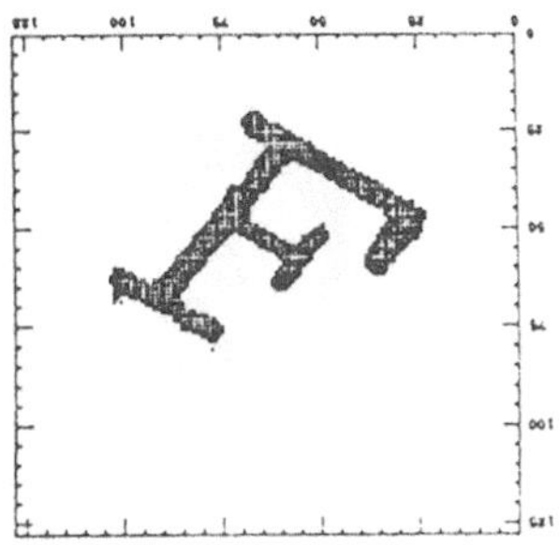

4

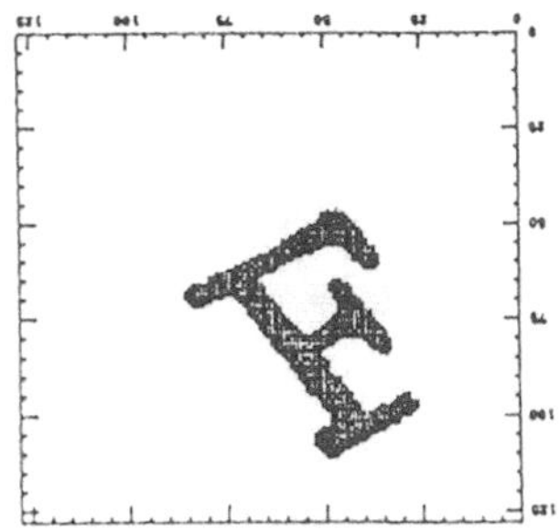

5

Picture	Hypothesis	Probability of Hypothesis	Energy of Hypothesis
F1	it is an A	0.0009967	5.9281058
F1	it is an E	0.0000012	12.6835003
F1	it is an F	0.9990017	-0.9819657
F2	it is an A	0.0136954	3.9498260
F2	it is an E	0.0032687	5.3824973
F2	it is an F	0.9830301	-0.3237556
F3	it is an A	0.0021422	5.9868760
F3	it is an E	0.0016594	6.2422500
F3	it is an F	0.9961985	-0.1552548
F4	it is an E	0.8462385	-0.0166720
F4	it is an F	0.1515211	1.7034043
F4	it is an G	0.0020051	6.0284510
F5	it is an A	0.0003445	8.3701258
F5	it is an E	0.8101026	0.6073851
F5	it is an F	0.1895521	2.0598819

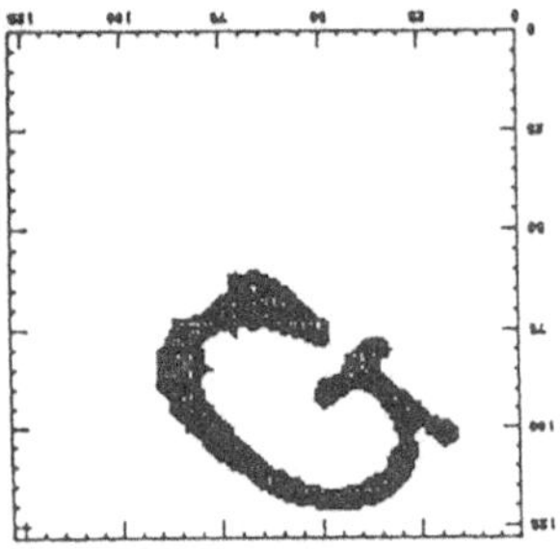

1

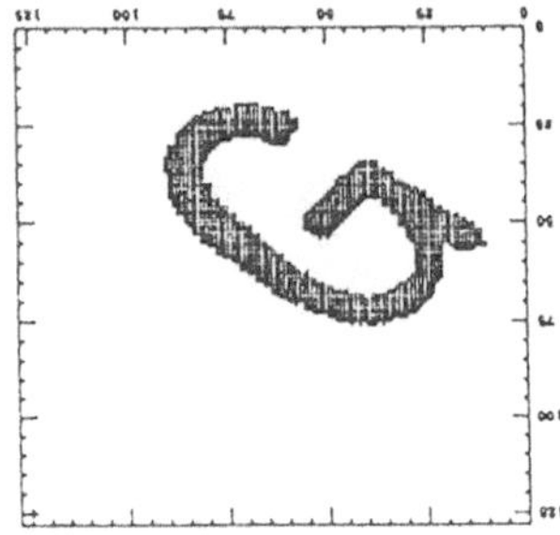

2

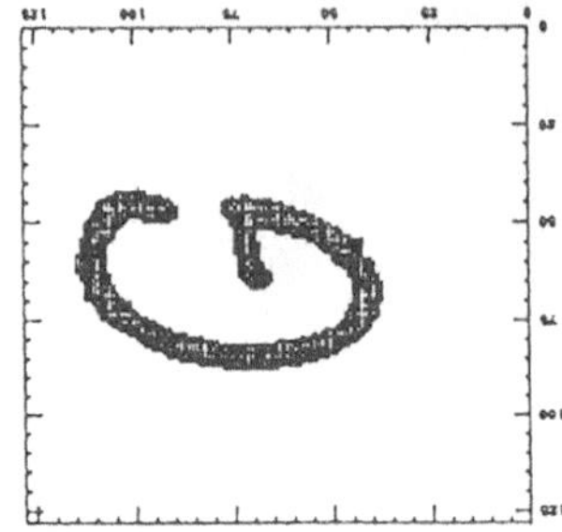

3

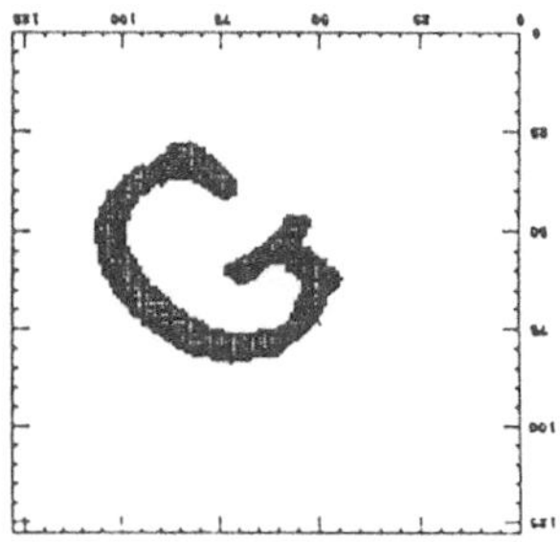

4

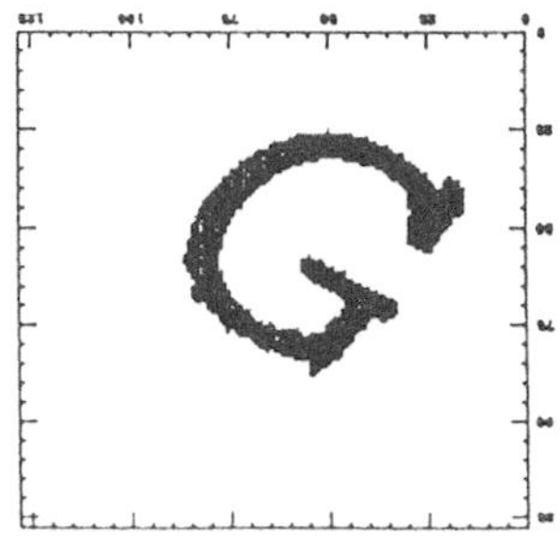

5

Picture	Hypothesis	Probability of Hypothesis	Energy of Hypothesis
G1	it is an C	0.1939279	1.1415343
G1	it is an E	0.2315701	0.9641383
G1	it is an G	0.5542700	0.0913689
G2	it is an C	0.2378499	0.9435763
G2	it is an E	0.1392609	1.4788672
G2	it is an G	0.6008883	0.0168073
G3	it is an C	0.3128236	0.1166506
G3	it is an E	0.0234408	2.7078128
G3	it is an G	0.6587726	-0.6280884
G4	it is an C	0.2702022	0.0306356
G4	it is an E	0.0150385	2.9191945
G4	it is an G	0.7069070	-0.9310930
G5	it is an C	0.3253015	-0.3527871
G5	it is an E	0.1510101	0.4146184
G5	it is an G	0.5172358	-0.8165337

Bibliography

[1] Edwin T. Jaynes, 'On The Rational of Maximum-Entropy Methods,' *Proc. of the IEEE* **70** (1982), 939–952.

[2] Alan F. Lippman, 'A Maximum Entropy Method for Expert System Construction,' *Ph.D. Thesis, Brown University* (1986).

[3] Stuart Geman, 'Stochastic Relaxation Methods for Image Restoration and Expert Systems,' To appear in: *Automated Image Analysis: Theory and Experiments, D.B. Cooper, R.L. Launer, and D.E. McClure, Eds.* New York: Academic Press.

[4] Donald Geman, Personal Communication (1985).

[5] Censor,Y., Elfving,T., Herman,G.T., Kuo,Y.H., and Lent,A., 'On The Relationship Between "MART" and Bregman's Algorithm for Entropy Maximization Over Linear Inequalities,' **In preparation**.

[6] Peter Cheeseman, 'A Method of Computing Maximum Entropy Probability Values for Expert Systems,' **Preprint** SRI International, Menlo Park, California.

[7] Gull,S.F. and Skilling,J., 'The Entropy of an Image', *Proceedings of the Second Workshop on Maximum Entropy and Bayesian Methods in Applied Statistics,'* **To appear**

[8] Metropolis,N., Rosenbluth,A.W., Rosenbluth,M.N., Teller,A.H., and Teller,E., 'Equations of State Calculations by Fast Computing Machines,' *J. Chem. Phys.* **21** (1953), 1087-1091.

STOCHASTIC RELAXATION METHODS FOR IMAGE RESTORATION AND EXPERT SYSTEMS*

Stuart Geman

Division of Applied Mathematics
Brown University
Providence, Rhode Island

Introduction

This is an introduction to "stochastic relaxation" (SR), a highly parallel computational algorithm for various inference and optimization problems. The presentation will be nontechnical and by example, highlighting possible applications to image processing and expert systems. Concerning image processing, SR algorithms have been implemented and numerous experiments have been performed. Although the pictures used so far are quite simple, the restorations and segmentations are extremely good, even at low signal to noise ratios. The application to expert systems is more speculative. SR seems especially well-suited for the combinatorial and statistical matters that arise in expert system problems, but we have not yet experimented in a real problem domain. For concreteness, our discussion will focus on medical diagnosis, an often-chosen prototype problem.

The common thread between these two applications of SR is our "Bayesian" formulation. This means that we formalize the relevant prior information, whether about real-world images, or about symptoms and diseases, as a probability distribution. For image processing, this distribution is on all possible configur-

* This paper was written in 1983. Since then, applications to tomography [11], texture segmentation [14] and classification [9], and character recognition [21] have been explored. The theoretical underpinnings of stochastic relaxation have been advanced considerably ([5],[12],[16],[18], for example), and various extensions have been introduced and analyzed ([10], [13], for example).

G. J. Erickson and C. R. Smith (eds.),
Maximum-Entropy and Bayesian Methods in Science and Engineering (Vol. 2), 265–311.

ations of pixel grey levels, locations of edge elements, and states of other image attributes that may be of interest for a particular problem. For medical diagnosis, the distribution is on the entire collection of medically-relevant features: diseases, tests, patient characteristics, and so-on. In either case, this *prior distribution* captures our knowledge or experience about what is likely and what is not. In short, it plays the role of a "knowledge base". Then, given a prior distribution and an "observation", we form the *posterior distribution*: the conditional distribution given what is observed. In image processing, the observation is of a degraded and possibly transformed (as in tomography) image. In medical diagnosis, it is the collection of signs and symptoms of the presenting patient. In principle, the posterior distribution contains the information necessary for optimal image restoration or, in medicine, for hypothesis formation and diagnosis.

There are legendary practical problems with the actual implementation of Bayesian methods. It is difficult to choose appropriate prior distributions in all but the most simple (low dimensional) problems. Our problems are very high dimensional: at least 250,000 (roughly 500x500 pixels) for image processing, and many thousands (e.g. all relevant medical features) for interesting expert systems problems. Yet we must choose probability distributions on these spaces that meaningfully capture the important real-world relationships. And, having constructed a suitable prior, there is still the host of computational problems that arise in trying to exploit the information implicit in the posterior distribution. Here again, the dimensionality is confounding. We will want to compute expectations or choose most likely states with respect to these posterior distributions, but direct evaluation is already impossible with one hundred dimensions, much less thousands or hundreds of thousands.

We shall exploit an analogy to statistical physics through which these practical problems are largely overcome. This analogy will suggest a class of prior distributions (namely, Gibbs distributions) that are easily constructed and especially suitable for application to image processing and to expert systems problems. *Furthermore, this analogy suggests a tractable computational scheme, stochastic relaxation, for the necessary manipulations with posterior distributions.* Under the analogy, pixel grey levels and possibly other image attributes (locations of edge elements, locations and scales of objects, etc.) play the role of the states of individual particles, such as atoms or molecules, in a real physical system. For the medical problem, the parallel is between the states of particles and the states of the various disease entities, possible test results, patient characteristics, and other medically important features. SR is the procedure of simulating the dynamics of the imagined physical system. Through the analogy, these dynamics are actually equivalent to the desired manipulations on the posterior distributions. And, because real physical systems execute a parallel dynamics, the resulting computational algorithm (SR) is parallel as well. All of this will of course be made more concrete in the sections that follow, first in the context of image processing and then for a proposed approach to expert systems.

We have benefited from the advice and encouragement of Ulf Grenander. Our work reflects the influence of his Theory of Patterns [15], as is especially evident by the analogy to statistical physics. Also, many other colleagues have contributed ideas and technical assistance. Among these, we thank in particular Eugene Charniak, Donald Geman, David Hoffman, Donald McClure and Vincent Mirelli.

1. Image Processing

The application of SR that we have explored in greatest depth is to image processing, specifically image restoration and boundary finding. An introductory discussion follows, minus many of the technical details. Our forthcoming paper [8] will more thoroughly discuss the relation to existing approaches and the underlying mathematics, and we also refer the reader there for the proofs of theorems.

A. Formulation

For now, we will concentrate on image restoration. Our formulation of the boundary finding problem fits the same general framework, as will be illustrated later when some simulation results are presented.

As is customary, we take as our starting point an observed digitized image $\vec{O} = \{O_{ij}\}\ 1 \leq i,j \leq n$, where O_{ij} is the grey level of the pixel at the lattice location (i,j). "n" is typically a power of 2, such as 256, 512, or 1024. We think of $\vec{O}$ as the result of a degradation of some "true" image $F = \{F_{ij}\}$:

$$\vec{O} = D(\vec{F}) \odot \vec{N}$$

with the following interpretation. D is a deformation mechanism, typically involving a convolution ("blurring") of $\vec{F}$ with some "point- spread function" H, and sometimes also a nonlinear pixel-by-pixel transformation. For example

$$D(\vec{F})_{ij} = \sqrt{(H^*\vec{F})_{ij}}$$

where

$$(1.A.1) \quad (H^*F)_{ij} = \sum_{|k|,|\ell| \leq 1} F_{i+k,j+\ell} H(k,\ell)$$

and

$$H(k,\ell) = \begin{cases} \frac{1}{2} & (k,\ell) = (0,0) \\ \frac{1}{16} & |k|, |\ell|, (k,\ell) \neq (0,0) \end{cases}$$

$\vec{N} = \{N_{ij}\}$ is a noise process, the prototype being independent Gaussian variables with common mean μ and variance σ^2. In our experiments we will focus on this Gaussian case. However, neither the Gaussian assumption nor that of independent components is necessary, and our methods are unchanged for a large class of noise statistics. Finally, the symbol $\odot$ represents a pixel-by-pixel operation, most typically addition or multiplication:

$$O_{ij} = D(\vec{F})_{ij} + N_{ij}, \quad \text{or} \quad O_{ij} = D(\vec{F})_{ij}\, N_{ij}\,.$$

These elements of the model, D, $\odot$, and N, are assumed known from the "physics" of the recording process (i.e. from the optics, the digitization procedure, etc.).

The problem is to estimate $\vec{F}$ from the observation $\vec{O}$.

B. Gibbs Priors for F ("Image Modelling")

We base our restorations on a *prior distribution* for the image process $\vec{F}$. This is the Bayesian framework, and it is particularly apt for image restoration. The possible images $\vec{F}$ constitute a very high-dimensional and complex parameter space, a space of possible restorations of an observed (degraded) image $\vec{O}$. In actual restoration problems, only a small fraction of these candidate images make for reasonable estimates. The rest grossly violate regularities that are characteristic of "real scenes". A properly constructed prior can focus the search for a restoration on the relatively few images consistent with real-world expectations.

But the benefits of Bayesian estimation are often only theoretical; for high dimensional parameter spaces it is notoriously difficult to construct reasonable prior distributions that lead to computationally feasible estimators. In our experiments, we have used a class of distributions, namely Gibbs distributions with "local energy functions", that appear to avoid these problems. By exploiting a connection to statistical physics, we can devise highly parallel algorithms for computations associated with these Gibbs distributions. Morever, their characterization in terms of an "energy function" provides a workable and intuitive framework for embodying real-world knowledge. So far we have attempted to capture only the most elementary sorts of prior information. But the resulting restorations are extremely good, albeit for some very simple pictures. And, the very same techniques extend, at least in principle, to the inclusion of more complex prior knowledge and hence to the analysis of more complex scenes.

A basic characteristic of real-world scenes is the high likelihood that grey levels at nearby locations will be nearly the same. We shall illustrate our approach by constructing a prior distribution that is consistent with this simple property. We first construct an "energy function" U, which assigns a value to each possible scene (i.e. to each possible realization of $\vec{F}$). U is endowed with the key property that it takes relatively small values for those scenes consistent with prior information. As we shall see, the Gibbs prior on $\vec{F}$ is then derived directly from U. The construction of U is based on a *neighborhood system* $\{G_{ij}\}$ $1 \leq i,j \leq n$, wherein G_{ij} is a collection of pixels that we think of as "directly interacting" with the pixel (i,j). For example,

(1.B.1) $\quad G_{ij} = \{(k,\ell):(k,\ell) \neq (i,j),\ |k-i| \leq 1,\ |j-\ell| \leq 1\}$

This example is a "nearest neighbor" system, including nearest diagonal neighbors. Let us suppose that the grey levels of $\vec{F}$ are discrete, for example $F_{ij} \in \{1,2,3,4,5\}$ $1 \leq i,j \leq n$. For every $x,y \in \{1,2,3,4,5\}$, we define a "bond energy" V(x,y). An example would be:

$$V(x,y) = \begin{cases} -1/3 & x = y \\ 1/3 & x \neq y \end{cases}$$

These bond energies, calculated among neighboring pixels, are combined to define the "energy of the picture" $\vec{f} = \{f_{ij}\}$ $1 \leqslant i,j \leqslant n$, $f_{ij} \in \{1,2,3,4,5\}$:

$$U(\vec{f}) = \frac{1}{2} \sum_{1 \leqslant i,j \leqslant n} \sum_{(k,\ell) \in G_{ij}} V(f_{ij}, f_{k\ell}) .$$

The idea is that U is small for pictures consistent with the properties we wish to capture, in this case pictures for which neighboring pixels tend to have common grey levels.

We construct the Gibbs prior from U. Imagine a real physical system with "particles" (perhaps atoms or molecules) at the sites (i,j) $1 \leqslant i,j \leqslant n$. Each particle can be in any of the "states" {1,2,3,4,5}, and this defines a *configuration* space for the entire system:

$$\Omega = \{\vec{f} : f_{ij} \in \{1,2,3,4,5\},\ 1 \leqslant i,j \leqslant n\}.$$

Recall that the energy function $U:\Omega \to R$ was constructed so that the lower energy configurations are the more expected configurations. A physical system with energy U would have an associated *Gibbs distribution*,

(1.B.2) $$\pi(\vec{f}) = P(\vec{F} = \vec{f}) = \frac{1}{Z} \exp\{-U(\vec{f})\} .$$

Here Z is the "partition function" (actually a constant),

$$Z = \sum_{\vec{f} \in \Omega} \exp\{-U(\vec{f})\},$$

making π a proper probability distribution on Ω. Under π, the lower energy configurations are the more likely configurations. A sample image from π is shown in Figure 2A. It has the property that nearby pixels are typically at the same grey level.

Of course the property of locally constant grey levels is very primitive. Real images have complex geometric properties: edges tend to maintain their curvature or their straightness; edges generally end in corners, or in some smooth transition to another "regular" edge; edges often define rectangles or ellipsoidal shapes, and these are typically surfaces of complex three dimensional objects; and so on. Even the near-constancy of grey levels is not truly typical. Most regions in real scenes are better described by a more or less regularly varying pattern of grey levels known as "texture". To differing degrees, these properties resist a purely local description, and this is a major difficulty in incorporating them into an image model. The property of locally constant grey levels was captured by an energy function comprising only pairs of neighboring variables. We shall see that this local structure is critical for the computational feasibility of our methods.

To avoid long-range interaction terms in the energy function, and the resulting computational difficulties, we propose to incorporate complex geometric properties by developing a multi-level hierarchical image model. We will illustrate the approach by a modification of the previously developed energy U to accommodate straight-running vertical and horizontal edges. The idea is to append the pixel process $\vec{F}$ with an unobservable line process $\vec{L}$. The components of $\vec{L}$ are 0-1 random variables indicating the absence or presence of an edge element:

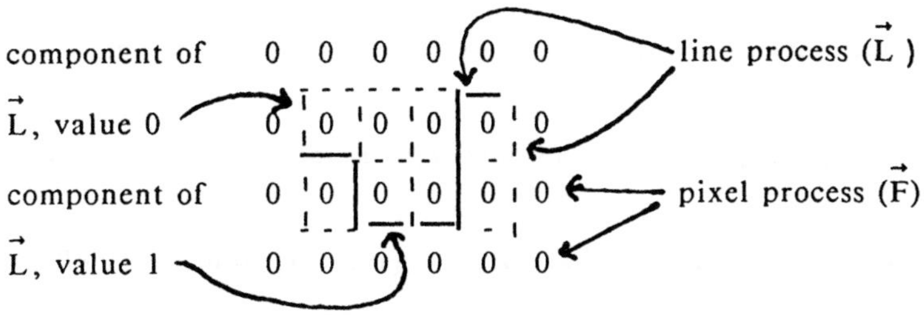

We want to capture the notions that a straight-running edge tends to continue running straight, that edges typically do not end "blindly", that corners are often right angled, and, most fundamentally, that pixels separated by an edge have no tendency to be at similar grey levels. For this, we define an energy U of the form $U(\vec{f},\vec{l}) = U_1(\vec{l}) + U_2(\vec{f},\vec{l})$. The pure line contribution, U_1, is minimized by line configurations that either define no edges at all, or contain unbroken curves with mostly straight edges, and occasional corners. (The details are layed out in [8].) U_2 is substantially the same as the pixel process defined earlier, except that there is no contribution from bonds that are "broken" by the presence of an intervening edge element ($l_{ij}=1$). The upshot is an energy function that is still built from strictly local contributions, albeit on an expanded configuration space. In our experiments (discussed later) each pixel site has a total of eight neighbors in its neighborhood system, comprising the four nearest pixel sites and the four surrounding line sites. Each line site also has eight neighbors: six are nearby line sites and the other two are the pixel sites on either side of the associated edge element.

The energy defines a Gibbs prior for the extended process (F,L):

$$\begin{aligned}\pi(\vec{f},\vec{\ell}) &= P(\vec{F}=\vec{f},\vec{L}=\vec{\ell})\\ &= \frac{1}{Z}\exp\{-U(\vec{f},\vec{\ell})\}\\ &= \frac{1}{Z}\exp\{-U_1(\vec{\ell}) - U_2(\vec{f},\vec{\ell})\}\end{aligned}$$

Later, we will present restorations based on this prior. Of course π defines a *marginal distribution* on the pixel process $\vec{F}$, obtained by summing over all allowed values of $\vec{L}$. It is easily demonstrated that this marginal distribution is *global* in nature: its characterization as a Gibbs distribution defines an energy function with terms that are simultaneously dependent on all components of $\vec{F}$. The multilevel model thus introduces long-range interactions into the *pixel* process, whereas it preserves the computational advantages of a locally-composed energy function.

We have begun a further development of this image model, adding "higher level" and spatially invariant geometric attributes such as line segments, smooth curves, and simple polygons with arbitrary scale and location. The pixel process $\vec{F}$ is further appended, and new attributes are thereby accommodated with energies built entirely from local contributions. But since we have not yet experimented with these more ambitious models, we shall forego detailed descriptions.

The more complex image models involve numerous unspecified parameters that dictate the detailed characteristics of typical images under the distribution π. Indeed, already for the simple model (1.B.2), which merely captures the property of locally constant grey levels, we have rather arbitrarily fixed the bond energies V(x,y) at +/- 1/3, depending on $x \neq y$ or $x=y$. When "1/3" is replaced by larger constants the typical regions in an image drawn from Ω under the distribution π are larger. Parameters such as these should be estimated from sample pictures, but there is not yet a well-developed and computationally feasible approach. The estimation problem is complicated by the nature of the observations, which are of $\vec{O}$ instead of $\vec{F}$, and further by the use of random variables, such as those constituting the line process $\vec{L}$, that are not directly related to the observed picture. There are some results to draw on (e.g. [1] and [23]), and we have done some preliminary work ourselves ([7]), but we are far from a coherent estimation theory.

C. Restoration

As described by the degradation model $\vec{O} = D(\vec{F}) \odot \vec{N}$, observations are of a noisy and deformed version of $\vec{F}$. Our main tool for recovering $\vec{F}$ is the *posterior distribution* of $\vec{F}$ given $\vec{O}$, or of $\vec{F}$ and any appended process, such as $\vec{L}$. For the degradations that arise in most image processing applications, the posterior distribution shares with the prior distribution a pivotal property: it is Gibbs with a locally-composed energy function. This can be is illustrated with the prototype model $\vec{O} = H^*\vec{F}+\vec{N}$, with H as in (1.A.1), $\vec{N} = \{N_{ij}\}$ iid zero mean Gaussian, and a prior on $\vec{F}$ alone, as in (1.B.2). The posterior distribution is

$$P(\vec{F}=\vec{f}\,|\,\vec{O}=\vec{o}) \tag{1.C.1}$$
$$= \frac{1}{\hat{Z}} \exp\{-U(\vec{f}) - \frac{1}{2\sigma^2} \sum_{i,j} (o_{ij} - (H^*\vec{f})_{ij})^2\}$$

$\hat{Z}$ is a new normalizing constant which, happily, will not concern us. The exponent in the posterior distribution defines a new, posterior, energy function $\hat{U}$:

$$\hat{U}(\vec{f}) = U(\vec{f}) + \frac{1}{2\sigma^2} \sum_{i,j} (o_{ij} - (H^*\vec{f})_{ij})^2 . \tag{1.C.2}$$

Like U, the posterior energy function, $\hat{U}$, is a sum of terms, each of which is only locally dependent upon the components of $\vec{f}$. In fact, the neighborhood system associated with $\hat{U}$ is simply

$$\mathcal{G}_{ij} = \{(k,\ell):(k,\ell) \neq (i,j),\ |k-i| \leq 2,\ |\ell-j| \leq 2\}\ ,$$

an extension of the original system (1.B.1) induced by the point spread function H. (We again refer to [8] for more details.) This local property of the posterior distribution is quite general. It is preserved with nonlinear deformations, with multiplicative and other noise mechanisms, and with certain models of $\vec{N}$ that allow for dependencies (namely, Markov random field models).

The posterior distribution is a powerful tool for image restoration, and, more generally, for scene analysis. We can, in principle, construct optimal Bayesian estimators for $\vec{F}$ (or $\vec{F}$ together with appended processes) given $\vec{O}$, such as the so-called MAP estimator, i.e. the image that maximizes the posterior distribution. Or, more subjectively, we can examine images sampled from the posterior distribution and select "reasonable" restorations. But conventional approaches to these operations involve prohibitive amounts of computation. MAP estimation of $\vec{F}$ illustrates this point. To be specific, consider the posterior distribution (1.C.1). The MAP estimator is the configuration $\vec{f}$ that achieves the minimum of (1.C.2). Realistic image models lead to complex energy functions U, and typically n^2 (the number of pixels) is of the order of 250,000. Minimization of a complex function such as (1.C.2) of this many variables is essentially impossible, and even the identification of near-optimal configurations is extremely difficult. Notice that a direct approach of trying every configuration is not feasible: there are roughly $M^{250,000}$ configurations, when M is the number of allowed gray levels in the image model. With regard to sampling, the story is the same: the very large configuration space renders the usual Monte Carlo techniques impractical.

D. Stochastic Relaxation

In Section 2 our discussion of expert systems will lead us to similar computational problems. We will again define high-dimensional Gibbs distributions with locally-composed energy functions, and again seek the most likely (=minimum energy) configurations. We will also want to evaluate certain expectations with respect to these distributions, and this is another computation confounded by the size of the configuration space and the complexity of the energy function. Let us formulate more carefully and more generally these computational tasks, and the key notion of locally-composed energy functions.

The general discussion is in terms of a *graph* G, with *sites* (or nodes) S, and *neighborhood system* $\{G_s\}$ $s \in S$. For each $s \in S$, G_s is the collection of sites connected to s in G. The neighborhood system is asssumed to be "local", meaning that for each $s \in S$, $|G_s|$ is very small compared to $|S|$. In our discussion of image restoration, S is the index set for the n components of the pixel process $\vec{F}$, as well as additional components in the case of multilevel models. In our image restoration experiments, $|G_s| = 8$ whereas $|S|$ ranges from 64^2 to 128^2. Associated with G is a random process indexed by S, $\vec{X} = \{X_s\}$ $s \in S$. $\vec{X}$, in the general case, plays the role of $\vec{F}$, or of $(\vec{F},\vec{L})$, in the image processing examples. Each component, X_s, of $\vec{X}$ has a *state space* (range)

Λ_s, and these determine the *configuration space* Ω :

$$\Omega = \{\vec{X} = \{X_s\}\ s\epsilon S:\ X_s \epsilon \Lambda_s \quad s\epsilon S\}.$$

We will simplify the discussion (and the mathematics) by assuming that Λ_s is finite for each $s\epsilon S$. The distribution on $\vec{X}$ is determined by an *energy function* $U:\Omega \rightarrow R$ that respects G in the sense that

$$\text{(1.D.1)} \qquad U(\vec{x}) = \sum_{C\epsilon \mathcal{C}} V_C(\vec{x}).$$

In this representation, $\mathcal{C}$ is the set of all *cliques*: collections $C \subseteq S$ of sites such that every pair of sites in C are neighbors. For each $C\epsilon\mathcal{C}$, $V_C(\vec{x})$ is independent of all x_s for which $s \notin C$. This is what we mean by a locally-composed energy function. It defines a *Gibbs distribution* on $\vec{X}$, with respect to G:

$$\pi(\vec{x}) = P(\vec{X}=\vec{x}) = \frac{1}{Z} \exp\{-U(\vec{x})\}$$

where $Z = \sum_{\vec{x}\epsilon\Omega} \exp\{-U(\vec{x})\}$.

The three computational tasks that we will now address are these:

(A) sample form π ;

(B) compute the mode(s) of π (=minimum energy state(s));

(C) compute expected values under π :

$$E[f] = \sum_{\vec{x}\epsilon\Omega} f(\vec{x})\, \pi(\vec{x}) \quad \text{for} \quad f: \Omega \rightarrow R.$$

In all interesting applications direct calculations are impossible: $|S|$ is large, $|\Omega|$ is enormous, and U is highly nonlinear.

We will outline the architecture for a parallel processing machine designed to make suitable approximations. (A version of this machine, specialized to binary state spaces, Λ_s, and proposed as a model of neural activity, is described by Hinton and Sejnowski in [17].) This architecture essentially conforms to the graph G. Let us imagine a simple processor placed at each site of this graph. The connectivity relation among the processors is described by the neighborhood system of the graph: processor s is connected to processor r if and only if r and s are connected in G. For the applications that we have in mind $|S|$ is large (roughly 250,000 for image processing) but the neighborhood sizes, and thus the number of connections to a given processor, are modest.

The state of the machine evolves by discrete changes, and it is therefore convenient to discretize time, say t=1,2,3,... . At time t, we describe the state of the processor at site s by a random variable $X_s(t)$, where $X_s \epsilon \Lambda_s$. The configuration at time t is $\vec{X}(t) = \{X_s(t)\}$ $s \epsilon S$. We begin with an arbitrary initial state $\vec{X}(0) \epsilon \Omega$, after which the configuration evolves because of state changes of the individual processors. Computation is asynchronous in the sense that each processor is driven by its own clock. Every processor is programmed to follow the same algorithm. At site s, for example, the "flow chart" looks like this:

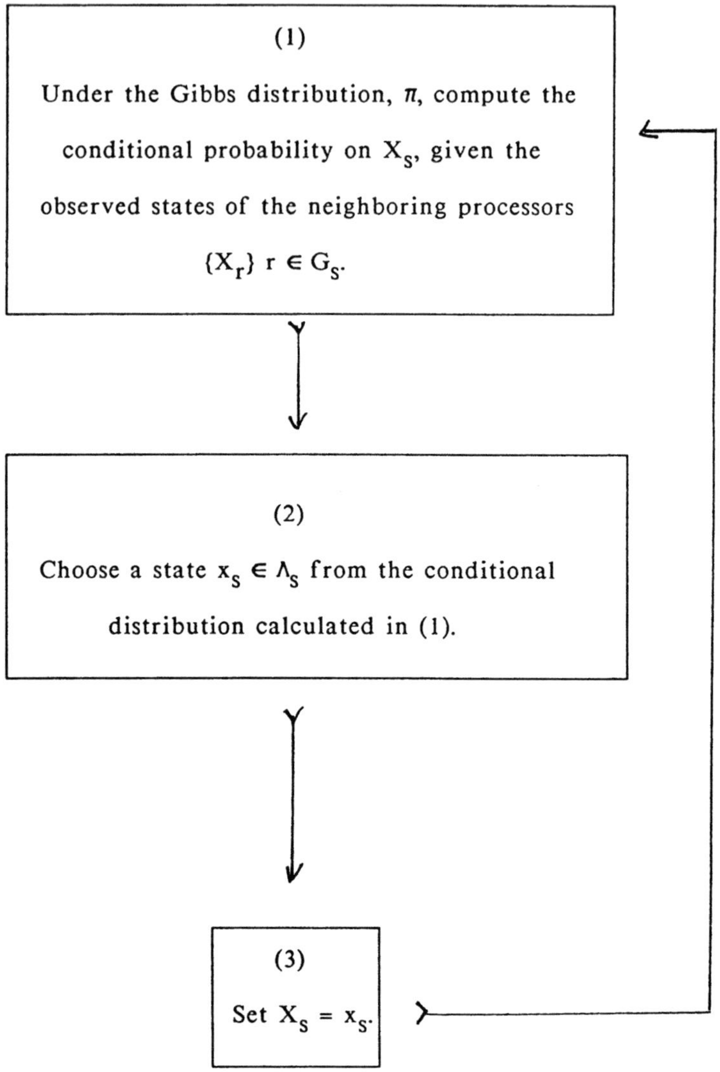
(1)
Under the Gibbs distribution, π, compute the conditional probability on X_s, given the observed states of the neighboring processors $\{X_r\}$ $r \in G_s$.
(2)
Choose a state $x_s \in \Lambda_s$ from the conditional distribution calculated in (1).
(3)
Set $X_s = x_s$.

Figure 1

Our convention for "time" is that $t \to t+1$ each time *any* processor changes state, i.e. performs step (3).

We stress the point that these computations are *local.* In particular, the conditional probability on X_s computed in step (1) depends only on the states of the neighboring sites, and these states are available at site s through the above-defined connectivity. Moreover, the calculation of this conditional probability is trivial. It depends only on those terms in the energy function whose associated cliques contain s, and it does not require evaluation of the partition function Z:

$$P(X_s=x_s \mid X_r=x_r,\ r \in G_s) \tag{1.D.2}$$

$$= \frac{\exp\left\{-\sum\limits_{\substack{C\in\mathcal{C}:\\ s\in C}} V_C(\vec{x})\right\}}{\sum\limits_{y_s\in\Lambda_s} \exp\left\{-\sum\limits_{\substack{C\in\mathcal{C}\\ s\in C}} V_C(\vec{y})\right\}}$$

where $y_r = x_r$ $r \neq s$. Recall that $V_C(\vec{x})$ depends only on x_s and neighbors of x_s, when $s \in C$.

The result of these computations is a random sequence of configurations $\vec{X}(1),\vec{X}(2),\ldots$ Shortly, we will discuss convergence and related properties, and connect these properties to the computational problems of sampling, finding modes, and computing expectations. But let us first discuss briefly the sense in which the proposed computation scheme is parallel, and relate this to a serial (conventional) algorithm. Computation is parallel in the sense that it is realized by simple and alike units operating largely independently. Units are dependent only to the extent that each must transmit its current state to its neighbors, the latter defined by the neighborhood system of the graph G. Most importantly, the amount of time required to update once the state value at every site is, in principle, independent of the number of sites. (Although, this time *will* depend on the size of the neighborhood systems.) An alternative and more conventional approach is to use one processor to "visit" sites successively. Upon arriving at a site, this processor must first load the local neighborhood relations and state values, perform the replacement, and then move on to a next site. The time required to refresh the state of every site then grows linearly with the number of sites.

The convergence properties of the random process $\vec{X}(1),\vec{X}(2),\ldots$ are essentially independent of the details of the sequence of site replacements. This sequence does not need to be cyclic, balanced (with respect to frequencies of replacements), or, in any other manner, regular. It is because of this flexibility that one can imagine performing the computations with independent, *asynchronous*, processors. And for this same reason, other, not fully parallel, architectures can be used. The graph G can be divided into subsets of sites with each subset "assigned" to a conventional machine (instead of a special purpose processor). Each machine can independently process its assigned sites; convergence is still guaranteed. Of course, the graph should be divided with an effort at minimizing the communication requirements among machines, which amounts to minimizing the number of neighboring sites that are placed in distinct subsets.

We now investigate the convergence properties of the random process $\vec{X}(t)$. The local computations made by the individual processors result in a sequence of changes of state at the individual sites. Let us denote by n_t ($n_t \epsilon S$) the t'th site to perform step (3) of the "local flow chart" (Figure 1). The distribution on $\vec{X}(t)$ can be expressed in terms of n_t and the distribution on X(t-1): for any x

$$P(\vec{X}(t) = \vec{x})$$

$$= P(X_{n_t} = x_{n_t} | X_r = x_r, r \epsilon G_{n_t}) P(X_r(t-1) = x_r,\ r \neq n_t, r \epsilon S)$$

where the conditional probability is calculated under the Gibbs distribution π , as in (1.D.2). Our most basic task is to sample from π . For this we want to assert that $P(\vec{X}(t)=\vec{x}) \to \pi(\vec{x})$, in which case $\vec{X}(t)$, for large t, essentially represents a sample from π . The only requirement is that we continue to visit every site:

Theorem *(relaxation). Assume that for each* $s \epsilon S$ *the sequence* $n_1, n_2, \ldots$ *contains* s *infinitely often. Then for every* $\vec{X}(0)$ *and every* $\vec{x} \epsilon \Omega$, $P(\vec{X}(t)=\vec{x}) \to \pi(\vec{x})$ *as* $t \to \infty$.

This is simply a modification of the well-known Metropolis algorithm [22] for sampling from a Gibbs distribution. The only complication in the proof is the arbitrary order of site replacements, $n_1, n_2, \ldots$. The Markov chain $\{\vec{X}(t)\}$, t=1,2,..., does not have stationary transition probabilities and the transition matrices do not commute. This precludes the usual Perron-Frobenius, algebraic, treatment. Nevertheless, the proof is straightforward; see [8] for details.

For computing expectations we can exploit the connection to statistical physics that is implicit in our use of the Gibbs distribution. We think of the configuration sequence $\vec{X}(1), \vec{X}(2), \ldots$ as a sequence of states of a physical system with energy function U. The relaxation theorem asserts that the approach to equilibrium is independent of initial conditions. In statistical physics one attempts to predict the observable quantities of a system in equilibrium from a description of the system's distribution. The observable quantities are time averages of functions of the configuration. Under the ergodic hypothesis, the ergodic theorem is in force and states that time averages approach the corresponding expectations (so-called phase averages) with respect to the equilibrium distribution. The analogue for our system is the statement that, in some suitable sense,

$$\text{(1.D.3)} \qquad \frac{1}{n} \sum_{t=1}^{n} f(X(t)) \xrightarrow{n\to\infty} E[f]$$

where expectation is with respect to the Gibbs distribution π. We have already made the observation that direct calculation of E[f] is not feasible. The limiting relation in (1.D.3) suggests that we approximate E[f] by a long-run time average of $f(\vec{X}(t))$. For most physical systems the ergodic hypothesis is extremely difficult to verify; statistical physics typically proceeds under the assumption that (1.D.3) is valid. Fortunately, for our system it is rather easy to directly establish ergodicity.

***Theorem** (ergodicity). Assume that there exists a τ such that for every* t=0,1,2,...

$$S \subseteq \{n_{t+1},\ldots n_{t+\tau}\}\ .$$

Then for every $\vec{X}(0)$ *and every function* f *on* Ω

$$\frac{1}{n} \sum_{t=1}^{n} f(\vec{X}(t)) \to E[f]$$

with probability one.

The added assumption on the site replacement sequence is of no practical importance, since the clocks of the various processors contributing to the computations will in fact have some common bound on their periods.

Finally, we address the problem of finding configurations $\vec{x}_0$ at which energy is approximately minimized, $U(\vec{x}_0) \simeq \min_{\vec{x} \in \Omega} U(\vec{x})$. For this, we follow Černý [2] and Kirkpatrick et. al. [19], and simulate "annealing". Annealing is a process of heating and then gradually cooling a physical system. Careful annealing yields low energy configurations. This can be understood by noting that low temperatures concentrate the Gibbs distribution at low energy states. The functional dependence of the Gibbs distribution on temperature (T) is well-known:

$$\pi_T(\vec{x}) = \frac{1}{Z_T} \exp\{-U(\vec{x})/T\}$$

where

$$Z_T = \sum_{\vec{x} \in \Omega} \exp\{-U(\vec{x})/T\}.$$

We can think of our earlier discussion as concerning this same distribution at fixed temperature T=1. When T is small, the likely configurations are the ones that nearly minimize U. The idea of Černý and Kirkpatrick is to minimize U by a simulated annealing: run the Metropolis algorithm for sampling from π_T while gradually lowering T. Kirkpatrick applies this to energies whose minimums correspond to the solutions of certain combinatorial optimization problems. We will apply it to finding nearly-maximum likelihood configurations, as in MAP image restoration.

The temperature is an easily controlled global parameter, and with the obvious modifications in step 1 of the local flow chart (Figure 1) we can replace T=1 by a decreasing sequence T(t) to simulate annealing. Processing is still parallel, and the resulting behavior is akin to that of a real physical system: If T is lowered to rapidly, $\vec{X}(t)$ is likely to converge to a state with energy far above the desired minimum. If T is lowered more gradually then the process $\vec{X}(t)$ concentrates at low energy configurations:

Theorem (annealing). Let π_0 *be the distribution uniformly concentrated on the minimum energy configurations:*

$$\pi_0(\vec{x}) = \begin{cases} \frac{1}{|\Omega_0|} & \vec{x} \in \Omega_0 \\ 0 & \textit{otherwise} \end{cases}$$

where $\Omega_0 = \{\vec{y} \in \Omega : U(\vec{y}) = \min_{\vec{x} \in \Omega} U(\vec{x})\}$. *Again assume that there exists a* τ *such that for every* $t=0,1,2,\ldots$

$$S \subseteq \{n_{t+1}, \ldots n_{t+\tau}\}$$

Then there exists a constant c such that if

(1) $T(t) \to 0$, *decreasing*,

(2) $T(t) \geqslant c/\log(t)$ $t=2,3,\ldots,$

then for every $\vec{X}(0)$ *and every* $\vec{x} \in \Omega$,

$$P(\vec{X}(t)=\vec{x}) \to \pi_0(\vec{x})$$

as $t \to \infty$.

In particular, $U(\vec{X}(t)) \to \min_{\vec{x} \in \Omega} U(\vec{x})$ in probability. We again refer to [8] for the proof.

Of course, for our experiments we must actually choose an "annealing schedule" $T(t)$. In our image restorations we have used $c/\log(1+k)$, where k is the number of iterations through all sites in G ($k \simeq t/|S|$) and where c=3 or 4. We do not yet know if these values for c are sufficiently large to guarantee the convergence asserted by the annealing theorem. Good results are obtained in reasonable time, but faster convergence may be possible. In fact, the basic mathematical questions concerning optimal relaxation techniques and annealing schedules, and rates of convergence, are unsolved.

E. Experiments

The results of three restoration and boundary finding experiments are presented in Figures 2,3, and 4. All of these involved additive Gaussian noise, with or without blur. We have obtained comparable results, that are presented in [8], in the presence of multiplicative noise and nonlinear deformations. As we have stressed, the computational burden is mostly determined by neighborhood sizes. Our experiments with more complex degradations involved the very same neighborhood systems that we used here, and hence they required essentially the same amounts of computation.

Of course, for the experiments the SR algorithm was run serially. We used a "raster scan" pattern of site replacements. In our discussion below, "sweep" refers to the visiting of each site exactly once, hence a sweep comprises $|S|$ local replacements, where $|S|$ is the number of components to the configuration. In each of the experiments, MAP configurations are approximated by the result of a simulated annealing, with annealing schedule $T(k)=c/\log(1+k)$, where k is the number of sweeps and c=3 or 4.

Figure 2A is a sample drawn from the Gibbs distribution (1.B.2), via SR after 200 sweeps. Figure 2B is the result of adding iid zero mean Gaussian noise, with $\sigma=1.5$. Figures 2C and 2D are obtained by annealing, under the posterior distribution conditioned upon 2B, after 25 and 300 sweeps respectively.

In Figure 3 we present the results of an experiment with a "hand- drawn" image, restored under the multilevel prior that includes a "line process", described earlier. Figure 3A is the original image. It contains 3 grey levels separated by 1 intensity unit. Figure 3B is again the result of adding iid mean zero Gaussian noise, this time with $\sigma=.7$. Figure 3C is the restoration of 3B by annealing, after 1,000 sweeps. For comparison, Figure 3D is the restoration of 3B using a different prior: namely, the one used in Figure 2, making no attempt to model edges.

The last series is from the experiment in boundary finding. Figure 4A is a part of a roadside picture taken from a moving car. As is apparent from the picture, the intensity histogram shows two prominent peaks. Zero mean Gaussian noise with standard error $\sigma=.5$, relative to the spacing between these peaks, was added to the original image to produce Figure 4B. The blur of Figure 4A was modelled by the convolution $H*\vec{F}$ as in (1.A.1), and thus the degraded image, Figure 4B, was modelled by $H*\vec{F}+\vec{N}$. Our goal was to identify the two boundaries of the original image. The line process was modified in such a way that each line element could be directed in any of three orientations (instead of just one), to better accommodate edges that are not partial to the vertical and horizontal. (See [8] for a detailed description of the resulting multilevel prior distribution.) The result of annealing, after 1000 sweeps, is shown in Figure 4C. Pixels that sit to the left of or above a line element, whatever its orientation, are indicated in black. The two main regions are circumscribed by an unbroken sequence of contiguous line elements.

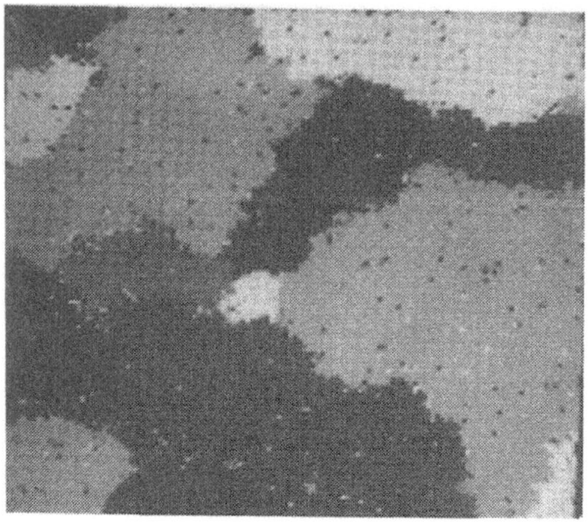

A. Sample from prior

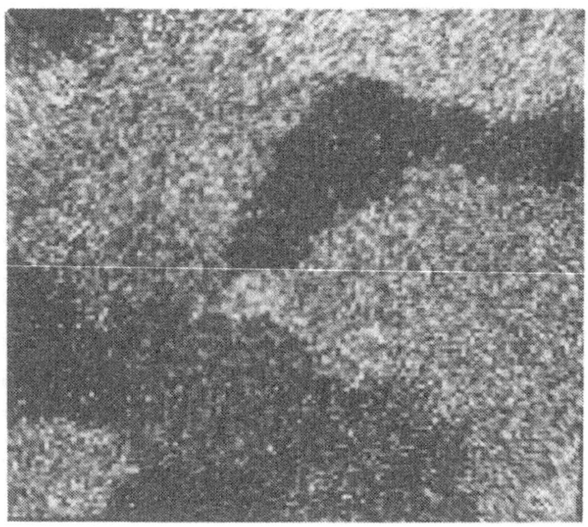

B. Sample plus noise

Figure 2

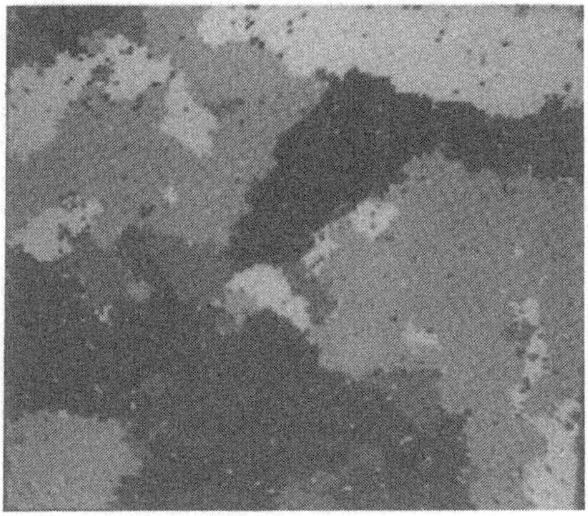

C. 25 sweeps of SR, with annealing

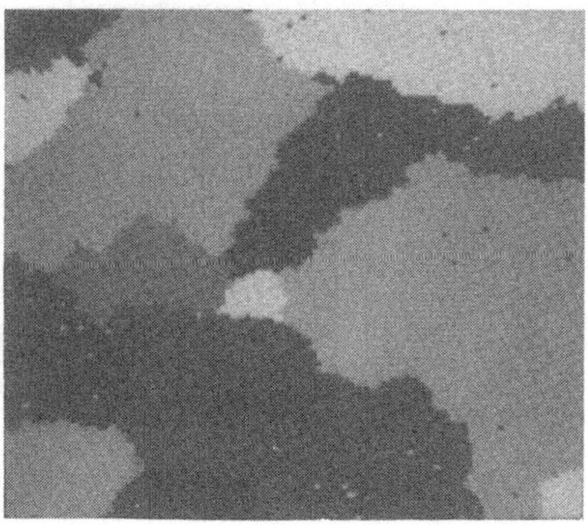

D. 300 sweeps of SR, with annealing

Figure 2

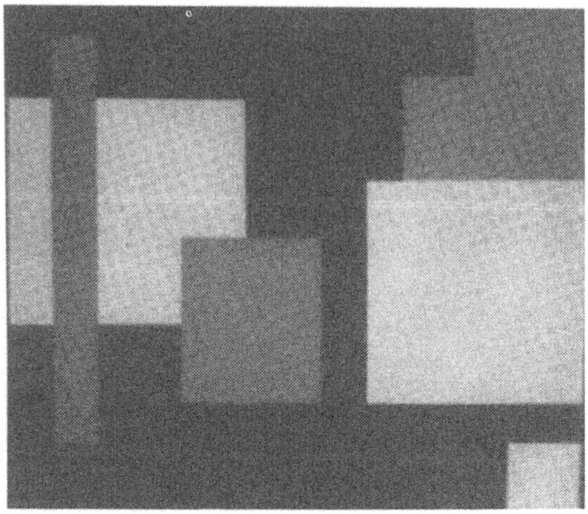

A. Hand-drawn image

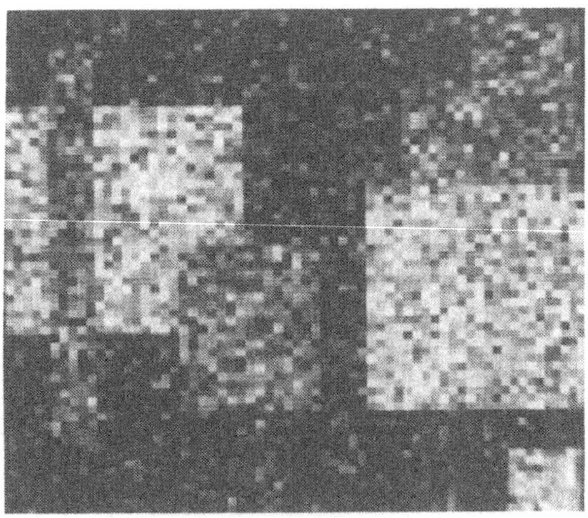

B. Hand-drawn image plus noise

Figure 3

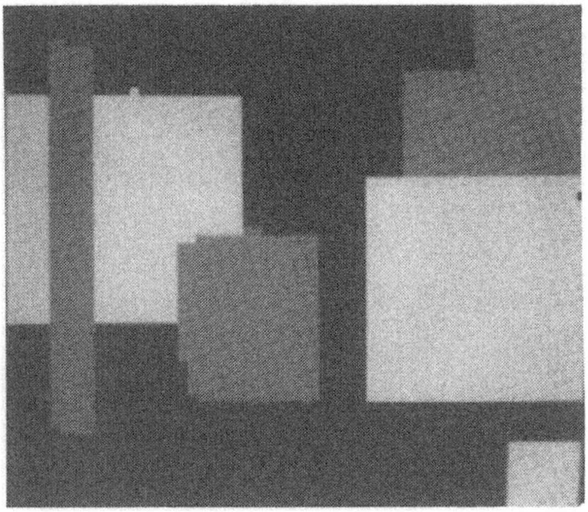

C. 1,000 sweeps of SR, with annealing, prior with line processes

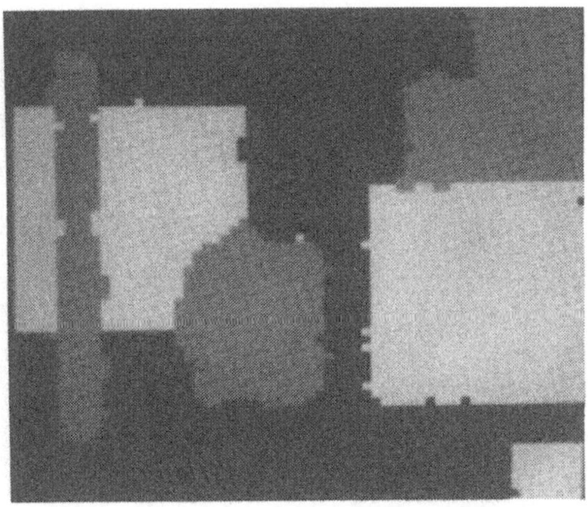

D. 1,000 sweeps of SR, with annealing, prior without line processes

Figure 3

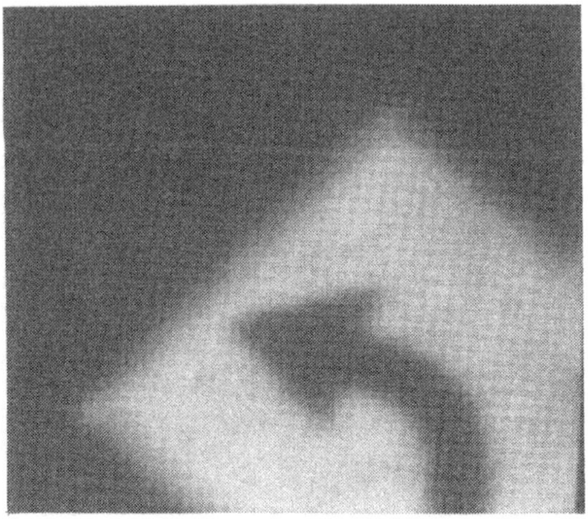

A. Road sign

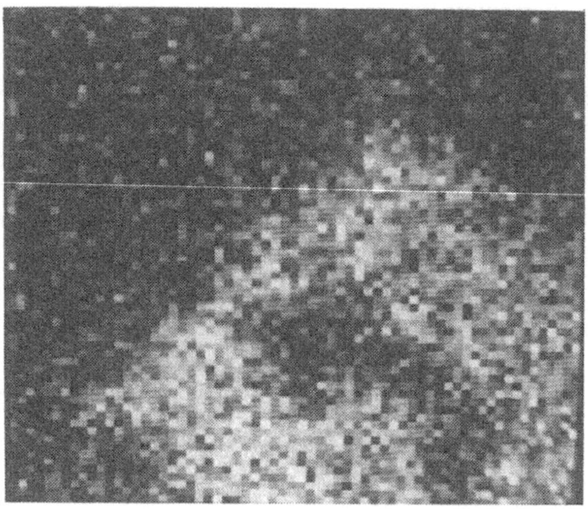

B. Scene plus noise

Figure 4

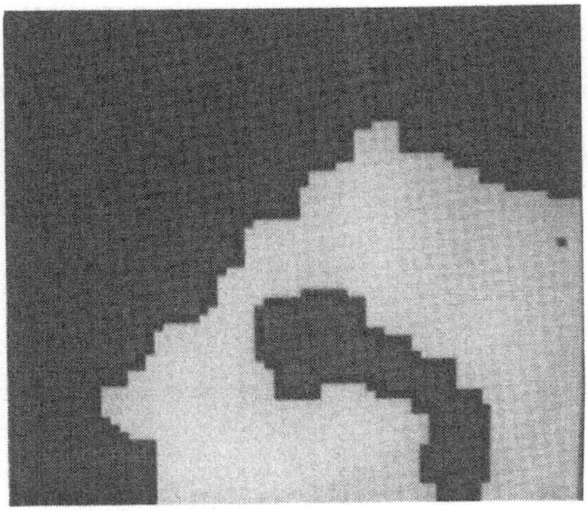

C. 1,000 sweeps of SR, with annealing, showing estimated locations of boundary elements

Figure 4

2. Expert Systems

Here we propose the application of SR, again within a Bayesian framework, to problems arising in the construction of expert systems. The discussion will necessarily be more superficial: we have not yet performed experiments, either with actual or with simulated data. Nevertheless, the Bayesian/SR technology seems to be especially well-suited for handling the *statistical issues* that arise in many expert systems problems.

Our discussion is of computer-aided medical diagnosis, a commonly discussed prototype problem involving the various issues of "reasoning under uncertainty", formalizing and manipulating "prior knowledge", the "nature of evidence", computational feasibility, etc. We recognize that the "statistical" aspect of this and other expert system problems is only part of the story. Systems of real practical value must be able to explain the reasoning behind a hypothesis or conclusion, and must often incorporate deterministic, or causal, reasoning. Still, at some level in its workings a diagnostic system confronts certain fundamentally statistical issues: given the observed signs and symptoms, and the known pathologies, what configurations of symptoms and diseases are most likely? Our chief goal is to develop a logical and tractable means for arriving at an answer based upon available medical knowledge, and to use this for developing hypotheses and for suggesting an efficient sequence of diagnostic steps.

For background we refer to Kulikowski [20] , which is an introduction to the practical aspects of computer-aided medical consultation and to existing systems and the various problems that they address. Also, we refer to a paper by Charniak [3], which argues for the value of a more formal, and in particular Bayesian, approach to medical diagnosis, a somewhat unusual position in the "AI" literature, and to Cheeseman [4], who proposes an approach very similar to ours.

For some background on expert systems in general, there is an excellent review article by Duda and Shortliffe [6].

A. Formulation

Our formulation is in terms of a large collection of patient and medical attributes, including signs and symptoms (as obtained from a medical history or diagnostic tests), disease entities, and patient characteristics such as age, sex, race, and so on. We think of these attributes as random variables $X_1, X_2, ... X_N$, with individual state spaces Λ_i $1 \leq i \leq N$. Concerning the attribute sex, for example, $\Lambda = \{M,F\}$, whereas age may be continuous, $\Lambda = [0,100]$, or perhaps categorized into clinically meaningful ranges, $\Lambda = \{[0,1], (1,3], ...\}$. (But, as in Section 1, we will focus our discussion on the discrete case.) A disease may have a binary state space $\Lambda = \{P,A\}$ (*present* or *absent*), or a more descriptive space of *intensities*, e.g. $\Lambda = \{0,1,2,3\}$. At this point we do not distinguish the different varieties of attributes (personal, laboratory, disease, etc.).

Medical information, as may be available from actual patient records or from less formal sources such as textbooks or expert diagnosticians, will be summarized in a "knowledge base". Of course, a *patient record* will supply only a *partial* list of attribute values; only a small fraction of the patient and medical attributes are observed in the course of treating any one individual. As for the less formal sources, the information may be of a variety of types. It may be available, for example, as modified "productions", of the form "if $X_1 = x_1$ and $X_2 = x_2$ then 'typically' $X_3 = x_3$ and $X_4 = x_4$ "; or as associations, such as "X_1 and X_2 are 'highly' correlated"; or as conditional probabilities, such as "70% of the individuals with $X_1 = x_1$ also have $X_2 = x_2$ "; etc. We postpone, for now, the details of the representation of this information in our proposed system.

Let us imagine that we encounter a patient presenting with definite values for some subset of the attribute variables. For example, we may be given the patient's age, sex, race, and other personal characteristics, a list of chief complaints, perhaps some of the patient's relevant medical history, and perhaps also the results of some diagnostic tests. In all cases the values of most attributes would be unknown. Based on these observed attributes, and based on the available body of medical information, we seek to:

(1) Estimate the likelihood of various unobserved attribute values, such as diseases or the results of candidate diagnostic procedures.

(2) Develop *hypotheses*. By these, we mean logical "pictures" of the patient's overall "profile", i.e. collections of signs, symptoms, diseases, etc. that are compatible given the observed attributes. The observed attributes will typically suggest several hypotheses.

(3) Select among the unobserved attributes those personal characteristics, elements of the patient history, or diagnostic tests that would best distinguish among the current hypotheses.

An additional goal is to accomplish these tasks with a system whose mechanisms and knowledge base can easily accommodate new medical information.

B. Maximum Entropy Prior Distribution

We shall adopt a Bayesian approach. Bayesian methods for automated medical diagnosis have been much discussed but generally agreed to be computationally intractable. For interesting problems (those involving numerous disease, sign and symptom entities), combinatorial difficulties associated with high dimensional joint probability distributions seem to force unreasonable simplifying assumptions, such as statistical independence of diseases, or of symptoms, or of symptoms *given* a disease. Here, we will employ stochastic relaxation to make the necessary statistical calculations; there will be no need for independence or related assumptions.

Medical knowledge, as represented in our knowledge base, will be used to construct a probability distribution on the medical attributes. This distribution, the *prior distribution*, is intended to embody the known implications, correlations, and so-on, among the signs, symptoms, diseases, and patient characteristics. The purpose of this section is to explain the construction of a particular prior distribution, the so-called *maximum entropy* prior. This distribution is "minimal" in the sense that it is consistent with the knowledge base while respecting no additional (and, presumably, unwarranted) constraints.

We begin our construction by reformulating, into probabilistic terms, the information within the knowledge base. Medical knowledge will be idealized as a collection of probability relations among attributes. If, for example, disease D_1 is found in 10% of the patients presenting with symptoms S_1 and S_2 then we would write, for the appropriate i,j, and k, $P(X_i=1|X_j=1,X_k=1)=.1$. If S_3 and S_4 are simultaneously present in 40% of the patients with disease D_2 and characteristic C_1 then $P(X_\ell=1,X_m=1|X_n=1,X_o=1)=.4$ for suitable l,m,n, and o.

These probabilistic statements, whether marginal probabilities, joint probabilities, correlations or conditional probabilities, permit a common and convenient representation. For example, the information $P(X_1 \in \mathcal{S})=p$ $(\mathcal{S} \in \Lambda_1)$ about some attribute variable X_1 can be expressed as

$$E[F(\vec{X})] = 0$$

by defining

$$F(\vec{X}) = \chi_{\mathcal{S}}(X_1) - p,$$

where $\chi_S(x)$ is 1 when $x \in S$, and 0 otherwise. Joint probabilities are handled in the same way. The statement $P(X_1 \epsilon S_1, X_2 \epsilon S_2, X_3 \epsilon S_3) = p$ $(S_1 \subseteq \Lambda_1, S_2 \subseteq \Lambda_2, S_3 \subseteq \Lambda_3)$ may be rewritten as E[F(X)]=0 by defining $F(\vec{X}) = \chi_{S_1}(X_1)\chi_{S_2}(X_2)\chi_{S_3}(X_3) - p$. This representation also applies to correlations or conditional probabilities; they can be represented as E[F(X)]=0 for suitably chosen F. The statement $E[X_1 X_2] = 1.8$, for example, is rewritten as E[F(X)] =0 by defining $F(X) = X_1 X_2 - 1.8$. Or, the conditional probability assignment

$$P(X_1 \epsilon S_1, X_2 \epsilon S_2 \,|\, X_3 \epsilon S_3, X_4 \epsilon S_4) = p$$

is the same as

$$P(X_1 \epsilon S_1, X_2 \epsilon S_2, X_3 \epsilon S_3, X_4 \epsilon S_4) - p\, P(X_3 \epsilon S_3, X_4 \epsilon S_4) = 0$$

and is therefore equivalent to E[F(X)]=0 when

$$F(\vec{X}) = \chi_{S_1}(X_1)\chi_{S_2}(X_2)\chi_{S_3}(X_3)\chi_{S_4}(X_4) - p\chi_{S_3}(X_3)\chi_{S_4}(X_4)$$

By these reformulations, we arrive at the following representation for the knowledge base:

$$\text{(2.B.1)} \qquad E[F_\alpha(\vec{X})] = 0 \qquad \alpha \in A$$

for some collection $\{F_\alpha\}$ $\alpha \in A$.

Unfortunately, the information in (2.B.1) constrains only "low-order" marginal statistics. The typical function F_α will depend on no more than four or five attributes, and most commonly just two. A patient, however, is likely to present as an observation of numerous attribute values, including personal characteristics, various signs and symptoms, and possibly even the already known status of certain diseases. Thus even the most basic questions, such as the probable state of a particular unknown disease given the presenting observations, are not directly computable from the information in (2.B.1). The low-order statistical constraints that constitute our knowledge base are insufficient to determine the distribution on unobserved attributes given a typical collection of observed attributes.

On the other hand, these conditional distributions could be directly computed (at least in principle) from the complete *joint* distribution on the attribute variables $\vec{X}=(X_1,\ldots X_N)$. Our plan here is to *construct* a particular distribution on $\vec{X}$ satisfying the various constraints of (2.B.1), and then to proceed as though this were the actual joint distribution. Recall that each attribute variable X_i takes values in an associated state space Λ_i. We will use Ω to designate the *configuration space* $\Omega = \Lambda_1 \times \Lambda_2 \times \ldots \Lambda_N$, which is the range space for $\vec{X}$. We seek a probability distribution π on Ω that satisfies the constraints in (2.B.1)[1], i.e.

$$\text{(2.B.2)}\quad \begin{aligned} &\text{(a)} \quad \pi(\vec{x}) \geqslant 0 \qquad \vec{x} \in \Omega \\ &\text{(b)} \quad \sum_{\vec{x}\in\Omega} \pi(\vec{x}) = 1 \\ &\text{(c)} \quad \sum_{\vec{x}\in\Omega} \pi(\vec{x})F_\alpha(\vec{x}) = 0 \qquad \alpha \in \mathcal{A}. \end{aligned}$$

Furthermore, to the extent possible, we would prefer that π be without information beyond what is explicitly justified by our knowledge base, and embodied in (c).

[1] We shall proceed under the assumption that at least one such distribution exists and that it can be chosen to be strictly positive: $\pi(x)>0\ \forall x \in \Omega$. In view of the subject origin of the constraints, it is possible that no such distribution exist. We believe that these cases will be amenable to very similar treatments.

To formalize this last requirement we bring in the *entropy* of a distribution ([24]):

$$H = - \sum_{\vec{x} \in \Omega} \pi(\vec{x}) \log \pi(\vec{x}).$$

H is minimum (H=0) for those distributions that concentrate on a single configuration: $\pi(\vec{x}_0)=1$ for some $\vec{x}_0 \in \Omega$. For these distributions no information is gained by an observation of $\vec{X}$, since the necessary state, $\vec{X}_0$, of X is already known a priori. The other extreme is the distribution which puts equal weight on *every* configuration $\vec{x} \in \Omega$, or, equivalently, the distribution under which all attributes are *independent* and *uniformly distributed* over their respective state spaces. Here, our prior knowledge is minimal, and the information about $\vec{X}$ gained in an observation of $\vec{X}$ is maximal. This uniform distribution achieves the maximum entropy ($H=-\log|\Omega|$). Among all distributions satisfying the constraints in (2.B.2)(c), we propose to choose the one maximizing H to serve as joint distribution on $\vec{X}$.

By a simple and well-known exercise in variational calculus, this constrained maximization is achieved by a unique π . This *maximum entropy* π can be written in the Gibbs form

$$\pi(\vec{x}) = \frac{1}{Z} \exp\{ \sum_{\alpha \in A} \lambda_\alpha F_\alpha(\vec{x})\} \qquad \vec{x} \in \Omega$$

where

$$Z = \sum_{\vec{x} \in \Omega} \exp\{ \sum_{\alpha \in A} \lambda_\alpha F_\alpha(\vec{x})\}$$

and where $\{\lambda_\alpha\}$ $\alpha\epsilon A$ are constants (Lagrange multipliers) determined by condition (2.B.2)(c). Substituting π above into (2.B.2)(c) yields:

(2.B.3) $$\sum_{\vec{x}\in\Omega} \exp\{ \sum_{\alpha\in A} \lambda_\alpha F_\alpha(\vec{x})\} F_\beta(\vec{x}) = 0 \qquad \beta\epsilon A \, .$$

Based upon this distribution, which will serve as a prior, we will perform the various tasks of computing conditional probabilities, generating hypotheses, and choosing diagnostic protocols.

The actual computation of the Lagrange multipliers $\{\lambda_\alpha\}$ $\alpha\epsilon A$ from the equations in (2.B.3) is not trivial. Any kind of direct approach is impossible since the summation in (2.B.3), which is over the entire configuration space Ω, involves an unmanageable number of terms. We propose to circumvent direct calculation by a Monte Carlo method based on SR, which will be briefly described later, in section D. For the time being, we will proceed under the assumption that the Lagrange multipliers are known.

C. Bayesian Approach to Diagnosis and Hypothesis Formation

We refer to the terminology and notation introduced in Section 1.D. We assign one site of the graph G for each attribute variable X_i, so that $S = \{1,2,\ldots N\}$. The neighborhood system, $\{G_s\}$ $s\epsilon S$, is determined by the constraint functions $\{F_\alpha\}$ $\alpha\epsilon A$: For any $s\neq r$, $1 \leq s,r \leq N$, s and r are neighbors ($r \in G_s$) if and only if there is an $\alpha\epsilon A$ such that $F_\alpha(\vec{X})$ depends upon both of the attributes X_s and X_r. We have already noted that most of the functions F_α will depend upon a small number (most commonly just two) of the attribute variables. It is also true that most pairs of variables, X_s and X_r, will not be connected by any statement in our knowledge base, and hence the corresponding sites, s and r, will not be neighbors in G. Thus the neighborhood system is relatively sparse, and the connectivity in a fully parallel realization of SR would be modest.

Observe that with this neighborhood system π is a Gibbs distribution with respect to G. Indeed, the "energy function" U is just

$$U(\vec{x}) = - \sum_{\alpha \in A} \lambda_\alpha F_\alpha(\vec{x}),$$

and this "respects G" in the sense that it admits the representation (1.D.1). (In fact, each $\lambda_\alpha F_\alpha$ is one of the "V_c" terms in (1.D.1).) Thus the relaxation, ergodicity and annealing theorems, set out in Section 1.D, are in force, allowing us to choose samples from π, compute expectations with respect to π, and find (approximately) most likely configurations under π.

Actually, we would rarely be interested in making calculations with the unconditional distribution π. It represents our *prior* knowledge about an imagined population of potential patients. It does not reflect the information about a particular patient that we would have at hand in an actual diagnostic situation. Instead, the relevant distribution is the *conditional* (or *posterior*) distribution, under π, on the unobserved attributes given the values of the observed attributes. It is not hard to see that the posterior distribution is again a Gibbs distribution, this time with respect to the *subgraph* of G that contains the sites associated with *unobserved* attributes. Furthermore, the SR procedure for this posterior Gibbs distribution is the same as for π, except that the variables associated with observed sites are held fixed at their observed values; all other sites execute the usual site replacement scheme. The analogue of the theorems on relaxation, ergodicity, and annealing all hold, with the posterior distribution playing the role of π.

We turn now to the three goals articulated in Section 2.A.

Concerning (1), let us suppose that X_i is an unobserved attribute, and that we wish to know the probability of some event depending upon X_i. X_i, for example, may be a binary variable with $X_i=1$ signifying "Disease i present" and $X_i=0$ signifying "Disease i absent". In this case we may wish to know the probability that $X_i=1$ *given the observed attribute values.* More generally, for some set $S_i \in \Lambda_i$ we seek $P(X_i \in S_i | \text{Observed attribute values})$. The calculation can be made by exploiting ergodicity (see Section I.D): defining $f(\vec{x}) = \chi_{S_i}(\vec{x}_i)$,

$$\frac{1}{n} \sum_{t=1}^{n} f(\vec{X}(t)) \rightarrow E[f(\vec{X}) | \text{Observed attribute values}]$$

$$= E(X_i \in S_i | \text{Observed attribute values})$$

as $n \rightarrow \infty$. Thus, conditional probabilities are calculated by recording the relative frequencies of associated events. Remember that the sites corresponding to observed attributes are fixed; during these calculations, the *posterior distribution* controls SR.

At first glance, hypothesis generation appears to be a straightforward extension of the computation of conditional probabilities. It would appear that one could examine individually the probability of each unobserved attribute, and then combine these to form an overall picture, or "hypothesis". Unfortunately, this simple approach is unjustified in all but the most unstructured settings. For example, the presenting signs and symptoms may suggest several disease entities, whereas it may be extremely unlikely that any two of these will exist together. Certain conglomerations of diseases, symptoms and characteristics "hang together"; they are recognizable as "syndromes" or "patterns". We want the likely patterns given the observed attributes.

These relatively common conglomerations of diseases, symptoms, and characteristics are represented by minima, local or global, in the energy surface.

We seek to identify the more prominent of these relative minima. Recall that the temperature-dependant Gibbs distribution is

$$\pi_T(\vec{x}) = \frac{1}{Z_T} \exp\{-U(\vec{x})/T\}$$

where Z_T is the normalizing constant. At lower temperatures the lower energy states dominate in the sense that all other states are increasingly unlikely. This, of course, is the basis of annealing for minimizing U. Thus a sample from π_T (or the appropriate conditional Gibbs distribution) at low temperature will most probability produce a state, $\vec{x}$, that is near a prominent minimum of U. We propose to generate hypotheses by low-temperature sampling from the posterior distribution. Of course, there are many ways in which this might be implemented. Putting details aside (these should await actual experiment), we might proceed roughly as follows:

(a) Beginning with a random configuration on the unobserved attributes, anneal from a relatively "high" to a relatively "low" temperature. (Our image processing experiments indicate that low temperature samples are most efficiently generated via annealing.)

(b) Execute SR at the low temperature while monitoring the total energy $U=U(\vec{X}(t))$.

(c) While in a relative minimum (as evidenced by near-constant energy) identify those attributes, among the unobserved attributes, whose values are strongly inclined toward one state. This may be done, for example, by demanding that the percentage of time spent in a particular state during the SR exceeds some threshold value, such as 80% or 90%.

(d) Re-initiate the procedure (go to (a)).

The result of each execution of (c) is a list of attributes and corresponding values that has the property that it is a likely scenario given the observed (presenting) attribute values. The more prominent of these "hypotheses" will presumably be identified after a modest number of iterations of the above procedure.

This procedure also suggests a means for choosing tests and questions that are likely to distinguish among the reasonable hypotheses. Step (c) will typically yield various test results and elements of a patient's history and personal characteristics that are expected (likely) to be associated with the active hypotheses (i.e. relative energy minima). Recall that these are *unobserved* attributes. Distinct hypotheses will suggest distinct values for these various attributes. A likely positive test result or particular patient characteristic under one hypothesis may be unlikely or less likely under other hypotheses. It is these attributes that will best select among the hypotheses, and this suggests that the corresponding elements of patient history or characteristics be ascertained and that the corresponding diagnostic tests be considered.

Observe that the knowledge base is easily supplemented to accommodate new constraints among existing attributes or the introduction of new attributes, as may represent new diagnostic procedures. In either case, the architectural change is modest, involving a small number of new connections among existing attribute sites, or an additional site with corresponding communication channels. There will be additional Lagrange multipliers, and the values of *all* multipliers will need to be recalculated. In this "rebooting" operation, the vast majority of the multipliers will undergo very little change of value, so that the old values should provide an excellent starting point.

D. Computation of the Lagrange Multipliers

We refer to Section 2.B, and recall that the specification of the maximum entropy prior is in terms of Lagrange multipliers $\vec{\lambda} = \{\lambda_\alpha\}\ \alpha \in \mathcal{A}$:

$$\pi(\vec{x}) = \frac{1}{Z} \exp\{ \sum_{\alpha\in\mathcal{A}} \lambda_\alpha F_\alpha(\vec{x})\} \tag{2.D.1}$$

Z is the normalizing constant, which will, of course, depend upon $\vec{\lambda}$. The values for the components, $\{\lambda_\alpha\}\ \alpha \in \mathcal{A}$, are to be chosen so that π satisfies the constraints that constitute the knowledge base. These have the form

$$\sum_{\vec{x}\in\Omega} \pi(\vec{x})F_\alpha(\vec{x}) = 0 \qquad \alpha \in \mathcal{A}. \tag{2.D.2}$$

Recall our "consistency assumption": there exists at least one (strictly positive) distribution, π , satisfying these constraints. For convenience, we shall also assume that the constraint functions, $F_\alpha: \Omega \to R$, are linearly independent.

Our scheme for computing $\vec{\lambda}$ is based upon a fortuitous relation between the normalizing constant Z (the "partition function") and $\vec{\lambda}$. To emphasize the dependency of Z upon $\vec{\lambda}$ we will write $Z(\vec{\lambda})$. $Z(\vec{\lambda})$ is of course just

$$\sum_{\vec{x}\in\Omega} \exp\{ \sum_{\alpha\in\mathcal{A}} \lambda_\alpha F_\alpha(\vec{x})\}$$

The key properties of $Z(\vec{\lambda})$ are summarized by:

Proposition

(1) $Z(\vec{\lambda})$ *is strictly convex.*

(2) *There is a unique* $\vec{\lambda} \in R^{|\mathcal{A}|}$ *such that* π *(defined by* (2.D.1)) *satisfies* (2.D.2). *Furthermore, this* $\vec{\lambda}$ *achieves the global minimum of* Z.

Thus the problem of choosing the Lagrange multipliers so that the knowledge-base constraints are satisfied is equivalent to minimizing the partition function, a convex function of $\vec{\lambda}$. We hasten to point out, however, that the partition function *cannot be directly computed* for any value of $\vec{\lambda}$, since it involves a summation over an astronomical number of terms.

Nevertheless, at any given $\vec{\lambda}$ we can approximate the *direction of the gradient* of the partition function with respect to $\vec{\lambda}$. The α component of this gradient is given by

$$(\nabla Z(\vec{\lambda}))_\alpha = \sum_{\vec{x}\in\Omega} F_\alpha(\vec{x}) \exp\{ \sum_{\beta\in A} \lambda_\beta F_\beta(\vec{x})\}$$

$$= Z(\vec{\lambda})E[F_\alpha(\vec{X})]$$

where E[ ] refers to expectation under the distribution π, and where π is defined by (2.D.1) using the current value of $\vec{\lambda}$. Thus $\nabla Z(\lambda)$ has the same direction as the vector whose α component is $E[F_\alpha(\vec{X})]$ for every $\alpha \in A$ (since $Z(\vec{\lambda}) > 0$). Observe that this latter vector can be approximated by time averages, using once again the ergodicity of SR (this time under the Gibbs distribution defined by the *current value* of $\vec{\lambda}$). The upshot is that we can employ one of the various gradient descent algorithms to locate the minimum of $Z(\vec{\lambda})$. The Lagrange multipliers, and hence the prior distribution, are thereby specified.

References

1. J. Besag (1974), "Spatial interaction and the statistical analysis of lattice systems," (with discussion), *J. Royal Statist. Soc.*, Series B, Vol. 36, 192-236.

2. V. Cêrný (1982), "A thermodynamical approach to the traveling salesman problem: an efficient simulation algorithm," Inst. Phys. & Biophys., Comenius Univ. Bratislava (preprint).

3. E. Charniak (1983), "The Bayesian basis of common sense medical diagnosis," *Proceedings of the 1983 Conference of the American Association for Artificial Intelligence*, M. Kauffman.

4. P. Cheeseman (1983), "A method of computing generalized Bayesian probability values for expert systems," *Proceedings of 8th International Joint Conference on Artificial Intelligence*, Karlsruhe, West Germany, 198-202.

5. T.-S. Chiang and Y. Chow (1987), "A limit theorem for a class of inhomogeneous Markov processes," Academia Sinica, Taipei, Taiwan (preprint).

6. R.O. Duda and E.H. Shortliffe (1983), "Expert systems research," *Science*, Vol. 220, 261-268.

7. D. Geman and S. Geman (1983), "Parameter estimation for some Markov random fields," Technical Report, Div. Appl. Math., Brown University.

8. S. Geman and D. Geman (1984), "Stochastic relaxation, Gibbs distributions, and the Bayesian restoration of images," *IEEE Trans. Pattern Anal. Machine Intell.*, 6, 721-741.

9. S. Geman and C. Graffigne (1987, "Markov random field image models and their applications to computer vision," *Proceedings of the International Congress of Mathematicians 1986,* Ed. A.M. Gleason, American Mathematical Society, Providence.

10. S. Geman and C.R. Hwang (1986), "Diffusions for global optimization," *SIAM J. Control and Optimization*, 24, 1031-1043.

11. S. Geman and D.E. McClure (1987), "Statistical methods for tomographic image reconstruction," Proceedings of the 46th Session of the International Statistical Institute, *Bulletin of the ISI*, Vol. 52.

12. B. Gidas (1985), "Non-stationary Markov chains and convergence of the annealing algorithm," *J. Stat. Phys.*, 39, 73-131.

13. B. Gidas (1986), "A renormalization group approach to image processing problems," Div. Appl. Math., Brown University (preprint).

14. C. Graffigne (1987), "Experiments in texture analysis and segmentation," Ph.D. Thesis, Div. Appl. Math., Brown University.

15. U. Grenander (1976,1978,1981), *Lectures in Pattern Theory*, 3 volumes, Springer-Verlag.

16. B. Hajek (1985), "Cooling schedules for optimal annealing," *Mathematics of Operation Research*, (to appear).

17. G.E. Hinton and T.J. Sejnowski (1983), "Optimal perceptual inference," *Proc. IEEE Conf. Comput. Vision Pattern Recognition*.

18. C.R. Hwang and S.J. Sheu (1986), "Large time behaviors of perturbed diffusion Markov processes with applications III: simulated annealing," Academia Sinica, Taipei, Taiwan (preprint).

19. S. Kirkpatrick, C.D. Gellatt, and M.P. Vecchi (1983), "Optimization by simulated annealing," *Science*, Vol. 220, 671-680.

20. C.A. Kulikowski (1980), "Artificial intelligence methods and systems for medical consultation," IEEE Trans. Pattern Anal. Machine Intell., 2, 464-476.

21. A. Lippman (1986), "A maximum entropy method for expert system construction," Ph.D. Thesis, Div. Appl. Math., Brown University.

22. N. Metropolis, A.W. Rosenbluth, M.N. Rosenbluth, A.H. Teller, and E. Teller (1953), "Equations of state calculations by fast computing machines," *J. Chem. Phys.*, 21, 1087-1091.

23. D.K. Pickard (1979), "Asymptotic inference for an Ising lattice III. Non-zero field and ferromagnetic states," *J. Appl. Prob.*, 16, 12-24.

24. C.E. Shannon and W. Weaver (1949), "*Mathematical Theory of Communication*," University of Illinois Press, Urbana.

TOWARDS A METHOD OF CORRECTING BIAS INTRODUCED BY THE USE OF THE ERROR FITTING METHOD IN CONJUNCTION WITH A MAXIMUM ENTROPY PROCESSING ALGORITHM

N.A. Farrow and F.P. Ottensmeyer

Ontario Cancer Institute and Department of Medical
Biophysics, University of Toronto,
500 Sherbourne Street, Toronto,
Ontario, Canada. M4X 1K9

G. J. Erickson and C. R. Smith (eds.),
Maximum-Entropy and Bayesian Methods in Science and Engineering (Vol. 2), 313–322.

INTRODUCTION.

High resolution dark field electron micrographs are corrupted by large amounts of noise arising mainly from the scattering of electrons by the thin carbon film that is used to support the samples that are to be imaged. Processing to produce an image from which a structure can be deduced has in the past relied on methods such as filtering with crystalline arrays, contrast enhancement and the averaging of a number of images of similar structures. These methods however are rather arbitrary in their application and often lead to a reduction in the spatial resolution of resultant images. A maximum entropy processing method applied to digitised micrographs has been investigated to see whether significant image enhancement could be achieved without either subjective decisions as to the degree of image processing or a reduction in the spatial resolution in the final image.

THEORY AND METHOD.

The electron microscope image, conventionally recorded on a photographic plate, may be digitised to form an N element array M, in which each element of the array represents a grey level corresponding to a single position in the image.

$$M = \{m_1, m_2, m_3, \ldots\ldots, m_N\} \qquad (1)$$

The image M will be corrupted by noise arising from the structures within the carbon support film and the poor electron statistics involved in taking a dark field electron micrograph. The image M will be modelled by an image P.

$$P = \{p_1, p_2, p_3, \ldots\ldots, p_N\} \qquad (2)$$

Values of the individual p_i that are to model the image M can be found by maximising the configurational entropy of the image P, thus selecting the most probable image,the entropy was defined.

$$S_p = -\sum_{i=1}^{N} p_i \log p_i \qquad i=1,2,\ldots\ldots N \qquad (3)$$

The entropy maximisation (Gull & Daniell 1980) is conventionally constrained to prevent the image attaining it's global maximum entropy, at which point all grey levels are equal, and to force the image P to be consistent with the data by constraining the x^2 distribution defined by

$$x^2 = \frac{1}{\sigma}\sum_{i=1}^{N} (m_i - p_i) \qquad (4)$$

to be equal to its expected value N. A second constraint applied when maximising the entropy of the model image P is that the total intensity in the image P must be the same as that in the original data image M.

The expression that will now let us determine the model image P by its maximisation is,

$$S = -\sum_{i=1}^{N} p_i \log(p_i) - \lambda_1 \chi^2 - \lambda_2 \sum_{i=1}^{N} p_i \qquad (5)$$

where λ_1 and λ_2 are Lagrangian multipliers, the values of which are determined during the iterative process of determining the p_i. The individual p_i are found by taking the appropriate partial differential of the above expression and finding it's root, by a Newton-Raphson iterative approach (Farrow & Ottensmeyer 1987) thus finding the maxima of S, S being an upwardly concave function. The value of the sum of the intensity values in the image P can then be found and the initial value of λ_2 adjusted; the process is then repeated until the sum of the p_i has converged on that of the m_i . The value of χ^2 can then be computed and λ_1 adjusted until the value of χ^2 has converged to N. It should be noted that the sum of the intensties in P and the χ^2 are not uniquely dependent on λ_2 and λ_1 respectively but the above iterative procedure nevertheless still allows convergence to be achieved for both of the constraints.

The χ^2 distribution and the retention of the same intensity in the model as in the data, are the two constraints used by most authors. It became clear however that constraining only the size of the value of χ^2 produced a distribution of residuals, differences between data and model values, that was intuitively incorrect. This was noticed early in the development of the above algorithm. It is illustrated in Fig. 1, a simple case in which eight pixel values are considered, one to simulate a high signal value and seven lower values simulating background values. Fig. 1 shows how the data would be modulated as the size of the standard deviation postulated as a noise on the data values is allowed to increase. It can be seen that for all values of the standard deviation, postulated as an error on the data, the point furthest from the global mean of the distribution, and consequently the global entropy maximum, moves a disproportionately large distance towards the mean and has a much greater residual than any of the other points. The χ^2 term is derived almost exclusively from this single large residual.

This is intuitively troubling. If each point in the image were to have been subject to a similar noise generating process one would expect that any process which is applied to remove the noise would remove similar amounts of noise from each of the points in the distribution and not as in the case illustrated in Fig. 1 to assign a large correction to a single point in the image whilst leaving the other points in the distribution essentially alone. This intuitive feeling may be formalised and realised in combination with the maximisation of configurational entropy by application of a similar procedure to that described by Bryan & Skillng (1980). The basis of this method of Error

Figure 1. The affect of changing the standard deviation of the uncertainty of the data values on the maximum entropy solution. The largest residual is assigned to the highest data value as it is brought towards the mean of the distribution.

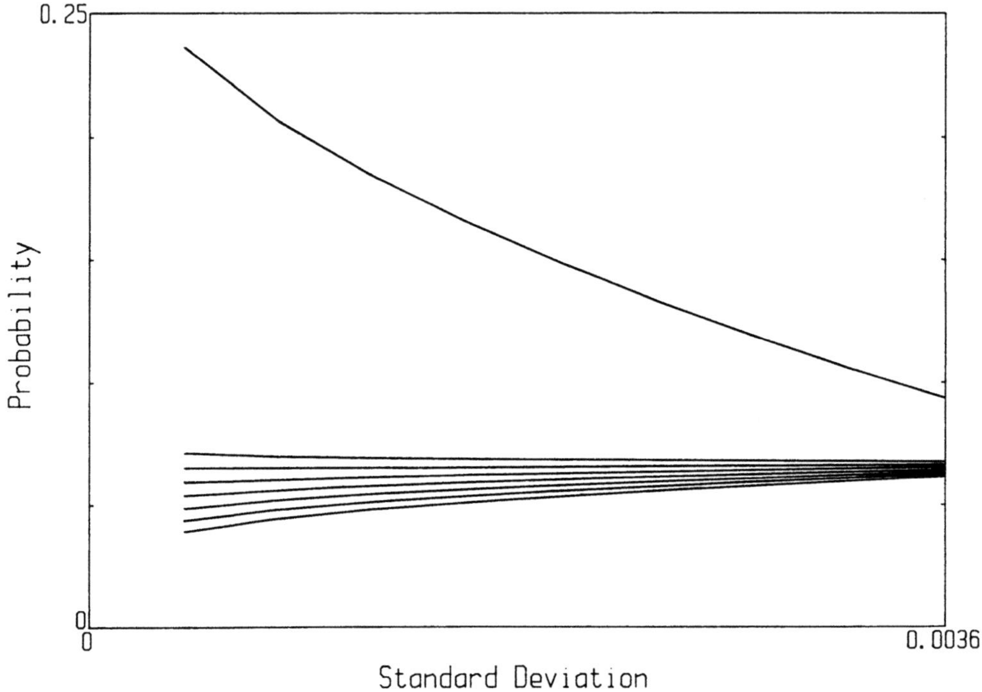

Fitting is described below.

N individual residual terms describing the relation of the model to the data at each point in the image are now defined,

$$u_i = \frac{1}{\sigma}(m_i - p_i) \tag{6}$$

The division by σ, the standard deviation of the noise in the image M, normalises the observed distribution of residuals to allow it to be compared to that which one would expect had gaussian noise been added to the original image and then exactly removed in the model. Expected residuals, ν, that would result from the removal of gaussian noise can be calculated from the following.

$$\nu_i = \phi^{-1}\left(\frac{i-0.5}{N}\right) \qquad i=1,\ldots.N \tag{7}$$

Where ϕ^{-1} indicates the inverse of the function ϕ defined by,

$$\phi(x) = \frac{1}{(2\Pi)^{1/2}} \int_{-\infty}^{x} \exp(-0.5u^2)du \tag{8}$$

To compare the actual residuals u_i obtained to those expected from the definition of ϕ, the exact unit normal distribution, N(0,1), the actual residuals are ordered such that,

$$u_{(1)} < _{(2)} < \ldots\ldots\ldots < u_{(N)} \tag{9}$$

where the parentheses indicate an ordering index. The actual residuals are then compared to those expected using the E^2 statistic where E^2 is defined as,

$$E^2 = \sum_{i=1}^{N} (u_{(i)} - \nu_i)^2 \tag{10}$$

It can be shown that the expected value of E^2 to aim at is log(log(N)) which indicates within a 95% confidence limit the likelihood of fitting of the actual residuals to the expected residuals from an N(0,1) distribution after the actual residuals have been normalised. The effect of replacing x^2 with E^2 is to force the deviations of the model from the data not only to be consistent with the size of the noise on the data but also to have the same pattern of residuals as would be generated by an N(0,1) distribution about mean zero.

Figure 2. A 4096 pixel, 16 gray level image electron microscope image of a nucleosomal core particle. (a) before processing. (b) after processing using an error fitting maximum entropy algorithm.

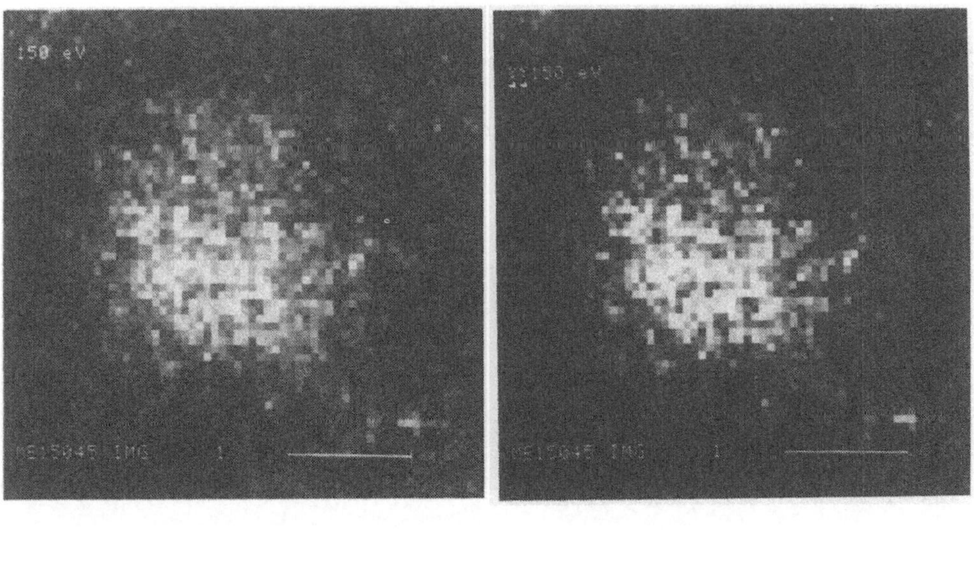

The solution for the individual p_i that comprise P can now proceed as before solving for the root of the N partial derivatives of the expression.

$$S = -\sum_{i=1}^{N} p_i \log(p_i) - \lambda_1 E^2 - \lambda_2 \sum_{i=1}^{N} p_i \quad (11)$$

The procedure used to maximise the above was to set the initial model P to be uniformally flat, to order the residuals to allow a first calculation with λ_1 and λ_2 equal to zero, to perform Newton-Raphson iterations to solve for each p_i, to recalculate and reorder the residuals and to procede as described for the solution of (5). The end point of the iteration was that at which the sum of the model is equal to the sum of the data and the value of E^2 is log(log(N)).

RESULTS.

The results of solving equation (11) where M was a digitised electron micrograph are shown in Fig. 2. The improvement in signal to noise ratio, measured over 100 pixels, is a factor of 1.6. This improvement was not as high as had been anticipated. One possible cause of this is that the noise in an electron micrograph, is markedly non-Gaussian, as shown by Andrews et al (1986). So we are modelling the noise incorrectly. The photograph digitized in Fig. 3a had pure gaussian noise added to give the image in Fig. 3b. This noisy image was then processed to yield the image shown in Fig. 3c. Again the improvement in image quality was not very significant despite the fact that the noise distribution was correctly modelled. Therefore even though it will be necessary to characterise the experimental noise distribution, ϕ, for electron micrographs this improved noise description is unlikely to improve significantly the level of image enhancement.

Figure 3. (a) 512X512 digitised photograph. (b) image after addition of gaussian noise. (c) image after processing with a maximum entropy error fitting algorithm.

Figure 4. (a) 512 point distribution of gaussian noise added to a rectangular function (b) distribution after processing with a maximum entropy error fitting algorithm. (c) the residuals between distributions (a) and (b).

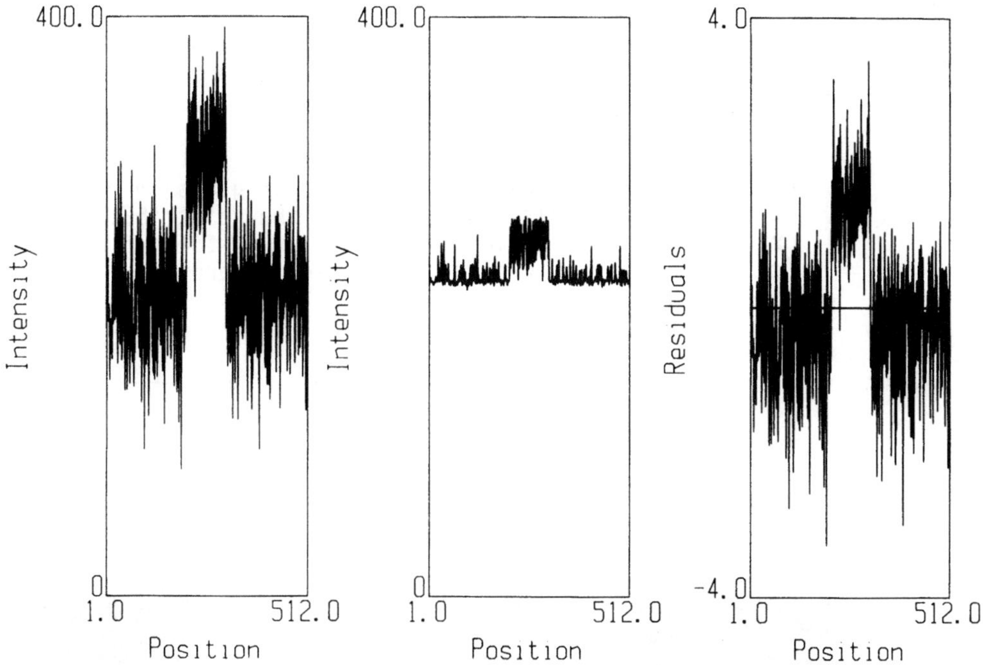

A key to the lack of enhancement is indicated by the three distributions of Fig. 4. This example attempted to model a cross section of a 512x512 digitised image in which the signal is elevated above that of the general image background. In this example the signal was represented in the central 100 of the total of 512 image values. The background was set to 200, the signal to 300, and the standard deviation of the noise was 40.0 giving a signal to noise ratio of 2.5. The distribution prior to processing is displayed in Fig. 4a, the model of this data distibution having maximised an expression equivalent to that of (11) is shown in Fig. 4b. It can be seen in Fig. 4b that the processing of the original distribution has led to a biassing of the values in the distribution: the noise in the lower levels of the distribution now lies almost entirely above a constant level whilst the noise in the signal lies mainly below some constant upper level. A visualisation as to why this artifact occurs can be seen if one plots the size of the residuals against their position, Fig 4c. It is seen that all the largest positive residuals are concentrated in the higher regions that are simulating the signal, whilst none of them lie in the background region even though it may contain some large positive deviations from the original background level. Figure 5 shows the effect of running the algorithm using error fitting on the data set which is comprised entirely of noise, Fig. 5a. The resulting model distribution is shown in Fig. 5b and the residuals plotted against position in Fig. 5c. If Fig. 4c is compared to Fig. 5c it is seen that the two distributions are quite different despite the fact that the noise was identical in both situations. Having a signal in the data set shown in Fig. 4a has concentrated all the large positive residuals in the region of the signal. The distribution of residuals shown in Fig. 5c is much closer to that which would be expected since the noise can be considered to be stationary. The residuals correspond to those expected on removal of stationary noise, unlike those shown in Fig. 4c.

Figure 5. (a) 512 point distribution of gaussian noise added to a uniform distribution. (b)the distribution after processing with a maximum entropy error fitting algorithm. (c) the stationary residuals between (a) and (b).

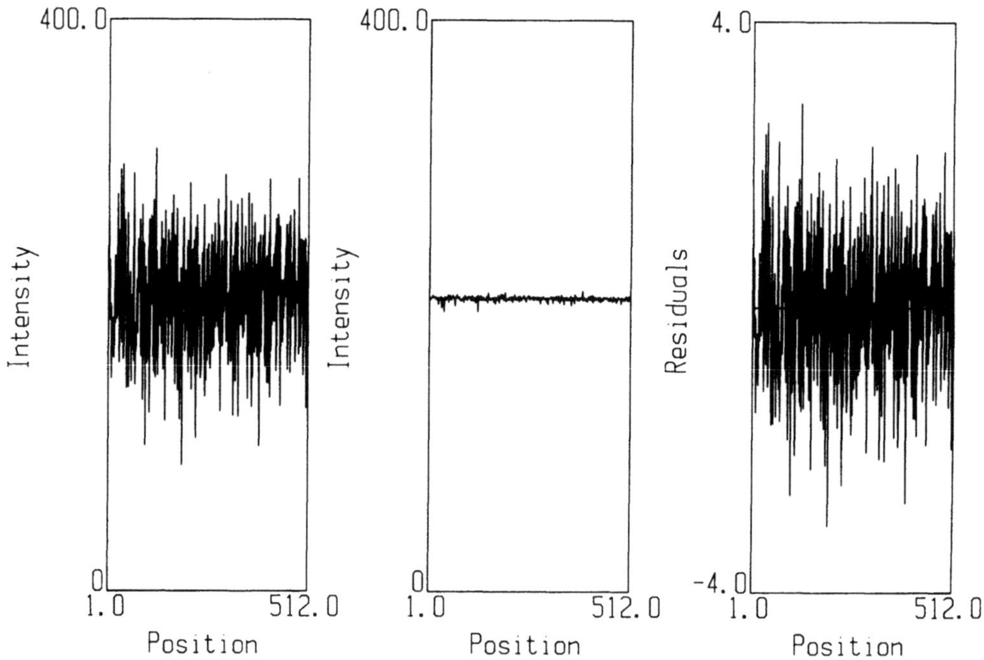

It would seem therefore that a better solution to the problem of selecting the best model for the data would be achieved if the spatial relationship of the residuals could be constrained to be more like that of Fig. 5c. In searching for a practically applicable constraint to bring about this condition it was noted that the lag 1 serial correlation, c_1, of the residuals defined by

$$c_1 = \frac{1}{\sigma_u} \sum_{i=1}^{N} (u_i - \bar{u})(u_{i+1} - \bar{u}) \tag{12}$$

has a value of ~512 for the distribution of fig(4c) whilst it is ~15 for the distributon of fig(5c). This indicates only a small degree of nearest neighbour correlation in the latter case. Thus it was thought that c_1 would provide a suitable parameter to force the distribution of residuals to be more like that of Fig. 5c regardless of the amount of signal in a distribution. The expression that is now to be maximised is,

$$S = -\sum_{i=1}^{N} p_i \log p_i - \lambda_1 E^2 - \lambda_2 \sum_{i=1}^{N} p_i - \lambda_3 c_1 \tag{13}$$

Figure 6. (a) 512 point distribution of gaussian noise added to a rectangular function. (b) the distribution after processing with a maximum entropy error fitting algorithm constraining C1 to be zero. (c) residual distribution between (a) and (b)

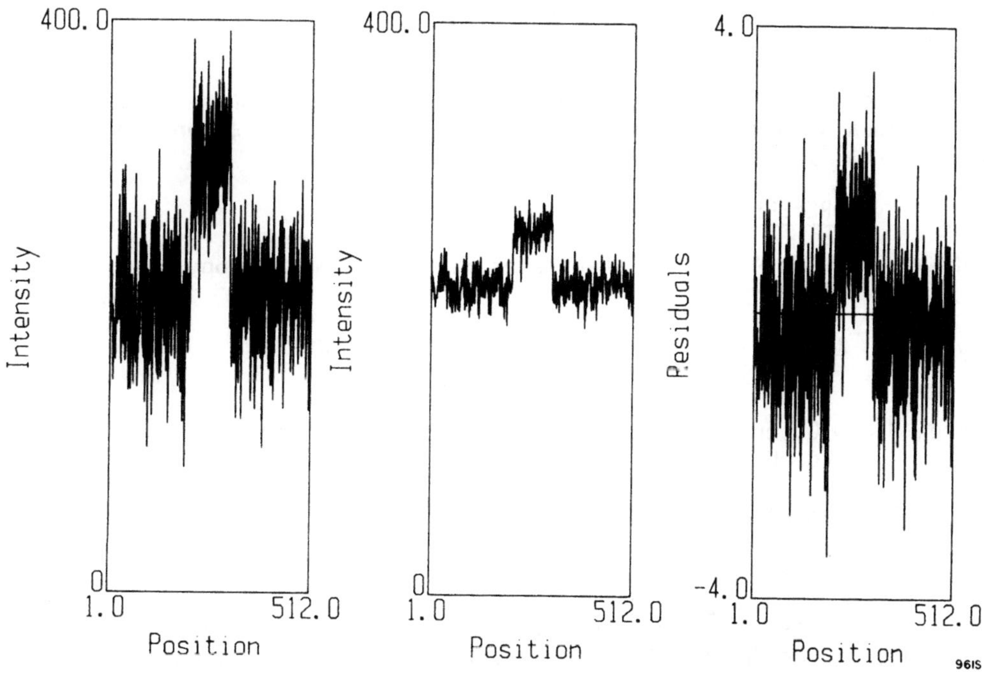

When including c_1 in the algorithm a number of simplifications can be made. As the residuals are constrained to be of an N(0,1) distribution, the mean of the residuals will be zero and the standard deviation will be unity. In addition for this data distribution the residuals at the extreme spatial positions can be accommodated by a wraparound of the distribution.

The algorithm now proceeds in a similar manner to that used for the maximization of (11) but now λ_3 is adjusted from it's initial zero value to give a value of c_1 close to zero. The solution for the same distribution as that in Fig. 4a is shown in Fig. 6a. It can be seen on comparison with Fig. 4b that the biassing artifact in the noise has been removed to a large degree but when the residuals are again plotted against position in the distribution, Fig. 6c, it is seen that the larger positive residuals are still concentrated in the elevated "signal" region of the distribution.

CONCLUSION.

It has been demonstrated that fitting only the value of χ^2 can lead to an anomalous distribution of residuals between the data and the final model. Error fitting was introduced to remove this problem and signal to noise ratio enhancement was demonstrated in both digitised images and in test distributions. A biassing artifact was noted, however. In an attempt to remove this artifact a constraint on the lag 1 serial correlation was added. The affect of this constraint was to reduce the extent of the artifact in the final model distribution. Examination of the residuals shows that the noise in the data is still not removed as we would require, knowing the noise to be stationary. It may be that using multiple constraints on different lag serial correlations will allow the data to be processed in a manner that forces the residuals to be stationary, i.e. the auto correlation function to be uniformally zero for all lags.

REFERENCES

Andrews D.W., Yu A.H.C. & Ottensmeyer F.P.; Ultramicroscopy 19,1986,1-14

Bryan R.K. & Skilling J.; Mon.Not.R.astr.Soc.(1980), 191, 69-79

Farrow N.A & Ottensmeyer F.P.; Image and Signal Processing, Ed. Hawkes P.W., S.E.M.Inc.,(1987), (in press)

Gull S.F & Daniell G.J; Nature vol 272, 20 April 1980, 686-690

ACKNOWLEDGEMENTS

This work was supported by grants from the National Cancer Institute of Canada, the Medical Research Council and the Ontario Cancer Research and Treatment Foundation. Neil Farrow is a recipient of a Connaught Scholarship.

Making Maximum Entropy Computations Easier By Adding Extra Constraints
(Extended Abstract)

Sally A. Goldman
Ronald L. Rivest

Laboratory for Computer Science
Massachusetts Institute of Technology
Cambridge, Massachusetts 02139

July 31, 1986

Abstract

This paper presents a new way to compute the probability distribution with maximum entropy satisfying a set of constraints. Unlike previous approaches, our method is integrated with the planning of data collection and tabulation. We show how *adding constraints* and performing the associated additional tabulations can substantially speed up computation by replacing the usual iterative techniques with a straight-forward computation. These extra constraints are shown to correspond to the intermediate tables used in Cheeseman's method. We also show that the class of constraint graphs that our method handles is a proper generalization of Pearl's singly-connected networks. An open problem is to determine a minimal set of constraints necessary to make a hypergraph acyclic. We conjecture that this problem is NP-complete, and discuss heuristics to approximate the optimal solution.

1 Introduction

Many applications require reasoning with incomplete information. For example, in artificial intelligence one may wish to develop expert systems that can answer

This research was supported in part by NSF Grant DCR–8006938. Sally Goldman received support from an Office of Naval Research Fellowship.

G. J. Erickson and C. R. Smith (eds.),
Maximum-Entropy and Bayesian Methods in Science and Engineering (Vol. 2), 323–340.

questions when given an incomplete model of the world. Having an incomplete model means that a question can have more than one answer consistent with the model. How can a system choose an answer to such a question? This paper discusses a technique based on the method of maximum entropy. After formally defining this problem, we discuss previously known methods for calculating the maximum entropy distribution. Then we present a new technique which makes maximum entropy computations easier by adding extra constraints. Finally, we compare our technique to previous methods.

2 Formal Problem Definition

In this section we formally define our problem. We begin by defining some notation. Let $V = \{A, B, C, \ldots\}$ be a finite set of binary-valued variables, or attributes. (The generalization to finite-valued variables is straightforward.) Consider the event space Ω_V defined to be the set of all mappings from V to $\{0,1\}$. (We call such mappings *assignments* since they assign a value to each variable in V.) It is easy to see that $|\Omega_V| = 2^{|V|}$. If $E \subseteq V$, we have Ω_V is isomorphic to $\Omega_E \times \Omega_{V-E}$; we identify assignments in Ω_E with subsets of Ω_V in the natural manner.

We are interested in probability distributions defined on Ω_V. We use the following convention throughout this paper. If $E \subseteq V$, we write $P(E)$ to denote the probability of an element of $\Omega_E \subseteq \Omega_V$. In other words, we specify only the variables involved in the assignments and not their values. For example

$$P(V) = P(A)P(B)P(C)\cdots \tag{1}$$

represents $2^{|V|}$ equations, stating that the variables are independent. (We do not assume equation (1) in this paper.) By convention, all assignments in an equation must be consistent. We also write $P(A)$ instead of $P(\{A\})$, $P(AB)$ instead of $P(\{AB\})$, and so on.

We use a similar convention for summations: $\sum_E$ stands for a summation over all assignments in Ω_E, when $E \subseteq V$. Using our conventions, we see that $Y \subseteq E \subseteq V$ implies that

$$P(Y) = \sum_{E-Y} P(E) \tag{2}$$

We are interested in probability distributions on Ω_V satisfying a given set of constraints. Let $E_1, \ldots E_m$ be distinct subsets of V. Let us suppose that for each i we are given the $2^{|E_i|}$ constraint values $\{P(E_i)\}$, and that these values are consistent (i.e. there exists at least one probability distribution on Ω_V which satisfies the given constraints). A common way of ensuring that the constraints are consistent is to derive the constraints by computing the observed marginal probabilities from a common set of data. (Using "experts" to provide subjective probability estimates is a well-known way of deriving a set of ***inconsistent*** constraints.) Note that equation (2) states that constraints on the $P(Y)$'s are implied by the constraints on the $P(E_i)$'s when $Y \subseteq E_i$. In general, there may be many probability distributions P satisfying the given constraints; in this case we are interested in that unique distribution P^* which maximizes the entropy

$$H(P) = -\sum_V P(V)\log(P(V)) \tag{3}$$

Motivation for this choice can be found in [Ja79,Ja82,JS83,Le59,SJ80,TTL84].

The maximum entropy distribution P^* is known to have a simple representation. For each ω in Ω_{E_i}, there are $2^{|E_i|}$ non-negative real variables $\alpha_{E_i}(\omega)$ (i.e., one variable per constraint), that determine P^* as follows. Let us write $\alpha_i(\omega)$ instead of $\alpha_{E_i}(\omega)$ for brevity, and omit the argument ω when it can be deduced from context. Now we may write simply

$$P^*(V) = \alpha_1\alpha_2\ldots\alpha_m. \tag{4}$$

Each element of Ω_V is assigned a probability which is the product of the appropriate α's where each α determines its argument from the assignment to V. This is known as a *log-linear* representation; it is an interesting fact that the maximum entropy distribution is the ***unique*** log-linear distribution of form (4) which satisfies the given constraints. (That is, to find the maximum entropy distribution satisfying the constraints, it suffices to find the log-linear distribution satisfying the constraints.)

Of course, in order for equation (4) to hold, the α's must be correctly computed. The problem of computing the maximum entropy distribution becomes the problem of computing the α_i's from the $P(E_i)$'s.

3 Previous ME Methods

Most existing methods for calculating the maximum entropy distribution are iterative. They typically begin with a representation of the uniform distribution and converge towards a representation of the maximum entropy distribution. Each step adjusts the representation so that a given constraint is satisfied. To enforce a given constraint $P(E_i)$, all of the elementary probabilities $P(V)$ relevant to that constraint are multiplied by a common factor. Because constraints are dependent, adjusting the representation to satisfy one constraint may cause a previously satisfied constraint to no longer hold. Thus one must iterate repeatedly through the constraints until the desired accuracy is reached. (We note that the implicit constraint — that the probabilities sum to one — must usually be explicitly considered here.) Examples of this type of algorithm are given in [Br59,Ch83,Cs75,Fi70,IK68,KK69]. Representing the probability distribution explicitly as a table of $2^{|V|}$ values is usually impractical. For this problem, it is most convenient to store only $\alpha_1, \alpha_2, \ldots \alpha_m$; this is a representation as compact as the input data, which represents the current probability distribution implicitly via equation (4). To represent the uniform distribution, every α is set to 1, except for the α corresponding to the requirement that entries of the probability distribution must sum to 1 — which is set to $2^{-|V|}$. To determine if a constraint is satisfied, one must sum the appropriate elements of the probability distribution Any particular element can be computed using equation (4). If the constraint is not satisfied, the relevant α is multiplied by the ratio of the desired sum to the computed sum. Thus in originally calculating the α's and later in evaluating queries it is necessary to evaluate a sum of terms, where each term is a product of α's. This sum is difficult to compute since it may involve an exponential number of terms.

Cheeseman [Ch83] proposes a clever technique for rewriting such sums in order to evaluate them more efficiently. For example

$$\alpha \sum_{A \ldots F} \alpha_{AB}\, \alpha_{ACD}\, \alpha_{DE}\, \alpha_{AEF}$$

is rewritten as follows. First, $\sum_{A \ldots F}$ is broken into six sums, each over one variable. Arbitrarily choosing the variable ordering $CDFEAB$ we obtain

$$\alpha \sum_B \sum_A \sum_E \sum_F \sum_D \sum_C \alpha_{AB} \alpha_{ACD} \alpha_{DE} \alpha_{AEF}.$$

Now each α is moved left as far as possible (it stops when reaching a sum over a variable on which it depends). The above sum then becomes

$$\alpha \sum_B \sum_A \alpha_{AB} \sum_E \sum_F \alpha_{AEF} \sum_D \alpha_{DE} \sum_C \alpha_{ACD}.$$

The sums are evaluated from right to left. The result of each sum is an intermediate *table* containing the value of the sum evaluated so far as a function of variables further to the left which have been referenced. For example after evaluating $\sum_C$ a table is kept containing α_{ACD} for Ω_{AD}. (As we shall see, in some cases the intermediate table is most efficiently represented as two or more smaller tables.) The variable ordering must be chosen carefully in order to take full advantage of this technique. A poor choice of variable ordering can yield a sum which is not much better than explicitly considering all $2^{|V|}$ terms; a good choice can dramatically reduce the work required. The choice of variable ordering which minimizes the cost of evaluating a sum can be viewed as a vertex ordering problem in a graph. This problem is very similar to the minimum fill-in problem encountered when performing Gaussian elimination on sparse symmetric matrices [RT78,RTL76]. Since the minimum fill-in problem has been proven to be NP-Complete [Ya81], we conjecture that this problem is as well.

Some alternative approaches to the iterative scheme have been proposed. One of the more interesting proposals is due to Geman [Ge86,Li86]. This method uses stochastic relaxation to simultaneously adjust the α's to meet all of the constraints, instead of satisfying one constraint at a time.

Some authors restrict their attention to probability distributions that can be easily worked on without resorting to the iterative methods needed to compute the maximum entropy distribution. These approaches usually restrict the kinds of constraints that might be supplied, and assume conditional independence explicitly as needed to force a unique result. This approach is taken by Chow and Liu [CL68] and Pearl [Pe85]. These methods construct a dependency tree where nodes represent variables and links represent direct dependencies; all direct influences on a node come from its parent. Here the set of all conditional probabilities of the form, $P(\text{child}|\text{parent})$, together with the probability distribution of the variable at the root, suffice to define a unique probability distribution. Pearl [Pe85] generalizes the tree condition to a network which has at most one undirected path between any pair of nodes (a *singly-connected network*).

One can view the contribution of the current paper as providing a synthesis of these two approaches, by showing how the difficulties of computing a maximum entropy distribution can be substantially alleviated by *enlarging* the set of constraints to be considered *before* the data is gathered and tabulated. With the enlarged set of constraints, the computation of the maximum entropy distribution has a simple form which generalizes the equation suggested by Pearl.

4 Using Acyclic Hypergraphs

Our approach is based on the work of Malvestuto [Ma85], who derived sufficient conditions for writing marginals of the maximum entropy formula as a product of easily calculated probabilities. We begin by describing how to model a set of attributes and associated constraints as a hypergraph. (A similar model was given by [EK83].) It is interesting to note that the work on the desirability of acyclic schemas first appeared in the database literature [BFMMUY81,BFMY83,TY82]. The attributes of the database replace the attributes in our problem, and the relations replace the constraint sets. Given that the database scheme is acyclic many problems are simplified

A hypergraph is like an ordinary undirected graph, except that each edge may be an arbitrary subset of the vertices, instead of just a subset of size two. We define the hypergraph $G = (\mathcal{V}, \mathcal{E})$ to contain a vertex for each variable, and a hyperedge for each constraint. For example the hyperedge $\{ABC\}$ corresponds to the constraint set $\{A, B, C\}$. We say that hyperedge X ***subsumes*** hyperedge Y if $Y \subseteq X$. It is important to observe that the constraints on a sub-hypergraph induced by restricting attention to a subset of the vertices can be inferred from the original hypergraph constraints.

A hypergraph is *acyclic* if repeatedly applying the following reduction steps gives the empty hypergraph (containing no edges and no vertices):

1. Delete any vertices which belong to only one hyperedge.

2. Delete any hyperedges which are subsumed by another hyperedge.

Graham's algorithm is the procedure of applying reduction steps 1 and 2 until either the empty set is reached, or neither can be applied [Gr79].

Before proceeding, we define some necessary notation regarding the above reduction procedure. Let $\mathcal{E}^{(0)} = \{E_1^{(0)}, \ldots, E_m^{(0)}\}$, where $E_i^{(0)}$ is the ith hyperedge of G. Let $E_i^{(k)} = Z_i^{(k)} \cup Y_i^{(k)}$, where $Z_i^{(k)}$ is the set of variables which appear only in $E_i^{(k)}$ and $Y_i^{(k)}$ is the set of variables which appear in at least one hyperedge other than $E_i^{(k)}$. Finally let $\mathcal{E}^{(i+1)}$ be the result of applying reduction step (1) and then (2) to $\mathcal{E}^{(i)}$. If G is acyclic then there exists an l such that $\mathcal{E}^{(l+1)} = \emptyset$.

When the hypergraph is acyclic, the maximum entropy distribution, $P^*(V)$, is given by: [Ma85]

$$P^*(V) = \left(\prod_{k=0}^{l-1} \frac{\prod_i P(E_i^{(k)})}{\prod_i P(Y_i^{(k)})} \right) \left(\prod_i P(E_i^{(l)}) \right) \tag{5}$$

Note that no α's are needed; the formula depends only on probabilities in the original input data (constraints). This formula is an immediate extension of the following theorem.

Theorem 1 [Ma85] *Given a decomposition,* $\mathcal{E} = \{E_1, \ldots, E_m\}$, *the maximum* *entropy distribution is given by the following.*

$$P^*(V) = \frac{P(E_1)\cdots P(E_m)}{P(Y_1)\cdots P(Y_m)} P^*(Y)$$

where $P^*(Y)$ is the maximum entropy distribution for the constraints $P(Y_1), \ldots,$ (Y_m).

Proof: From the marginal constraints we have the following

$$\begin{aligned} P(E_i) &= \sum_{V-E_i} \alpha_1 \cdots \alpha_m \\ &= \alpha_i \sum_{V-E_i} \prod_{j \neq i} \alpha_j \end{aligned} \tag{6}$$

Similarly we have,

$$\begin{aligned} P(Y_i) &= \sum_{Z_i} \sum_{V-E_i} \alpha_1 \cdots \alpha_m \\ &= \left(\sum_{Z_i} \alpha_i \right) \left(\sum_{V-E_i} \prod_{j \neq i} \alpha_j \right) \end{aligned} \tag{7}$$

Let $\beta_i = \sum_{Z_i} \alpha_i$. Combining equations (6) and (7) from above gives:

$$\alpha_i = \frac{P(E_i)}{P(Y_i)} \beta_i \tag{8}$$

Now writing $P^*(V)$ in its product form we get

$$\begin{aligned} P^*(V) &= \alpha_1 \cdots \alpha_m \\ &= \frac{P(E_1)\cdots P(E_m)}{P(Y_1)\cdots P(Y_m)} \beta_1 \cdots \beta_m \end{aligned} \tag{9}$$

We want to show that $\psi(Y) = \beta_1 \cdots \beta_m$ is $P^*(Y)$, the maximum entropy distribution for the constraints $P(Y_i)$. To do this, it is suffices to prove that the marginal constraints hold.

$$\begin{aligned} P^*(E_i) &= \sum_{V-E_i} \left(\prod_j \frac{P(E_j)}{P(Y_j)} \right) \psi(Y) \\ &= \sum_{V-E_i} \psi(Y) \frac{P(E_i)}{P(Y_i)} \prod_{j \neq i} \frac{P(E_j)}{P(Y_j)} \\ &= \frac{P(E_i)}{P(Y_i)} \sum_{Y-Y_i} \left(\psi(Y) \prod_{j \neq i} \frac{1}{P(Y_j)} \sum_{Z-Z_i} \left(\prod_{j \neq i} P(E_j) \right) \right) \end{aligned} \tag{10}$$

where $Z = Z_1 \cup \cdots \cup Z_m$, so that $V = Y \cup Z$. Now, since the Z_j's are disjoint,

$$\begin{aligned}\sum_{Z-Z_i}\prod_{j\neq i} P(E_j) &= \prod_{j\neq i}\sum_{Z-Z_i} P(E_j) \\ &= \prod_{j\neq i}\sum_{Z_j} P(E_j) \\ &= \prod_{j\neq i} P(Y_j) \qquad (11)\end{aligned}$$

Substituting equation (11) into equation (10) gives:

$$P^*(E_i) = \frac{P(E_i)}{P(Y_i)} \sum_{Y-Y_i} \psi(Y)$$

However since $P^*(E_i) = P(E_i)$ we get $P(Y_i) = \sum_{Y-Y_i} \psi(Y)$, so $\psi(Y)$ satisfies the constraints $P(Y_i)$. ∎

5 A New ME Method

In this section we present a new procedure for calculating the maximum entropy distribution. The main advantage of our procedure is that it avoids the iteration previously required by providing a direct formula for the desired answer. The major disadvantage is that the method cannot ordinarily be applied if the data is already tabulated and the constraints already derived; the method requires that one "plan ahead" and tabulate additional constraints when processing the data.

Equation (5) allows one to avoid iteration when calculating the maximum entropy distribution for schemas having acyclic hypergraphs. What should one do for *cyclic* hypergraphs? Our method is based on the observation that *a hypergraph can always be made acyclic by adding hyperedges.* (This is trivial to prove, since at worst a hypergraph can be made acyclic by adding the hyperedge containing all vertices.) For example, the hypergraph:

$$(\mathcal{V}, \mathcal{E}) = (\{ABCDEF\}, \{\{AB\}, \{ACD\}, \{DE\}, \{AEF\}\})$$

becomes acyclic when the hyperedge $\{ADE\}$ is added. Thus by adding additional constraints (edges) the maximum entropy calculation can be simplified so that no iteration is required. Here is a summary of how our method works:

1. We begin with a set of variables (attributes) and a set of constraint groups deemed to be of interest. (Cheeseman [Ch84] discusses a learning program which uses the raw data to find a set of significant constraints. Edwards and Kreiner [EK83] also discuss how to choose a good set of constraints.) Here a "constraint group" is a set of variables; the intent is that during data-gathering there will be one table created for each constraint group, and the observed events will be tabulated once in each table according to the values of the attributes in the constraint group. For example, if $\{A, B, C\}$ is a constraint group of three binary valued attributes, then there will be a table of size 8 used to categorize the data with respect to these three attributes. This will give rise to 8 constraints on the maximum-entropy distribution desired, one for each of the eight observed probabilities $P(ABC)$.

2. Construct the corresponding hypergraph $G = (\mathcal{V}, \mathcal{E})$, where there is one vertex for each variable and one hyperedge corresponding to each constraint group.

3. Perform Graham's algorithm on G, and let G' denote the resulting hypergraph. If G' is the empty hypergraph, then G is acyclic, and the following step is skipped.

4. Find a minimal set $\mathcal{X}$ of additional hyperedges (constraint groups) which can be added to G' to make it acyclic. Note that any original edges subsumed by edges in $\mathcal{X}$ are eliminated.

5. Collect data for the expanded set $\mathcal{E} \cup \mathcal{X}$ of constraints [1].

6. Apply equation (5) to calculate individual elements of the maximum entropy distribution. If sums of elements are desired, use Cheeseman's summation technique, choosing a good variable ordering. Note that instead of having a summation over a product of α's, here the summation is over a product of probabilities which are equivalent in form to the α's.

6 Possible Problems With Our Method

In this section we consider possible inefficiencies of our method. First, it may be necessary to add "large" hyperedges containing many vertices in order to make

[1]Our method is unusual in that it extends the set of tables (constraints) used to tabulate the data. To fill in the entries of a new table, the raw data must still be available in step (5). Thus steps 1-4 may be considered to be "planning" steps.

the hypergraph acyclic. For example, to make the complete undirected graph (containing all hyperedges of size two) acyclic, one must add the "maximum" hyperedge containing all vertices. Since the size of the table corresponding to a hyperedge is an exponential function of the size of the hyperedge, adding large hyperedges creates a problem. Furthermore, the table corresponding to the maximum hyperedge is itself the probability distribution that we are estimating, so the above situation is clearly undesirable. This kind of behavior depends on the structure of the hypergraph; hypergraphs which are "highly connected" will tend to require the addition of large hyperedges. However, when the graph is highly connected other computational techniques seem to "blow up" as well.

Second, because of our method's unique approach, we have a unique concern. Recall that since the data is tabulated *after* adding the additional constraints; steps 1-4 of our algorithm must be performed while the source of the constraints (i.e., the raw data) is still available. If the added hyperedges are too large, there may not be enough data to calculate meaningful statistics. Tabulating 100,000 data points in a table of size $\approx$ 1,000 will give reasonable estimates, while tabulating them in a table of size $\approx 1,000,000$ will not.

7 Comparison with Cheeseman's Method

We now compare the tables (constraints) added by our technique, with the intermediate tables used in Cheeseman's method. We demonstrate, by means of an example, that these are the same, except that Cheeseman's tables are half the size, since they are are already summed over the variable being eliminated.

When evaluating a sum one can visualize an imaginary "scan line" moving from left to right across the hypergraph, where all variables to the left of the scan line have already been summed over.

- According to the position of a scan line the vertices of the graph may be divided into three parts ("eliminated", "boundary", and "unseen") as follows:

    1. $V_E = \{v \in V \mid v \text{ is left of the scan line } \}$
    2. $V_B = \{v' \in V - V_E \mid \exists v \in V_E : (v, v') \in E \}$
    3. $V_U = V - (V_E \cup V_B)$

- Let $G_1, G_2, \ldots, G_l$ be the connected components of the subgraph induced by $V_E \cup V_B$.

- Let $\Gamma(G_i)$ denote the set of vertices of V_B in G_i.

- Let $\Gamma'(G_i)$ be that subset of $\Gamma(G_i)$ consisting of the vertices adjacent to some vertex in V_U or V_B.

Theorem 2 *Let $X_1, X_2, \ldots, X_l$ be a collection of subsets of V_B. Then $\{X_j\}$ is sufficient and necessary for Cheeseman's algorithm if*

$$(\forall i)(\exists j) \mid \Gamma'(G_i) \subseteq X_j. \tag{12}$$

Now we will look at the following example:

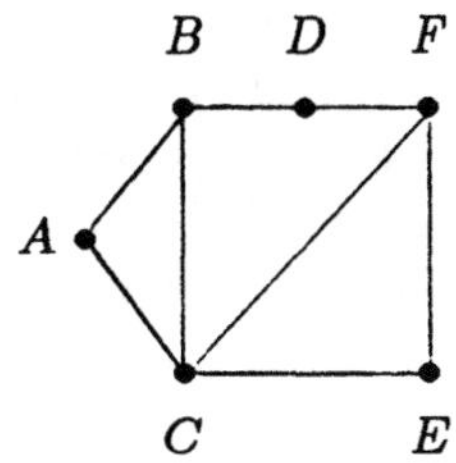

To explain this example we introduce new notation, where the variable sets for the intermediate tables are put in parenthesis above the summation sign. Consider the vertex ordering $DEFCAB$; given such a vertex ordering, theorem 2 specifies the temporary tables needed.

$$P^*(\emptyset) = \sum_{AB} \alpha_{AB} \overset{(AB)}{\sum_{C}} \alpha_{AC}\alpha_{BC} \overset{(BC)}{\sum_{F}} \alpha_{CF} \overset{(BF,CF)}{\sum_{E}} \alpha_{CE}\alpha_{EF} \overset{(BF)}{\sum_{D}} \alpha_{BD}\alpha_{DF}$$

(This summation is only being used for explanatory purposes. Since the elements of the probability distribution sum to 1, we know that $P^*(\emptyset) = 1$.) When evaluating $\sum_D$ it will be necessary to keep a table with the value of $\alpha_{BD}\alpha_{DF}$ for Ω_{BF}. When evaluating $\sum_E$ as well as keeping the table for B and F, another table with all combinations of C and F must be created. Below are the temporary tables used by Cheeseman's method.

$$BF, CF, BC, AB$$

Now we look at the hyperedges which must be added to make a hypergraph acyclic.

- Let $v_1, \ldots, v_n$ be an ordering of the vertices.

- Let $G_i = \begin{cases} G & \text{if } i = 0 \\ G_{i-1} + \phi(v_i) - \{e \mid v_i \in e\} & \text{otherwise} \end{cases}$

- Let $\phi(v_i) = \{v \mid v \text{ is adjacent to } v_i \text{ in } G_{i-1}\}$.
- Let $\Phi(v_i) = \begin{cases} \emptyset & \text{if } \phi(v_i) = \emptyset \\ \phi(v_i) \cup v_i & \text{otherwise} \end{cases}$

Theorem 3 *The set of all non-empty $\Phi(v_i)$ is necessary and sufficient to make a hypergraph, G, acyclic.*

Now we return to our example. Theorem 3 defines a set of additional hyperedges that make our hypergraph acyclic. First, the hyperedge BFD eliminates vertex D. (Hyperedge BFD subsumes hyperedges BD and DF and thus eliminates them. Now D is only in hyperedge BDF and so is eliminated, leaving hyperedge BF.) Second, hyperedge CFE eliminates vertex E. Now vertex F is only in hyperedges BF and CF so BCF eliminates it. Finally, hyperedge ABC eliminates the remaining vertices. Thus the following additional hyperedges will make our example graph acyclic (in parenthesis is the variable eliminated by adding the edge):

$$BF(D), CF(E), BC(F), AB(C)$$

Ignoring the variables in parentheses, these are identical to Cheeseman's tables. However there are important differences between these methods.

First, in terms of time complexity, Cheeseman's method specifies an iterative approximation of the αs, whereas our method requires no such iteration. So, if Cheeseman's method requires 10 iterations on the average, our method should yield an average speed-up of a factor of 10.

Second, in terms of space complexity, both methods use approximately the same amount of space. However, our method adds what might be called "permanent" edges, since they correspond to tabulations of the raw data. Note, however, that new edges may subsume and thus eliminate original edges, so the space required by our method may not be quite as great as it first appears. In Cheeseman' method the tables exist only temporarily during the course of the computation, and not all such tables may be needed at the same time.

And finally, in terms of the "precomputation" needed, both methods need to compute a vertex ordering to use. We observe that a good summation ordering is a good ordering for eliminating vertices. So the problem of choosing the hyperedges to make a graph acyclic seems comparable to the problem of choosing an optimal summation ordering.

8 Comparison to Pearl's Work

In this section we will compare our work to that of Pearl, and show that the formula we use is a proper generalization of Pearl's. Pearl's technique depends on the Bayesian network being singly-connected. A Bayesian network is a directed acyclic graph; such a network is said to be *singly-connected* if it has no *undirected* cycles (i.e., no cycles if we ignore the directions of the edges).

We begin by proving that a singly-connected network is a special case of an acyclic hypergraph. Then we show that both Pearl's and Malvestuto's formulas yield the same estimated probability distribution for a singly-connected network. Given a network N, we define a corresponding hypergraph G as follows. The vertices of G are the nodes of N. For each node x in N we create a hyperedge in G, consisting of the corresponding vertex and the vertices corresponding to all immediate predecessor of x in N.

Theorem 4 *If a Bayesian network N is singly-connected, then the corresponding hypergraph G is acyclic.*

Proof: Since N is singly-connected, it contains no undirected cycles, by the definition of singly-connected. Since an acyclic undirected graph is a tree or forest, N must contain some node s with degree at most one. The corresponding vertex in G is contained in exactly one hyperedge and so can be eliminated by reduction step 1. Finally, we must show that the reduced network N' so obtained corresponds to the reduced hypergraph G' that remains after eliminating the vertex corresponding to s. When s is a source in N (or a sink which is the second to last node) this correspondence is obtained immediately. In the remaining cases, the correspondence holds only after applying reduction step 2 to eliminate the hyperedge remaining after s is eliminated. (This hyperedge contains only the parent of s.) The network N' is singly-connected, and so by induction every node in G can be eliminated. Thus G is acyclic. ∎

Theorem 5 *A Bayesian network is not necessarily singly-connected if the corresponding hypergraph is acyclic.*

Proof: We prove this by means of an example. The following Bayesian network is not singly-connected since there is an undirected cycle.

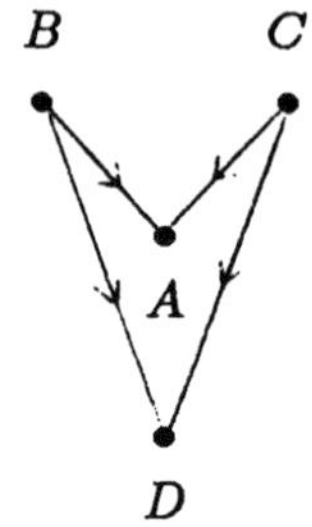

The hypergraph corresponding to the above network is acyclic as shown by the reduction below:

$$\{B, C, ABC, BCD\}$$
$$\Downarrow$$
$$\{BC\}$$
$$\Downarrow$$
$$\emptyset$$

■

Thus Pearl's condition of the network being singly-connected is a special case of having an acyclic hypergraph. So when given a singly-connected network, our technique will apply without adding any constraints.

Now we show (by example) that for singly-connected networks, Pearl's and Malvestuto's formulas for the estimated probability distribution are equivalent. We use the following notation for stating Pearl's formula.

- $\{f_x\}$ are the fathers (immediate ancestors) of node x in the network.
- $\mathcal{R}$ is the set of roots (sources).

Pearl's equation is

$$P^*(V) = \left(\prod_{x \in \mathcal{R}} P(x)\right) \left(\prod_{x \notin \mathcal{R}} P(x|\{f_x\})\right) \tag{13}$$

We use the following example to compare Pearl's and Malvestuto's formulas.

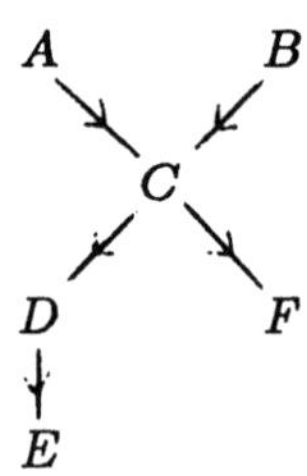

Pearl's formula gives:

$$\begin{aligned} P^*(A\ldots F) &= P(A)P(B)P(C|AB)P(D|C)P(F|C)P(E|D) \\ &= P(A)P(B)\frac{P(ABC)}{P(AB)}\frac{P(CD)}{P(C)}\frac{P(CF)}{P(C)}\frac{P(DE)}{P(D)} \end{aligned}$$

By definition A and B must be independent since otherwise there would be a link connecting them. Thus $P(AB) = P(A)P(B)$ and the above becomes

$$P^*(A\ldots F) = P(ABC)\frac{P(CD)}{P(C)}\frac{P(CF)}{P(C)}\frac{P(DE)}{P(D)} \tag{14}$$

To apply Malvestuto's formula, it is necessary to perform Graham's algorithm. (Elements of $Y_i^{(0)}$ are underlined.)

$$\begin{gathered} \{AB\underline{C}, \underline{CD}, \underline{C}F, \underline{D}E\} \\ \Downarrow \\ \{CD\} \\ \Downarrow \\ \emptyset \end{gathered}$$

Applying equation (5) gives:

$$P^*(A\ldots F) = \frac{P(ABC)}{P(C)}\frac{P(CF)}{P(C)}\frac{P(DE)}{P(D)}P(CD) \tag{15}$$

As shown by equations (14) and (15) the probability distribution given by Pearl's and Malvestuto's formulas are the same. So Pearl's formula is a special case of Malvestuto's formula. Thus our method is a proper generalization of the method given by Pearl. That is, not only does our technique give the same results as Pearl's when the network is singly-connected, but our formula applies (without adding any hyperedges) to cases where Pearl's does not.

9 Conclusions and Open Problems

We have presented an efficient algorithm for calculating the maximum entropy distribution given a set of attributes and constraints. Using a hypergraph to model the attributes and constraints, we show the benefits of making the corresponding hypergraph *acyclic*. We then show how to make a hypergraph acyclic by adding hyperedges (constraints). We have shown that our technique is at least as efficient as Cheeseman's method, and that our technique generalizes Pearl's method for singly-connected networks.

An open problem is how to choose the best set of hyperedges which will make a hypergraph acyclic; we conjecture that this problem is NP-complete. If so, then step (4) of our method cannot be done efficiently. One can use heuristics (such as a minimum-degree heuristic) to approximate the optimal answer, or maybe there is a pseudo-polynomial time algorithm in the size of the contingency tables corresponding to the edges of the optimal hypergraph.

We intend to try our technique on some realistic examples. Our goal is to determine if the size of the hyperedges will remain within reasonable limits for such realistic examples. We expect that in practice our new method will give substantial improvements in running time.

Finally, we will study the effects on accuracy of keeping tables which may be larger than the original tables.

References

[BFMMUY81] Berri, C., R. Fagin, D. Maier, A. Mendelzon, J.D. Ullman and M. Yannakakis, "Properties of Acyclic Database Schemas," in *Proc. 13th Annual ACM STOC* (1981), 355–362.

[BFMY83] Beeri, C., R. Fagin, D. Maier and M. Yannakakis, "On the Desirability of Acyclic Database Schemas," *J. ACM*, **30**,3 (1983), 355-362.

[Br59] Brown, D.T., "A Note on Approximations to Discrete Probability Distributions," *Information and Control*, **2** (1959), 386–392.

[Ch83] Cheeseman, P.C., "A Method For Computing Generalized Bayesian Probability Values For Expert Systems," in *Proc. Eighth International Conference on Artificial Intelligence* (August 1983), 198–202.

[Ch84] Cheeseman, P.C., "Learning of Expert Systems From Data," in *Proc. IEEE Workshop on Principles of Knowledge Based Systems* (1984), 115–122.

[CL68] Chow, C.K. and C.N. Liu, "Approximating Discrete Probability Distributions With Dependence Trees," *IEEE Trans. on Info. Theory*, **IT-14**,3 (May 1968), 462–467.

[Cs75] Csiszár, I., "*I*-Divergence geometry of probability distributions and minimization problems," *Annals of Probability*, **3**,1 (1975), 146–158.

[EK83] Edwards, D., and S. Kreiner, "Analysis of contingency tables by graphical models," *Biometrika* **70**,3 (1983), 553–565.

[Fi70] Fienberg, S.E., "An Iterative Procedure For Estimation In Contingency Tables," *The Annals of Mathematical Statistics*, **41**,3 (1970), 907–917.

[Ge86] Geman, S., "Stochastic Relaxation Methods For Image Restoration and Expert Systems," In Cooper, D.B., R.L. Launer, and E. McClure, editors, *Automated Image Analysis: Theory and Experiments*, New York: Academic Press, (to appear).

[Gr79] Graham, M.H., "On the Universal Relation," *University of Toronto Technical Report* (1979).

[IK68] Ireland, C.T., and S. Kullback, "Contingency tables with given marginals," *Biometrika* **55**,1 (1968), 179–188.

[Ja79] Jaynes, E.T., "Where Do We Stand On Maximum Entropy," In Levine and Tribune, editors, *The Maximum Entropy Formalism*, M.I.T. Press, (1979).

[Ja82] Jaynes, E.T., "On the Rationale of Maximum-Entropy Methods," *Proceedings of the IEEE*, **70**,9 (September 1982), 939–952.

[JS83] Johnson, R.W., and J.E. Shore, "Comments and corrections to 'Axiomatic derivation of the principle of maximum entropy and the principle of minimum cross-entropy'," *IEEE Trans. Inform. Theory* **IT-29**, 6 (Nov. 1983), 942–943.

[KK69] Ku, H.H. and S. Kullback, "Approximating Discrete Probability Distributions," *IEEE Trans. on Info. Theory*, **IT-15**,4 (July 1969), 444–447.

[Le59] Lewis, P.M., "Approximating Probability Distributions to Reduce Storage Requirements," *Information and Control*, **2** (1959), 214–225.

[Li86] Lippman, A.F., "A Maximum Entropy Method for Expert System Construction," PhD thesis, Brown University, Division of Applied Mathematics, (May 1986).

[Ma85] Malvestuto, F.M., "Approximating Discrete Probability Distributions: Easy and Difficult Cases," Unpublished Manuscript, (December 1985).

[Pe85] Pearl, J., "Fusion, Propagation and Structuring in Bayesian Networks," *University Of California, Los Angeles Dept. of Computer Science Technical Report CSD-850022 R-42*, (April 1985).

[RT78] Rose, D.J. and R.E. Tarjan, "Algorithmic Aspects of Vertex Elimination in Directed Graphs," *SIAM Journal Applied Math*, **24** (1978), 176–197.

[RTL76] Rose, D.J., R.E. Tarjan, and G.S. Lueker, "Algorithmic Aspects of Vertex Elimination on Graphs," *SIAM Journal Comput.*, **5**,2 (June 1976), 266–283.

[SJ80] Shore, J.E., and R.W. Johnson, "Axiomatic Derivation of the Principle of Maximum Entropy and the Principle of Minimum Cross-Entropy," *IEEE Trans. Inform. Theory*, **IT-26**,1 (Jan. 1980), 26–37.

[TTL84] Tikochinsky, Y., N.Z. Tishby, and R.D. Levine, "Consistent inference of probabilities for reproducible experiments," *Physical Rev. Letters* **52**, 16 (16 April 1984), 1357–1360.

[TY82] Tarjan, R.E. and M. Yannakakis, "Simple Linear-Time Algorithms to Test Chordality of Graphs, Test Acyclicity of Hypergraphs, and Selectively Reduce Acyclic Hypergraphs," *SIAM J. Comp.*, **13**,3 (August 1984), 566–579.

[Ya81] Yannakakis, M., "Computing the Minimum Fill-in is NP-Complete," *SIAM Journal Alg. Disc. Meth.*, **2**,1 (March 1981), 77–79.

Image Restoration and Reconstruction using Entropy as a Regularization Functional

Ali Mohammad-Djafari and Guy Demoment
Laboratoire des Signaux et Systèmes (CNRS-ESE-UPS)
Plateau du Moulon, 91190 Gif-sur-Yvette, France.

Abstract

Entropy has been widely used to solve inverse problems, especially when a complete set of data is not available to obtain a unique solution to the problem. In this paper we show that a Maximum Entropy (ME) approach is appropriate for solving some inverse problems arising at different levels of various image restoration and reconstruction problems :

- Reconstruction of images in X ray tomography,

- Reconstruction of images in ultrasound or microwave diffraction tomography (in the case of Born or Rytov approximation) from either Fourier domain data or directly from diffracted field measurements;

- Restoration of positive images by deconvolution when data is missing.

Our contribution is twofold :

i) Entropy has been introduced by various approaches: combinatorial considerations, probabilistic and information theory arguments, etc. But in real situations, we also have to take into account the presence of noise on the data. To do this, a χ^2 statistics is added to the entropy measure. This leads us to a novel interpretation of entropy as a particular choice of regularizing functional.

ii) In some of the above-mentioned applications, the object to be restored or reconstructed is a complex valued quantity. Consequently, we extend the definition of the entropy of an image which is considered as a function of $\mathbb{R}^2$ to $\mathbb{C}$.

The discussion is illustrated with some simulated results in X ray and diffraction tomography which allow a comparison between ME methods and some classical ones to be made. The computational cost and practical implementation of the algorithm are discussed.

G. J. Erickson and C. R. Smith (eds.),
Maximum-Entropy and Bayesian Methods in Science and Engineering (Vol. 2), 341–355.

1. Inverse problems considered

Entropy has been widely used to solve inverse problems, especially when a complete set of data is not available to obtain a unique solution to the problem. In this paper we show that a Maximum Entropy (ME) approach is appropriate for solving some inverse problems arising at different levels of various image reconstruction and restoration problems.

In all these problems we have to solve an integral equation of the first kind in the form

$$g(r,\phi) = \iint_D h(x,y,r,\phi)\, f(x,y)\, dx\, dy \qquad (1)$$

where (x,y) and (r,ϕ) are the coordinates of a point in a 2-D space, D is a closed region in this espace, g is called the **image** or **projections**, f is the **objet** and h is the **point-spread function** (PSF) or **kernel** of the imaging system. Some of the problems we are interested in are:

a) Reconstruction of images in X ray tomography:

In this case we have

$$h(x,y,r,\phi) = \delta(r - x \cos \phi + y \sin \phi) \qquad (2)$$

and the integral equation (1) becomes

$$g(r,\phi) = \iint_D f(x,y).\delta(r - x \cos \phi + y \sin \phi)\, dx\, dy = \int_L f(x,y)\, dl \qquad (3)$$

which is the Radon transform of $f(x,y)$. In this equation L is a straight line defined by $r=x \cos \phi + y \sin \phi$ and $g(r,\phi)$ is a projection of $f(x,y)$ on a line perpendicular to lines L.

b) Reconstruction of images in diffraction tomography:

If we make some approximations (such as the Born or Rytov ones) in the wave equation to linearize it [1-3], we have a relation between the diffracted field measured on a line $g(r,\phi)$ and the diffracting object $f(x,y)$ in the form of integral equation (1) in which h is given by

$$h(x,y,r,\phi) = -\frac{j}{4} H_0^{(1)} \left[jk_0\sqrt{(x_0-x')^2+(r-y')^2} \right] \exp[-jk_0 |x_0-x'|] \qquad (4)$$

where x_0 is a constant, $x'=x \cos \phi + y \sin \phi$, $y'= -x \sin \phi + y \cos \phi$ and $H_0^{(1)}$ is a first kind Hankel function.

c) Fourier synthesis problem :

In this case we have

$$g(r,\phi) = \iint_D f(x,y) \exp\{-j[x\, T_1(\Omega,\phi)+y\, T_2(\Omega,\phi)]\}\, dx\, dy \qquad (5)$$

where $T_1(\Omega,\phi)$ and $T_2(\Omega,\phi)$ are known functions which define a set of algebraic contours in (ω_x,ω_y) space given by:

$$\begin{pmatrix}\omega_x \\ \omega_y\end{pmatrix} = \begin{pmatrix}T_1(\Omega,\phi) \\ T_2(\Omega,\phi)\end{pmatrix} = \begin{pmatrix}\cos\phi & -\sin\phi \\ \sin\phi & \cos\phi\end{pmatrix}\begin{pmatrix}\beta(\alpha) \\ \alpha=\Omega\end{pmatrix} \qquad (6)$$

This problem is raised in many tomographic imaging applications when one tries to solve the inversion problems in the Fourier domain. For example in X ray tomography if one uses the FT of the projections as the data we have $\beta(\alpha)=0$ and the contours are the straight lines given by $\omega_x = -\Omega \sin\phi$ and $\omega_y = \Omega\cos\phi$. In diffraction tomography also one has a relation between the 1-D FT of the diffracted field and the 2-D FT of the object, when it is illuminated by a plane wave, and if we use these data we have $\beta(\alpha) = -k_0+\sqrt{k_0^2-\alpha^2}$ and the contours are the semi-circles with rays k_0 and centered on $\omega_x = -k_0\cos\phi$ and $\omega_x = -k_0 \sin\phi$.

This last problem is raised in many other imaging applications such as RMN tomography, radio-astronomy, synthetic aperture radar imaging, interferometry, etc.

2. Difficulties

In all these problems one disposes of $g(r,\phi)$ for a finite discrete set of ϕ_j, $j=1,\ldots m$ and the function $f(x,y)$ to be reconstructed is positive and has limited support. Solving these integral equations is in general an **ill-posed** problem and one must study not only the mathematical properties of **existence** and **uniqueness** of the solution, but also its **stability** with respect to errors on the data. When discretized, they give rise to the resolution of an under determined and ill-conditioned linear system of equations.

One must note also that these problems are under-determined even if we dispose of a continuum of data samples, i.e. in the case of a continuous problem. This is due to the fact that we dispose of the projections for a finite discrete set of ϕ_j.

To find a unique and stable solution to these problems, one must regularize the solution by introducing any prior information that one has about the solution. In these tomography applications one knows by the physical properties that the solution is positive and has limited support. Some information is also known about the noise on the data. In general it is assumed that the PSF is known exactly and that the noise is independent of the data and is just additive. The knowledge about the noise is often limited to its mean and its energy. In some applications these two quantities can be measured.

If all we know is limited to this information, the ME principle can help us to solve the problem. First, if we only have the mean and the energy of the noise, this principle tells us that the most non-committal hypothesis is that the noise is a Gaussian one. We can now use a χ^2 statistics to determine the feasible solutions, i.e. the solutions which are consistent with data. This can be done by defining a soft constraint $\chi^2 \leq$ cte on the solution, as is done by Daniell and Gull [4], Skilling [5,6,7], Wernecke and d'Addario [8], etc. But in general in many inverse problems the data are not sufficient to define a unique solution even if they are without noise. So the ME principle can be used to choose from among these feasible solutions one which has its entropy maximal. When the noise exists, the set of feasible solutions is just enlarged, but again one can choose the solution of maximum entropy.

The idea of using the ME principle to solve inverse problems is not new [9, 10], but we think that our contribution is a novel interpretation of the entropy as a particular choice of regularization functional in solving the inverse problem of Fourier synthesis arising in X ray or diffraction tomography in which the solution must be positive. We will now discuss this idea.

3. Regularization and the principle of ME.

When the integral equation (1), (3) or (5) are discretized one has to solve a system of linear equations of the form

$$\mathbf{d} = \mathrm{A}\,\mathbf{f} + \mathbf{b} \qquad (7)$$

where **d** is a vector containing the data, **f** a vector containing the unknown parameters of the problem, A a known matrix and **b** a vector containing the noise terms which are supposed to be Gaussian with zero mean and variance σ^2. In this system, A is in general an under-determined and ill-conditioned matrix.

One can define a regularized solution to this problem by minimizing a functional

$$J_\alpha(\mathbf{f}) = \| \mathrm{A}\,\mathbf{f} - \mathbf{d} \|^2 + \alpha\,\Omega(\mathbf{f}) \qquad \alpha>0 \qquad (8)$$

where $\| \,.\, \|^2$ is a norm in the space of observations to measure the residual error $\mathrm{A}\,\mathbf{f} - \mathbf{d}$ of the solution, and $\Omega(\mathbf{f})$ is a norm in the space of the parameters which controls the a priori information on the solution. When this a priori information can be translated by a quadratic regularization functional $\Omega(\mathbf{f})$, one has an explicit solution to the problem. This is the case, for example, when the information tells us that the solution is smooth and infinitely differentiable. But if the information is in the form of constraints such as positivity, which is the case in our applications, one must find another functional.

In the Bayesian approach to ME in image reconstruction, as explained first by Jaynes [11-14] and developed by other authors [5-8], one obtains the posterior probability

$$p(\mathbf{f} \mid D,\sigma,I_0) = \text{cte.} \exp[\, N\,(H-\lambda Q)] \quad \text{with} \quad \lambda = \frac{1}{N\sigma^2} \tag{9}$$

in which **f** is a given scene, D the set of data, σ^2 the variance of the noise, I_0 the prior information that all the scenes are equiprobable [14], N the total number of particles in image, H the entropy of the scene $\mathbf{f}=\{f_1,\ldots,f_n\}$ given by

$$H = -\sum_{j=1}^{n} f_j \, \mathrm{Log} f_j \tag{10}$$

and, finally, Q is given by

$$Q = \frac{1}{2}\sum_{i=1}^{m} \frac{1}{\sigma^2} \Big| d_i - \sum_{j=1}^{n} A_{ij} f_j \Big|^2 = \| A\,\mathbf{f} - \mathbf{d} \|^2 \tag{11}$$

Now maximizing $p(\mathbf{f} \mid D,\sigma,I_0)$ is equivalent to maximizing $H-\lambda Q$, or equivalent to minimizing

$$J_\alpha(\mathbf{f}) = Q(\mathbf{f}) + \alpha\,[-H(\mathbf{f})] \quad \text{with} \quad \alpha = \frac{1}{\lambda} = N\sigma^2 \tag{12}$$

Comparing this function to be maximized with the function given in (9) permits us to give another interpretation to this method. As one can see, we can consider $-H(\mathbf{f})$ as a regularization functional and can conclude that when the prior information on the solution of an inverse problem is given by a constraint of positivity, one can use the entropy of the solution as a regularization functional.

We must now verify the existence and uniqueness of the solution to this minimization problem. For this, it is not difficult to show that there is a number $\alpha_1>0$ for which $J_\alpha(\mathbf{f})$ with $0<\alpha<\alpha_1$ and $\mathbf{f}>\mathbf{0}$ has a unique minimum [2].

The major drawback to this choice for $\Omega(\mathbf{f})=-H(\mathbf{f})$ is that $H(\mathbf{f})$ is not a quadratic function of **f** and one has no explicit solution. However, the method can be implemented by an iterative method. There are many ways for solving this minimization problem. We have used a conjugate gradient technique and made some modifications to improve its performance and to adapt it to our applications.

In some of the above-mentioned applications, the objects to be restored or reconstructed are complex valued quantities. Consequently, we extend the definition of the entropy of an image which is considered as a function of $\mathbb{R}^2$ to $\mathbb{C}$.

4. Definition of the entropy of a complex valued quantity

When the image is a discrete real and positive quantity, as is the case of optical image restoration or X ray tomography reconstruction problems, the entropy of an image $\mathbf{f}=[f_1,\ldots,f_j,\ldots,f_n]$ is expressed either as

$$H_1(\mathbf{f})=\sum_{j=1}^{n} \operatorname{Log} f_j \quad \text{or} \quad H_2(\mathbf{f})=-\sum_{j=1}^{n} f_j \operatorname{Log} f_j \tag{13}$$

In the case of ultrasounds or microwave tomography, the object to reconstruct may be either a real quantity (sound speed in ultrasound or induced current densities inside the object) or a complex quantity (propagation constant or complex permittivity of the object). In this latter case we must define the entropy of an object considered as a function of $\mathbb{R}^2$ to $\mathbb{C}$. In this case the entropy can be defined in the two following ways:

i) if the real f_j^R and imaginary f_j^I parts of the object f_j can be considered as two positive and independent quantities, the entropy is simply defined as

$$H_1(\mathbf{f})=\sum_{j=1}^{n} (\operatorname{Log} f_j^R + \operatorname{Log} f_j^I) \quad \text{or} \quad H_2(\mathbf{f})=-\sum_{j=1}^{n} (f_j^R \operatorname{Log} f_j^R + f_j^I \operatorname{Log} f_j^I) \tag{14}$$

This is the case, for example, in microwave tomography, where the image to be reconstructed represents the distribution of the complex permittivity of an object illuminated with a monochromatic wave [1-3]. Actually, for a single and fixed frequency, the real part and the imaginary part of the complex permittivity of an object represent two different characteristics of the object and can be considered as independent quantities.

ii) if we are only interested by the module of f_j, the entropy is defined as

$$H_1(\mathbf{f})=\sum_{j=1}^{n} \operatorname{Log} |f_j| \quad \text{or} \quad H_2(\mathbf{f})=-\sum_{j=1}^{n} |f_j| \operatorname{Log} |f_j| \tag{15}$$

This is the case when the image represents the distribution of one axial coordinate of the induced current densities in the illuminated object.

5. Algorithm

We have implemented a general algorithm to solve the following problem:

Given $d_i = \sum_{j=1}^{n} A_{ij} f_j + b_i \sigma_i \quad i=1,\dots,m,$ estimate **f** by maximizing $J(\mathbf{f}) = H(\mathbf{f}) - \lambda Q(\mathbf{f})$,

where H(**f**) is the entropy expression given by one of the equations (14) to (16), and Q(**f**) is a quadratic expression of **f** given by

$$Q(\mathbf{f}) = \frac{1}{2} \sum_{i=1}^{m} \frac{1}{\sigma_i^2} | d_i - \sum_{j=1}^{n} A_{ij} f_j |^2 = [A\,\mathbf{f} - \mathbf{d}]^t D\, [A\,\mathbf{f} - \mathbf{d}] \tag{16}$$

with $$D = \mathrm{diag}[\frac{1}{\sigma_i^2}, \dots, \frac{1}{\sigma_i^2}] \tag{17}$$

λ is a constant which must be chosen so that the solution $\mathbf{f}^*$ satisfies the constraint $Q(\mathbf{f}^*) \leq M$. The optimum choice of λ needs a more complicated and time-consuming algorithm [6,7]. In our algorithm we choose it empirically at the initialization of the algorithm. Some discussion about this choice in different applications are given in [12].

The algorithm follows these sequences:

1) An initial estimate is calculated using $f_n^{(0)} = \frac{1}{m} \sum_{i=1}^{m} d_i A_{ij}^* = \frac{1}{m} A^{*t} \mathbf{y}$

where A_{ij}^* is the conjugate complex of A_{ij}. In the case of the Fourier synthesis problem $\mathbf{f}^{(0)}$ is the inverse FT of the data with all the lacking data set to zero.

2) At any iteration k :

a) apply the positivity constraint, i.e.

if $f_j \leq 0$ then $f_j = \varepsilon$ with ε a small positive value.

b) calculate $J(\mathbf{f}) = H(\mathbf{f}) - \lambda Q(\mathbf{f})$ and its gradient $\nabla J(\mathbf{f}) = \nabla H(\mathbf{f}) - \lambda \nabla Q(\mathbf{f})$

c) deviate the gradient;

if $f_j \leq 0$ then $f_j = \varepsilon$ and if $\frac{\partial J(\mathbf{f})}{\partial f_j} > 0$ then $\frac{\partial J(\mathbf{f})}{\partial f_j} = 0$

d) calculate a new estimate of f_j by the Conjugate Gradient.

e) Some criteria are calculated and tested to ensure the normal execution of the algorithm. For example

$$E_1=\frac{\sum |d_i-\tilde{d}_i|^2}{\sum |d_i|^2}=\frac{[\mathbf{d}-\tilde{\mathbf{d}}]^{*t}[\mathbf{d}-\tilde{\mathbf{d}}]}{\mathbf{d}^{*t}\mathbf{d}}=\frac{\|\mathbf{d}-\tilde{\mathbf{d}}\|}{\|\mathbf{d}\|}<\varepsilon_1 \qquad \text{with } \tilde{\mathbf{d}}=A\,\mathbf{f}$$

or

$$E_2=\frac{\|\mathbf{f}^{(k)}-\mathbf{f}^{(k-1)}\|}{\|\mathbf{f}^{(k-1)}\|}<\varepsilon_2$$

can be used to stop the algorithm, and

$$\text{Test}=\frac{1}{2}\left|\frac{\nabla H}{|\nabla H|}-\frac{\nabla Q}{|\nabla Q|}\right|$$

is used to verify the parallelism between the two components of the function to be maximized.

The algorithm takes in general about 10 to 30 iterations to give a satisfying result. In each iteration one needs to calculate J and its gradient. For this one needs to calculate $\tilde{\mathbf{d}}=A.\mathbf{f}$ and $A^t.[\mathbf{d}-\tilde{\mathbf{d}}]$. To calculate a new estimate of **f** by conjugate gradient one needs to calculate J on two other points. So in each iteration one needs to go, 3 times, from image space to projection space, and, once, come back to image space. Thus the algorithm needs about 30 to 100 times more cpu time than a usual linear interpolation method.

6. Simulations

We have applied this method to the solution of some inverse problems arising at different levels of various image reconstruction problems in X ray and diffraction tomography. In this section we give some of these results.

6.1- X ray tomography:

In these simulations we chose a mathematical phantom of the same form as the phantom used by Shepp and Logan [15] as an object. We calculated the projections by considering the participation of a pixel in a ray to be 0 or 1 depending on the position of its centre in the ray, and considered 3 cases:

i) In the first case we used 12 projections uniformly distributed around the object,

ii) in the second case we limited the number of projections to 8, and

iii) in the third case we limited not only the number of projections to 5 but also limited them in angle to 90°. This means that we calculated the projections at 45°,67.5°,90°,

112.5° and 135°.

Given these data we proceeded to reconstruct directly from them. To compare these results with a linear method and to analyse the effect of the entropy expression, we made the reconstructions by our algorithm in the three following cases:

i) Normal procedure of the algorithm (we call it **ME**);

ii) Keeping the entropy term to zero but applying the positivity constraint. This can be considered as an iterative least squares (LS) method with the positivity constraint which is regularized by choosing a criterai to stop the iterations before the algorithm diverges (we call it **LSP**);

iii) Keeping the entropy term to zero without applying the positivity constraint. This can be considered as an iterative least squares method (we call it **LS**).

Figure 1 shows these results with the original and noisy data. These results are obtained after 20 iterations.

6.2- Diffraction tomography:

In these simulations we limited ourselves to the Fourier Synthesis part of the problem. To simulate the diffraction tomography imaging we used three different objects and calculated their FT on semi-circles. Then, given these data, we made the reconstruction in the same three cases as in A.

For these simulations we considered two cases:

i) In the first case we used 16 projections uniformly distributed around the object,

ii) in the second case we limited the number of projections to 8, and their distribution in angle to 90°.

Figure 2 shows these results. In these figures, Np is the number of projections and, θ is the angle restriction.

6.3- Positive Image restoration by deconvolution:

In these simulations we used a phantom (object) which is blurred by a PSF to obtain a blurred image. We added some noise to this image and used our algorithm to restore the original object either from the just blurred or the blurred and noisy images (figure3). Figures 4 and 5 show these results. In figure 4 we show a comparison of the results obtained by our **ME** algorithm with the results obtained by two linear Kalman filtering methods developed in our laboratory. In figure 5 we show the results of restoration in the following situations: **a**) using all the data in image; **b**) using only 1/2 of the data; **c**) using only 1/4 of the data; and **d**) using only 1/9 of the data (one row over three and one column over three).

6.4- Conclusions on the simulations

In all these simulations one can see that the results obtained by the ME method have a greater spatial resolution and a better precision in amplitudes. If we keep the entropy term to zero but apply the positivity constraint on the solution at each iteration we obtain results which are as good as those obtained by ME when the data are not noisy. But one must note that in general a LS method does not give a unique solution even if we constrain the solution to be positive. In the presence of noise these results are not so good and there are many abnormal bright points on the results. More important is that if we continue the iterations the algorithm diverges. Finally, when one just uses an LS method, the algorithm diverges more rapidly than the LSP method.

In the case of deconvolution results one can also see that the linear methods of Kalman filtering give results which are very smooth and their spatial resolution poor. This is not surprising because in these methods one uses a regularization functional which assumes that the solution must be smooth.

7. Conclusions

Entropy can be used as a regularization functional to solve inverse problems in which the solution must be positive, and especially when a complete set of data is not available to obtain a unique solution.

We discussed this idea and illustrated it with some simulated results in X-ray tomography, diffraction tomography and positive image restoration.

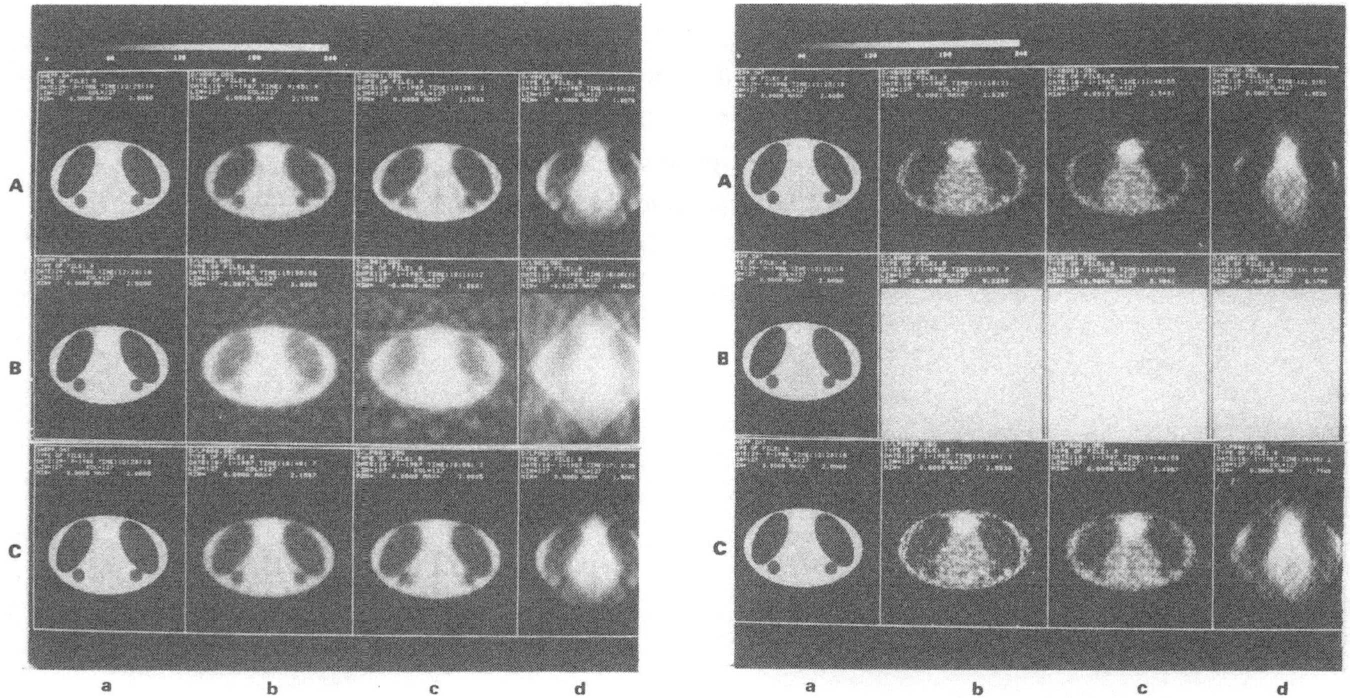

Figure 1: Reconstruction from limited projections in X ray tomography:

a) original objects, **b**) reconstructions when Np=12 and θ=180°, **c**) reconstructions when Np=8 and θ=180°, and **d**) reconstructions when Np=5 and θ=90°.

A: Maximum Entropy (**ME**), **B:** Least Squares (**LS**),

C: Least Squares with positivity constraint (**LSP**)

Left: Data without noise, **Right:** Noisy data

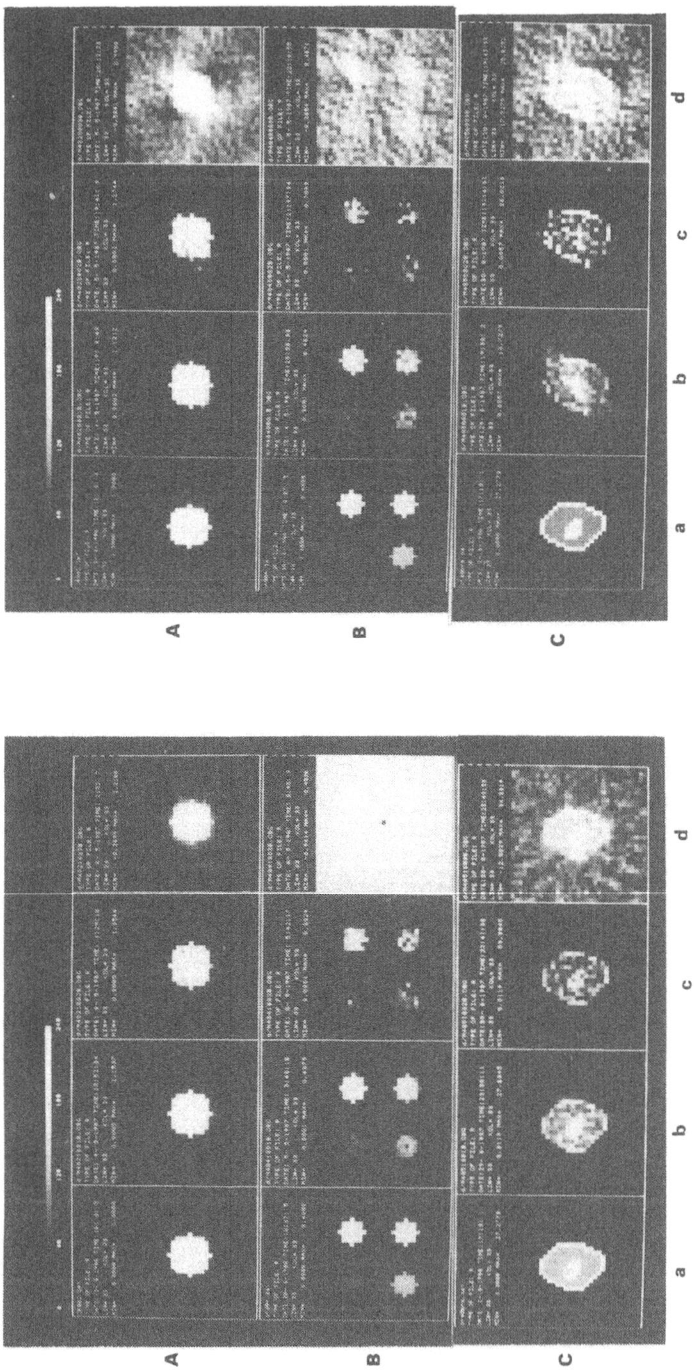

Figure 2: Reconstruction from the FT of the diffracted field in diffraction tomography:
a) original objects, **b)** reconstructions by **ME,** **c)** reconstructions by **LSP** and **d)** reconstruction by **LS.**

Left: Np=16 and θ=360°, **Right**: Np=8 and θ=90°.

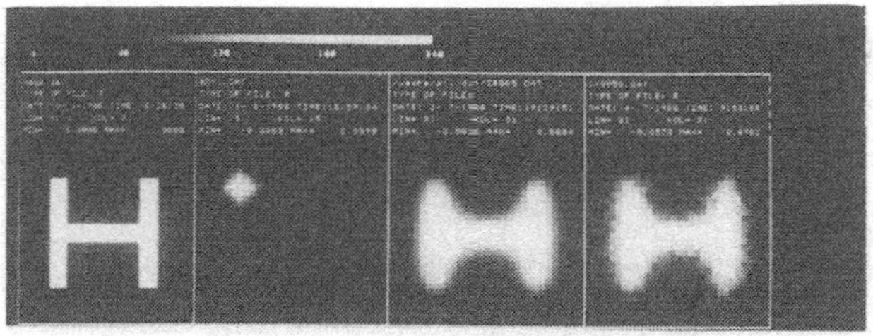

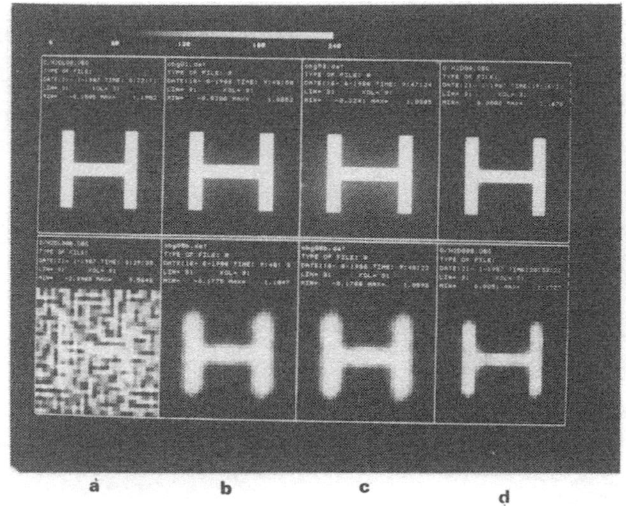

Figure 3: Positive image restoration:

a) original image, **b**) PSF, **c**) blurred image and **d**) blurred and noisy image.

Figure 4: Positive image restoration:

a) Deconvolution by **LS**, **b**) Deconvolution by an optimal Kalman filter,

c) Deconvolution an sub-optimal Kalman filter and **d**) Deconvolution by **ME**.

A: Data without noise, **B:** Noisy data

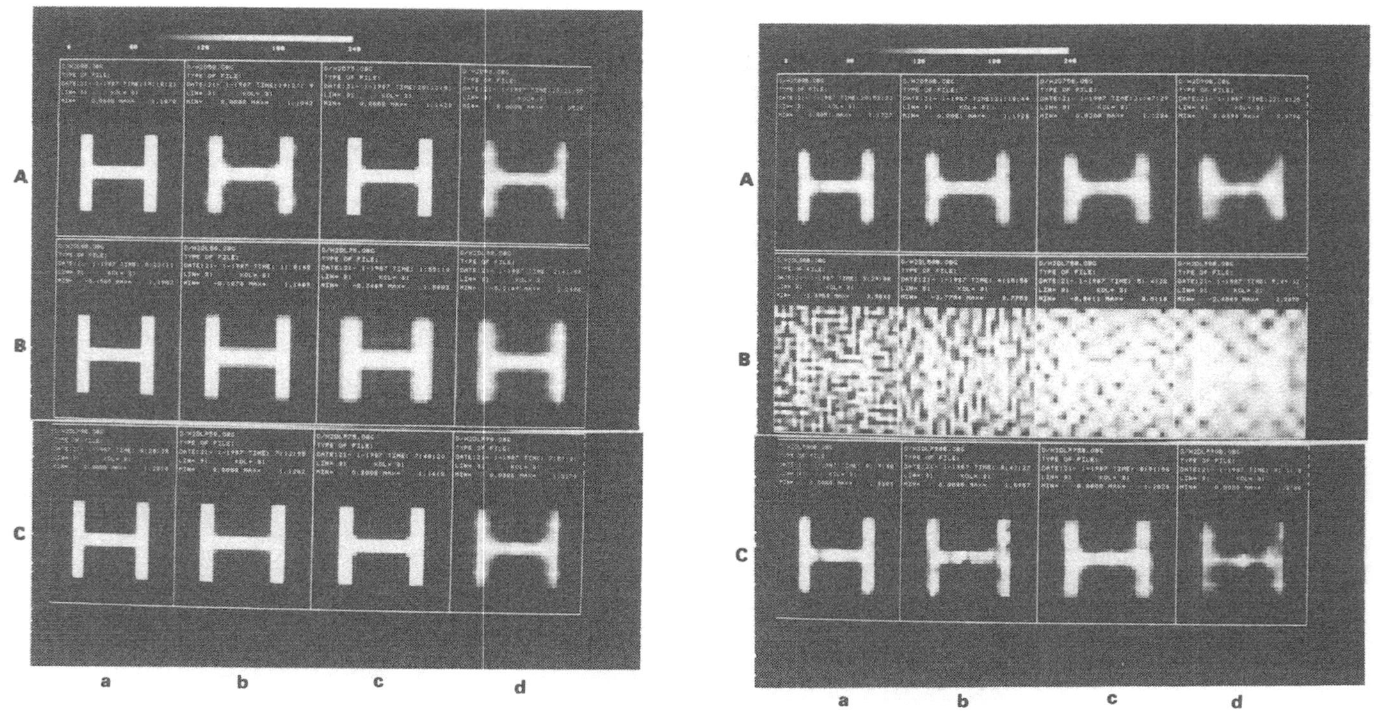

Figure 5: Restoration of positive images with missing data:
a) Restoration with all the data, **b**) Restoration with 1/2 of the data,
c) Restoration with 1/4 of the data and **d**) Restoration with 1/9 of the data.

A: Restorations by **ME**, **B:** Restorations by **LS**, **C:** Restorations by **LSP.**
Left: Data without noise, **Right:** Noisy data

8. References

[1] A. Mohammad-Djafari and G. Demoment, “Maximum entropy diffraction tomography,” Proc. of ICASSP, **Tokyo**, pp:34.7.1-34.7.4, 1986.

[2] A. Mohammad-Djafari and G. Demoment, “Tomographie de diffraction et synthèse de Fourier à maximum d'entropie,” Revue Phys. Appl., Vol. **22**, pp:153-167,1987.

[3] A. Mohammad-Djafari and G. Demoment, “Maximum entropy Fourier synthesis with application to diffraction tomography,” Applied Optics, Vol.**26**, No. 10, 1987.

[4] G.J. Daniell and S.F. Gull, "Maximum Entropy algorithm applied to image enhancement," Proc. Inst. Electr. Eng., **127E**, 170, 1980.

[5] J. Skilling and S.F. Gull, "Maximum entropy method in image processing," IEE Proc., **131**, 646, 1984.

[6] S.F. Burch, S.F. Gull and J. Skilling, "Image restoration by a powerful maximum entropy method," Comput. Vision & Image Proc., **23**, 113-128, 1984.

[7] J. Skilling and S.F. Gull, “The Entropy of an Image,” SIAM-AMS Proceedings, **14**, 167-189, 1984.

[8] S.J. Wernecke and L.R. D'Addario, "Maximum Entropy Image Reconstruction," IEEE Trans. Comput., **C-26**, 351, 1977.

[9] Hunt B.R. , “Bayesian Methods in Nonlinear Digital Image Restoration,” *IEEE Trans.*, **C-26**, pp. 219-229, 1977.

[10] Frieden B. , “Statistical Models for the Image Restoration Problem,” *Comp. Graph. & Imag. Proc.*, **12**, pp. 40-59, 1980.

[11] E.T. Jaynes, "On the Rationale of Maximum-Entropy Methods," Proc.IEEE, **70**, 939, 1982.

[12] E.T. Jaynes, "Where do we go from here?," in Maximum-entropy and Bayesian Methods in Inverse Problems, Ray Smith and Grandy eds. (1985).

[13] E.T. Jaynes, "Bayesian methods: general backgrounds," in Maximum-entropy and Bayesian Methods in Applied Statistics, J.H. Justice eds. (1986).

[14] E.T. Jaynes, "Monkeys, Kangaroos and N," in Maximum-entropy and Bayesian Methods in Applied Statistics, J.H. Justice eds. (1986).

[15] L.AL Shepp and B.F. Logan, "Fourier reconstruction of a head section," IEEE Trans. on Nucl. Sci., **NS-21**, 21-31, 1974.

APPLICATION OF LIKELIHOOD AND ENTROPY FOR TOEPLITZ CONSTRAINED COVARIANCE ESTIMATION

Michael I. Miller [†]
Department of Electrical Engineering
Electronic Systems and Signals Research Laboratory
Washington University in St. Louis
St. Louis, Missouri 63130

1. Abstract

For the class of likelihood problems resulting from a complete-incomplete data specification in which the complete-data **x** are nonuniquely determined by the measured incomplete-data **y** via some many-to-one set of mappings **y**=**h**(**x**), it is shown that the density which maximizes entropy is identical to the conditional density of the complete data given the incomplete data which would be derived via rules of conditional probability. It is precisely this identity between the *maxent* density and the conditional density which results in the fact that maximum-likelihood estimation problems may be solved via an iterative joint-maximization of the sum of the entropy plus expected log-likelihood. It is demonstrated that for the problem of spectrum estimation from finite data sets, this view results in the derivation of the maximum-likelihood estimates of the Toeplitz constrained covariance parameters via an iterative maximization of the likelihood function.

2. Introduction

There has recently been a tremendous increase in the application of maximum-entropy techniques to constraint problems with nonunique solutions[1-5]. The rational, as first proposed by Jaynes[6] is that of all candidates consistent with a set of constraints the maximum-entropy *(maxent)* solution is the one which occurs with greatest multiplicity. The success of the entropy function is due to the property that the candidate solutions are concentrated strongly near the maxent one; solutions with appreciably lower entropy are atypical of those specified by the data[7]. The fact that entropy methods have been successful for the solution of underdetermined inference problems suggests that these methods may play an important role in the solution of maximum-likelihood (ML) parameter estimation problems. In particular, problems encompassed by a *complete-incomplete* data specification, in which the measured data specifies many possible complete data sets over which the estimates may be obtained via maximization of the likelihood seem particularly well suited for entropy techniques. For the problems examined in this paper, a function is estimated which parameterizes a known probability density; the actual data (denoted as the *complete-data)* described by the density are not observed. Rather, observations consist of data (denoted as *incomplete-data*) which nonuniquely specifies the complete-data via some set of many-to-one mappings.

Our motivation is that we have been working on problems in image reconstruction and spectrum estimation in which parameters are estimated from measurements which are both noisy, i.e. samples of a stochastic process, as well as incomplete[8-13]. For the former, images are reconstructed which are the intensity of a Poisson process; due to errors introduced by the measurement device, the data do not uniquely specify the point-process. In the spectrum estimation problem, the Toeplitz covariance of a Gaussian random process is estimated from finite measurements of a stationary process. For both problems, ML techniques are an obvious choice for generating the parameter estimates; however, the fact that the measured data do not uniquely determine the underlying stochastic processes suggests that entropy may play a key role. In fact, both entropy and likelihood approaches have been applied for the solution of these problems[1-5, 8, 14, 15].

[†] This work was supported by the NSF via a Presidential Young Investigators Award Grant No. ECE-8552518.

G. J. Erickson and C. R. Smith (eds.),
Maximum-Entropy and Bayesian Methods in Science and Engineering (Vol. 2), 357–362.

The classical entropy formulation is stated as follows. Given a prior density f, the maxent density to which we refer is the maxent density q maximizing the entropy

$$E(q,f) = -\int_D q(\mathbf{x})\log\left[\frac{q(\mathbf{x})}{f(\mathbf{x})}\right] d\mathbf{x} \ , \tag{1a}$$

subject to the constraints $\mathbf{H}$ fixing mean values of the observation function given by

$$\mathbf{H} = \int_D q(\mathbf{x})\mathbf{h}(\mathbf{x})d\mathbf{x} \ . \tag{1b}$$

The maxent density $\hat{q}$ is

$$\hat{q}(\mathbf{x}) = \frac{\exp(\mathbf{v}^t\mathbf{h}(\mathbf{x}))\, f(\mathbf{x})}{Z(\mathbf{v})} \ , \qquad \text{for } \mathbf{x}\in D \ . \tag{1c}$$

The Lagrange multiplier vector $\mathbf{v}$ is chosen so that $\hat{q}$ satisfies the constraints of (1b) over its support set D, with t denoting matrix transpose. Alternatively the negative of E has been called the I-divergence, K-L number, or cross-entropy between q and f [16, 17].

The problem we set up is similar to that which generates $\hat{q}$ of (1c) in that we assume a prior density $f(\mathbf{x};\phi)$ describing the complete-data $\mathbf{x}$, parameterized by some function ϕ, and observations $\mathbf{y} = \mathbf{h}(\mathbf{x})$ where $\mathbf{h}(\mathbf{x})$ is a many-to-one vector mapping from the complete data observations. We are interested in finding the maxent density $\hat{q}$ although we do not assume that the observations $\mathbf{y}$ provide moment constraints on $\mathbf{h}$. For the estimation problems in which we are involved the set of observations $\mathbf{h}(\mathbf{x}_i)$ for $i=1,\cdots,N$ is small so that the average of $\mathbf{h}(\mathbf{x}_i)$ may not be close to its expectation. To anticipate our results on the use of the entropy function for the finite observation problem, we show that by viewing the incomplete data $\mathbf{y}$ as restricting the domain over which the maxent density is defined, rather than as a moment constraint on $\mathbf{h}(\mathbf{x})$, the density maximizing $E(q,f)$ is identical to the conditional density derived via formal rules of conditional probability. This equivalence results in the fact that a large class of ML problems may be posed as a joint-maximization of the entropy function.

3. Generation Of The Complete-Incomplete Data Model

We begin by defining the underlying probability space $\left[\Omega,\sigma(\Omega),P\right]$, with sample points ω in Ω, events in the sigma field $\sigma(\Omega)$ of subsets of Ω, and probability measure P. Then we define the *complete-data* random variable X as a measurable function so that $X:(\Omega,\sigma(\Omega))\rightarrow(\mathbf{D},\sigma(\mathbf{D}))$, with the probability of an event $B\in\sigma(\mathbf{D})$ given by $P_X(B)=P\{\omega:X(\omega)\in B\}$. We shall for the entire paper *assume* that $P_X()$ is absolutely continuous with density $f(\mathbf{x};\phi)$. The family of densities $f(\mathbf{x};\phi)$ parameterized by ϕ we term the *complete-data* densities. We say that we are given *incomplete-data* if instead of observing $\mathbf{x}$ in $\mathbf{D}$, only the sample $\mathbf{y}$ is available, where $\mathbf{y} = \mathbf{h}(\mathbf{x})$ for some measurable m-dimensional vector mapping $\mathbf{h}$; this mapping is, in general, many to one, so $\mathbf{x}$ is not uniquely specified by $\mathbf{y}$. Thus the incomplete data $\mathbf{y}$ results from the existence of $m+1$ sample spaces, the complete data space $\mathbf{D}$ and the m incomplete data spaces $\mathrm{Y}_1,\mathrm{Y}_2,\cdots,\mathrm{Y}_m$. We denote the product space describing the incomplete data vector $\mathrm{Y} = \mathrm{Y}_1 \times \mathrm{Y}_2 \times \cdots \times \mathrm{Y}_m$.

The complete data $\mathbf{x}$ are a particular realization from $\mathbf{D}$, and the incomplete observed data $\mathbf{y}$ are a particular realization from Y. Therefore, the many-to-one mapping $\mathbf{h}(\mathbf{x})$ taking $\mathbf{D}$ to Y specifies the subset $\mathbf{D}(\mathbf{y})\subset\mathbf{D}$ in which the complete data $\mathbf{x}$ is an element, with $\mathbf{D}(\mathbf{y})$ given by the following relation:

$$\mathbf{D}(\mathbf{y}) = \{\, \mathbf{x} : \mathbf{h}(\mathbf{x}) = \mathbf{y} \,\} \ . \tag{2}$$

The family of densities $g(\mathbf{y};\phi)$ describing the incomplete data are derived according to the following relation:

$$g(\mathbf{y};\phi) = \int_{\mathbf{D}(\mathbf{y})} f(\mathbf{x};\phi)d\mathbf{x} \ . \tag{3a}$$

The conditional density of $\mathbf{x}$ given $\mathbf{y}$ is then

$$k(\mathbf{x}\,|\,\mathbf{x}\in \mathbf{D}(\mathbf{y}),\phi) = \frac{f(\mathbf{x};\phi)}{\int_{\mathbf{D}(\mathbf{y})} f(\mathbf{x};\phi)d\mathbf{x}} \qquad \text{for } \mathbf{x}\in \mathbf{D}(\mathbf{y}) \; ; \tag{3b}$$

$$= 0 \qquad \text{for } \mathbf{x}\notin \mathbf{D}(\mathbf{y}) \; .$$

4. Maximum-Likelihood Posed as a Joint-Maximization of the Entropy Function

By incorporating the complete-incomplete model of the measurements $\mathbf{y}=\mathbf{h}(\mathbf{x})$ into the entropy formalism of (1) it follows directly that the density maximizing entropy $E(q,f)$ is identical to the conditional density $k(\mathbf{x}\,|\,\mathbf{x}\in \mathbf{D}(\mathbf{y}),\phi)$. Following the maximum-entropy approach given by (1a,b) we find the density q maximizing $E(q,f)$ subject to the constraints determined by the data $\mathbf{y}$? Since moment values on $\mathbf{h}$ do not exist, the solution of the incomplete data problem is different then that stated in (1). The data $\mathbf{y}$ determines the domain $\mathbf{D}(\mathbf{y})$ as given by (2) over which the complete-data are defined. Therefore, rather than specifying moment constraints on $\mathbf{h}(\mathbf{x})$ the data $\mathbf{y}$ specifies the domain D over which the maxent density has support, with the constraint given by $\int_{\mathbf{D}(\mathbf{y})} q(\mathbf{x})d\mathbf{x} = 1$. From Jensen's inequality, the density q maximizing entropy $E(q,f)$ subject to the support constraint

$$\int_{\mathbf{D}(\mathbf{y})} q(\mathbf{x})d\mathbf{x} = 1 \tag{4}$$

becomes

$$\hat{q}(\mathbf{x}) = \frac{f(\mathbf{x};\phi)}{\int_{\mathbf{D}(\mathbf{y})} f(\mathbf{x};\phi)d\mathbf{x}} \qquad \text{for} \quad \mathbf{x}\in \mathbf{D}(\mathbf{y}) \; . \tag{5}$$

This is precisely the conditional density denoted as $k(\mathbf{x}\,|\,\mathbf{x}\in \mathbf{D}(\mathbf{y}),\phi)$ in (3).

Now it follows directly that the MLE $\hat{\phi}$ may be posed as a joint-entropy maximization. The MLE is obtained by maximizing the log-likelihood of the incomplete data $\log g(\mathbf{y};\phi)$. Applying (3a,b) the log-likelihood of $\mathbf{y}$ is given by

$$\log g(\mathbf{y};\phi) = \log f(\mathbf{x};\phi) - \log k(\mathbf{x}\,|\,\mathbf{x}\in \mathbf{D}(\mathbf{y}),\phi) \; , \tag{6}$$

and evaluating the expectation of $\log g(.)$ in (6) with respect to the conditional density $k(\mathbf{x}\,|\,\mathbf{x}\in \mathbf{D}(\mathbf{y}),\phi)$ yields the following log-likelihood to be maximized:

$$\log g(\mathbf{y};\phi) = \int_{\mathbf{D}(\mathbf{y})} k(\mathbf{x}\,|\,\mathbf{x}\in \mathbf{D}(\mathbf{y}),\phi)\log f(\mathbf{x};\phi)d\mathbf{x} - \int_{\mathbf{D}(\mathbf{y})} k(\mathbf{x}\,|\,\mathbf{x}\in \mathbf{D}(\mathbf{y}),\phi) \log k(\mathbf{x}\,|\,\mathbf{x}\in \mathbf{D}(\mathbf{y}),\phi)d\mathbf{x} \; . \tag{7}$$

Since the density maximizing $E(q,f)$ is precisely the conditional $k(\mathbf{x}\,|\,\mathbf{x}\in \mathbf{D}(\mathbf{y}),\phi)$, the log-likelihood to be maximized becomes the following:

$$\log g(\mathbf{y};\phi) = \max_{\{q\}} \left\{ - \int_{\mathbf{D}(\mathbf{y})} q(\mathbf{x})\log\left[\frac{q(\mathbf{x})}{f(\mathbf{x};\phi)}\right] d\mathbf{x} \right\} \; ,$$

with q a density over $\mathbf{D}(\mathbf{y})$; that is $\int_{\mathbf{D}(\mathbf{y})} q(\mathbf{x})d\mathbf{x}=1$. The MLE is then simply $\hat{\phi}\leftarrow \underset{\{\phi\}}{argmax}\left\{ \max_{\{q\}} E(q,f(\phi))\right\}$. It follows that the MLE $\hat{\phi}$ is given by the following coupled maximizer equations:

$$\hat{\phi} \leftarrow \underset{\{\phi\}}{argmax}\left\{ \int_{\mathbf{D}(\mathbf{y})} \hat{q}(\mathbf{x})\log f(\mathbf{x};\phi)d\mathbf{x}\right\} \tag{8a}$$

$$\hat{q} \leftarrow \underset{\{q\}}{denmax}\left\{ -\int_{D(y)} q(\mathbf{x})\log\left[\frac{q(\mathbf{x})}{f(\mathbf{x};\hat{\phi})}\right]d\mathbf{x}\right\} . \tag{8b}$$

In (8a) "argmax" means the MLE $\hat{\phi}$ is the argument which maximizes the expectation of $\log f(\mathbf{x};\phi)$; the notation denmax means the density $\hat{q}$ maximizes $E(q,f)$.

Because of the equivalence between the conditional and maxent densities, the incomplete-data log-likelihood is simply the joint-maximum with respect to q,ϕ of the entropy function $E(q,f(\phi))$. This results in the estimation problem being expanded to what appears to be a larger problem in which both the parameters ϕ as well as density q must be estimated; this reformulation gives rise to an iterative solution of (8) by maximizing (8a) with respect to ϕ with $\hat{q}$ fixed, and then maximizing (8b) with respect to q with $\hat{\phi}$ fixed. This is precisely the alternating maximization sequence we shall use for the constrained Toeplitz covariance spectrum estimation problem, and is an instance of an expectation-maximization sequence first described by Dempster et al.[18].

5. Maximum-Likelihood Estimation of Toeplitz Constrained Covariances

As first shown by Burg, when given exact values of some finite set of the autocovariances of a stationary time-series, but no values of the actual time series, the maximum entropy density becomes a multivariate Gaussian constructed by maximizing the entropy function subject to covariance constraints[4,7,19]. For the spectrum estimation problem we now address, samples from the actual time series do exist, whereas the autocovariance values themselves are unknown. In order to apply the entropy theory, many investigators have generated second order statistics from the time series, from which moment constraints are assumed[4,20]. We instead derive maximum-likelihood estimates of the autocovariance parameters under the Gaussian model determined by the maximum-entropy rule. This involves the complete-incomplete data set-up in which the incomplete data correspond to the finite length measured sequence and the complete data correspond to the infinite duration sequence.

We assume that we are given an observed series $\{y_0,\cdots,y_{G-1}\}$ of length G, which is part of a larger stationary, zero-mean Gaussian periodic time-series of period $N>G$ where $y_0=y_N, y_1=y_{N+1},\cdots$, and we wish to find the maximum-likelihood estimates of the Toeplitz covariance matrix $\mathbf{K}$ of the N-periodic sequence. The choice of the finite period of length N is arbitrary, and we can allow it to grow when modeling stationary non-periodic processes. We solve for the maximum-likelihood estimates $\hat{\mathbf{K}}$. The proper choice for the *complete-data* becomes the entire set of samples $\{y_0,\cdots,y_{G-1},y_G,\cdots,y_{N-1}\}$ from the N-periodic time-series. We denote this complete-data by the N-dimensional vector $\mathbf{y}_N$, consisting of the given G-dimensional vector $\mathbf{y}_G$ corresponding to the set $\{y_0,\cdots,y_{G-1}\}$, augmented by the $N-G$ dimensional vector $\mathbf{y}_A$ corresponding to the set $\{y_G,y_{G+1},\cdots,y_{N-1}\}$. The many-to-one function $\mathbf{h}$ mapping the complete data $\mathbf{y}_N$ to the incomplete data $\mathbf{y}_G$ deletes all time series points corresponding to the augmented vector of length $N-G$, and is given by

$$\mathbf{y}_G = \mathbf{h}(\mathbf{y}_N) . \tag{9}$$

The complete-data density is

$$f(\mathbf{y}_N;\mathbf{K}) = (2\pi)^{-\frac{N}{2}} \det^{-\frac{1}{2}} |\mathbf{K}| \exp(-\frac{1}{2}\mathbf{y}_N^{\dagger}\mathbf{K}^{-1}\mathbf{y}_N) , \tag{10}$$

with $\dagger$ denoting hermitian transpose, det $|\ |$ denoting matrix determinant, and $\mathbf{K}^{-1}$ matrix inverse. We rewrite (10) as

$$f(\mathbf{y}_N;\mathbf{v}) = = \frac{\exp(-\sum_{k=0}^{N-1} v_k [\sum_{m=0}^{N-1} y(m)y^*(<m+k>_N)])}{\int\exp(-\sum_{k=0}^{N-1} v_k [\sum_{m=0}^{N-1} y(m)y^*(<m+k>_N)])d\mathbf{y}_N} , \tag{11}$$

where $< >_N$ denotes *modulo* N, and * denotes complex conjugate. Equation 11 is derived by defining the $N \times N$ circulant matrix $\mathbf{V}=\frac{1}{2}\mathbf{K}^{-1}$ whose entries are $\mathbf{V}_{ij}=v_{j-i}$ with $v_{j-i}=v^*_{i-j}$, $v_{j-i}=v^*_{N-(j-i)}$. Since $\mathbf{V}$ is related to the inverse of $\mathbf{K}$, finding the maximum-likelihood estimate of $\mathbf{V}$ yields the maximum-likelihood estimates of $\mathbf{K}$. Applying (8a) to (11) yields the following equation which the maximum likelihood estimates of the Toeplitz covariances must satisfy:

$$\hat{\mathbf{K}}(\tau) = \frac{1}{N}\sum_{m=0}^{N-1} E\{y(m)y^*(<m+\tau>_N) \mid \mathbf{y}_G,\hat{\mathbf{K}}\} \ . \tag{12}$$

The conditional expectation of (12) is with respect to the maximum-entropy density of (8b). Iteratively maximizing (12) using an iterative maximization of (8a) and (8b) yields the sequence of iterates $\{\mathbf{K}^{(p)}(\tau);p=1,2,\cdots\}$ given by

$$K^{(p+1)}(\tau) = \frac{1}{N}\sum_{m=0}^{N-1} E\{y(m)y^*(<m+\tau>_N) \mid \mathbf{y}_G,\mathbf{K}^{(p)}\} \ . \tag{13}$$

We have proven in another paper[21], that all of the limit points of the algorithm of (13) are stable, and satisfy the necessary maximizer conditions for a maximum-likelihood estimator. We have also shown that the performance of the ML estimator under the Toeplitz constraint is by far superior to conventional lag product estimators, and can from small segments of correlated Gaussian periodic processes (ones in which the spectra are fairly concentrated) yield estimates "as good as" estimators derived from complete data corresponding to full periods of the process.

References

1. S.F. Gull and G.J. Daniell, ''Image reconstruction from incomplete and noisy data,'' *Nature*, vol. 272, p. 686, 1978.
2. S.F. Gull and G.J. Daniell, ''The maximum entropy algorithm applied to image enhancement,'' *Proc. of the IEEE*, vol. 5, p. 170, 1980.
3. J. Skilling and R.K. Bryan, ''Maximum entropy image reconstruction,'' *MNRAS*. submitted 1982
4. J.P. Burg, *Maximum entropy spectral analysis,* Univ. Microfilms No. 75-25, Stanford University, Stanford,California, 1975.
5. J.E. Shore, ''Minimum cross-entropy spectral analysis,'' *IEEE Trans. on Acoustics,Speech and Signal Processing*, vol. ASSP-29, No.2, pp. 230-237, April 1981.
6. E.T. Jaynes, ''Information theory and statistical mechanics,'' *Phys. Rev.*, vol. 106, pp. 620-630, 1957.
7. E.T. Jaynes, ''On the rationale of maximum-entropy methods,'' *Proc. of the IEEE*, vol. 70, pp. 939-952, September 1982.
8. D. L. Snyder and D. G. Politte, ''Image reconstruction from list-mode data in an emission tomography system having time-of-flight masurements,'' *IEEE Trans. on Nuclear Science*, vol. NS-30, pp. 1843-1849, 1983.
9. D.L. Snyder and M.I. Miller, ''The use of sieves to stabilize images produced with the EM algorithms for emission tomography,'' *IEEE Trans. on Nuclear Science*, vol. NS-32, October 1985.
10. M. I. Miller, D. L. Snyder, and T. Miller, ''Maximum likelihood reconstruction for single photon emission computed tomography,'' *IEEE Trans. on Nuclear Science*, vol. NS-32,No.1, pp. 769-778, February 1985.
11. M.I. Miller, ''Algorithms for removing recovery related distortion from auditory-nerve discharge patterns,'' *J.Acoust. Soc. Am.*, vol. 77, pp. 1452-464, 1985.

12. M.I. Miller, N. Karamanos, and W.E. Bosch, "E-M algorithms for estimating parameters from single-memory Markov point-processes having a multiplicative intensity," *23rd Annual Allerton Conference*, University of Illinois, Urbana, Illinois, October 1985.

13. M.I. Miller, K.B. Larson, J.E. Saffitz, D.L. Snyder, and L.J. Thomas, Jr., "Maximum-likelihood estimation applied to electron-microscope autoradiography," *Journal of Electron Microscopy Techniques*, 1985.

14. J.P. Burg, D.G. Luenberger, and D.L. Wenger, "Estimation of structured covariance matrices," *Proc. of the IEEE*, vol. 70,9, pp. 963-974, September 1982.

15. L. A. Shepp and Y. Vardi, "Maximum-likelihood reconstruction for emission tomography," *IEEE Trans. on Medical Imaging*, vol. MI-1, pp. 113-121, 1982.

16. S. Kullback, in *Information Theory and Statistics*, Wiley,Dover, New York, 1959,1968.

17. I. Csiszar, "I-divergence geometry of probability distributions and minimization problems," *Ann. Prob*, vol. 3, pp. 146-158, 1975.

18. A. D. Dempster, N. M. Laird, and D. B. Rubin, "Maximum likelihood from incomplete data via the EM algorithm," *J. of the Royal Statistical Society*, vol. B,39, pp. 1-37, 1977.

19. A. van den Bos, *IEEE Trans. Inform. Theory*, vol. IT-17, pp. 493-494, 1971.

20. in *Modern Spectrum Analysis*, ed. D.G. Childers, New York: IEEE Press, 1978.

21. M.I. Miller and D.L. Snyder, "The Role of Likelihood and Entropy in Incomplete-Data Problems: Applications to Estimating Point-Process Intensities and Toeplitz Constrained Covariances," *Proceedings of the IEEE*, July 1987.

THE CONCEPT OF EPOCH ENTROPY IN COMPLEX SYSTEMS

K. L. Ngai

A. K. Rajagopal

S. Teitler
Naval Research Laboratory, Washington, D.C. 20375-5000

The concept of epoch entropy for relaxation processes is introduced and used to discuss relaxations in complex systems. Empirical rules for relaxation of many such systems are used in explicit evaluations of the epoch entropy. It is shown how by appropriate choices of scale factors the value of the epoch entropy may be the same for different conditions of relaxation. The relation of the empirical rules and the evaluations of epoch entropy to some existing models of relaxation are indicated.

I. INTRODUCTION

By way of introduction, we develop the concept of epoch entropy for relaxation in this first section. In the next section, after noting the empirical characteristics of relaxation in a large class of complex systems, we estimate the epoch entropy associated with relaxations in such systems. We also show how with appropriate choices of scale factors, the epoch entropy for relaxation of a particular quantity under different conditions can be made to have the same value. In the third, concluding section, we indicate how a class of existing models of relaxation in complex systems, some of which anticipated the empirical relations, provide a rationale for choosing scale factors in such a way as to maintain the value of the epoch entropy.

We turn first to the concept of epoch entropy. As shown by Bottcher and Bordewijk (1978), the relaxation function $\Phi(t)$ characterizing a stochastic relaxation process which is unity at the initial time and zero at infinite time may be interpreted as a survival probability with corresponding cumulative probability $P(t) = 1 - \Phi(t)$. A dimensionless time $\bar{t}$ may be defined by dividing the measured time by a scale factor with the dimensions of time. Then the derivative of the cumulative probability with respect to $\bar{t}$ is a dimensionless probability density $f(\bar{t})$. The dimensionless entropy over the epoch of the relaxation or the epoch entropy is then defined by

$$S_T = - \int_0^\infty d\bar{t}\; f(\bar{t}) \log f(\bar{t}) \qquad (1.1)$$

G. J. Erickson and C. R. Smith (eds.),
Maximum-Entropy and Bayesian Methods in Science and Engineering (Vol. 2), 363–369.

The epoch entropy is a differential entropy which is neither necessarily nonnegative nor invariant under a transformation of epoch parameter $\bar{t}$. However, these properties, rather than being drawbacks, add a flexibility that can lead to interesting physical interpretations. For the moment, let us consider the formal effect of a transformation of epoch parameter from $\bar{t}$ to $\tilde{z}$.

Here we use a tilde over the z to indicate that the scale factor used in forming the dimensionless $\tilde{z}$ may be different from the one used in forming $\bar{t}$. We write

$$\tilde{z} = \tilde{z}(\bar{t}) \tag{1.2}$$

and find

$$S_Z = S_T + \int d\bar{t} f(\bar{t}) \log (d\tilde{z}/d\bar{t}) \tag{1.3}$$

when $\tilde{z}$ and $\bar{t}$ share the same scale factor, it is clear that S_Z and S_T will be different for a nontrivial transformation $\tilde{z}(\bar{t})$. However, when $\tilde{z}$ and $\bar{t}$ have appropriately chosen different scale factors, the values of S_Z and S_T can be made to coincide. We shall consider such a possibility in the next section as part of evaluations and a discussion of epoch entropy for relaxation in complex systems.

II. RELAXATION IN COMPLEX SYSTEMS

When considering relaxation phenomena, one typically thinks of linear exponential decay

$$\Phi_E = \exp - W_o t = \exp - \bar{W}_o \bar{t} \tag{2.1}$$

Here W_o and $\bar{W}_o$ are respectively the dimensional and dimensionless constant relaxation rate. However, in many instances, the observed decay is not a linear exponential but rather a fractional exponential form Φ_K associated with the name Kohlrausch.

$$\Phi_K = \exp - (W^*z)^{1-n} = \exp - (\tilde{W}^*\tilde{z})^{1-n}, \qquad 0<n<1 \tag{2.2}$$

Here we have used $\tilde{z}$ as the dimensionless epoch parameter both to indicate that $\tilde{z}$ may be related as shown below to $\bar{t}$ by an epoch parameter transformation and that different scale factor may be employed.

It has now been established by Ngai et al (1986) that many relaxation phenomena in complex systems obey the following three empirical rules. For a given relaxing quantity, there is a critical time t_c such that:

(a) If a measured relaxation occurs for $t/t_c \ll 1$, the observed relaxation function is of the form $\Phi_E(t)$;
(b) If a measured relaxation occurs for $t/t_c \gg 1$, the observed relaxation function is given by a Kohlrausch

form $\Phi_K(t)$;
(c) The relaxation rates W_o and W^* are related by

$$W_o t_c = (W^* t_c)^{1-n}, \quad 0<n<1 \tag{2.3}$$

This means that when $\Phi_E(t)$ is observed, the relaxation process has run its course by the time t_c is reached, or in other words $W_o t_c \gg 1$. On the other hand, when $\Phi_K(t)$ is observed, there is only negligible relaxation before t_c and $W_o t_c \ll 1$ or by Eq. (2.3) $W^* t_c \ll 1$. Elsewhere (Ngai et al 1986), we have called such behavior respectively exponential and Kohlrausch dominated relaxation. Thus t_c plays the role of infinite time when the relaxation function is $\Phi_E(t)$, and the role of zero time when the relaxation function is $\Phi_K(t)$. We can therefore extend the time domain to $[0,\infty]$ in the evaluation of the epoch entropy for the empirical situation (a) and (b) above.

For situation (a) when the relaxation function has the form $\Phi_E(t)$, the epoch entropy is evaluated to be

$$S_E = \bar{W}_o \int_o^\infty d\bar{t} \{\bar{W}_o \bar{t} - \log \bar{W}_o\} \exp(-\bar{W}_o \bar{t}) = 1 - \log \bar{W}_o \tag{2.4}$$

Similarly for situation (b), the epoch entropy is evaluated to be

$$S_K = 1 - \log \tilde{W}^* - \log (1-n) - n\gamma/(1-n), \qquad 0<n<1 \tag{2.5}$$

Here $\gamma = 0.57721566...$ is the Euler constant. It is clear that both S_E and S_K depend on the respective time scale factors chosen to make $\bar{W}_o$ and $\tilde{W}^*$ dimensionless. For generality and later purposes, we allow the scale factors to be different, i.e., τ_E and τ_K. Then the difference in the two entropies may be expressed as

$$S_K - S_E = \log (\tau_E/\tau_K) + \log \{[W_o/W^*(1-n)] \exp - [n\gamma/(1-n)]\} \tag{2.6}$$

This difference is in general nonvanishing and depending on the empirical value of n and choice of τ_E/τ_K may be positive or negative. However, we note that if we put the choice of τ_E/τ_K aside, the difference is predominantly positive since by Eq. (2.3) $W_o/W^*>1$ for all $0<n<1$. Thus there is a mild paradox. The distribution $f_K(\tilde{z})$ corresponds to a maximum entropy distribution subject to two constraints; namely, averages of both $\log \tilde{z}$ and fractional monomial $(\tilde{z})^{1-n}$ are taken as given. On the other hand, the linear exponential distribution corresponds to a maximum entropy distribution subject only to the first moment as given. Thus a maximum entropy subject to two constraints may be greater than one subject to only one constraint. This of course is not really a paradox because the theorem (Levine 1980) specifying that increasing constraints reduces the maximum entropy requires the new constraints must be added to the already existing ones in a hierarchal manner. In the above example, the one constraint determining S_E is not included in the constraints determining S_K so the theorem does not apply.

Another formal point already indicated above is S_K may be obtained from S_E by a transformation of epoch parameter. Thus consider

$$\tilde{z} = (\bar{W}_o \bar{t})^{1/(1-n)} / \tilde{W}^* \qquad (2.7)$$

Then use of Eq. (1.3) will once again provide the equivalent of Eq. (2.6). Again the actual evaluation of the difference between S_K and S_E depends on the ratio W_o/W^* consistent with Eq. (2.3), the empirical parameter n, and the choice of scale factors τ_E and τ_K. An interesting possibility is to choose τ_E and τ_K in such a way that the values of S_E and S_K coincide. Using Eq. (2.6), we are then led to the condition

$$\frac{\tau_E}{\tau_K} = (1-n) \exp [n\gamma/(1-n)] \, W^*/W_o, \qquad S_E = S_K \qquad (2.8)$$

We may renormalize the τ_K scale factor by defining

$$\tau_K^* = (1-n) \exp [n\gamma/(1-n)] \tau_K \qquad (2.9)$$

Then τ_e/τ_K^* is just the inverse ratio of the corresponding constant relaxation rates. But the constant relaxation rates W_o and W^* may be viewed as the "natural" reciprocal scale factors that cause the respective arguments of the relaxation functions to be dimensionless. These scale factors may be viewed as natural since a change in scale factor does not change the form of the relaxation function. Indeed the respective relaxation functions then take the form.

$$\Phi(\bar{t}) = \exp - \bar{t} \qquad (2.10a)$$

$$\Phi_K(\tilde{z}) = \exp - \tilde{z}^{1-n} \qquad (2.10b)$$

Of course, if τ_E is taken to be W_o^{-1} and τ_K^* is taken to be W^{*-1}, then τ_E and τ_K^* are related by the equivalent of Eq. (2.3)

$$t_c/\tau_E = (t_c/\tau_K^*)^{1-n}, \quad 0<n<1 \qquad (2.11)$$

Further insight into nature of scaling in relaxation processes may be obtained by considering the instantaneous decrement rate that enters into some existing models of relaxation in complex systems. We discuss this in the next section.

III. MODELING CONSIDERATIONS

To provide context for our discussion here, we introduce

the instantaneous decrement rate $\lambda(t)$.

$$\lambda(t) = f(t)/\Phi(t) \tag{3.1}$$

For $\Phi_E(t)$, $\lambda_E(t) = W_o$ while for $\phi_K(z)$, $\lambda_K(z) = (1-n)W^{*(1-n)}z^{-n}$. The corresponding dimensionless instantaneous decrement rates are

$$\bar{\lambda}_E(\bar{t}) = \bar{W}_o \tag{3.2}$$

$$\tilde{\lambda}_K(\tilde{z}) = (1-n)\tilde{W}^{*(1-n)}\tilde{z}^{-n} \tag{3.3}$$

It may be noted that in general the expectation of the logarithms of the instantaneous decrement rate is simply related to the differential entropy.

$$E[\log \bar{\lambda}(\bar{t})] = 1-S \tag{3.4}$$

This follows because

$$\int_o^\infty d\bar{t} f(\bar{t}) \log [\Phi(\bar{t})] = - \int_o^1 d\Phi \log \Phi = -1 \tag{3.5}$$

This means our evaluations of the epoch entropy in the previous section was tantamount to evaluations of the expectation of the logarithm of the instantaneous decrement rate. These evaluations among other things depended on the respective constant relaxation rates W_o and W^*. When the scale factors are chosen such that $\tau_E = W_o^{-1}$ and $\tau^*_K = W^{*-1}$, $E[\log \lambda_E]$ is zero trivially and $E[\log \tilde{\lambda}_K(\tilde{z})]$ is made to have value zero.

The instantaneous decrement rate has been particularly emphasized by Ngai and coworkers et al (1986) in their development of models of relaxation in complex systems. In such models, the decrement rate is constant up to a time t_c when, by a physical mechanism described in the particular model, it becomes time dependent so that overall

$$\lambda(t) = \begin{cases} W_o, & t/t_c<1 \\ (1-n)W_o(t/t_c)^{-n}, & t/t_c>1 \end{cases} \tag{3.6}$$

In these models, the dressing of W_o at $t = t_c$ leads to a Kohlrausch form for the relaxation function with W^* predicted to have the form given in Eq. (2.3). Furthermore if W_o is such that $W_o t_c >> 1$, then the observed relaxation function has the form $\Phi_E(t)$. Thus all the empirical features cited in section II are encompassed in these models. The discontinuity in the decrement rate at $t = t_c$ indicated in Eq. (3.6) is the onset of the effect of complexity in the relaxation. In the region beyond $t = t_c$

$$\frac{d\lambda}{\lambda} = -n\frac{dt}{t}, \quad t>t_c \tag{3.7}$$

so that changes in the decrement rate are independent of time scale for the Kohlrausch situation. The scaling feature of Eq. (3.7) holds also when n = 0 which corresponds to the linear exponential situation. Thus the form of both the linear exponential and Kohlrausch relaxation should be independent of scale factors. This lends credibility to the choices of W_o^{-1} and W^{*-1} as the time scale factors respectively for the linear exponential and the Kohlrausch situations. With such choices, the value of the epoch entropy for relaxation is always unity and relaxation functions take their canonical form as in Eqs. (2.10).

Further support for such a viewpoint is provided by experiments in which the constant relaxation rate in a particular relaxation regime is shifted by variation of thermodynamic variables such as the temperature T. Often the form of the relaxation function remains unchanged. When this is the case, the relaxation is frequently termed "thermorheological simple" in phenomenological and experimental descriptions. Many relaxation processes though not all are thermorheological simple at least in some limited temperature range. On varying T, the relaxation process is the same except for a shift in the effective relaxation rate. It is gratifying to be assured that the epoch entropy continues to remain unchanged with temperature if the effective relaxation rate is chosen to be the inverse scale factor in the evaluation of the epoch entropy.

It is tempting to reverse this development and require as a physical principle that the value of the epoch entropy to be unity in all cases and most particularly when the relaxation process is observed to be thermorheologically simple. Then $\tau_E = W_o^{-1}$ and $\tau_K^* = W^{*-1}$ are respectively the <u>only</u> choice for linear exponential and fractional exponential relaxations. In this way, the property of thermorheological simplicity is directly obeyed.

K. L. Ngai and A. K. Rajagopal were partially supported by ONR Contracts N0001487WX24039 and N0001487WX24028 respectively.

References

Bottcher, C.J.F. & Bordewijk, P. (1978) Theory of Electric Polarization Vol. II, Ch. 8. New York: Elsevier

Levine, R.D., (1980). An Information Theoretical Approach to Inversion Problems, J. Phys. A: Math. Gen. 13, 91-108.

Ngai, K.L., Rendell, R.W., Rajagopal, A.K., and Teitler, S. (1986). Three Coupled Relations for Relaxations in Complex Systems. Annals New York Academy Sci., 484, 150-184

A GENERAL THEORY OF INHOMOGENEOUS SYSTEMS

S. A. Trugman
Theoretical Division
Los Alamos National Laboratory
Los Alamos, NM 87545
and
Department of Physics
Princeton University
Princeton, NJ 08544

Abstract. Inhomogeneous systems are modeled using maximum entropy methods so that only explicitly stated properties are taken into account, with no arbitrary assumptions. Sandstones are considered as an example. A phase diagram describes the evolution of one inhomogeneous system into another. Methods from statistical mechanics, including Monte Carlo simulation, series expansions, and the renormalization group should be useful as well for other maximum entropy problems involving many degrees of freedom.

We consider the application of maximum entropy methods[1] to the study of inhomogeneous systems.[2] Typical examples of inhomogeneous systems include cermets (fine mixtures of insulating ceramic and conducting metal), sandstones, vycor (a porous glass network), granular films, precipitates from solution, biological tissues, etc.[3] Both the structural and transport properties of inhomogeneous systems will be considered.

The problem of inhomogeneous systems forms connections between maximum entropy methods and those of recent statistical mechanics that are not apparent in other maximum entropy work. These methods of statistical mechanics, including Monte Carlo simulation, series expansions, the renormalization group, and exact solutions should be useful in many other maximum entropy studies.

The first section of this paper is a brief review of previous work on inhomogeneous systems. The second section describes the present theory (see also Ref. (2)), and the third contains a summary and conclusions.

I. PREVIOUS MODELS FOR INHOMOGENEOUS SYSTEMS

The most widely studied model for inhomogeneous systems is the percolation model.[4] In its simplest form, the percolation model requires that the sites of an infinite lattice (for example cubic) be independently occupied, each with a probability p. Two occupied sites

G. J. Erickson and C. R. Smith (eds.),
Maximum-Entropy and Bayesian Methods in Science and Engineering (Vol. 2), 371–379.

belong to the same (connected) cluster if there is a path between them on occupied nearest neighbor sites.

Some of the key results of percolation theory are the following: (1) The fraction of sites belonging to the infinite cluster is proportional to $(p - p_c)^\beta$ as $p \to p_c^+$. Here, p_c is the percolation threshold (equal to .59 for a square lattice), and β is a critical exponent (equal to 5/36 in two dimensions). For $p < p_c$, no infinite cluster exists. (2) The largest diameter of finite clusters, with exponentially rare exceptions, is $\xi \sim |p - p_c|^{-\nu}$ where ν is another critical exponent (equal to 4/3 in two dimensions).[5] This formula applies as p approaches p_c from either side. (3) If occupied nearest neighbor sites are connected by a resistor of resistance R, the conductivity of a large block is proportional to $(p - p_c)^t$ as $p \to p_c^+$. The static exponents β and ν, and the transport exponent t are universal, in that they do not depend on the type of lattice (although they do depend on the spatial dimension). The percolation threshold p_c is not universal.

Percolation is perhaps the simplest example of a continuous phase transition. Others include the appearance of magnetism as the temperature is lowered through the Curie point, and the gas-liquid critical point. [Discontinuous transitions, such as ice melting, are perhaps more familiar.] Continuous transitions have unusual properties, including fluctuations on all length scales. To understand these fluctuations, consider an ordinary gas at atmospheric density. The density of gas in a box 1Å on an edge at an instant in time would not be representative of the density of the entire system (it would contain either one atom or zero). A box of edge larger than about 100Å would, however, be representative. At a continuous transition, any finite box, no matter how large, is too small to be representative. This is a statement that there are fluctuations on all length scales. The structure of the system is self-similar. If one defines the spatial dimension D of the system in a standard way (the Hausdorf dimension), one finds that D is not an integer. These objects are called fractals.[6] For example, the Hausdorf dimension of the infinite percolation cluster at $p = p_c$ is 1.896.

Percolation is an adequate model for some inhomogeneous systems, but it fails to describe the properties of others. Percolation fails, for example, as a model of sandstones. Sandstones empirically have a percolation threshold at or near zero; fluid can be forced through the pores of a sandstone even when the pore volume fraction is less than 0.01.[7] In contrast, percolation models have the property that when $p < p_c$ (p_c is generally near 0.3 in three dimensions), the pore space becomes disconnected.

Historically, the response to this failure has been to invent specialized models that have a percolation threshold at zero. One model, proposed by Sen et al.,[8] starts with randomly packed spheres, puts smaller spheres in the interstices, yet smaller spheres in the remaining interstices, etc. A second model, proposed by Wong et al.,[9] begins with a cubic lattice of pipes. A pipe is chosen at random, and its cross section reduced by a constant factor (e.g. 1/2). Another pipe is chosen at random (perhaps the same one), and its cross section reduced. As the process is continued, a zero threshold model eventually results. A third model has been proposed by the author and Weinrib.[10] It is a type of continuum percolation model in which a random surface is created and then flooded to a constant depth. A two-dimensional realization is shown in Fig. (1).

Fig. (1) The figure is a realization of a continuum percolation problem that has a percolation threshold at zero (see Ref. 10). (b) The black network connects arbitrarily distant points, in spite of the fact that its volume fraction is small. (a) The volume fraction is essentially zero, corresponding to the width of a line, and yet arbitrarily distant points remain connected.

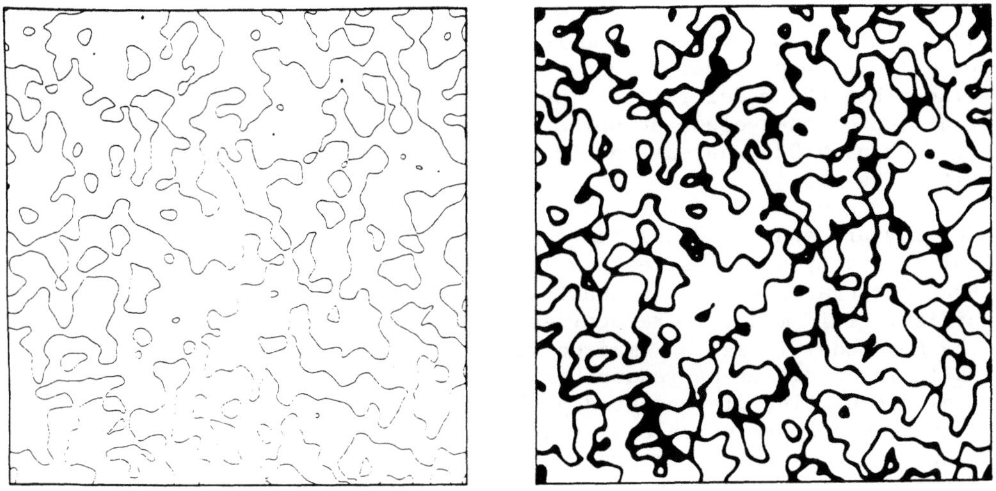

(a) (b)

A problem with these zero threshold models for sandstones is that they seem ad hoc and arbitrary. It is not clear which, if any of them, forms an adequate model for sandstones. The problem is compounded by the fact that many other zero threshold models could clearly be proposed.

These problems lead to the following question: Is it possible to start from an explicit list of given properties (e.g. $p_c = 0$) and to construct a unique model containing no additional arbitrary assumptions? This question appears to be wildly underdetermined and perhaps even ill-posed, but it does in fact have an answer in the context of maximum entropy theory.

II. MAXIMUM ENTROPY THEORY OF INHOMOGENEOUS SYSTEMS

To implement the maximum entropy method, one assumption is required, which is the prior probability.[1] Perhaps the simplest way to dispose of this problem is to assume that the system has no structure on a length scale smaller than b. Depending on the system, b may correspond to the size of a grain, the size of an atom, etc. We will in fact make a somewhat stronger assumption and put the system on a lattice (for example cubic) of lattice constant b. The fundamental cubes of the lattice are occupied by either material A or material B. It is hoped that by universality the long distance properties will not be affected by the particular choice of lattice. The lattice contains a finite but large number of sites N.

One may now ask, making no further assumptions (other than a two component lattice system), what is <u>the</u> generic inhomogeneous system? There are 2^N different configurations. Maximum entropy theory assigns an equal probability to each. (To do otherwise would imply that additional arbitrary assumptions are being made.) This model can be recognized as the percolation model with p = 1/2. Note that p was unspecified, and that the maximum entropy method chose p = 1/2.

The generic model can be compared to the particular system of interest. We will again consider a sandstone to have a specific example. One finds that the model fails. The most conspicuous failure is that the fraction of the volume occupied by the pores of a particular sandstone is perhaps 0.02, whereas the fraction occupied in the model (the A sites) is 0.5.[11]

One may proceed by adding the constraint that the model must have a volume fraction of A sites equal to q = 0.02. Now all of the 2^N configurations that satisfy the constraint p = q are taken with equal probability. This model is recognized as percolation, but now with p = q. The new model is again compared with the system of interest. It now fails for a different reason. The pore space of the model at q = .02 consists of tiny disconnected pockets, whereas that of a sandstone is a connected network capable of transporting fluid an arbitrary distance. A straightforward way to proceed is to add a second constraint: The A sites must form a single connected cluster. All configurations satisfying both constraints are now taken with equal weight. This model is <u>not</u> equivalent to percolation. It is the generic zero threshold model.[12] It correctly models some properties of sandstones. If it is found not to accurately model other important properties of sandstones, the response would be different from that discussed in Section I. Previously, one would hire more theorists to construct new models, hoping that at least one would work. In the present formulation, one returns to the experimentalist and informs him that there is an important property of the system that has not yet been stated, and that a new model cannot be constructed until the property is specified.

It is sometimes convenient to specify the Lagrange multiplier conjugate to a constraint, rather than imposing the constraint explicitly. These two procedures are essentially equivalent for large systems. (A familiar example is the study in statistical mechanics of either the microcanonical or the canonical ensemble.) Now all of the 2^N configurations are allowed, with a relative weight W given by

$$W = \exp(-\lambda_s n_s - \lambda_c n_c) \quad , \tag{1}$$

where n_s is the number of sites and n_c the number of connected clusters. The Lagrange multipliers λ_s and λ_c are adjusted until $\langle n_s \rangle$ and $\langle n_c \rangle$ have the desired values.

Any set of constraints and conjugate multipliers may be used, but they must all be expicit.[13] The nature of the inhomogeneous system changes continuously with the vector (λ_s, λ_c). One would like to understand the properties of an (infinite) inhomogeneous system at a given point (λ_s, λ_c), and in particular to know whether there are special points at which the structure changes qualitatively. (Special points do exist, as will be discussed below.)

The problem is to determine the macroscopic properties from the microscopic weights of equation (1). Several methods from statistical mechanics are useful, and may indeed help to understand other problems that are analyzed by maximum entropy methods. Only a brief description is given; references are provided for further information.

A system whose properties are not understood is, if possible, often first studied in mean field theory.[14] Mean field theory is simpler to implement than the methods described below, and will often predict correct qualitative features. Mean field theory does not, however, treat fluctuations correctly and does not quantitatively describe the nature of singularities (it gives incorrect critical exponents unless the number of spatial dimensions is large). The methods that follow, in contrast, do treat fluctuations correctly.

One generally useful method is Monte Carlo simulation.[15] From a starting configuration, one generates a proposed new configuration (for example by choosing a fundamental cube at random and changing the substance that occupies it.) The new configuration is accepted with probability W_2/W_1 if its weight W_2 is less than that of the previous configuration W_1. It is accepted with probability one if $W_2 \geq W_1$. It has been shown that this algorithm will generate an ensemble in which configurations appear with weight W. Some preliminary Monte Carlo simulations for the inhomogeneous systems weight (Eq. 1) have been performed.[16]

The remaining methods have general applicability, but they are particularly useful for studying continuous transitions. One may perform series expansions in one or more of the Lagrange multipliers about $\lambda = 0$ or about $\lambda = \pm\infty$, known respectively as high and low temperature series. These series, which in practice contain on the order of twenty or fewer terms, can be analyzed with Padé or partial differential approximants.[17]

The renormalization group is a method in which the system is coarse-grained in steps.[18] For the inhomogeneous system problem, one version of the renormalization group would choose a single large cube to represent the configuration of several fundamental cubes. This process is repeated again and again, until the long distance behavior is apparent.

Methods of finite size scaling extract the properties of an infinite system by solving for the behavior of strips of increasing width.[19] Monte Carlo renormalization group is a hybrid of those two methods discussed above.[20]

These methods should be useful for broad classes of maximum entropy problems involving many degrees of freedom.

We return to the question of the nature of an infinite inhomogeneous system as a function of (λ_s, λ_c). One can draw a phase diagram with lines to separate regions of different qualitative behavior (see Fig. (2)).

Fig. (2) The phase diagram for inhomogeneous systems is shown as a function of p and λ_c. The connected and disconnected phases are labeled, with ordinary percolation on the line $\lambda_c = 0$. A region with antiferromagnetic (checker board) order is labeled AF. The shaded area is one of phase separation. This area would contract to a line if λ_s, rather than p, were plotted on the x-axis.

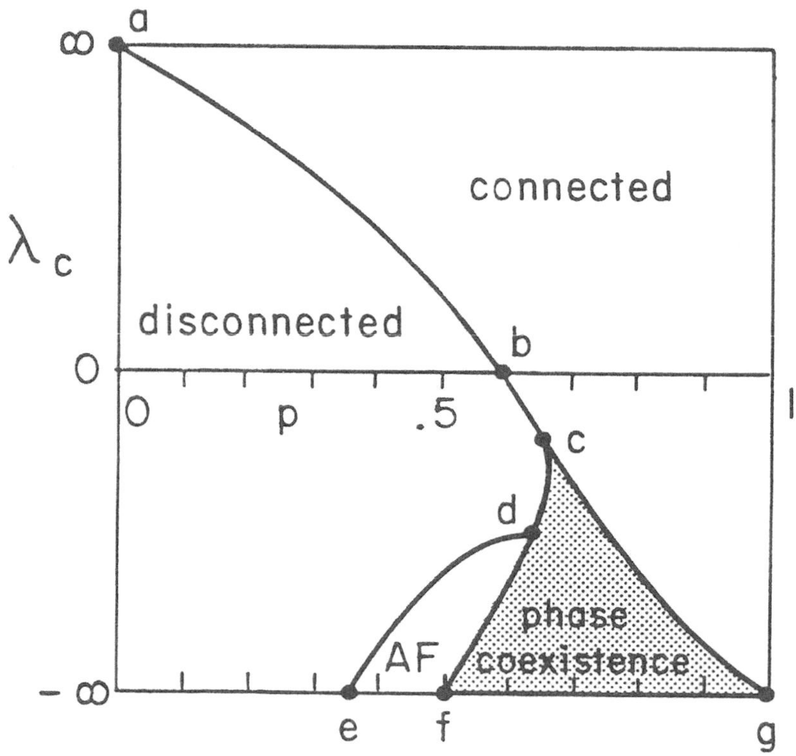

Some aspects of the phase diagram are speculative, and some are well understood. The vertical axis is the Lagrange multiplier λ_c controlling clustering. When $\lambda_c \to \infty$ the system has one connected cluster, and when $\lambda_c \to -\infty$, the system has as many clusters as possible. On the horizontal axis the volume fraction p of component A is plotted, which is somewhat more convenient than the conjugate Langrange multiplier λ_s.

We first discuss the well understood regions of the diagram. In the upper left corner, labeled "a", the configurations are identical to those of a well studied problem known as lattice animals (see Ref. 4). There is a cluster connecting opposite faces of the system even when a vanishing fraction p of the sites are occupied. As discussed previously, the line $\lambda_c = 0$ is identical to ordinary percolation. There is a second order phase transition at $p = p_c$, where large fluctuations occur in the cluster size distribution and in the conductivity. The static critical exponents are known exactly in two dimensions and approximately in three dimensions.[4,5] An infinite connected cluster is present for $p \geq p_c$.

The line $\lambda_c = -\infty$ is also understood. For $p \leq 1/2$ the system can be shown to be equivalent to the lattice gas with infinite nearest neighbor repulsion. The equivalence follows from the fact that for $p \leq 1/2$ the maximum number of clusters $n_c = n_s$ is formed if nearest neighbor occupation is forbidden. The known properties of the lattice gas[21] allow the following description: For small p, the A sites form a dilute, nearly ideal gas. As p increases, the hard-core repulsion becomes more important until a continuous phase transition occurs with "checker-board" or "antiferromagnetic" long-range order for $p > p_2$. (If A sites are identified with spin up and B sites with spin down, the system has antiferromagnetic long-range order.) The point $p = p_2$ is labeled e in Fig. 2, with $p_2 = 0.371$ for a square lattice.[21] At $p = p_1 = \frac{1}{2}$, labeled f, the entropy per site is zero and only the two equivalent antiferromagnetic ground states are allowed. If p is fixed between p_1 and 1 and the limit $\lambda_c \to -\infty$ taken, the system is no longer equivalent to a lattice gas with short-range repulsion. In this region, the system becomes macroscopically inhomogeneous and phase separates into an antiferromagnetic phase and a connected phase (one in which all sites are occupied). The phase separation can be seen most clearly by induction on the number of B sites. The region $p_1 < p \leq 1$ is thus one of phase coexistence, characteristic of a first-order phase transition (see Fig. 3).

The remaining aspects of the phase diagram are speculative, although the diagram is constrained by several physical principles and guided by a tentative mapping to the Potts model.[2] The lines a-c and e-d mark continuous phase transitions with large scale (fractal) fluctuations and critical exponents. To the right of line a-c-f, it is possible to travel an arbitrarily large distance on a single connected cluster of A sites. The region just below and to the right of point "a" in the connected phase is expected to provide the best model of sandstones available in this plane.

A quantity called the order parameter indicates which phase occurs; it is zero on one side of a continuous phase transition and nonzero on the

Fig. (3) The diagram illustrates the phase separation that occurs between points f and g of Fig. (2), at $\lambda_c = -\infty$. The checkerboard phase floats in a connected phase. If one moves slightly upward in the phase diagram ($\lambda_c > -\infty$), the checkerboard develops defects and some uncolored sites detach and form a dilute gas in the connected region.

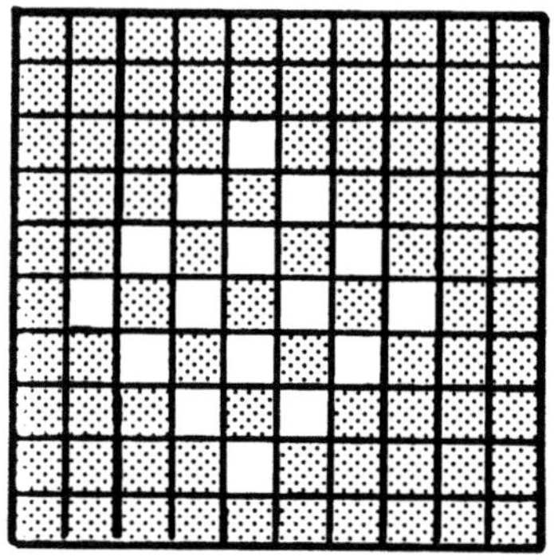

other. The fraction of A sites in the largest connected cluster, denoted P_∞, is an order parameter that is nonzero to the right of a-c-f, and indicates that the fluid contained in A sites can move an arbitrarily large distance. There is a second order parameter, m_s (the staggered magnetization), that is nonzero in the antiferromagnetic, or checkerboard phase.

III. SUMMARY AND CONCLUSIONS

A general model for the structural and classical transport properties of inhomogeneous systems has been proposed. It requires as input an explicit list of specified properties of the system, and makes no further arbitrary assumptions. The behavior of an inhomogeneous system with two specified properties (the number of sites and the number of connected clusters) has been determined in part, although further Monte Carlo studies will be required to resolve remaining questions.

The ideas and methods of statistical mechanics should apply to many maximum entropy problems with a large number of degrees of freedom. In particular, there should be special points where the properties (including the entropy) are not smooth functions of the Lagrange multipliers and qualitatively new properties arise. These special points, corresponding to continuous and discontinuous phase transitions, can be plotted in a phase diagram. Order parameters can be used to indicate the phase. Methods from statistical mechanics, including mean field theory, Monte Carlo, the renormalization group, and series analysis should be generally useful.

REFERENCES

1. E. T. Jaynes: Papers on Probability, Statistics and Statistical Physics, edited by R. D. Rosenkrantz, (D. Reidel Pub., Boston, 1983).
2. S. A. Trugman, Phys. Rev. Lett. 57, 607 (1986).
3. Electrical Transport and Optical Properties of Inhomogeneous Media, edited by J. C. Garland and D. B. Tanner, AIP Conf. Proc. No. 40, (American Institute of Physics, New York, 1978).
4. D. Stauffer, Phys. Rep. 54, 1 (1979).
5. F. Y. Wu, Rev. Mod. Phys. 54, 235 (1982).
6. B. B. Mandelbrot, Fractals: Form, Chance and Dimension (Freeman, San Francisco, 1977).
7. Physical Properties of Rock, ed. by K. H. Hellwege, (Springer, New York, 1982).
8. P. N. Sen, C. Scala, and M. H. Cohen, Geophysics 46, 781 (1981).
9. Po-zen Wong, J. Koplik, and J. P. Tomanic, Phys. Rev. B 30, 6606 (1984).
10. S. A. Trugman and A. Weinrib, Phys. Rev. B 31, 2974 (1985).
11. The maximum entroy method predicts a very sharp value for the volume fraction of A, equal to 1/2 with fluctuations of the order $N^{-1/2}$. The prediction, although sharp, is completely wrong. This situation can arise quite generally in large systems, in contrast to the incorrect notion that a sharp maximum entropy prediction will be correct and that a broad one may not be.
12. One could consider an alternate formulation in which there must be at least one "infinite" cluster (a cluster that connects opposite faces), rather than requiring that all A sites form a single cluster. The alternate formulation can be shown to be uninteresting for large systems. The system will simply run a string between opposite faces to remove the constraint, and then do precisely what it would have done with no constraint present.
13. Some facts are known when a Lagrange multiplier conjugate to the conductivity is used; see Ref. (2).
14. R. K. Pathria, Statistical Mechanics, (Pergamon, Elmsford, NY 1972).
15. K. Binder, in Monte Carlo Methods in Statistical Physics, edited by K. Binder (Springer, Berlin, 1979), Vol. 7.
16. G. Dodge, Junior Paper, Princeton University (unpublished).
17. M. E. Fisher and R. M. Kerr, Phys. Rev. Lett. 39, 667 (1977).
18. Phase Transitions and Critical Phenomena, ed. by C. Domb and M. S. Green (Academic, New York, 1976), Vol. 6.
19. M. P. Nightengale, Proc. K. Ned. Akad. Wet. Ser. B 82, 235 (1979), and J. Appl. Phys. 53, 7927 (1982).
20. R. H. Swendson, Phys. Rev. Lett. 52, 1165 (1984).
21. L. K. Runnels, in Phase Transitions and Critical Phenomena, edited by C. Domb and M. S. Green (Academic, New York, 1972) Vol. 2.

MAXIMUM ENTROPY AND CRACK GEOMETRY IN GRANITIC ROCKS

P.M. Doyen
The Rock and Borehole Geophysics Project, Stanford University, Stanford,CA 94305

Abstract. The maximum entropy method is used to infer the dimensions of cracks in granitic rocks from measurements made under confining pressure of their hydraulic permeability and of their electrical conductivity. The fluctuations of crack dimensions are characterized by a crack spectrum which gives the frequency of occurrence of cracks as a function of their cross-sectional length and aspect ratio in the absence of stress. First, using an effective medium approximation, the variations of the transport coefficients resulting from crack deformation under pressure are expressed by averaging the pressure-dependent crack electrical or hydraulic contributions over the unknown crack spectrum. Next, the maximum entropy crack spectrum in Westerly granite is determined from laboratory measurements of the transport coefficients. The predicted spectrum is comparable to the distribution of dimensions estimated from direct observations of cracks with a scanning electron microscope. The main features of the calculated spectrum are interpreted from a model of growth of the microfissures during cooling of the granite.

Introduction

The study of the electrical conductivity and permeability of water-saturated granitic rocks is important to understand the transport of current and that of fluids in the earth's crust. In laboratory experiments, rock samples saturated with a conducting brine are confined under different levels of hydrostatic pressure to restore the conditions prevailing in situ at different depths in the crust. Figures 1 and 2 (after Brace et al., 1965, 1968) show respectively the d.c. electrical conductivity σ^* and the hydraulic permeability κ^* of Westerly granite measured at room temperature as a function of confining pressure p from 0 to 10^3 MPa. The large decrease of the transport coefficients by factors of 10^2 to 10^3 is typical of granitic rocks. As the minerals forming the granite are essentially insulating at room temperature, the variations of the

G. J. Erickson and C. R. Smith (eds.),
Maximum-Entropy and Bayesian Methods in Science and Engineering (Vol. 2), 381–393.

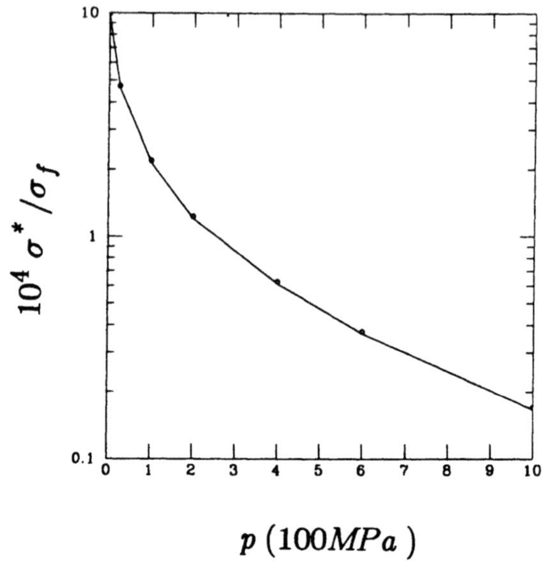

Figure 1. Pressure dependence of the d.c. electrical conductivity of Westerly granite saturated with a conducting electrolyte of conductivity σ_f (after Brace et al., 1965). The o's are the experimental data used in the inversion.

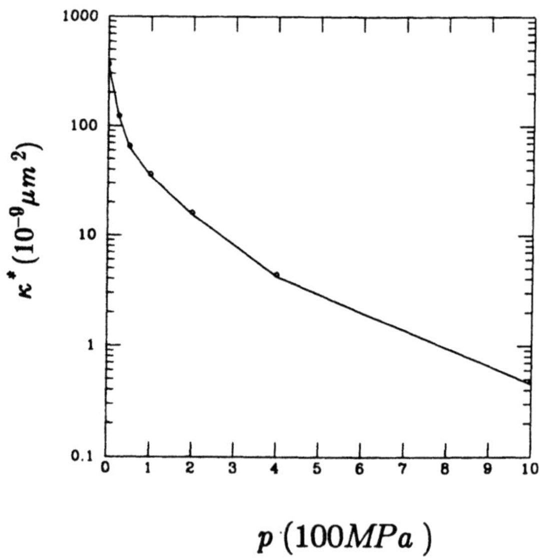

Figure 2. Hydraulic permeability of Westerly granite measured as a function of pressure (after Brace et al., 1968). The o's are the experimental data used in the inversion.

transport properties with pressure are controlled by the connectivity and shape of the fluid-filled pore space. The porosity of granite reflects the presence of sheetlike microcracks occurring at the boundaries between the mineral grains. At low confining stresses, although the porosity of granite is usually very small (less than 1%), the large density of cracks and their large specific surface area guarantee that the crack assembly is interconnected and that the conducting fluid percolates across the rock. When a rock sample is compressed, the closure of flat cracks creates nonconducting paths in the pore space and a decrease of the transport coefficients results. However, at pressure as high as 10^3 MPa, the conductivity of the electrolyte-saturated rock remains several orders of magnitude greater than the conductivity of the dry rock. At high pressure, the fluid phase is still continuous and fills a residual subnetwork of stiff cracks which remain open.

In this study, the variations with pressure of the transport coefficients of granite are interpreted from the elastic deformation and closure of cracks. In the first section, we introduce a *crack spectrum* which specifies the variability of crack dimensions in the absence of stress. Then, using an effective medium approximation (EMA), the d.c. conductivity and permeability are expressed at each pressure in terms of the spectrum of crack dimensions and in terms of the pressure-dependent crack electrical and hydraulic conductances. In the second section, the inverse problem is addressed. For Westerly granite, the crack spectrum representing a condition of maximum entropy (ME) is inferred from the pressure-dependent macroscopic transport coefficients. The inverted spectrum is then compared with the distribution of crack dimensions estimated from scanning electron micrographs of the granite. Finally, the ME spectrum is interpreted from a simple model of crack growth.

Transport properties of granite under pressure

The pore space of granite is modelled by a topologically disordered network of interconnected tubular cracks with approximately constant length, l, but variable cross-sectional dimensions (figure 3). In the absence of stress, the crack cross-sectional dimensions are described by two variable parameters: the half length and the half width, c and αc respectively. The aspect ratio α measures the ellipticity of the crack cross-section. The fluctuations of crack dimensions are characterized by a crack spectrum $n(\alpha, c)$ which gives the frequency of occurrence of cracks as a function of α and c.

Locally, at the scale of a crack, the flow of current is controlled by the crack electrical conductance g. This local conductance is proportional to the fluid conductivity σ_f, to the crack cross-sectional area s and to l^{-1} (see figure 4). The conductance of a crack decreases as the confining pressure p is raised. This variation reflects the shortening of the crack cross-sectional dimensions under pressure. We assume that the cracks have a regular cross-sectional shape

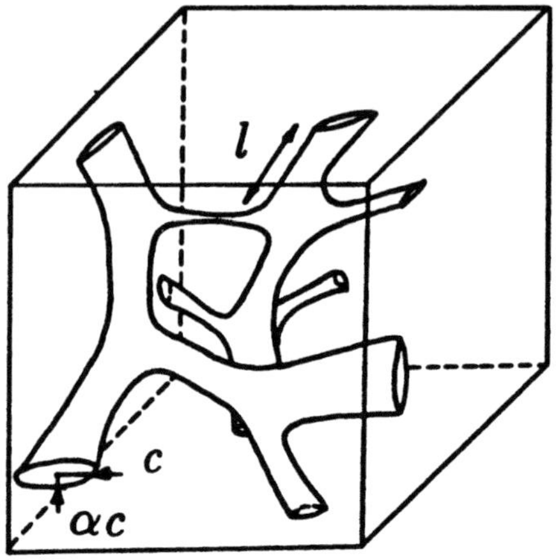

Figure 3. In the absence of stress, the pore space is represented by a fully interconnected network of tubular cracks with approximately constant length, l, but variable cross-sectional half length, c, and half width, αc. The cracks deform elastically under pressure at a rate determined by the aspect ratio in the absence of stress, α.

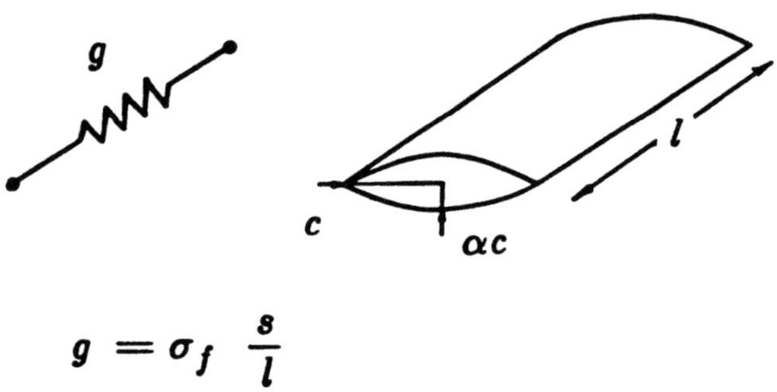

Figure 4. The flow of current in a crack filled with an electrolyte of conductivity σ_f is modelled by an equivalent electrical conductance g.

as shown in figure 4 and that they deform elastically under pressure. We also assume that the tubular cracks close uniformly over their length and we neglect the change of l with p. The crack electrical conductance at pressure p can then be calculated from the cross-sectional dimensions c and αc in the absence of stress, i.e., $g = g(p, \alpha, c)$. Figure 5 shows that the aspect ratio in the

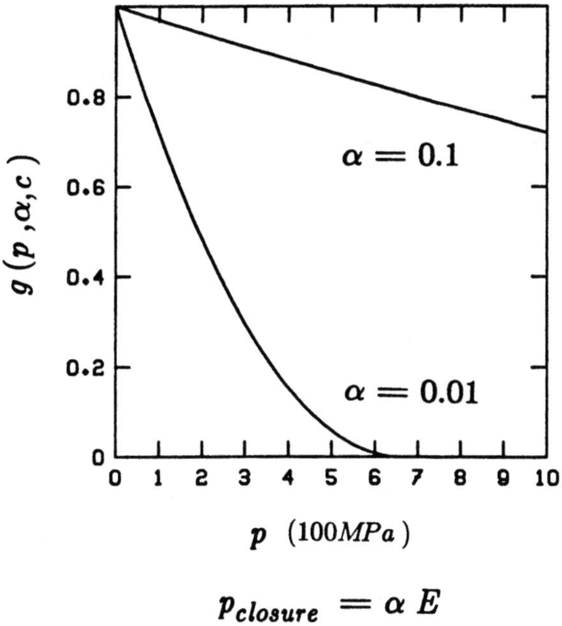

Figure 5. The electrical conductance g decreases when the confining pressure p is raised. This decrease refects the elastic closure of the crack under pressure. The rate of decrease of the conductance and the closure pressure are determined by the aspect ratio α of the crack in the unstressed state.

unstressed state, α, controls the rate of decrease of g and determines the pressure $p_{closure}$ at which a crack closes and ceases to conduct. A thin crack ($\alpha<<1$) closes at low pressure while a more equidimensional cavity ($\alpha\approx1$) requires very high pressure of the order of the Young's modulus E of the crystalline matrix to deform substantially. It can be shown that the dependence of the closing pressure on the product αE is a general result, independent of the exact crack shape. It follows that the inverted spectrum of crack dimensions is relatively insensitive to the selected crack shape in the model (Doyen, 1987).

At the scale of a laboratory sample containing many cracks, the flow of current in the electrolyte saturating the crack network is controlled by the spectrum $n(\alpha,c)$ which specifies the crack dimensions in the absence of stress. If this spectrum is measured experimentally, a good estimate of the d.c. electrical conductivity can be obtained from an effective medium approximation (EMA) (Kirkpatrick, 1973). At each pressure p, the conductivity σ^* can be calculated implicitly from the following self-consistency condition:

$$\sum_{\alpha,c} n(\alpha,c) \frac{l^* \sigma^*(p) - g(p,\alpha,c)}{g(p,\alpha,c) + (\frac{\bar{z}}{2} - 1)\, l^* \sigma^*(p)} = 0 \tag{1}$$

where $\bar{z}$ is the average coordination number of the crack network and l^* is some characteristic length of the topologically disordered crack assembly (Doyen, 1987). In the EMA, the electrical interactions between cracks are approximately accounted for by considering that each crack is embedded in a homogeneous crack assembly which models the average effect of all the other cracks. The kernel in (1) represents the conductivity perturbation resulting from the introduction of a crack with conductance $g(p,\alpha,c)$ in the homogeneous network where all the cracks have the same conductance $l^* \sigma^*(p)$. The self-consistency condition (1) requires that the average perturbation vanishes when the average is taken over the spectrum $n(\alpha,c)$. The pressure-dependent permeability $\kappa^*(p)$ can be calculated in a similar way from the crack spectrum by considering the crack hydraulic instead of electrical conductances (Doyen, 1987).

Maximum entropy crack spectrum in Westerly granite

In the previous section, the pressure-dependent transport coefficients were calculated by averaging the crack contributions over the crack spectrum $n(\alpha,c)$. In this study, the spectrum is unknown and it must be estimated from measurements of κ^* and σ^* made at different confining pressures. To each measurement of the permeability or of the conductivity corresponds a linear constraint like (1) on the unknown spectrum. The problem is that there is an infinite number of crack spectra that may give rise to the same response of the rock under pressure. Among the spectra consistent with the macroscopic averages, we select the one which has maximum entropy. Introducing Lagrange's multipliers λ_p for the linear constraints (1), the well-known formal solution for the ME spectrum $n(\alpha,c)$ is (Jaynes, 1957):

$$n(\alpha,c)\ \alpha\ \exp\left[-\sum_p \lambda_p \frac{l^* \sigma^*(p) - g(p,\alpha,c)}{g(p,\alpha,c) + (\frac{\bar{z}}{2} - 1)\, l^* \sigma^*(p)}\right] \tag{2}$$

where the summation extends over all pressures p at which the conductivity σ^* is measured. In practice, additional terms appear in the exponential (2) when constraints corresponding to the permeability are included in the inversion.

The maximum entropy (ME) crack spectrum in Westerly granite is now inferred from measurements of the transport coefficients under pressure. The circles in figures 1 and 2 show the experimental measurements of $\sigma^*(p)$ and $\kappa^*(p)$ used in the inversion. In addition to the permeability and conductivity data, measurements of the compressibility of the rock under pressure are also incorporated in the inversion. (See Doyen, 1987 for more details.) Figures 6a-d show the evolution of the ME spectrum when the number of experimental constraints is progressively increased, the data $\sigma^*(p)$ and $\kappa^*(p)$ being added in the inversion in order of decreasing pressure. This sequence of crack spectra illustrates clearly the criterion of maximum uniformity underlying the maximum entropy method. With increasing number of constraints, the range of crack parameters compatible with the data narrows down, the entropy decreases monotonically and the crack spectrum departs more and more from the original log-uniform distribution (figure 6a) which represents the state of null information. The relative entropy s given in figure 6 is defined as the ratio between the entropy of the crack spectrum and that of the original uniform distribution. In figure 6b, all the experimental data correspond to pressures $p \geq 200MPa$ and the ME spectrum is still uniform for all cracks with $ln(\alpha) \leq \ln(200/E) \approx -2.5$. This uniformity expresses simply equal uncertainty about the frequency of occurrence of cracks closing at $p_{closure} \approx \alpha E \leq 200MPa$. Indeed, the transport coefficients measured at a certain pressure level are not sensitive to the dimensions of the cracks that are closed at that pressure. In figure 6b, the maximum entropy method gives then naturally equal frequency of occurrence to all the thin cracks and expresses the fact that their dimensions cannot be distinguished from the given data. Going from figure 6b to 6d, new modes develop in the spectrum at smaller α values. This evolution underlines the opening of the thinner cracks as the pressure is reduced. When the low pressure information is taken into account, a large decrease of the entropy arises and the crack spectrum becomes more sharply defined. The large decrease of s reflects the constraining nature on the spectrum of the large variations of σ^* and κ^* at low stresses.

The ME spectrum is now compared with direct experimental measurements of crack dimensions. Using scanning electron micrographs (SEM) of Westerly granite, Hadley (1976) estimated the crack length and aspect ratio distribution from a population of 344 cavities in an area of about 1 mm^2. Figure 7 shows a perspective contour of this experimental crack spectrum. Figure 8a shows a perspective contour of the inverted spectrum represented in figure 6d. Several remarkable similarities exist between the two spectra. The most striking common feature is the negative correlation existing between the crack parameters. The two distributions tend to be symmetrical and concentrated about a line inclined at 45 degrees, $\log \alpha c = \log \alpha + \log c \approx cst$. The main modes of the ME spectrum (labelled 1 and 2 in figure 8a) occur at the edges of this diagonal feature. Mode 1 is also observed in the experimental spectrum. Mode 2 corresponds to very long flat cracks. Only the flank of this second maximum is apparent in the experimental distribution. As the area of the micrograph studied by Hadley (1976) is only $1mm$ in diameter, the very long cracks corresponding to this second mode may have been missed. The maxima in the experimental spectrum are less pronounced than in the calculated spectrum. There are two reasons for this difference. First, the SEM study was

Figure 6a-6d. Evolution of the maximum entropy crack spectrum in Westerly granite as the number of constraints is increased. s is the relative entropy of the spectrum. The pressure range indicated is that of the experimental data used in each inversion. The underlying grid has $(N_\alpha = 65)$ x $(N_c = 40)$ nodes.

$s = 1.0$ (a)

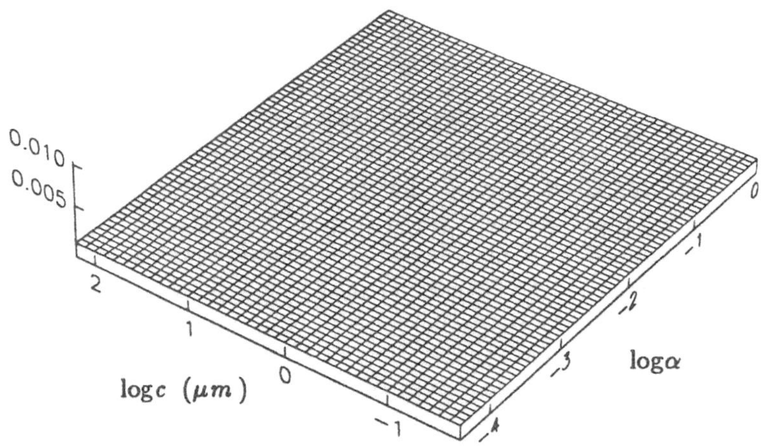

$s = 0.94$ $p \geq 200$ MPa (b)

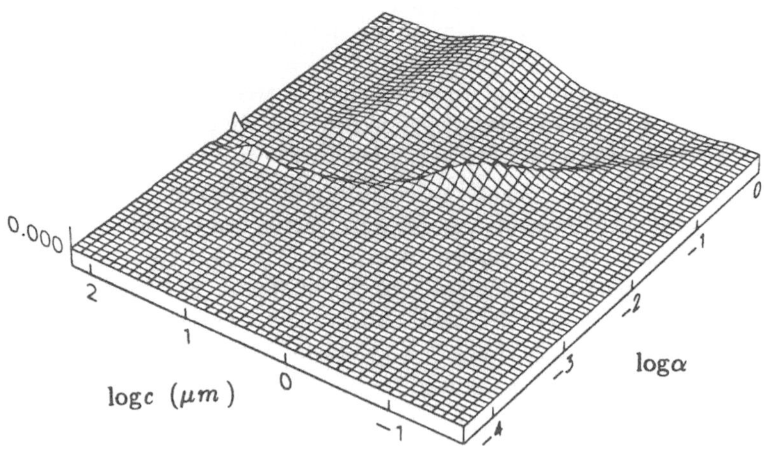

$s = 0.90 \quad p \geq 25\ MPa$ (c)

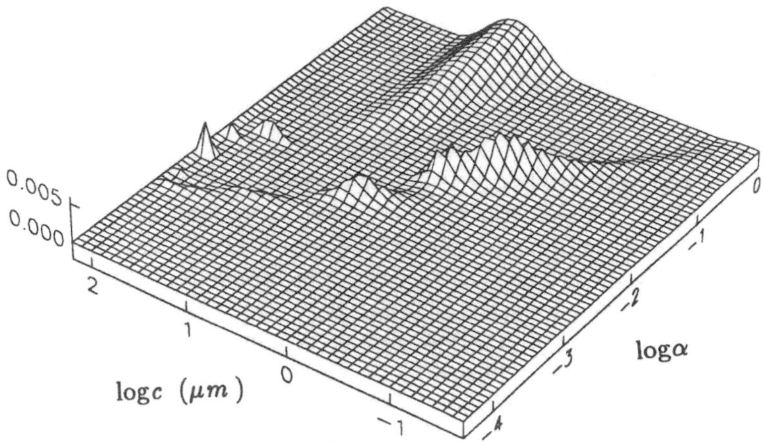

$s = 0.75 \quad p \geq 0\ MPa$ (d)

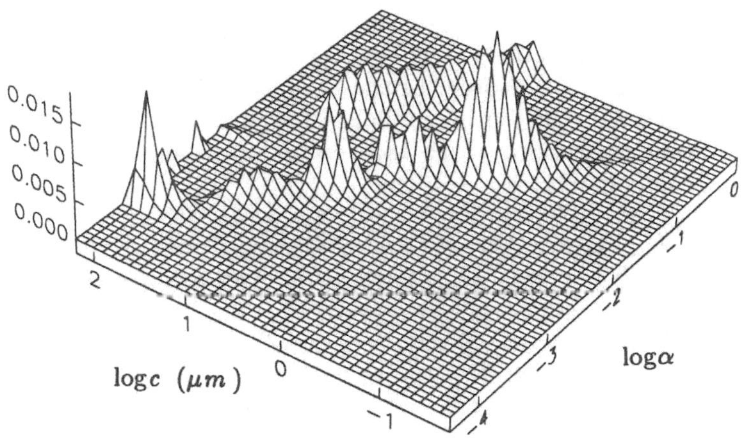

based on a collection of only a few hundred cracks. If a great number of cracks were counted in Westerly granite, the magnitude of the maxima would be enhanced. Second, as a consequence of the small number of cracks involved in Hadley's work, the crack parameter space was partitioned into much fewer classes. The SEM spectrum is then naturally smoother.

The clustering of the two crack spectra about a diagonal line indicates the existence of an approximately constant crack width $w = 2\alpha c \approx 0.1\mu m$ in Westerly granite. This clustering can easily be understood from a simple model of crack formation in granitic rocks. SEM observations (Sprunt et al., 1974) show that microcracks in Westerly granite occur mostly at the boundaries between quartz grains and other minerals. The thermal expansivity contrast existing between quartz and other grains is believed to be responsible for the creation of microfissures. The microcrack growth is driven by intergranular stresses that are produced by the cooling which arises when granite, formed at depth in the earth, is progressively uplifted to the surface (Nur et al., 1970; Bruner, 1984). The occurrence of a constant crack width, w, suggests that a crack, initiated by the separation of two grains, propagates along the grain boundary without changing its aperture. It can be shown that the crack aperture is determined by the average grain diameter in the granite (Doyen, 1987). It is now clear that the diagonal trend in the crack spectrum represents the path of evolution of cracks during cooling as they extend along grain boundaries and coalesce with other cracks (figures 8a and 8b). The initial small microfissures have lengths of a few micrometers close to the main mode in the spectrum (mode 1 in figure 8). They then possibly grow from the tips to a maximum length $2c$ of the order of the grain diameter (mode 2 in figure 8). As not all the cracks extend during cooling, the spectrum shows cracks at different stages of evolution ranging from short round openings (type 1) to very long sheetlike cracks (type 2).

Figures 1 and 2 show that at low confining stresses, the decrease of the transport coefficients is very pronounced while at pressures exceeding 100 or 200 MPa, the decrease appears more gradual. The dramatic decrease of the transport coefficients at low pressure is controlled by the closure of the long flat cracks which are the most conducting. The constancy of the crack aperture implies in particular that the crack electrical conductance is inversely proportional to the aspect ratio ($g \approx w^2/\alpha \approx cst/\alpha$). The sheetlike cracks with $\alpha << 1$ are squeezed closed at low pressure and the most conducting paths in the crack system are destroyed. At higher pressure, the closure of rounder and less conducting cavities affect less the transport properties.

Conclusion

The maximum entropy method is useful in obtaining information about the pore space microstructure from measured bulk properties of rocks.

Figure 7. Contouring of the experimental spectrum measured by Hadley (1976) in Westerly granite. The original data divided into 9x7 classes were interpolated on a 65x40 grid to allow comparison with the inverted spectrum.

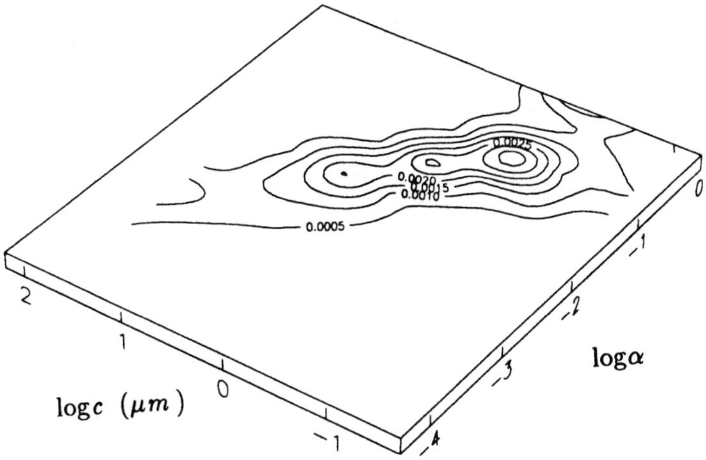

Figure 8a-b. (a) Contouring of the inverted crack spectrum in Westerly granite. This figure represents the same spectrum as figure 6d. The contouring interval is 0.005. (b) During cooling of the granite, cracks of type 1 may grow, merge with others and produce very long cracks of type 2 extending over a large portion of the grain boundary.

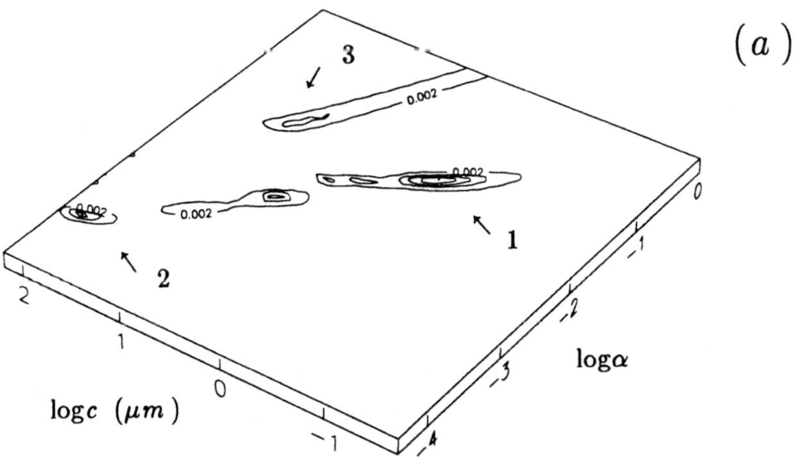

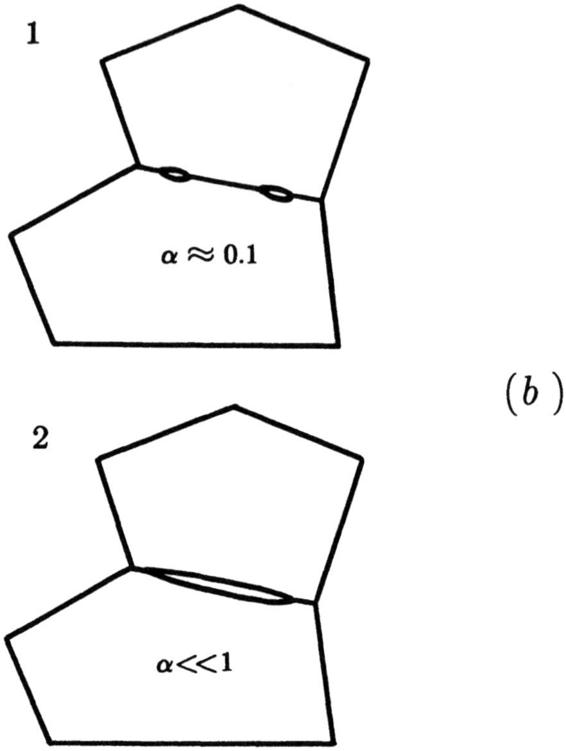

The maximum entropy crack spectrum is the most conservative estimate of the actual distribution of crack dimensions in the sense that it is the most uniform distribution compatible with the data. Departure from uniformity indicates the sensitivity of macroscopic measurements to a particular range of crack dimensions. The ME crack spectrum in Westerly granite was inferred from measurements made under confining pressure of compressibility, conductivity and permeability. It was found in good agreement with the spectrum estimated from SEM observations. The wide variability of crack length and aspect ratio contrasts with the existence of an approximately constant crack aperture in the cavity population. The form of the crack spectrum underlines the path of propagation of the microfissures as they extended along grain boundaries during cooling of the granite. The large sensitivity of Westerly granite to confining pressure stems from the closure of the long thin cracks which are the most conducting and compliant cavities in the rock.

Acknowledgments

This work was supported by the grant #DE-FG-03-86ER 1301 from the Office of Energy Research, Division of Engineering, Mathematics & Geosciences, U.S. Department of Energy. We thank A. Journel, A. Nur, R. Blankenbecler, D. Ellis and R. Ehrlich for helpful discussions and comments.

References

Brace, W.F., A.S. Orange, and T.R. Madden, The effect of pressure on the electrical resistivity of water-saturated crystalline rocks, *J. Geophys. Res.*, *70*, 5669-5678, 1965.

Brace, W.F., J.B. Walsh, and W.T. Frangos, Permeability of granite under high pressure, *J. Geophys. Res.*, *73*, 2225-2236, 1968.

Bruner, W., Crack growth during unroofing of crustal rocks: effects of thermoelastic behavior and near-surface stresses, *J. Geophys. Res.*, *89*, 4167-4184, 1984.

Doyen, P., Crack geometry in igneous rocks, a maximum entropy inversion of elastic and transport coefficients, to appear in the *J. Geophys. Res.*, 1987.

Hadley, K., Comparison of calculated and observed crack densities and seismic velocities in Westerly granite, *J. Geophys. Res.*, *81*, 3484-3494, 1976.

Jaynes, E.T., Information theory and statistical mechanics, *Phys. Rev.*, *106*, 620-630, 1957.

Kirkpatrick, S., Percolation and conduction, *Rev. Mod. Phys.*, *45*, 574-588, 1973.

Nur, A., and G. Simmons, The origin of small cracks in igneous rocks, *Int. J. Rock Mech. Min. Sci.*, *7*, 307-314, 1970.

Sprunt, E.S., and W.F. Brace, Direct observation of microcavities in crystalline rocks, *Int. J. Rock Mech. Min. Sci. Geomech. Abstr.*, *11*, 139-150, 1974.

MAXIMUM ENTROPY ANALYSIS OF LIQUID DIFFRACTION DATA

John H. Root, P. A. Egelstaff, B. G. Nickel
Physics Department, University of Guelph,
Guelph, Ontario, Canada, N1G 2W1

Abstract:

A maximum entropy method for reducing truncation effects in the three dimensional inverse Fourier transform of S(q) to g(r) is described in detail. The failure of a naive approach, maximizing the entropy of g(r) subject to a chi-square statistic on the S(q) data is demonstrated. Subsequently, a method is described which yields better results, by restricting the spaces on which g(r) and S(q) are defined, and allowing a tail to form beyond the original range of the data. Our algorithm is tested using a PY hard sphere structure factor as model input data. An example using real data on the structure of light and heavy water is presented. It is seen that the maximum entropy method can greatly enhance our ability to distinguish physically meaningful structure from that arising from noise and the truncation effects of traditional Fourier transform methods.

Introduction

The pair correlation function, g(r), is of central importance in modern theories of the liquid state, and links the microscopic world of intermolecular potentials to the macroscopic world of pressure, compressibility, and other thermodynamic quantities. It is related to the probability that two sites (e.g. atoms or molecular centres) will be found separated by distance r. It is not a probability distribution itself, since it approaches a finite value of unity as $r \to \infty$, and hence is not normalizable. In a diffraction experiment on a liquid, one measures the intensity of scattered particles (e.g. photons or neutrons) as a function of scattering angle, Θ. After a number of correction and calibration procedures, this can be reduced to a structure factor, S(q), which is related, by Fourier transformation, to g(r). Here, q is the magnitude of the scattering vector, λ is the wavelength of the incident radiation, ρ is the fluid particle density and:

$$q = 4\pi/\lambda \sin(\Theta/2) \tag{1}$$

$$S(q) = 1 + 4\pi\rho/q \int_0^\infty dr\, r \sin(qr)\, [g(r) - 1] \tag{2}$$

G. J. Erickson and C. R. Smith (eds.),
Maximum-Entropy and Bayesian Methods in Science and Engineering (Vol. 2), 395–407.

Equation (2) has been integrated over angular variables, given that g(r) depends on the magnitude of r only.

What is desired is to obtain g(r) from experimental data on the structure factor. However, inverse Fourier transformation of (2) is not possible because there is a limit of q, denoted Q, beyond which data either cannot be collected (eg. $\theta < \pi$), or is of low quality, rendering it useless. Simply calculating:

$$g(r) = 1 + 1/(2\pi^2\rho r) \int_0^Q dq\ q \sin(qr)\ [S(q) - 1] \tag{3}$$

incurs a truncation error unless S(q) = 1 for all values of q > Q. In figure 1a, we show the inverse Fourier transform (IFT) of some structure factor input data, E(q), calculated for the PY hard sphere fluid with reduced density $\rho\sigma^3 = 0.5$, according to a standard formula. (For the PY examples in this paper, we shall use q and r for the dimensionless units $q\sigma$ and r/σ commonly used in liquid state literature.) E(q) is calculated on a grid of points with spacing Δq=0.2, to a maximum value Q = 14, and thus is a discrete, highly truncated sample of the complete structure factor, but is otherwise a set of perfect data. We also show, for comparison, the IFT of a data set which is identical, except that Q is extended to a range such that truncation errors are minimized. The IFT of E(q) exhibits rippling throughout its range, and spurious structure like this prevents a diffractionist from making reliable conclusions about small details in the correlation functions of real fluids. (In liquid diffraction studies, the shape of g(r) is as important as the positions and magnitudes of peaks.) In figure 1b we show the forward Fourier transform, S(q), of the IFT of E(q), including some points calculated for q > Q. We see that S(q) = 1 for q > Q. This corresponds to the biased estimate of a tail on the input E(q), implied by calculating g(r) at values of r less than π/Q in figure 1a.

Naive Maximum Entropy Solution

One might hope that a Maximum Entropy (ME) method could be readily applied to the problem of truncation effects in the IFT of structure factor data. We begin, following the lead of ME image reconstructors [Frieden, 1972; Gull and Daniell, 1978; Skilling and Bryan, 1985], by creating an entropy expression suitable for g(r), our underlying distribution function, and maximizing it subject to a mathematical constraint expressing "fit" to the structure factor data. For the entropy of g(r) we write:

$$H = - \sum_{i=1}^{N} 4\pi r_i^2 \Delta r\ g(r_i) \ln(g(r_i)/e) \tag{4}$$

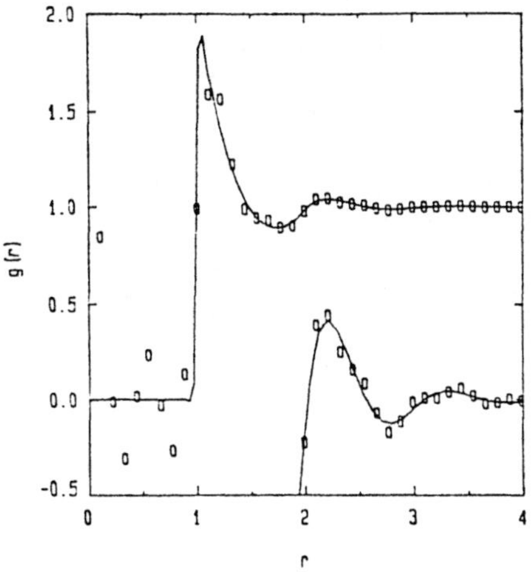

Fig 1a The solid line shows the IFT of the PY E(q) where Q=70. The circles are the IFT where Q=14. The inset is a X10 expansion for r > 2.

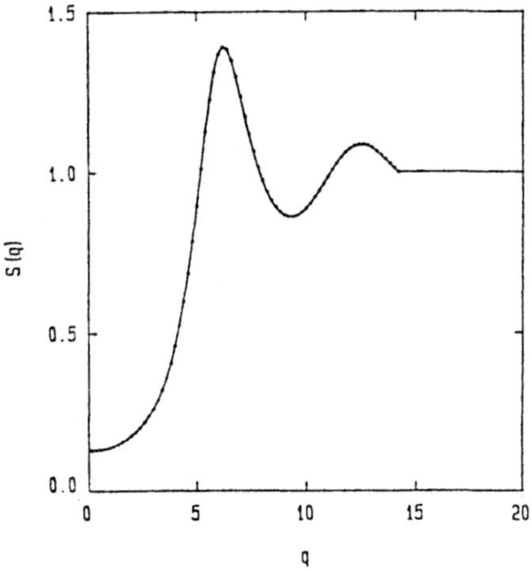

Fig 1b The solid is the FT of the IFT where Q=14 and the dots are the original E(q).

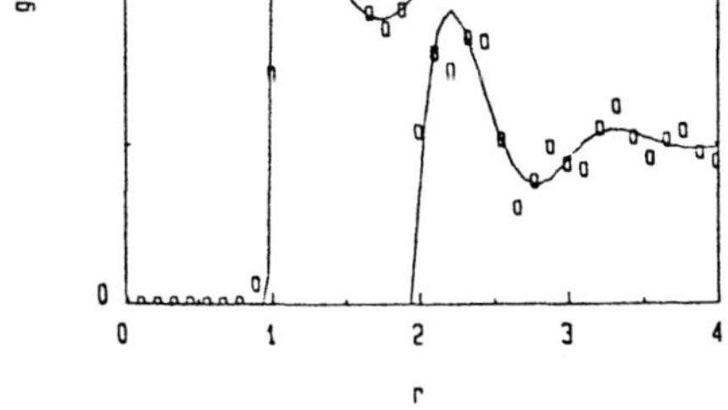

Fig 2a The line is the IFT of the PY E(q) where Q=70. The circles show the "naive" MEM g(r). The inset is a X10 expansion for r>2.

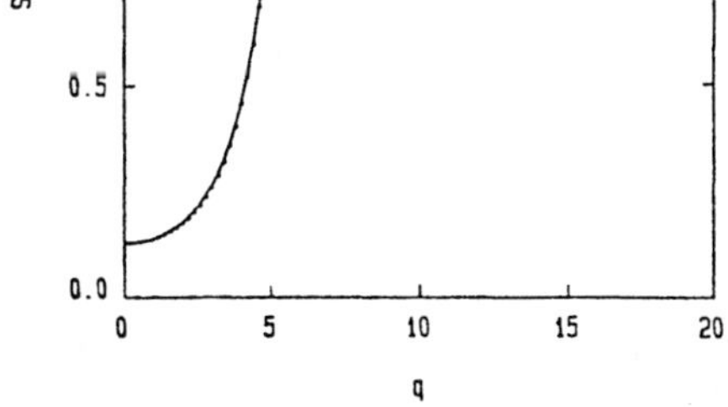

Fig 2b The line is the FT of the "naive" MEM g(r), and the dots are the original data.

This form incorporates the fact that g(r) is a centrosymmetric function in three dimensional space where volume elements of equal magnitude are the reasonable states over which our distribution will be defined. In the absence of data (ie constraints), maximizing this entropy with respect to g(r) yields g(r) = 1 for all values of r. This is the most sensible prior estimate of g(r) for a fluid, given only structure factor information as a constraint, since it corresponds to the correlation function for an ideal gas of point particles. We rule out the use of prior estimates which force g(r) to be zero for low values of r, as expected in an ideal liquid, preferring to take the view that the data measured on a liquid system will produce the properties of a liquid without such forcing, if the algorithm will allow it to do so. Any deviations of g(r) from expected behaviour are either due to lack of information in the data, or systematic errors therein, and thus are useful guides regarding the improvement of an experiment.

The entropy of g(r) (4) is maximized subject to the commonly used chi-square constraint expressing "fit" to the input E(q), of the output S(q), obtained from g(r) through equation (2):

$$1/N \sum_{k=1}^{N} (E(q_k) - S(q_k))^2/\sigma_k^2 = 1 \tag{5}$$

Division by the point-by-point variance of the data (if available) allows freedom to use data of quality that varies over the range of q studied. We can solve this maximization problem, using an undetermined Lagrange multiplier, A, to obtain an expression for the g(r) with maximum entropy that fits the structure factor data, E(q), as judged by the reduced chi-square statistic:

$$g(r) = \exp\left[A \sum_{k=1}^{N} ((E(q_k) - S(q_k))/\sigma_k^2)(\sin(q_k r)/q_k r)\right] \tag{6}$$

All that remains is to solve equation (6) for g(r), using various values of A until constraint (5) is satisfied. This is readily accomplished with a Newton-Raphson numerical method, which is appropriate given the typical numbers of data points involved in a liquid diffraction experiment, and given the assurance that the solution to the system of equations (2, 5, 6) is unique [Skilling and Bryan, 1985].

Before we plunge into numerical computation, let us study equation (6) and make some notes.

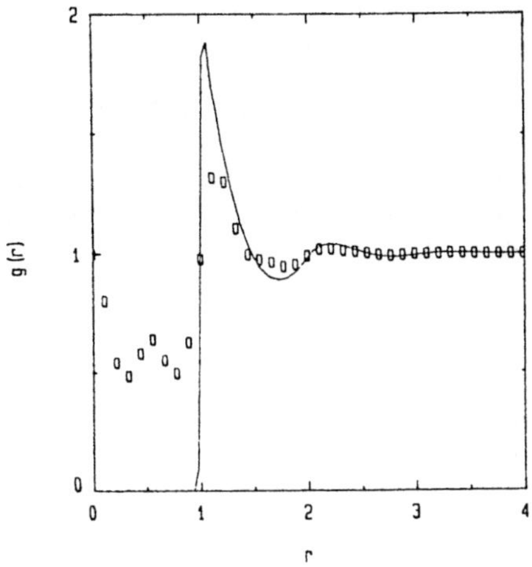

Fig 3a The line is the IFT of the PY E(q) where Q=70. The circles show the MEM g(r) when R = $2\pi/\Delta q$.

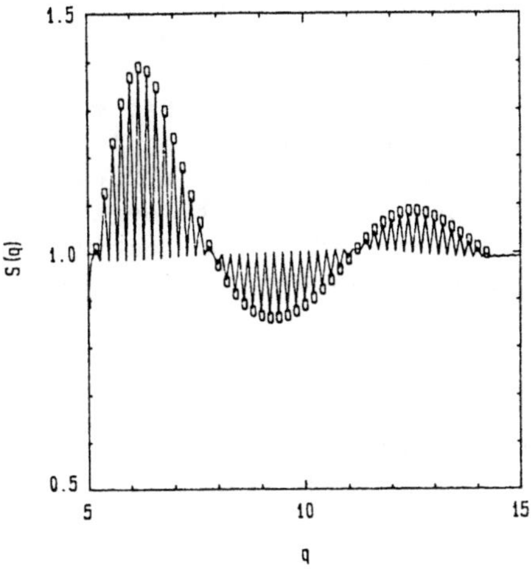

Fig 3b The line is the FT of the MEM g(r) where the range is doubled, and the circles show the original data, E(q).

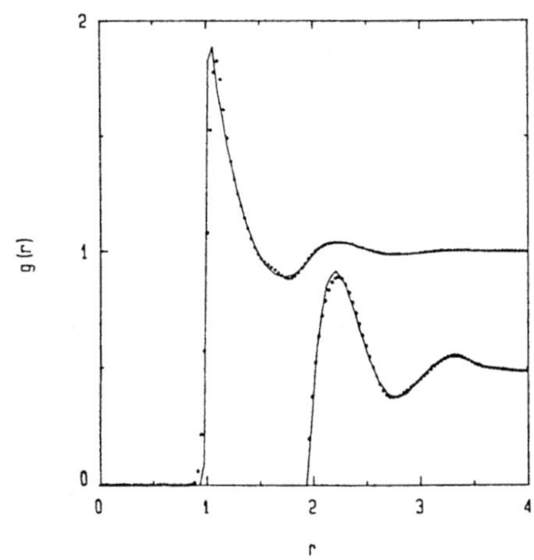

Fig 4a The line is the IFT of the PY E(q) where Q=70. The dots show the GRIT g(r). The inset is a X10 expansion for r>2.

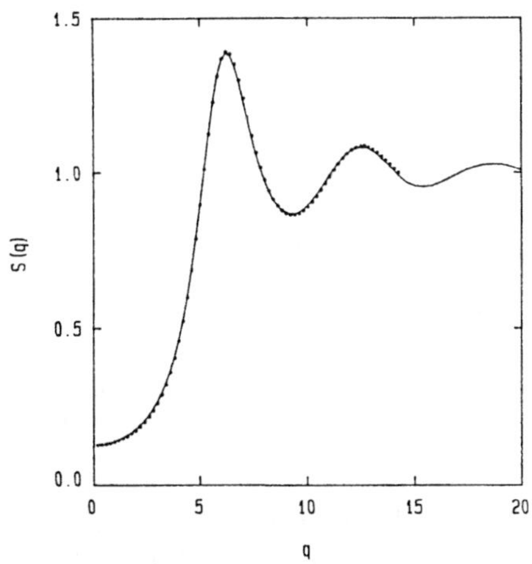

Fig 4b The line is the FT of the GRIT g(r) and the dots are the original E(q).

a) We are encouraged by the fact that at large r, the ME g(r) approaches the expected asymptotic value of unity.

b) It seems good that g(r) is strictly positive for all r, which must be the case for an ideal fluid. Of course, in a real experiment, unknown systematic errors in the data may require some of the low r values of g(r) to be negative if the data is to be "fit" within the predetermined error estimates. Since the ME g(r) cannot access negative values, this can result in spurious structure in other regions of r.

c) The contributions to the summation which ultimately give rise to the structure in g(r), are weighted by 1/q. This may be an improvement over the IFT, where contributions to the structure in g(r) are weighted by q and data quality decreases as q increases.

d) There remains in the ME g(r) a truncation error, since the summation still does not extend to infinite Q.

e) There is no indication in equations (5, 6) that S(q) is analytic, with continuous derivatives at all values of q. Thus, the ME g(r) of equation (6) need not correspond to an S(q) that smoothly interpolates between the input data points, nor smoothly extrapolates beyond Q.

f) There is probably a fundamental flaw in applying the ME principle to g(r) which is not a normalizable distribution on its own.

The points d, e and f are important negative features of the "naive" ME solution to the problem of IFT truncation error. Point b will be a problem for any ME solution if we are interested in applications to real data with unknown systematic errors.

To demonstrate the unacceptability of the "naive" ME solution (2, 5, 6), we show some graphical examples. Figure 2a shows the solution of equation (6) which fits the same PY E(q) as we used for the IFT in figure 1a. For comparison, we show the "correct" IFT also as in figure 1a. The ME g(r) has more spurious structure than was obtained in the IFT of the truncated E(q). In figure 2b we show the Fourier transform of the ME g(r), including some values of S(q) for q > Q. The input data has been fit, but the behaviour of S(q) at Q is nonsensical, in view of our additional knowledge that S(q) is continuous, with continuous derivatives at all values of q. Of course, as noted, this knowledge is not a part of the constraint (5), so we should not be surprised if the "least-biased" g(r) does not correspond to the "smoothest" extrapolation of S(q). Nevertheless, the ME g(r) is less acceptable than the IFT, at the moment.

In figure 2, we calculated the ME g(r) on the same grid of points as we did for the IFT, with the range of r, denoted R, limited to a value of $\pi/\Delta q$. In reality, g(r) has an infinite range, and the ME principle contains no reason for limiting it. In figure 3a we show the ME g(r) obtained when $R = 2x\pi/\Delta q$. By doubling the range of r we have halved the amplitude of g(r)-1. This nonsensical result may arise from the fact that g(r) is not a normalizable distribution. If R tends to infinity, as it should, g(r) will contain negligible structure, even though the constraint (5) is satisfied. In figure 3b we show the Fourier transform of the ME g(r) in figure 3a. At the data points E(q), the fit of S(q) is good, but between the data points, the ME S(q) tends to return to the base level of unity, rather than interpolate smoothly as we would prefer to see. It is possible that this unacceptable behaviour could be corrected by the use of a constraint containing functional information on S(q). (Chi-square treats every data point as an independent entity, averages the fit over the entire data set, and hence is too "blunt" to obtain a completely satisfactory ME solution to the truncation problem.) The search for such a constraint (or combination of constraints) is a matter of current study.

Patching the Naive Maximum Entropy Solution

Even though the foregoing "naive" Maximum Entropy solution exhibited unacceptable behaviour, we have pursued the method, since the underlying philosophy is so attractive to those of us who want to make the least biased conclusions about the "message" present in physical data. A few modifications to the above solution were tested to see if improvements in the ME results could be achieved. In essence, we have applied reasonable additional constraints on the maximization of the entropy of g(r). These constraints are incorporated into the numerical procedure used to solve the system of equations (2, 5, 6) rather than being cast into the form of Lagrange multipliers.

The resulting algorithm for reducing the effects of truncation error on the inverse transformation of E(q) data to g(r) is described in the appendix to this paper. Beyond the range of the data, Q, we define two more limits: Q_1 is about one additional wavelength beyond Q and Q_2 is about one additional wavelength beyond Q_1. Pseudo-data is created for the region $[Q, Q_2]$ by smoothly joining the current solution for S(q) for $Q_1 < q < Q_2$ to the input data E(q) with a polynomial fit to E(q) for the last wavelength before Q and the far region $[Q_1, Q_2]$. As numerical solution of (2, 5, 6) proceeds, the difference function (E(q) - S(q)) in (6) smoothly approaches zero at Q_2, thus reducing the truncation error of note (d). Furthermore, extending the data with a polynomial near Q is an indirect way to assert that S(q) is analytic for $q \geq Q$, as required to address note (e). We restrict the range of g(r), requiring $R \leq \pi/\Delta q$. This is the allowed range of data in r-space, given a discrete set of points separated by Δq in q-space, according to the rules of Fourier series [Lado, 1971]. The effect is to circumvent the normalizability problem of note (f). It is important to ensure that R is greater than the range of correlations in g(r), or one obtains a Fourier truncation error on the forward Fourier transform in equation (2). This may require interpolation of the data to reduce Δq. Interpolation is also required if there are gaps in E(q) grid of spacing Δq which would otherwise yield results like those shown in figure 3. Preparing E(q) for analysis by interpolation is an assertion that S(q) is an analytic function, again addressing note (e), and so is a reasonable operation. (See also the notes in the appendix.)

In figure 4a, we compare the result of our algorithm, GRIT, to the "correct" IFT for long range Q. The input data was the same as for all other truncated cases shown so far. We see that spurious structure is greatly reduced in comparison to the IFT of figure 1a and the "naive" ME solution of figure 2a. The remaining spurious structure is related to the order of the polynomial splice, the distance $[Q,Q_1]$, the density of E(q) points, and the level to which S(q) is required to fit E(q). We have not optimized these conditions for this figure. The height of the main peak is underestimated by our ME algorithm, but this is expected, since the Maximum Entropy priciple strongly restores g(r) to its baseline of 1 unless there is strong information in E(q) to the contrary. The truncated data set lacks the long q range information necessary to sharpen the first cusp to its true shape and height. Note that the low r behaviour of our ME g(r) came out to be nearly zero, without forcing the prior estimate of g(r) to be so. In figure 4b we compare the original truncated E(q) and the reconstructed S(q) which is the forward Fourier transform of the ME g(r). In the region $q > Q$, S(q) shows the behaviour expected of an analytic function and is greatly improved over similar regions in figures 1 through 3.

Achieving success in the inverse transformation of PY hard sphere data is sufficient to demonstrate that our algorithm is robust for all practical cases in liquid diffraction data analysis. The PY hard sphere E(q) was an "acid test" because of the discontinuity in g(r), and resulting long range oscillations in S(q). Real liquid systems have continuous correlation functions and so, in principle, are easier to analyze.

A Note on Systematic Errors

Unfortunately, in all practical cases, there is a new difficulty to face -- the presence of systematic errors in E(q) of unknown magnitude. The results of not knowing, for instance, that we have underestimated the fluid density, ρ, by 3% are illustrated in figure 5a. To fit the data, with the lower density, we must have $g(r) < 0$ for $r < 1$. Since, the ME g(r) cannot attain such values, there is no solution to equation (6) such that constraint (5) is also satisfied. The best reduced chi-square we can obtain is 9 (if the same error estimates on E(q) are used here as in the previous examples), and spurious structure is pronounced. The fact that we cannot find a ME solution that satisfies the constraint, indicates that there are additional physical effects not accounted for in our analysis [Jaynes, 1978], and thus we are alerted to the presence of systematic errors in our data set.

However, it is frightening to contemplate the possibility that a

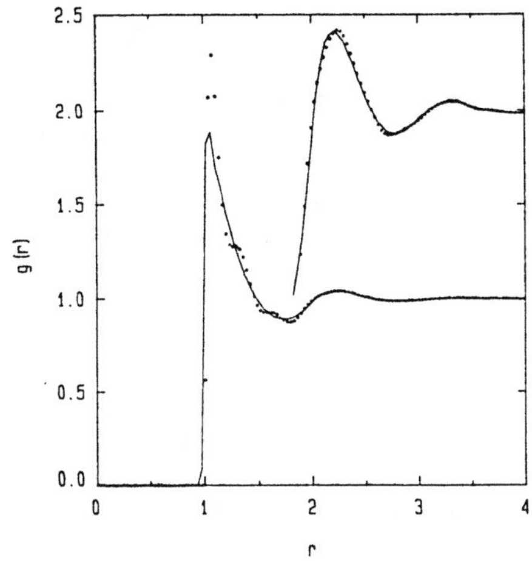

Fig 5a The line is the IFT of the PY E(q) where Q=70. The dots show the GRIT g(r) where the density is underestimated by 3%.

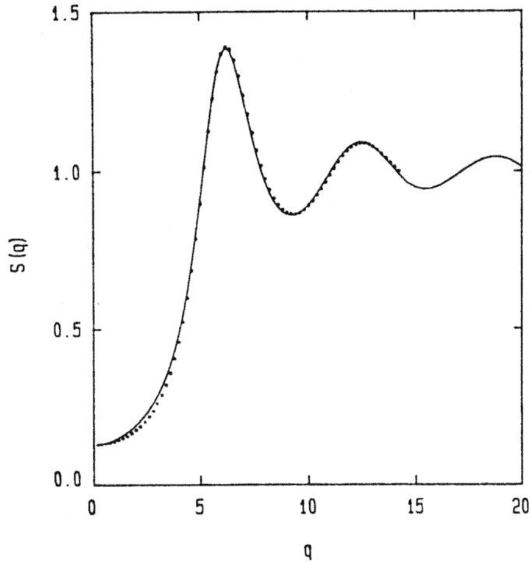

Fig 5b The line is the FT of the GRIT g(r). The dots are the original E(q) which cannot be fit due to the systematic error.

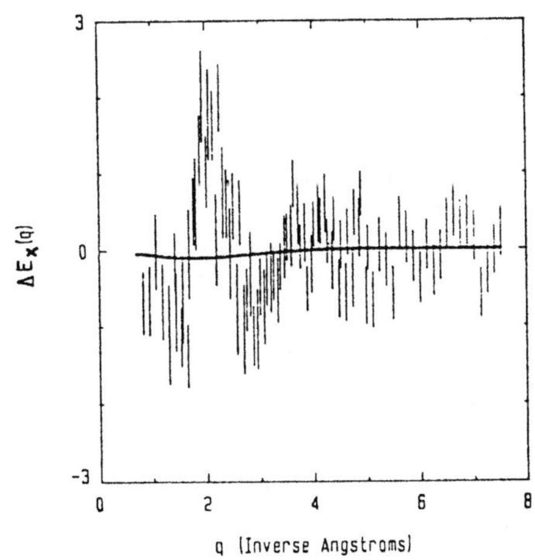

Fig 6a The error bars show our experimental $\Delta\mathcal{E}_X(q)$, indicating uncertainties due to counting statistics. The solid line is $\Delta\langle F^2\rangle$ calculated with free molecule data.

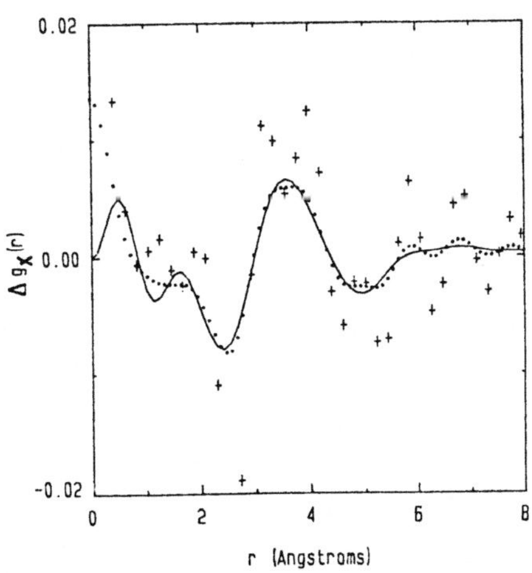

Fig 6b The crosses show the IFT of our data. The dots show the GRIT $\Delta g_X(r)$ obtained using all of the experimental data. The solid line is the GRIT transform using only data for $q < 5\ Å^{-1}$.

systematic error in the data, in conjunction with the necessary positivity of g(r) might not be so severe as to prevent constraint (5) from being satisfied, but to result in spurious structure in g(r). The best solution to this problem is to gain a lot of knowledge enabling the minimization of systematic errors. In the meantime, our approach is to widen the error estimates on the data until the entropy of g(r) "levels out". That is, by studying the variation of the entropy of g(r) with the average level of σ_q, one observes a change in the behaviour as g(r) positivity begins to contribute to the structure of the correlation function. We widen $\langle\sigma\rangle$ to slightly larger than this point, so the structure of g(r) predominantly arises from constraint (5), rather than the combination of systematic errors and the g(r) positivity condition. This seems to be the most conservative treatment of the data, in line with the underlying philosophy of Maximum Entropy methods. We are aware that some ME deconvolution algorithms use the positivity condition to produce "super-resolution", but remain wary of trusting E(q) data to the extent that the structure so generated is reliable.

Application to Real Data

A recent thesis [Root, 1986] studied the difference in the structure factors of light and heavy water. From a classical point of view, there should have been no measureable difference, because the forces between the molecules are identical. However, accounting for quantum effects in the molecular motions, there was a predicted slight difference for the two fluids [Kuharski and Rossky, 1985]. The experimental data contained low systematic errors due to the design of the diffractometer, but the statistical noise on the difference between the two structure factors ($\Delta E_X(q)$) was relatively large, and, the data set was truncated. We began work on our Maximum Entropy algorithm because such conditions could easily result in spurious structure in the difference in correlation functions, $\Delta g_X(r)$, leading us to make false conclusions about the small effect studied.

In figure 6a we show the actual data collected. Interpolation provided values for $\Delta E_X(q)$ to reach the known asymtotic value of −0.09 at q=0. To run this data with the present algorithm, a value of 1 was added to all values of $\Delta E_X(q)$ and subtracted from all values of the resulting $\Delta g_X(r)$. Thus, in this case, the g(r) positivity boundary was never encountered during the numerical procedure. In figure 6b we show the IFT of the data which exhibits a lot of spurious structure. It is difficult to say where the nodes of the difference are, and whether or not there is a significant feature near r = 1.7Å. Two ME $\Delta g_X(r)$s are also shown, one using the complete set of data, and the other using only the data of higher quality for values of q < 5Å^{-1}. In both cases, the spurious structure is greatly reduced, and our ability to distinguish the presence of "real" structures is improved. The loss of amplitude in the ME differences is due to the wide error bars on the data. This is the least biased $\Delta g_X(r)$ consistent with the data, and to increase the amplitude in this function would require more certain information in $\Delta E_X(q)$.

Summary

We have described the truncation problem arising in inverse Fourier transformation of liquid diffraction data to obtain the underlying pair correlation function. A straightforward application of a Maximum Entropy method using reduced chi-square as a single constraint on the maximization of the entropy of g(r) was shown to yield an unsatisfactory solution to the truncation problem. Modifications to this method approximately addressed the truncation effect remaining in the ME solution, the need to express the analyticity of S(q), and the need to express g(r) as a normalizable distribution. A very good solution resulted, indicating that future efforts of a more rigorous sort would indeed be worthwhile. We noted that unknown systematic errors in a data set might still result in spurious detail arising in a ME g(r). We presented an example of the use of our algorithm on real data in which it was important to distinguish the real physical structural effects from those arising due to noise, and truncation in the data.

In conclusion, we are optimistic that a rigorous ME method eventually will be developed to reduce truncation effects in the analysis of liquid diffraction data, as we have done with the approximate solution described in this paper. To do this, a mathematical way to introduce the analyticity of S(q) as a constraint on the maximization of the entropy of g(r) must be invented. The truncation error in the ME g(r) and the normalization question must also be addressed for this eventual solution to be successful.

Appendix: GRIT

This is a Maximum Entropy algorithm to minimize the truncation error incurred on Inverse Fourier Transformation of a structure factor of limited range, Q.

1.0 Input E(q), $\sigma(q)$ and control parameters: FT prefactor, initial A, Δq, ratio Q_2/Q, number of points to used to calculate polynomial, and range of data $[Q, Q_L]$ to be replaced with splice. (See notes)

2.0 Initialization

2.1 Establish grid of r values for g(r) with $\Delta r = \pi/Q_2$.

2.2 Set E(q) = 1 and set $\sigma(q) = \sigma(Q)$ for $Q < q \leq Q_2$.

2.3 Set initial guess: S(q) <-- E(q).

3.0 Smoothly join tail to E(q) by replacing E(q) for $Q < q < Q_2$ with current values of S(q) and then replacing E(q) for $Q < q < Q_1$ with polynomial fit to points at either end of the region.

4.0 Find $S(q)$ for current tail and value of A.

4.1 Calculate $g(r)$ from equation 2.6.

4.2 $F(q_j) = S(q_j)-1 - 4\pi\rho/q_j \sum_i \pi/Q_2\, r_i \sin(q_j r_i)\, [g(r_i)-1]$

4.3 If $F(q_j)$ is small for all j, jump to step 5.0

4.4 $M_{jj'} = \partial F(q_j)/\partial S(q_{j'})$ and invert it giving $M_{jj'}^{-1}$.

4.5 Improved guess: $S(q_j) \leftarrow S(q_j) - M_{jj'}^{-1} \cdot F(q_{j'})$

4.6 Repeat 4.1 - 4.5 until a solution is found.

5.0 Calculate $S2 = 1/N_Q \sum (E(q_k) - S(q_k))^2 / \sigma_k^2$ for $q_k < Q$.

5.1 If S2 is within preset boundaries ($1 +/- 1/\sqrt{N_Q}$) or if it has not changed much from previous value, jump to step 6.0.

5.2 If it is too high, increase A. Otherwise, decrease A.

5.3 Jump to step 3.0.

6.0 Output the final $S(q)$, and corresponding $g(r)$.

Notes:

1) We have used this algorithm to perform transformations of truncated $g(r)$ data (from computer simulations) to $S(q)$, and the procedure works very well in this type of problem, as well. The only modification to the algorithm is to input the FT prefactor as $1/2\pi^2\rho$ and the program will put this wherever the prefactor $4\pi\rho$ was used in lines 4.2 and 4.4. One mentally interchanges vector labels: g <--> S and r <--> q, but the program itself does not distinguish between correlation functions and structure factors. Since $S(q)$ does not equal zero for a range of low q values, the inconsistency between low σ and $S(q)$ positivity is less often encountered than it is in the transformation of $E(q)$ to $g(r)$. However, since oscillations in $S(q)$ often have a fairly long range, one needs to be careful to ensure that the spacing of points in r-space is sufficiently fine that Q truncation is minimized. (See note 4 below.)

2) If you guess the right value of A at the outset, the algorithm is powerful enough to find a solution immediately (if one exists). Unfortunately, such a solution may contain the effects of polynomial fitting errors (arising as discontinuities at the junction points Q and Q_1) and the original biased guess of the tail for $E(q)$. It is usually best to start A low enough that a number of tail iterations occur before the procedure finishes.

3) The spacing of data points obtained from a real experiment is usually less than the q-resolution of the diffractometer, and is normally much finer than the structure under observation. The q-resolution function of an instrument varies with the value of q, generally in shape and magnitude. In our algorithm, we have not

attempted to treat the problem of q-resolution, though if one had knowledge about the machine response as a function of q, an obvious procedure would be to forward-convolute the FT of g(r) in line 4.2 (as well as the term "1") before calculating the difference function F(q). This would be better than attempting to deconvolute the instrument response function, using traditional methods before starting the procedure.

4) The input value of Δq in line 1.0 of GRIT can be chosen larger than the spacing of data points, and it is advantageous to do so, since this effectively groups neighbouring points, and introduces the desired functional nature of S(q) into the algorithm. However, Δq should not be chosen so large that the range of g(r): $R = \pi/\Delta q$ is less than the distance to which significant correlations extend. If this occurs, there is a Fourier truncation error on the forward FT in line 4.2.

References

FRIEDEN, B. R., J. Opt. Soc. Am. **62**, 511 (1972).
GULL, S. F., DANIELL, G. J., Nature **272**, 686 (1978).
JAYNES, E. T., "Where Do We Stand on Maximum Entropy?" The Maximum Entropy Formalism, R. D. Levine and M. Tribus ed. MIT Press, Cambridge Mass. and London (1978).
KUHARSKI, R. A., ROSSKY, P. J., J. Chem. Phys. **82**, 5164 (1985).
LADO, F., J. Comput. Phys. **8**, 417, (1971).
ROOT, J. H., Quantum Effects in the Structure of Water, PhD Thesis, University of Guelph, 1986.
SKILLING, J., BRYAN, R. K., Mon. Not. R. Astr. Soc. **211**, 111 (1984).

RANDOM ARRAY BEAMFORMING

KEITH H. NORSWORTHY, AND PAUL N. MICHELS

BOEING ELECTRONICS HIGH TECHNOLOGY CENTER

ABSTRACT

Conventional random array beam forming methods give sidelobe responses with magnitudes near [-10 log N] db, where N is the number of receiving elements in the array, and it is found that conventional shading methods are ineffective in reducing the mean sidelobe magnitude. As a consequence, a low sidelobe requirement translates into a need for many receivers in the array (i.e. 100 receiver elements for a mean sidelobe of -20 db).

This presentation shows that the high sidelobes of random arrays arise principally from the mishandling of 'absent' data. New random array beamforming methods are described that lead to greatly improved sidelobe characteristics. The new beam forming methods are applied in the element-separation/cross correlation domain (as opposed to the conventional spatial/signal domain) and care is taken to ensure that 'absent' signals are not (unintentionally) assumed to have zero amplitudes.

Computer simulation results are presented for a partially filled uniform line-array, but the method is shown to be equally applicable for random arrays with linear, planar, or volumetric spatial extent.

1.0 INTRODUCTION

This paper describes signal processing methods for improving the directional response characteristics of an array of passive receiver elements when the spatial distribution of the elements is random. The work was performed as part of a Boeing independent research and development (IR&D) program which addressed potential submarine detection and tracking systems. Figure 1-1 illustrates the deployment of a set of acoustical receiver sonobuoys which sense the acoustic vibrations originating at the target. The target direction is determined by beamform processing of the element signals.

THE SIXTH ANNUAL WORKSHOP ON MAXIMUM ENTROPY AND BAYESIAN METHODS IN APPLIED STATISTICS
SEATTLE UNIVERSITY 8/5/86 - 8/8/86.

G. J. Erickson and C. R. Smith (eds.),
Maximum-Entropy and Bayesian Methods in Science and Engineering (Vol. 2), 409–420.

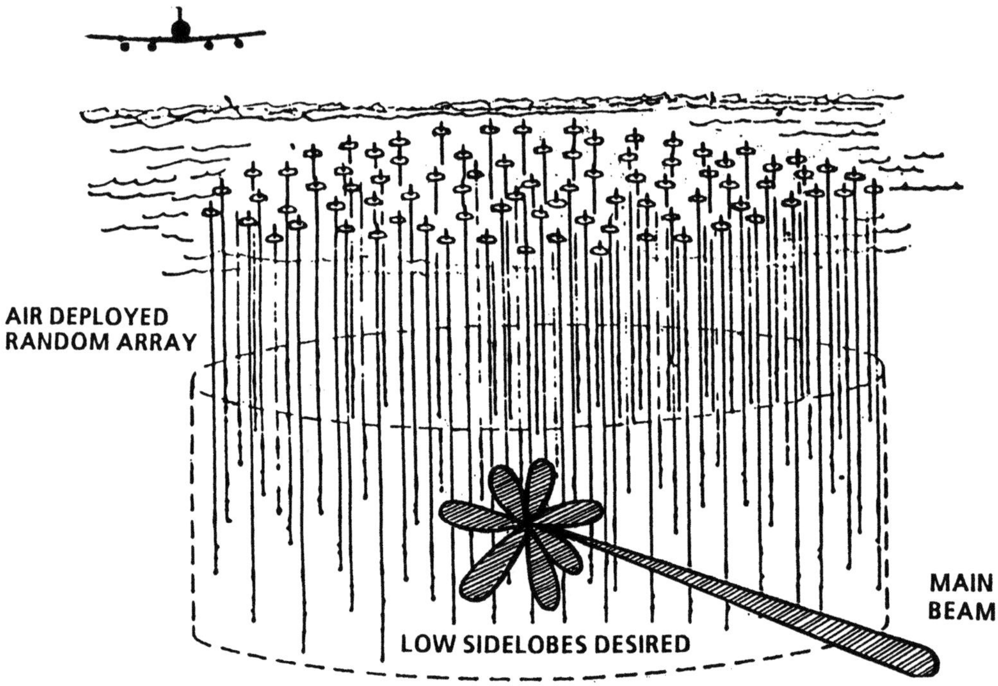

Figure 1-1. Submarine Detection and Tracking System.

The objective of the beamforming process is to develop a directional pattern which has maximum responsivity in the beam steer direction and minumum response in all other directions. For arrays of uniformly spaced receiving elements, methods are well known for suppressing the sidelobe response by factors in the range of 30 to 60 dB (ref 1,2). When the spacing of the elements are not uniform, however, the conventional sidelobe suppression methods are no longer effective and the degree of suppression achievable then becomes dependent on the number of elements employed. Steinberg (ref 1,3) shows that the mean sidelobe response for an N element random array is [-10 Log N] db and that the peak sidelobe response is 8 dB greater. Thus, for a 30 element random array, the mean sidelobe level is about -15 dB. In order to achieve a sidelobe level of -30 dB, to correspond with the performance achievable with a uniformly spaced array of 30 elements, the number of elements needed in the random array has been estimated as 1000. The increases in cost associated with the hardware and processing for this larger number of elements provides the incentive for the new sidelobe suppression described in this paper which improves sidelobe suppression without increasing the number of elements.

The new beamforming method is applied in the element-separation cross correlation domain in a manner which exploits the available sensor information content.

For purposes of clarifying the fundamental issues, a conventional line array is analyzed first and the analysis is then extended to similar line arrays where some of the receiving elements are absent. Discussion is then extended to line, planar, and volumetric arrays in which the receiver element positions are totally random.

Computer simulation results are presented for a partially filled uniformly spaced line array and the method is shown to be equally applicable to completely random arrays with linear, planar, or volumetric spatial extent.

2.0 CONVENTIONAL LINE ARRAY BEAMFORMING PROCESSING

The sensor array can be made responsive to signals from a selected direction by compensating each sensor signal to correct for the propagation time difference between the source and the individual sensors. The sum of the compensated signals is maximum in the selected direction and is low for sources at other directions. The ideal beamformer employs time delay compensation. However, if the signal is narrow band, the time delays are well approximated by corresponding phase shifts. A wide band signal can be processed by phase shift beamforming if the signal is first filtered into a set of narrow bands and separate beamforming is done in each band. Furthermore, the process of forming the entire set of independent directional beams is mathematically equivalent to a Fourier Transform process from the space (element location) dimension to the direction of arrival (angle) dimension. (ref. 4).

Figure 2-1 illustrates three different but equivalent methods of conventional line array beamforming:

METHOD 1: Spatial Fourier transformation of the receiver signal vectors.
METHOD 2: Fourier transformation of the spatial correlation function.
METHOD 3: Solution of simultaneous equations.

In each case, the beamforming algorithm achieves a responsivity pattern which is maximized in the main lobe direction and has a low response in all other (sidelobe) directions. The low sidelobe responses result from the alignment of the component signal phases into cancelling patterns and these sidelobes can be further reduced by known Fourier amplitude shading techniques.

3.0 LINE ARRAY WITH MISSING RECEIVER ELEMENTS

3.1 APPLICATION OF METHOD 1, FOURIER TRANSFORM OF SPATIAL VECTORS

When some of the receiving elements are missing (or inoperative), see Figure 3-1, the orthogonality of the Fourier components is upset and the low sidelobe characteristic is no longer achieved. This results because the phase distributions in the sidelobe signals are poorly cancelled when the contributions of the missing sensors are not present (ref. 1, 3). This condition also renders the normal sidelobe reduction shading methods ineffective since the needed compensation cannot be achieved in all beam directions. Figure 4-1 curve B shows typical results when the Fourier Transform of spatial vectors (Method 1) is applied to the array with missing elements.

3.2 FOURIER TRANSFORM OF SPATIAL CORRELATION FUNCTION

Some performance improvement is achievable by using Method 2 employing the spatial correlation method.

Working in the spatial correlation/"element-separation" domain has two benefits

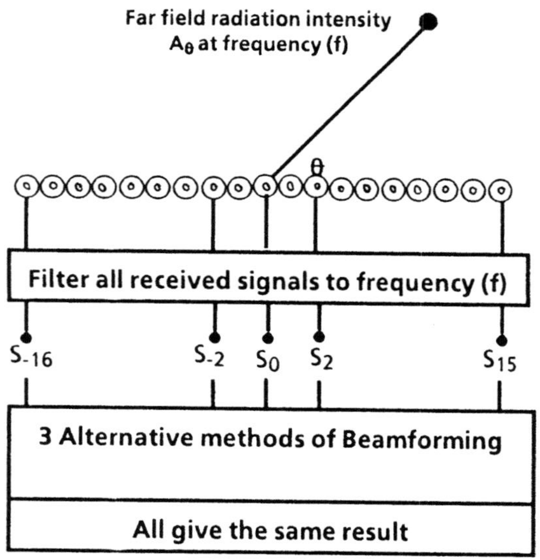

Note: In all cases, array shading (eg.Hanning)can be used to modify sidelobe characteristics.

$$S(n) = \Sigma A(\Theta) e^{-j[2\pi ft + \varnothing(n)]} \qquad (1)$$

where $\varnothing(n) = \frac{2\pi nd \cos\Theta}{\lambda}$

•To prevent spatial frequency Nyquist folding $d \leq \lambda/2$

Method No. 1 Spatial Fourier transform

$$F(k) = 1/N \sum_{n=-N/2}^{(N/2)-1} S(n) e^{\frac{-i2\pi Kn}{N}}$$

N = number of receiving elements

Method No.2 Fourier transform of Spatial Correlation Function

$$C(\Delta) = \Sigma \overline{S(n) S(n+\Delta)}$$

$$P(K) = 1/(2N-1) \quad \Sigma \quad C(\Delta) \cos\left(\frac{2\pi K\Delta}{N}\right)$$

Method No. 3:

Solution of simultaneous linear equations relating the amplitudes of uncorrelated far-field radiant intensities to the amplitude and phase of each correlation coefficient $C(\Delta)$.

Figure 2-1. Conventional Line Array Beamforming.

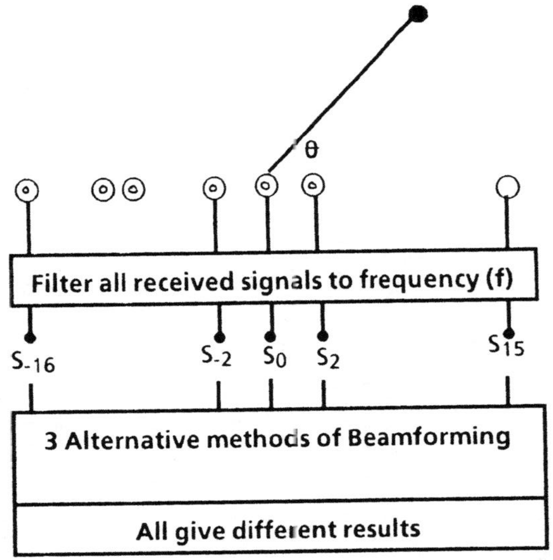

Previous Array with some receivers absent.
Beamforming methods now give different results.

Method No. 1 Spatial Fourier transform

$$F(k) = \frac{1}{(N-a)} \Sigma \, S(n)e^{\frac{-i2\pi Kn}{N}}$$

summation is over (N-a) signals only
This method is standard in the industry and leads to very poor sidelobe amplitudes that are not reduced by shading.

Method No. 2

A) Gives identical result to method 1 when

$$C(\Delta) = \Sigma \, \overline{S(n)S(n+\Delta)}$$

summed over all pairs with separation Δ.

$$P(K) = 1/(2N-1) \quad \Sigma \, C(\Delta) \cos\left(\frac{2\pi K\Delta}{N}\right)$$

B) Much better result if $C(\Delta)$ is normalized (divide by the number of pairs at Δ and multiply by the number of pairs if arrays were full). **However, sidelobes are still not good if some Δ values are absent.**

Method No. 3

Solution of simultaneous linear equations relating the amplitude of uncorrelated far-field radiant intensities to the amplitude and phase of each correlation coefficients $C(\Delta)$.

Figure 3-1. Thinned Line Array Beamforming.

1) The number of absent separation values is much less than the number of absent spatial positions, thereby reducing the seriousness of the absent data problem, see table I.

2) The range of feasible shading functions is expanded. A shading function in the spatial domain always can be transformed to an equivalent shading function in the element-separation domain , but the reverse is not true.

Having recognized the potential advantages of working in the correlation function/element-separation domain we define the following beamforming algorithm.

Step 1 Compute the spatial correlation function $C(\Delta) = \Sigma \; \overline{S(n)\, S(n+\Delta)}$ -------------(1)

Step 2 Normalize the spatial correlation function by one of the following two processes to get $C'(\Delta)$

$$\text{A)} \quad C'_A(\Delta) = C(\Delta)$$
$$\text{B)} \quad C'_B(\Delta) = C(\Delta)/R(\Delta)$$

Where $R(\Delta)$ is a ratio equal to the number of receiver element pairs with separation (Δ) divided by the number of element pairs with separation (Δ) if the array were full. However, if $R(\Delta) = 0$ then $C'_B(\Delta)$ is set equal to zero.

Step 3 Perform the Fourier transform of the (Even Functioned) normalized correlation function.

$$P(K) = 1/(2N-1) \; \Sigma \; C'(\Delta) \, \mathrm{Cos}\, 2\pi kn/N \text{ ---------- (2)}$$

When dealing with planar, or volumetric, arrays one can encounter high amounts of array '"thinning" which cause many of the separation values to be absent. For those circumstances the simultaneous equation beam forming method, described in the next section, is recommended

3.3 SIMULTANEOUS EQUATION BEAMFORMING

When the number of missing elements becomes too great, the quality of the direct interpolation used in Method 2 will eventually lead to performance degradation similar to that of Method 1. It then becomes more preferable to replace interpolation by the simultaneous equation beamforming method.

This method forms the separate spatial correlation coefficients ($C(\Delta)$ of equation 1) and equates each to the sum of elemental correlation terms arising from each unknown far-field source $P(k)$ at angle θ.

$$C(\Delta) = \sum_{K=0}^{N-1} P(k) e^{-j2\pi d\Delta \mathrm{Cos}\,\theta_k} \text{ --------------------------- (3)}$$

where Pk is the signal power from a far-field source at angle θk and (d) is the elemental element spacing in wavelength units. For convenience, we select our far-field angles to correspond to the decrete Fourier transform parameter values K. This gives N discrete beam angles.

Table 1. Spatial Position of Receiving Elements (With 20 of 32 Absent).

Spatial Position	Receiving Element Present or Absent
0	Present
~~1~~	(Absent)
2	Present
~~3~~	(Absent)
~~4~~	(Absent)
5	Present
~~6~~	(Absent)
~~7~~	(Absent)
~~8~~	(Absent)
9	Present
~~10~~	(Absent)
11	Present
~~12~~	(Absent)
~~13~~	(Absent)
14	Present
~~15~~	(Absent)
16	Present
~~17~~	(Absent)
~~18~~	(Absent)
~~19~~	(Absent)
~~20~~	(Absent)
21	Present
~~22~~	(Absent)
~~23~~	(Absent)
~~24~~	(Absent)
25	Present
~~26~~	(Absent)
27	Present
~~28~~	(Absent)
~~29~~	(Absent)
30	Present
31	Present

Element Separation (Δ)	Number of Element Pairs
0	12
1	1
2	4
3	3
4	3
5	5
6	3
7	3
8	0 ← Absent
9	5
10	2
11	4
12	2
13	1
14	4
15	1
16	5
17	1
18	1
19	2
20	2
21	2
22	2
23	1
24	0 ← Absent
25	3
26	1
27	1
28	1
29	1
30	1
31	1

If there are (X) absent separations we will have (N-X) linear equations in (N) unknowns. Furthermore, the 'confidence factor' for each equation can be related to the number of element pairs that contributed to the C(Δ) value considered and, if desired, shading can be included in the equation weighting process. Using standard computer programming a "least mean square error" solution can be derived for the under determined equation set and this is taken as the sensed angular distribution of far field sources.

4.0 COMPUTER SIMULATION RESULTS FOR FILLED AND PARTIAL FILLED LINE ARRAYS.

To illustrate the relative performances of the above defined beamforming methods a 32 element line array was evaluated; firstly with all elements present and secondly with 20 of the 32 elements absent. Table I indicates which spatial positions of the array were absent and also shows that, for the selected configuration, only 2 of the 32 separation distances were absent from the analysis.

4.1. SIMULATION RESULTS FOR THE SPATIAL FOURIER TRANSFORM (METHOD #1)

With the line arrays steered broadside (θ = 90degrees) the sidelobe response patterns are as shown in Figure 4-1. In all cases the equivalent of Hanning shading is applied. Curve 1 shows the desired low sidelobe values of a filled array, and curve 2 shows the less than satisfactory sidelobe pattern when 20 of the 32 elements are absent.

4.2 SIMULATION RESULTS FOR THE FOURIER TRANSFORM OF THE SPATIAL CORRELATION FUNCTION (METHOD #2).

Again with the line array steered broadside the beamforming sidelobe responses for processing methods A and B (described in Section 3.2) are depicted in Figure 4-2. As expected, curve A has the same high sidelobes as in Figure 4-1 curve 2, whereas the new curve B has improved sidelobes.

5.0 EXTENSION OF BEAMFORMING METHODS TO FULLY RANDOM ARRAYS.

The above analysis has treated a configuration in which the positions of the receiving elements are considered to be uniformally spaced along a line with some of the elements inoperative (or absent). If the receivers are free to take up random locations the processing methods should be modified slightly to allow for the non-integer values of the elements positions and separations.

For beamforming method # 1 it has been usual to perform the Fourier summation over the elements without concern for the specific randomness of position. Sidelobe patterns are found to be poor with the mean sidelobe level equal to approximately - 10 log N db, (ref 3).

For the spatial correlation beamforming method (method #2) we suggest that the array separation values for a hypothetical uniform array ($d = \lambda/2$) be defined and the correlation function 'count normalization' be performed in bands (± d/2) centered on those hypothetical separations. As explained before, when processing signals from thinned planar or volumetric array the simultaneous equation method is likely to be preferred, and in those cases the correlation function is again conveniently 'count normalised' in bands around the element separations of a hypothetical uniform planar or volumetric array.

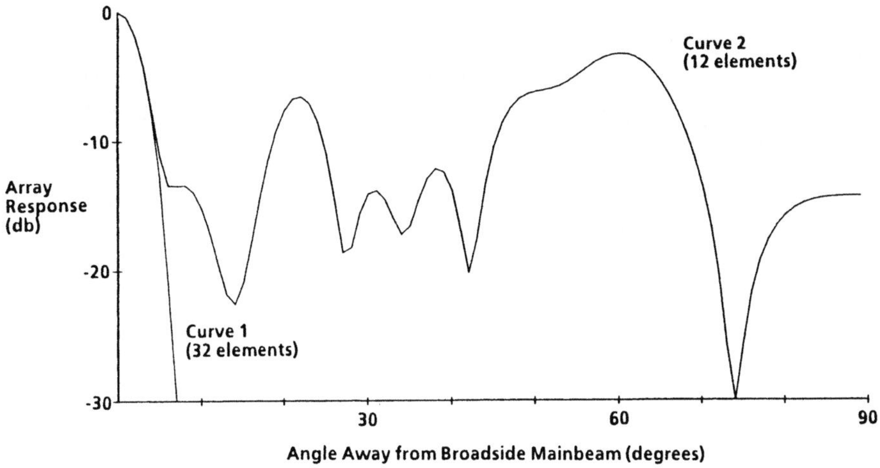

Figure 4-1. Simulation Results for Spatial Fourier Transform (Method 1).

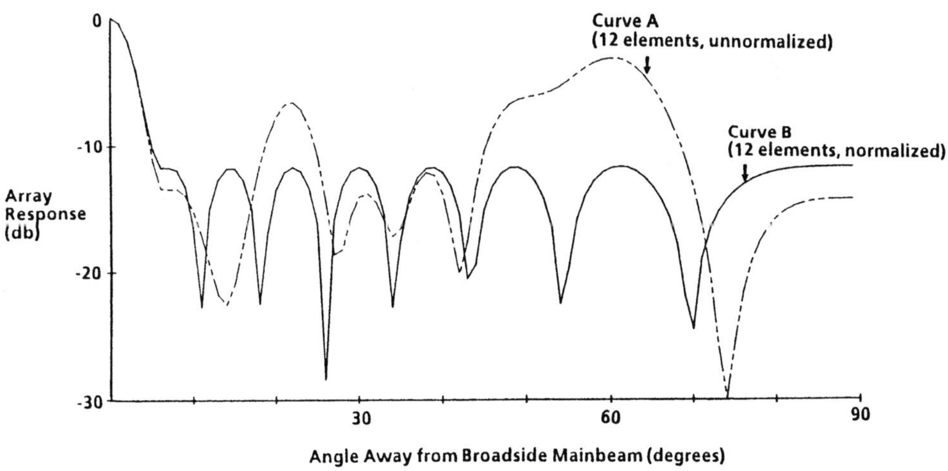

Figure 4-2. Simulation Results for Fourier Transform of Spatial Correlation Function (Method 2).

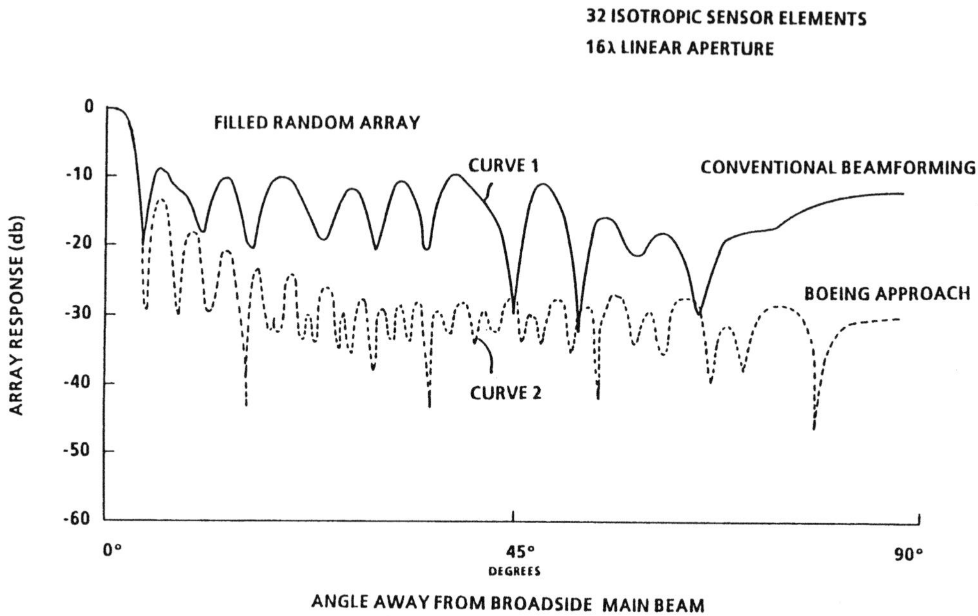

Figure 5-1. Sidelobe Pattern for Random Line Array (Small Aperture).

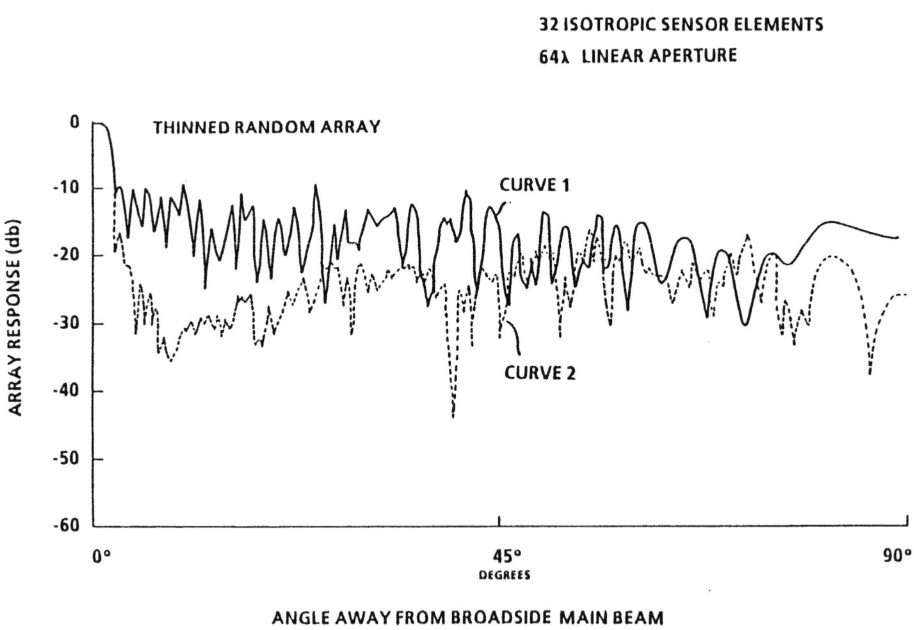

Figure 5-2. Sidelobe Pattern for Random Line Array (Large Aperture).

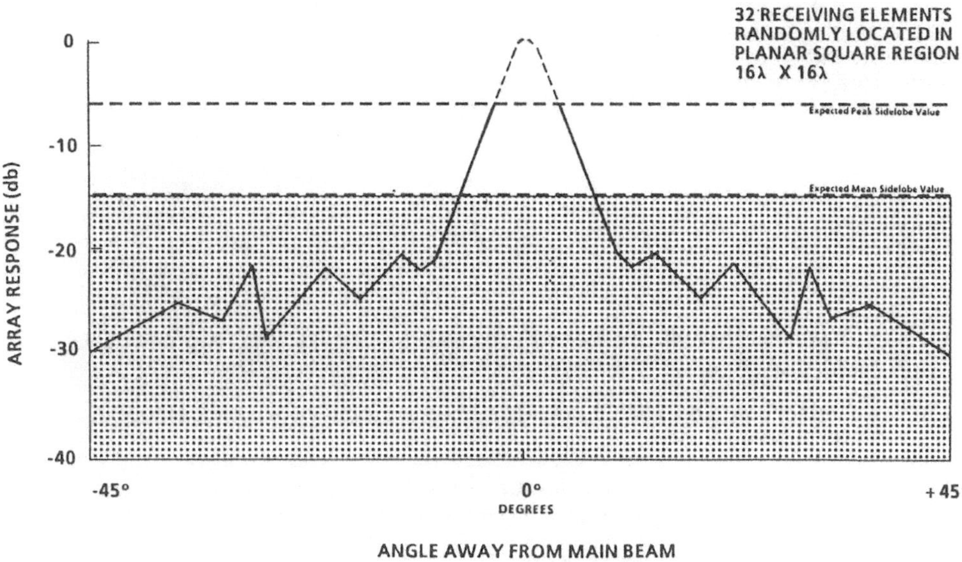

Figure 5-3. Sidelobe Pattern for Random Planar Array (Highly Thinned).

Computer simulation results for a random line array are shown in Figures 5-1 and 5-2 and for a random planar array in Figure 5-3. As expected, the results are generally in agreement with those derived for the partially filled line array reported in sections 4.1and 4.2.

The two curves in Figure 5-1 are for a line array in which 32 receiving elements are randomly distributed over a distance of 16λ, and the two curves in Figure 5-2 are for a similar 32 elements distribution over a distance 64λ. In each case straight-forward discrete Fourier summation (curves1) is compared with count normalised and interpolated correlation function processing (Curves 2). It is seen that excellent sidelobe performance is obtained for the 16λ array, but the 64λ array has high amplitude sidelobes at angles widely separated from the beamsteer angle because the interpolation process does not support good estimation of high frequency Fourier components.

The curve in Figure 5-3 shows results for a planar array with 32 receiver elements randomly distributed over a planar square 16λ x16λ . The processing method used was the (correlation coefficient) simultaneous equation method. It is seen that, despite substantial array thinning, good sidelobe characteristics are obtained.

6.0 CONCLUSIONS

The application of Fourier Transform techniques to random array beamforming involves processing characteristics which differ in key respects from the more usual time series signal waveform analysis. The combination of spatial and time processing presents peculiar challenges in the treatment and formulations of constraints and performance criteria in terms of information content (or lack thereof). It is anticipated that the incorporation of entropy and cross-entropy concepts will lead to improved beamforming results for future random arrays. The significant performance improvements achieved by the innovations described in this paper serve only as an illustration of the opportunities that still exist.

REFERENCES

(1) B.D. Steinberg, 'Principles of Aperture and Array System Design: including Random and Adaptive Arrays', John Wiley, 1976, New York

(2) R.B. Blackman and J.W. Tukey, 'The Measurement of Power Spectra', New York: Dover, 1959

(3) B.D. Steinberg, 'The Peak Sidelobe of the Phased Array Having Randomly Located Elements', IEEE Trans. Antennas & Propag., AP-20, March, 1972

(4) D.E. Dudgeon, 'Fundamentals of Digital Array Processing', Proc. IEEE, vol. 65, pp. 898-904, June 1977

DECISION MAKING WITH BARELY ANY INFORMATION: THE ROLE OF MIXED STRATEGIES

P. B. Kantor
OCLC, Dublin, Ohio and
Tantalus Inc., Cleveland, Ohio*

M. J. Kantor
Dept of Mathematics
University of Michigan

1. Introduction

At the 6th annual Maximum Entropy Workshop Carlos Rodrigues presented an amusing problem in decision theory. In this problem the normal application of expected utility theory leads to an unacceptable conclusion. We restate the problem, and the "paradox." We then give an "almost but not quite exactly unlike" formulation in which the paradox is resolved. We then pursue the new formulation to show how adopting a mixed strategy makes it possible to "certainly" gain while playing the game. We next show that in a suitable limit the "expected gain" goes to zero. Finally, we speculate on the possible application of these ideas to situations with "great uncertainty."

2. The Two Envelope Problem.

You play a game. You are offered your choice of either of two envelopes, and are told (correctly) that each envelope contains a check made out to you. You are also told, correctly, that one of the checks is twice as large as the other. You are permitted to select one envelope, and look inside it. Let us suppose that you find a check for $20. You are now given the following choice: you may either keep the $20 check, or exchange it for the one in the other envelope.

You reason as follows. "I chose at random, so the chance that I chose the smaller envelope is 0.5. So the chance that the contents of the other are larger is 0.5. The expected value is then 0.5x$10+0.5x$40 = $25. This is more than I have. So, by the expected utility theory, I should exchange for the other envelope(!!).

But this seems absurd. How can it be that, once you look into the envelope, you become "convinced" that you could have done better, when, in effect you chose at random. In fact, you don't even have to look into the envelope. You can simply say: Let the amount in the envelope be Y. Then the expected value is 0.5xY/2+0.5x2Y=1.25Y. So you should swap, without even looking.

*Permanent Address

G. J. Erickson and C. R. Smith (eds.),
Maximum-Entropy and Bayesian Methods in Science and Engineering (Vol. 2), 421–426.

Bayesians seek to escape this problem by imposing a prior distribution on the contents of the envelope, inverse to the size of the check. This solution fails if, for example, you must play a game in which the sum in one envelope is (approximately) the square of the sum in the other, or the two sums differ by a fixed amount.

Essentially we propose that the puzzle arises from an incorrect passage to an infinite limit. We describe a finite analog, and discuss its properties in some detail.

3. The deck of cards problem. A finite analog

Consider a deck of cards each of whose members has one number on one face, and double that number on the other side. Suppose further that, except for the lowest and highest numbers in the deck, each number appears on exactly two cards. Admitting fractions, we can, without loss of generality represent the cards as multiples of the smallest face value, as:

1	2	4	8	....	N/2
2	4	8	16		N

In this version of the game, you are shown one side of one card (remember, you don't know the actual lowest and highest values) and asked whether you would like to exchange it for the other side.

We might repeat the steps of the argument given above, to show that you should change. But of course that result must be wrong. [For example, we could run a money pump by showing the opposite sides of a card to two different people, and getting them to both pay us to receive whatever was on the other side of the card.]

A more careful calculation shows that the expected value of this game is a collapsing sum. There are 2xLog N faces that might be presented to you. We assume that each has equal probability, which we represent by $q=1/[2*\mathrm{Log}_2 N]$. The expected gain by always choosing the other side [we call this strategy {1}] is given by the sum:

$$
\begin{aligned}
E(\{1\}) = \;& q && \text{[if you see "1"]}\\
&+2q - q && \text{[if you see 2]}\\
&+4q - 2q && \text{[if you see 4]}\\
&\;\dots\dots\\
&+qN/2 - qN/4 && \text{[if you see N/2]}\\
&+0 - qN/2 && \text{[if you see N]}
\end{aligned}
$$

The expected value is positive as we sum the series, right up until the last term, which is only negative. We then recognize that this is a collapsing series whose sum is zero. Hence, as long as the game is played with a finite deck, no matter what the top and bottom values, the expectation value of the strategy "take the other side" will be zero.

This is reasonable. When we consider the two envelope game we may think of it as an infinite limit of the deck game. We contend that the correct way to calculate the value of that limit is to pass to the limit of finite values. Since the expectation value is zero for all finite decks, the limit is also zero.

4. A Mixed Strategy.

Somewhat to our surprise, there is more to be said. We approached the question, up until now, by supposing that there is a definite correct decision. However, our opponent (Nature) is using a random strategy in choosing which envelope to give us (!). So perhaps we can do better with a random strategy.

Let $\ldots n_{-1}, n_0, n_1, n_2, \ldots n_k, \ldots$ be a sequence of positive real numbers, extending infinitely in both directions, and having the property that:

$$2n_k > n_{k+1} > n_k.$$

and:

$$\lim_{k \to \infty} n_{-k} = 0$$

$$\lim_{k \to \infty} n_k = \infty$$

The points n_k divide the positive real line into an infinite set of non-overlapping intervals, which cover it, except for the point at 0. Because the upper point of each interval is less than twice the lower point, whatever value shows on the face that we look at, the value on the other face cannot lie in the same interval.

Next we form a sequence of decreasing numbers, all lying in the interval (0,1), and extending infinitely in both directions:

$$\ldots p_{-1}, p_0, p_1, p_2, \ldots p_k, \ldots$$

Now we adopt a mixed strategy. If the face we see lies in the interval whose upper endpoint is n_k, we will, with probability p_k, choose the other side.

This modifies the expected value calculated above so

that the terms of the series do not quite cancel. Call this strategy {p}.

$$
\begin{array}{lll}
E(\{p\}) = & qp_{J(1)} & \text{[if you see "1"]} \\
 & +2qp_{J(2)} - qp_{J(2)} & \text{[if you see 2]} \\
 & +4qp_{J(2)} - 2qp_{J(2)} & \text{[if you see 4]} \\
 & \ldots\ldots & \\
 & +qp_{J(N/2)} - qp_{J(N/2)} & \text{[if you see N/2]} \\
 & +0 - qp_{J(N)} N/2 & \text{[if you see N]}
\end{array}
$$

$$=q(p_{J(1)}-p_{J(2)})+2q(p_{J(2)}-p_{J(4)})+\ldots+qN/2(p_{J(N/2)}-p_{J(N)}).$$

Since the sequence $\{p_k\}$ is strictly decreasing, every term in this sum is positive, and so the sum is positive. Hence, by following any such mixed strategy we can expect to make some money by playing the game, above and beyond what is on the first face presented to us.

5. A potentially realistic example.

Suppose we are charged with protecting a basket of eggs from a marauding seagull who will drop two stones, one of which is twice as large as the other. We want to minimize the damage to the basket, which damage is assumed proportional to the size of the stone. The small stone does damage X. Our shield is made of a light material, and will be completely destroyed, whichever stone it stops. We have sharp eyes, and can judge the size of the stone as it falls. BUT, we do not know how big the stones are to begin with. [We also don't know how big the gull is.] We see a stone falling towards us. What should we do?

Clearly we'd rather stop the bigger stone. Our simplest strategy is to select an arbitrary cut point C. If the first stone is larger than C, stop that stone. If not, stop the other. If C falls between the sizes of the two stones, we do well, holding the expected damage to X. If it does not fall between the sizes, we expect damage 3X/2. Note that softening our strategy, using probabilities p_L, (p_S) to determine whether we stop the first stone if it is larger (smaller) than C, only weakens us. The expected gain drops to:

$$3X/2 + X(p_S - p_L)/2.$$

So the expected loss is minimized by the pure strategy $p_L=1$, $p_S=0$.

But, of course, if our cut does not divide the large stone from the small, we gain nothing by any strategy. We can then move to a strategy similar to our discussion of the game with the deck of cards. A sequence of cutpoints will ensure us that we have a different strategy for the large and small stones. But, since all the probabilities must lie between zero and one, we cannot make any of the steps (that is, the drop in probability of responding, as we move from the high to the low side of the interval dividers) large.

6. Are we crypto-Bayesians?

From the perspective of the sea gull game, we see that the key to a mixed strategy is to be good at guessing how big the stones are likely to be. A typical distribution for the values of {p} (remember now that we prefer to act if the value seems large) is: stop the first stone if it is larger than x, with probability:

$$p(x) = \{1+\tanh[a*\log(x-x_o)]\}/2.$$

The parameter x_o is the "best location for the cut. The parameter a determines how sharply our strategy changes across that cut. So, if we adopt some Bayes' prior for the size of the stone we will conclude that a specific distribution is best.

BUT, without any such assumption, we are still assured that all strategies of the form described in the deck game will improve our performance compared to, e.g. simply stopping the first stone.

Can we believe in a uniform distribution? No. At least not if we want to implement it, because this would require that {p} rise uniformly from 0 to 1 over the real line, and there is no function that does this. Another way of saying this is that, as the size of the deck becomes infinite, if we believe the distribution is uniform, our strategy eventually drops in value to 0.

7. So what?

It is not clear that this discussion has any importance beyond the present examples. However, we (and others to whom we have described it) have found it surprising that in a reasonable finite model of the envelope problem, one can have a mixed strategy which "makes money." [Compared to just keeping the first envelope.] It suggests that one can gain some of the benefits of specifiying a prior distribution without actually specifying that prior.

In conclusion, we have accomplished the following desirable goals:

(1) We have removed the paradox.

(2) We have discovered a mixed strategy with non-zero value.

(3) We have provided an effective distinction between the case where one looks in the envelope and the case where one does not. The mixed strategy requires that one look into the envelope. Thus, the value of the game is improved if and only if we get more information than is provided in the rules of the game alone. The question of measuring the amount of information represented by the number on the check is beyond the scope of this paper. [cf KANTOR,77]. The question of the value of that information seems to depend upon the prior distribution of the size of the checks or stones, if one believes in such things.

(4) We have accomplished all of this without claiming to adopt any belief about the "distribution in Nature" of the size of checks, stones or sea gulls. Conceivably, this becomes important in real conflict situations, where one cannot, with any real conviction, place a prior distribution on the array of threats.

8. Acknowledgements.

We thank Dr. Frederick Kantor, of Kantor Laboratories, New York, for an insightful remark about the vanishing value of the strategy if we believe that the distribution of threats is uniform and the range becomes infinite. One of us (PBK) thanks OCLC and its director of Research, Dr. Martin Dillon, for hospitality during the period when these results were written up.

9. Literature cited

RODRIGUEZ[1986], Carlos. "title?" in Proceedings of the Sixth Annual Workshop on Maximum Entropy and Bayesian Methods, ed. C. Ray Smith and Gary Erickson. Reidel. Year?

KANTOR[1977], Frederick W. Information Mechanics. John Wiley and Sons. NY

COMPARISON OF BAYESIAN AND DEMPSTER'S RULES IN EVIDENCE COMBINATION+

Yizong Cheng
Department of Computer Science, University of Cincinnati, Cincinnati, OH

R.L. Kashyap
School of Electrical Engineering, Purdue University, West Lafayette, IN 47907

Abstract. Suppose we have a hypothesis H and two evidential variables A and B. Suppose the conditional probabilities P(H|A) and P(H|B) are known to lie in specified intervals. We compare the intervals for the combination P(H|A,B) given by the Bayesian method and the Dempster's method. We show that the width of the interval given by the Dempster rule is narrower than that of Bayes. We discuss the one-sided cases, i.e., when we are given bounds on P(H|A) and P($\bar{H}$|B), $\bar{H}$ being the negation of H. Finally, we mention the sensitivity of the combination rules to small deviations of the members, especially when these values are near zero.

1. INTRODUCTION

Evidence is used to support ***hypotheses***. Evidence is better described by its impact on the belief on hypotheses than by its physical reality, especially in the context of evidence combination. Also, we often restrict ourselves to ***independent pieces of evidence***. Independence is a subtle concept and usually can only be defined by its consequences on combination. According to our point of view, independence is the most expected situation under ignorance of dependence. Because the meaning of independence is never unique, under different belief representation systems and different implications of independence, the combination rules are different. There are many ***ad hoc*** rules of combination available. In this paper, two major families of combination rules, that of Bayesian probability and that of Dempster's theory, are considered.

In section 2, we discuss the intervals for the combined probability P(H|A,B) in terms of the members P(H|A) and P(H|B) when the latter are specified by intervals, using both the Bayesian and methodology. In section 3, we consider the one-sided cases, i.e., when the intervals of the members are of the form $[0,\alpha]$ or $[\beta,1]$, i.e., one of the bounds is 0 or 1. In section 5, we discuss the non-robustness or sensitivity of the two combination formulae.

2. TWO-SIDED INTERVAL-VALUED CASES

The problem of combining evidence can be stated as follows: given probabilities P(H|A) and P(H|B), where H is a hypothesis, and A and B are two independent pieces of evidence, find P(H|AB). Following the Bayesian

+This work was partially supported by the Office of Naval Research under contract N00014-85K-0611 and the National Science Foundation under the grant IST-84-05052.

G. J. Erickson and C. R. Smith (eds.),
Maximum-Entropy and Bayesian Methods in Science and Engineering (Vol. 2), 427–433.

Theorem, or the definition of conditional probabilities, under the assumptions of conditional independence of evidence given both H and $\bar{H}$, the formula for combination is

$$P(H|AB) = \frac{P(H|A)P(H|B)}{1-P(H|A)-P(H|B)+P(H|A)P(H|B)/P(H)} \frac{1-P(H)}{P(H)} \tag{1}$$

If there is no prior information about the preference between H and $\bar{H}$, a reasonable assignment is P(H) = 0.5. In this case, Equation (1) becomes

$$P(H|AB) = \frac{P(H|A)P(H|B)}{1-P(H|A)-P(H|B)+2P(H|A)P(H|B)} \tag{2}$$

Denoting P(H|A), P(H|B), and P(H|AB) by α, β, and γ, respectively, we have

$$\gamma = \frac{\alpha\beta}{1-\alpha-\beta+2\alpha\beta} \tag{3}$$

Since

$$\frac{\partial\gamma}{\partial\alpha} = \frac{\beta(1-\beta)}{(1-\alpha-\beta+2\alpha\beta)^2} > 0 \tag{4}$$

γ is an increasing function of both α and β. Hence, when P(H|A) and P(H|B) take interval values $[a_1,a_2]$ and $[b_1,b_2]$, respectively, P(H|AB) will take interval value

$$P(H|AB) = \left[\frac{a_1b_1}{1-a_1-b_1+2a_1b_1}, \frac{a_2b_2}{1-a_2-b_2+2a_2b_2}\right] \triangleq [s,t] \tag{5}$$

In the very year, 1764, when Bayes proposed his theorems for conditional probabilities (Shafer, 1982), Lambert proposed his rule to combine independent evidence (Shafer, 1978). The major difference between these two approaches is that for Bayes, two pieces of evidence being independent means the probability of their conjunction is equal to the product of two individual probabilities, while for Lambert, it means the probability of the hypothesis that they support together is the product of those supported by only one of them. Although both of these approaches have convincing interpretations from frequency counting, Lambert's approach needs ***renormalization***.

Dempster's rule of combination (Dempster, 1967), is based on the same frequentist interpretation as Lambert's. For the case of one hypothesis and its negation, they are the same. Lambert's (and also Dempster's) result for two-side interval-valued probability combination is

$$P(H|AB) \in \left[\frac{a_1b_2+a_2b_1-a_1b_1}{1-a_1-b_1+a_1b_2+a_2b_1}, \frac{a_2b_2}{1-a_1-b_1+a_1b_2+a_2b_1}\right] \triangleq [u,v] \, . \tag{6}$$

We will show that the interval given by the Dempster rule is narrower or crisper than that given by Bayesian method.

Theorem 1. $s \le u \le v \le t$.
Proof. We first prove $v \le t$ and this is equivalent to

$$a_2 + b_2 - 2a_2b_2 \ge a_1 + b_1 - a_1b_2 - a_2b_1 .$$

This is true because

$$(a_2-a_1) + (b_2-b_1) - (a_2-a_1)b_2 - (b_2-b_1)a_2$$

$$= (a_2-a_1)(1-b_2) + (b_2-b_1)(1-a_2) \ge 0.$$

To prove $s \le u$ we only need to notice that the interval values for $P(\overline{H}|AB)$ are
[1-t, 1-s] and [1-v, 1-u], respectively, according to (5) and (6). Hence, $(1-t) \le (1-v)$, or $v \le t$.

■

Theorem 2. $v-u \le \min\{a_2-a_1, b_2-b_1\}$.
Proof. From (6) we know that

$$v-u = \frac{(a_2-a_1)(b_2-b_1)}{1-a_1-b_1+a_1b_2+a_2b_1}$$

Thus, in order to prove $v-u \le b_2-b_1$, we only need to prove

$$\frac{a_2-a_1}{1-a_1-b_1+a_1b_2+a_2b_1} \le 1$$

This is true because

$$(1-a_1-b_1+a_1b_2+a_2b_1) - (a_2-a_1) = (1-a_2)(1-b_1)+a_1b_2 \ge 0.$$

■

Similarly, one can prove $v-u \le a_2a_1$.

Other properties are:

1. If $s=t$ the $u=v$. The reverse is not true.
2. If $a_1=a_2$ or $b_1=b_2$, then $u=v$.
3. If $a_1=0$ or $b_1=0$, the $s=0$. Similarly, if $a_2=1$ or $b_2=1$, then $t=1$. This is not true for u and v.

3. ONE-SIDE INTERVAL-VALUED CASES

The lower bound of an interval value of probability is usually interpreted as "degree of belief." And the results after combination are also given in (5) or (6) by letting some of a_1 and b_1 take the extreme values. When two pieces of evidence give lower bounds a and b to a hypothesis, the combined lower bound from Bayesian approach (5) is

$$P(H|AB) \geq \frac{ab}{1-a-b+2ab} \tag{7}$$

and that from Dempster's approach (6) is

$$P(H|AB) \geq a+b-ab \tag{8}$$

The latter was proposed by James Bernoulli in 1713 (Shafer, 1979), although based on different philosophy than Lambert's and Dempster's.

Another situation is that $a_2=1$ and $b_1=0$. In this case, one piece of evidence is in favor of the hypothesis and the other is in favor of its opposite. Let $P(H|A) \geq a$ and $P(\overline{H}|B) \geq c$, i.e., we have one positive evidence favoring H, and a negative evidence favoring $\overline{H}$. From (5) and (6), we have:

$$\text{Bayesian method:}\quad P(H|AB) \in [0,1] \tag{9}$$

$$\text{Dempster:}\quad P(H|AB) \in \left[\frac{a-ac}{1-ac}, \frac{1-c}{1-ac}\right] \tag{10}$$

Notice that the new lower bounds in both (9) and (10) are less than the lower bounds in (5) and (6), i.e., $0 \leq s$ and $\frac{a-ac}{1-ac} \leq u$. In other words, there is some sort of cancellation of the positive and negative evidence. In the Bayesian approach (9), no matter how weak the negative evidence is, the positive evidence is completely cancelled, while in Dempster's approach (10), no matter how strong the negative evidence is, as long as it has not reached 1, the positive evidence cannot be cancelled. If the quantity a is small, the combined lower bound is small, whatever may be c. This means that a weak lower bound does not give much information for the hypothesis. A better representation for positive and negative evidence is to let the lengths of the intervals be less than one half. However, by Theorem 2, the combined result will be a narrower interval around 0.5, and there is no cancellation back to the interval for total ignorance, [0,1].

The situation of giving lower bounds for two mutually exclusive hypotheses from two pieces of evidence is similar. If the lower bounds of $P(H_1|A)$ and $P(H_2|B)$ are given as a and c with $H_1 \cap H_2 = \phi$, then the lower bound for $P(\overline{H}_1|B)$ is also c and the combination results are still (9) and (10).

4. POINT-VALUED CASES

Let us consider cases when intervals shrink into points, or

$$a_1=a_2=a \quad \text{and} \quad b_1=b_2=b$$

In this case, both (5) and (6) become

$$P(H|AB) = \frac{ab}{1-a-b+2ab} \tag{11}$$

For two mutually exclusive hypotheses H_1 and H_2, $P(H_2|B)=c$ implies $P(H_1|B) \in [c,1]$. This corresponds to the case of

$$a_1=a_2=a, \quad b_1=0, \quad \text{and} \quad b_2=1-c.$$

Now (5) becomes

$$P(H|AB) \in \left[0, \frac{a(1-c)}{a+c-2ac}\right] \tag{12}$$

for Bayesian approach and (6) becomes

$$P(H|AB) = \frac{a-ac}{1-ac} . \tag{13}$$

For n mutually exclusive and complete hypotheses, both Bayesian and Dempster's approaches give

$$P(H_i|AB) = \frac{P(H_i|A)P(H_i|B)}{\sum_{i=1}^{n} P(H_i|A)P(H_i|B)} \tag{14}$$

Equation (11) is its special case when n=2.

5. INDEPENDENT POOL AND SENSITIVITY TO ZERO

Equation (14) is called an ***independent opinion pool*** by Berger (1985). One of its properties is ***reinforcement*** [Berger (1985)] or ***unidirectionality*** [Dempster (1982)]. The property is such that if every piece of evidence provides the highest degree of belief on the same hypothesis, then the belief on that hypothesis after evidence combination is even higher.

Another property of independent pools, the sensitivity of the combination to small deviations of the individual members when these values are near zero or one, is illustrated by Zadeh's paradox (Zadeh, 1984 and Prade, 1985). The combination formula (14) from both Bayesian and Dempster's approaches give entirely different results for a slight change of probabilities from 0 to 0.01:

	H_1	H_2	H_3
A	0.9	0.1	0
B	0	0.1	0.9
Combined	0.00	1.00	0.00

	H_1	H_2	H_3
A	0.89	0.1	0.01
B	0.01	0.1	0.89
Combined	0.32	0.36	0.32

The following is a study on the sensitivity to zero. First let us consider the case of two alternatives with point-values. When $P(H_1|B)$ is relatively small comparing with $1-P(H_1|A)$,

$$\frac{P(H_1|A)P(H_1|B)}{1-P(H_1|A)-P(H_1|A)+2P(H_1|A)P(H_1|B)} \sim \frac{P(H_1|A)}{1-P(H_1|A)}\, P(H_1|B). \tag{15}$$

For example, if $P(H_1|B)$ changes from 0 to 0.05 and $P(H_1|A) = 0.9$, then $P(H_1|AB)$ will change from 0 to 0.32.

The more alternatives the hypothesis has, the more sensitive to zero (14) will be. Suppose we have uniformly distributed probabilities for hypotheses other than the first one from both sources and α and β ar still used to denote the probabilities for the first hypothesis. Then, from (14) we have the combined probability for the first hypothesis as

$$P(H_1|AB) = \frac{P(H_1|A)P(H_1|B)}{P(H_1|A)P(H_1|B)+(n-1)\left[\dfrac{1-P(H_1|A)}{n-1}\;\dfrac{1-P(H_1|B)}{n-1}\right]}$$

$$= (n-1)\,\frac{P(H_1|A)P(H_1|B)}{1-P(H_1|A)-P(H_1|B)+nP(H_1|A)P(H_1|B)} \tag{16}$$

where n is the number of mutually exclusive hypotheses under consideration. When $n \rightarrow \infty$, the limit for (16) is 1. For small n and small $P(H_1|B)$, (16) becomes approximately

$$P(H_1|AB) \sim (n-1)\,\frac{P(H_1|A)}{1-P(H_1|A)}\,P(H_1|B). \tag{17}$$

For example, if $P(H_1|A) = 0.9$ and $P(H_1|B)$ changes from 0 to 0.05, then $P(H_1|AB)$ will change from 0 to 0.49.

6. *CONCLUDING REMARKS*

In this paper, we have compared two approaches of combining evidence, Bayesian and Dempster's, for one or more than one hypothesis, for two-sided or one-sided interval-valued cases, and for point values. In general, Dempster's approach gives narrower interval-valued results and for point values they turn out to give the same result. The result is so-called independent opinion pool and has the problem of sensitivity to zero. Regarding the one-sided interval-valued case, there is the partial cancellation of positive and negative evidence. In general, one-side intervals are not appropriate representations for positive or negative evidence. A better way is to consider the difference or ratio between $P(H|A)$ and $P(H)$. This has been done in expert system MYCIN in the definition of measures of belief and disbelief, and the formula for certainty factors can better handle the presence of both positive and negative evidence.

References.

[1] J. Berger (1985). *Statistical Decision Theory.* 2nd ed. New York: Springer-Verlag.

[2] A.P. Dempster (1967). "Upper and lower probabilities induced by a multivalued mapping." *Ann. Math. Statist.* **38**, pp. 325-339.

[3] A.P. Dempster (1982). Comment on Shafer's "Lindley's Paradox." *J. the American Statistical Association,* **77**, pp. 339-341.

[4] H. Prade (1985). "A computational approach to approximate and plausible reasoning with applications to expert systems." *IEEE Trans.* **PAMI-7**, pp. 260-283.

[5] G. Shafer (1978). "Non-additive probabilities in the work of Bernoulli and Lambert." *Archive for History of Exact Sciences,* **19**, pp. 309-370.

[6] G. Shafer (1982). "Bayes's two arguments for the rule of conditioning." *The Annals of Statistics,* **4**, pp. 1075-1089.

[7] L.A. Zadeh (1984). Review of Shafer's "A Mathematical Theory of Evidence," *The AI Magazine,* **5**, pp. 81-83.

SUBJECT INDEX

adaptive filter, 20, 21, 30
autocorrelation/autocovariance 130, 131, 158, 215, 322
Bayesian analysis/methods 19, 427–433
Bayes' theorem 30, 244, 428
beamforming 409–420
Boltzmann principle 105, 107, 108
Burg's method (see maximum–entropy spectral analysis)
coherent states 111, 112, 113, 115, 118, 121, 123–125
complexion function 105, 107, 108
constraint 90, 101, 122, 130–133, 159, 163, 164, 244–248, 254, 255, 301–303, 307, 308, 314, 323–332, 344, 345, 347, 349, 350, 359, 360, 366, 374, 375, 396, 398, 401
crops 181, 185, 209, 216
cross entropy (see relative entropy)
data bases 183, 185, 328
differential entropy 83–88
directed divergence (see relative entropy)
DNA 147–150, 153
economy 181–183, 186, 209, 210, 215, 224, 225, 230
entropy 51, 53, 65, 74, 84–87, 90, 94, 96, 101, 106, 111–119, 125, 128, 130–134, 153, 161, 162, 166, 167, 176, 302, 313, 325, 340, 341, 343, 344, 345, 347, 349, 357–359, 361, 363, 365, 395, 398
epoch entropy 363–368
expert systems 243, 253–255, 265–267, 277, 296, 297
extra terrestrial intelligence 1
Gaussian distribution 132–137, 167, 269, 288, 315, 317, 327, 357, 360, 361
genetic 147
Heisenberg uncertainty 121, 123
ignorance (see complete ignorance)
image processing/restoration/enhancement 99, 265–269, 276, 285, 314, 341, 342, 350, 363
inductive inference 19, 20, 28, 30
inhomogeneous systems 371, 374, 375, 376, 378
instar 21
Kullback–Leibler information 83–88
Kullback–Leibler number (see relative entropy)
Lagrangian multiplier 94, 130, 136, 186, 187, 188, 210, 244–246, 303, 307–310, 358, 374, 375, 376, 377, 381, 398
learning law 28, 29
likelihood function 355–359
Lorentzian 133, 134, 137
maximum entropy (see principle of maximum entropy)

maximum–entropy 19, 34, 38, 92, 100, 119, 122, 123, 127, 128, 155, 160–164, 167, 171, 175, 177, 186, 217, 243, 245, 246, 249, 255, 298, 299, 302, 321, 323–325, 327, 328, 337, 357, 365, 371, 373, 375, 377, 379, 381–393, 395, 396, 401, 402, 403, 405, 421
maximum–entropy spectral analysis (MESA) 128–131, 134–144, 308, 313
neural networks 19–26
neurons 22, 27
nonparametric models 33
normal distribution (see Gaussian distribution)
optimum detection 7
percolation 371–374, 377
phase 155, 157, 158–167, 171, 172, 173, 174, 178, 181, 217, 220
phase space 51, 59
phase transition 377, 378
posterior probability (distribution) 265, 266, 275, 276, 304, 305, 345, 363
power spectrum 128, 131
principle of maximum entropy 34, 39, 53, 65, 89, 108, 121, 266, 269, 289, 299, 303, 309, 343, 346
prior information/knowledge 266, 269, 289, 299, 303, 309, 343, 345
probability measures 33
protein 147, 148, 155–165, 167, 171
quantum mechanics 51, 52, 59, 65, 88, 89, 106, 110
Radon 4, 6, 171, 175, 176, 177, 342
relative entropy 34, 35, 83, 84, 163
relaxation 361–365
SETI 1
spectral analysis 127–129, 187–189, 355
statistical mechanics 371, 375, 378
superradiance 112
thermorheological 366
time series 130, 131, 224–227, 355, 358, 359
truncation 395, 396, 400, 401, 402, 403
variational principle 89–101
Wigner formulation (of quantum mechanics) 51

Volume 1: Foundations

CONTENTS

PREFACE ix

HOW DOES THE BRAIN DO PLAUSIBLE REASONING?
E.T. Jaynes 1

THE RELATION OF BAYESIAN AND MAXIMUM ENTROPY METHODS
E.T. Jaynes 25

AN ENGINEER LOOKS AT BAYES
Myron Tribus 31

BAYESIAN INDUCTIVE INFERENCE AND MAXIMUM ENTROPY
Stephen F. Gull 53

EXCERPTS FROM BAYESIAN SPECTRUM ANALYSIS AND PARAMETER ESTIMATION
G. Larry Bretthorst 75

DETECTION OF EXTRA-SOLAR SYSTEM PLANETS
E.T. Jaynes 147

STOCHASTIC COMPLEXITY AND THE MAXIMUM ENTROPY PRINCIPLE
Jorma Rissanen 161

THE AXIOMS OF MAXIMUM ENTROPY
John Skilling 173

UNDERSTANDING IGNORANCE
C.C. Rodriguez 189

MAXIMUM ENTROPY CALCULATIONS ON A DISCRETE PROBABILITY SPACE
P.F. Fougere 205

QUANTUM DENSITY MATRIX AND ENTROPIC UNCERTAINTY
R. Blankenbecler and M.H. Partovi 235

INFORMATION-THEORETICAL GENERALIZATION OF THE UNCERTAINTY PRINCIPLE
A.J.M. Garrett 245

TIME, ENERGY, AND THE LIMITS OF MEASUREMENT
M.H. Partovi and R. Blankenbecler 249

ON A DETECTION ESTIMATOR RELATED TO ENTROPY
R.N. Madan 257

THE EVOLUTION OF CARNOT'S PRINCIPLE
E.T. Jaynes 267

A LOGIC OF INFORMATION SYSTEMS
N.C. Dalkey 283

METHODOLOGICAL PRINCIPLES OF UNCERTAINTY IN INDUCTIVE MODELLING: A NEW PERSPECTIVE
G.J. Klir 295

COMPARISON OF MINIMUM CROSS-ENTROPY INFERENCE WITH MINIMALLY INFORMATIVE INFORMATION SYSTEMS
N.C. Dalkey 305

Subject Index 313

Of Related Interest

Maximum-Entropy and Bayesian Methods in Inverse Problems

Edited by

C. Ray Smith and W. T. Grandy, Jr.

Department of Physics and Astronomy,
The University of Wyoming, Laramie, Wyoming, U.S.A.

This volume is the outcome of two workshops entitled "Maximum-Entropy and Bayesian Methods in Applied Statistics" and presents contributions by renowned authorities in many different fields. The purpose of these workshops was to bring together leading scientists whose research involved using the Principle of Maximum Entropy in a wide range of different applications in order to pool the experience gained and to identify common problems in need of solution. The result is stimulating and informative and provides many directions for further progress.

Audience

"Maximum-Entropy and Bayesian Methods in Inverse Problems" will be of great interest to mathematicians, physicists, geophysicists, electrical engineers, economists, and those working in communication and information theory and many other aspects of signal processing.

ISBN 90–277–2074–6 FTP 14

Of Related Interest

Maximum-Entropy and Bayesian Spectral Analysis and Estimation Problems

Edited by

C. Ray Smith

U.S Army Missile Command, Redstone Arsenal, Alabama, U.S.A.

and

Gary J. Erickson

Department of Electrical Engineering,
Seattle University, Seattle, Washington, U.S.A.

This volume contains 20 contributions by leading researchers from different fields who critically examine maximum-entropy and Bayesian methods in science, engineering, signal processing, medical physics, and other disciplines.
This is a sequel to the volume "Maximum-Entropy and Bayesian Methods in Inverse Problems', published by Reidel in 1985.

Audience

This book will be of interest to probability theorists, statisticians, electrical engineers, communication engineers, computer scientists, physicists, mathematicians, biologists, geophysicists, and those working in medical imaging.

ISBN 90–277–2579–9 FTP 21

GPSR Compliance

The European Union's (EU) General Product Safety Regulation (GPSR) is a set of rules that requires consumer products to be safe and our obligations to ensure this.

If you have any concerns about our products, you can contact us on ProductSafety@springernature.com

In case Publisher is established outside the EU, the EU authorized representative is:

Springer Nature Customer Service Center GmbH
Europaplatz 3
69115 Heidelberg, Germany

Batch number: 10607320

Printed by Printforce, the Netherlands